11/11/97

The Magnetic Field
of the Earth

Paleomagnetism, the Core,
and the Deep Mantle

This is Volume 63 in the
INTERNATIONAL GEOPHYSICS SERIES
A series of monographs and textbooks
Edited by RENATA DMOWSKA and JAMES R. HOLTON

A complete list of books in this series appears at the end of this volume.

The Magnetic Field of the Earth

Paleomagnetism, the Core, and the Deep Mantle

RONALD T. MERRILL
Department of Geophysics
University of Washington
Seattle, Washington

MICHAEL W. McELHINNY
Gondwana Consultants
Hat Head, New South Wales
Australia

PHILLIP L. McFADDEN
Australian Geological Survey Organisation
Canberra, Australia

ACADEMIC PRESS

San Diego   London   Boston   New York   Sydney   Tokyo   Toronto

The cover of this book shows the magnetic field for the Glatzmaier–Roberts dynamo at a time when the field is not undergoing a reversal (unpublished figure supplied to us by Glatzmaier, 1996). A change in color of the field lines indicates a change in the direction of the radial component of the magnetic field. On close inspection, the tangential cylinder is apparent in both the magnetic field (cover) and in the velocity field (Fig. 9.7).

This book is printed on acid-free paper.

Copyright © 1996, 1983 by ACADEMIC PRESS

Academic Press, Inc.
525 B Street, Suite 1900, San Diego, California 92101-4495, USA
http://www.apnet.com

Academic Press Limited
24-28 Oval Road, London NW1 7DX, UK
http://www.hbuk.co.uk/ap/

Library of Congress Cataloging-in-Publication Data

Merrill, Ronald T.
    The magnetic field of the earth : paleomagnetism, the core, and
  the deep mantle / by Ronald T. Merrill, Michael W. McElhinny,
  Phillip L. McFadden.
        p.   cm. -- (International geophysics series)
    Includes bibliographical references and index.
    ISBN 0-12-491245-1
    1. Geomagnetism.   2. Paleomagnetism.   3. Dynamo theory (cosmic
  physics)   I. McElhinny, M. W.   II. McFadden, Phillip L.   III. Title.
  IV. Series.
  QC816.M47   1996
  538'.72--dc20   96-28566
                                             CIP

PRINTED IN THE UNITED STATES OF AMERICA
96   97   98   99   00   01   QW   9   8   7   6   5   4   3   2   1

Contents

Chapter 4 The Recent Geomagnetic Field: Paleomagnetic Observations

Chapter 5 **Reversals of the Earth's Magnetic Field**

Chapter 6 **The Time-Averaged Paleomagnetic Field**

Chapter 8 **Introduction to Dynamo Theory**

Chapter 9 **Dynamo Theory**

Chapter 10 **The Magnetic Fields of the Sun, Moon, and Planets**

Chapter 11 **Examples of Synthesis**

Preface

Our knowledge of the Earth's magnetic field has developed over many centuries, originating from some of the earliest scientific investigations. The first scientific treatise ever written is generally recognized as being the geomagnetic text *Epistola de Magnete* written by Petrus Peregrinus in 1269. This was followed more than three centuries later by the classical work *De Magnete* by William Gilbert in 1600. Analysis of historical records was then placed on a firm mathematical footing by Gauss in 1838. The development of paleomagnetic studies in the second half of this century enabled the historical records of the Earth's magnetic field to be extended back not only to archeological times but to nearly the whole of geological time. One of the most significant discoveries of paleomagnetism has been that the Earth's magnetic field has reversed its polarity many hundreds of times. Over the past 15 years there has been a major resurgence of interest in geomagnetism, one result of which has been the formation of the international body SEDI (Study of the Earth's Deep Interior). This has now brought together scientists from many disciplines in geophysics, because it is widely recognized that our understanding of the Earth's deep interior (the core and deep mantle) is improved by integration of studies of the origin of the Earth's magnetic field (dynamo theory) with those from paleomagnetism and other disciplines such as seismology, gravity and high-pressure mineral physics.

This book is our view of the current status of the geomagnetic perspective of this integrated approach. Topics involved in studies of the Earth's magnetic field range from the intricate observations of geomagnetism, archeomagnetism, and paleomagnetism to the complex mathematics of dynamo theory. A fundamental problem in an integrated approach to geomagnetism has been the gulf between observationalists and theoreticians. We have tried to present the material in such a way that, hopefully, each group will understand what the others are about! Two of us (R.T.M. and M.W.M.) first attempted such an integration when we published a book on the Earth's magnetic field in 1983. The subject has developed so much since then that the three of us have now combined to bring the topic up to date. So while the general structure of this book is similar to the previous one, it is a completely new book with more than 60% new material. Some of the ideas put forward in the first book have been superseded by new developments. Our aim is to provide a broad overview of the present state of geomagnetism and paleomagnetism and the use of that knowledge to improve our understanding about processes and properties of the Earth's interior.

After a brief historical introduction, the foundations of geomagnetism and paleomagnetism are given in chapters 2 and 3. Chapters 4, 5, and 6 describe, explain, and summarize the current state of archeomagnetic and paleomagnetic studies. Chapter 5, in particular, reflects the concentration of work over the past decade on reversals and reversal transitions, and is now the largest chapter. Chapter 7 describes much of the seismology, mineral physics and fluid mechanics that provides a basic framework within which geodynamo theory and geomagnetic and paleomagnetic data must be interpreted. Chapters 8 and 9 deal with dynamo theory, the leading theory for explaining the origins of planetary and solar magnetic fields. Chapter 10 presents an overview of the magnetic fields of the Sun, our Moon, and the planets. Chapter 11 shows some examples of how integration of material from earlier chapters can be used to gain insight into what goes on in the Earth's deep interior. It should be noted that some important topics within geomagnetism and paleomagnetism, such as tectonics, magnetic induction, and magnetospheric physics, are only mentioned briefly. Chapters 7, 8, and 9 require a knowledge of electricity and magnetism and of vector calculus to a third year university level.

So many friends and colleagues have helped us in our efforts to write this book that it would be impractical to thank them all here. However, special thanks for discussions, reviews, or direct material contribution go to Charlie Barton, Jeremy Bloxham, Joe Cain, Pierre Camps, John Connerney, Vincent Courtillot, Gary Glaztmaier, Helen McFadden, Scott Merrill, Nancy Merrill, Andrew Newell, Neil Opdyke, and Hongbo Zheng, but responsibility for errors or inconsistencies lies with us. We have been carrying out joint research for more than a decade. Through many discussions and arguments we have reached certain common viewpoints that may not be shared by others. We have tried to represent conflicting views in this book, but clearly we will not always have been successful. Nevertheless, we hope you will find this book to be balanced, profitable, and enjoyable, and that you will share our enthusiasm for this integrated approach to one of the world's oldest sciences.

Seattle, Hat Head, and Canberra   **R.T. Merrill**
June 1996   **M.W. McElhinny**
                                                                   P.L. McFadden

Chapter One

HISTORY OF GEOMAGNETISM AND PALEOMAGNETISM

1.1 Discovery of the Main Magnetic Elements

1.1.1 The Magnetic Compass

The attractive properties of lodestone (magnetite) were known to both the Chinese and the Greeks in ancient times. The earliest observations on magnets are supposed to have been made by the Greek philosopher Thales in the sixth century B.C. The Chinese literature between the third century B.C. and the sixth century A.D. is full of references to the attractive power of the magnet, but there is nothing as far back in time as Thales (Needham, 1962). The properties were explained by Thales in animistic terms. In China the most usual name for a magnet was *ci shi*, the "loving stone," which may be compared with the French word *aimant*.

The discovery of the magnetic compass was an event of immense importance in science. It represents the very first dial and pointer instrument to be invented and is the forerunner of all those self-registering instruments that play so important a part in modern scientific investigation. The sundial was much older but in this case it was only a shadow, not the instrument itself, that moved. The windvane too was older but in all ancient forms there was no provision for precise readings on a circular graduated scale.

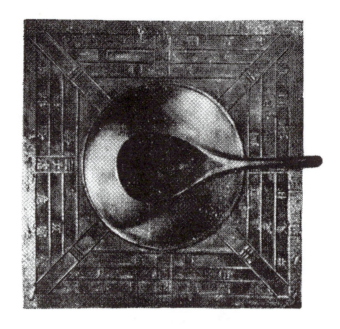

Fig. 1.1. Model lodestone spoon (*shao*) and bronze earth plate of the *shi* constructed by Wang Chen-To (1948). This is probably the earliest form of magnetic compass. North is to the left. Reproduced from Needham (1962) with the kind permission of Cambridge University Press.

The earliest known form of magnetic compass was invented by the Chinese by at least the first century A.D., and probably as early as the second century B.C. (Needham, 1962). The compass comprised a lodestone spoon rotating on a smooth board, a model of which has been constructed by Wang Chen-To (1948) (Fig. 1.1). Since the natural magnetization of such a spoon would establish itself in the direction of the main axis, this simple compass would automatically point north–south irrespective of the magnetization of the piece of rock from which it was cut. The spoon was probably given a thermoremanent magnetization by heating and cooling through the Curie Point while set in the north–south direction. The greater magnetization produced this way had the effect of overcoming the drag between spoon and board. The disadvantage of this frictional drag was overcome during succeeding centuries by the invention of floating, dry pivoted, and silk or other fiber suspended versions.

The compass arrived in Europe during the twelfth century A.D., where the first reference to it was made in 1190 by Alexander Neckham, an English monk of St. Albans. By what means it reached Europe is unknown, the most commonly accepted view being that it came by the sea route in the hands of the Arabs (Benjamin, 1895). A land route is still, however, a possibility because the earliest Arab references to the compass are all somewhat later than the European ones and no Indian reference of any significance has been discovered (Needham, 1962). The Chinese regarded the south pointing property of the magnet as fundamental, whereas mariners' compass needles were regarded as north pointers in Europe. This difference is of considerable significance because of the influence it had in Europe on theories of magnetic directivity.

The early thirteenth century philosophy of magnetic directivity proposed that the compass needle pointed toward the pole star. Since, unlike other stars, the pole star was fixed, it was concluded that the lodestone with which the needle was rubbed obtained its "virtue" from this star. In the same century the idea of polar lodestone mountains was propounded. The pole star gave its virtue to the lodestone mountains, which in turn imparted their virtue to the needle, which then directed itself toward the pole star. For the first time, theories of magnetic directivity had descended from the heavens to the Earth (Smith, 1968).

The idea of the universality of the north–south directivity of the compass was first questioned by Roger Bacon in 1266. A few years later Petrus Peregrinus (Peter the Wayfarer), an Italian scholar from Picardy, questioned the idea of polar lodestone mountains. Lodestone deposits occurred in many parts of the world, so why should the polar ones have preference? However, Peregrinus went further than logical argument and performed a remarkable series of experiments with spherical pieces of lodestone. These are reported in his *Epistola de Magnete* written in 1269. Although written in 1269 and widely circulated in Europe during the succeeding centuries, the *Epistola* was not published in printed form

Fig. 1.2a. The floating compass described in Peregrinus' *Epistola* and the reconstruction by Bertelli in 1868. An oval magnet (*magnes*) is placed in a bowl that floats in water in a larger transparent vessel. The top of the inner vessel is graduated so that 0° corresponds with east (*ortus*), 90° with south (*meridies*), 180° with west (*occasus*) and 270° with north (*septentrio*). These directions are with respect to the axis of the magnet, each pole (*polus lapidis*) of which Peregrinus had previously determined. A ruler with erect pins (*regula cum erectis stylis*) rests on the edges of the graduated bowl, and is used to determine the azimuth of the Sun or Moon. The rule is turned until the shadows of the pins fall longitudinally down it. The magnetic azimuth of the Sun (or Moon) is then the angle between the ruler and the magnetic axis. The wooden strip (*lignum gracile*) is used to determine the fiducial line across the larger vessel. *Note*: The style and precise details shown by Peregrinus' illustrations vary slightly from manuscript to manuscript. These are from the first printed edition of 1558. From Smith (1970a).

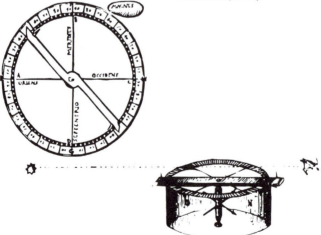

Fig. 1.2b. Peregrinus' pivot compass and the reconstruction by Bertelli in 1868. The iron magnet is placed through the pivot which rotates in a transparent vessel (cf. the fixed pivot of the modern compass). The ruler with erect pins serves the same purpose as in the floating compass. The vessel is graduated as before. *Oriens*, east; *occidens*, west. The lodestone shown here was used to magnetize the iron needle. From Smith (1970a).

under Peregrinus' name until 1558 (Smith, 1970a). He defined the concept of polarity for the first time in Europe. He discovered magnetic meridians and showed several ways of determining the positions of the poles of a lodestone sphere, each of which demonstrated an important magnetic property. He thus discovered the dipolar nature of the magnet, that the magnetic force is strongest and vertical at the poles, and became the first to formulate the law that like poles repel and unlike poles attract. The *Epistola* bears a remarkable resemblance to a modern scientific paper. Peregrinus used his experimental data from which to draw conclusions, unlike his contemporaries who sought to reconcile facts with preexisting speculation. His work is probably the first scientific treatise ever written. The second part of the *Epistola* contains descriptions of two types of magnetic compass, one floating and one pivoted. Peregrinus' version was the first pivoted compass to have been described in Europe (Fig. 1.2).

1.1.2 Declination, Inclination, and Secular Variation

The first observation of magnetic declination appears to have been made in China by the Buddhist astronomer Yi-Xing about A.D. 720 (Needham, 1962). However, knowledge of declination did not travel to Europe with the compass. Although the Chinese records reveal at least nine measurements of declination from A.D. 720–1280 (Fig. 1.3), the first measurement in Europe appears to be

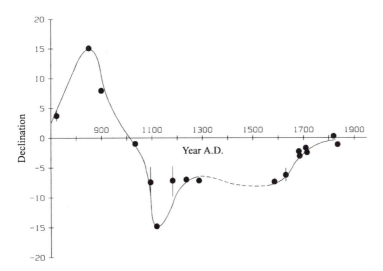

Fig. 1.3. Variation of magnetic declination with time in China between A.D. 720 and 1829. Error bars represent the range within which the declinations lie as quoted by Chinese texts. After Smith and Needham (1967).

that of Georg Hartmann, a Vicar of Nurenberg, made in Rome in about 1510. It has often been stated that Christopher Columbus first discovered declination in the European region during his first voyage to the West Indies in 1492. This claim now appears to be untrue (Chapman and Bartels, 1940), because there is strong evidence that declination was at least known in Europe by the early 1400s, even if precise measurements were not recorded for another century.

Magnetic inclination was discovered by Georg Hartmann in 1544, as mentioned in a letter of that year. His letter was only found in the Konigsberg archives in 1831, so his discovery had no influence on studies of the Earth's magnetism. It was rediscovered by Robert Norman, an English hydrographer, in 1576. There appears to be no record of any discovery of inclination by the Chinese.

In 1634 Henry Gellibrand, Professor of Astronomy at Gresham College, first discovered that magnetic declination changed with time. The observations in London on which Gellibrand based his conclusion were as follows:

Date	Observer	Declination
16 Oct 1580	William Borough	11.3° E
13 June 1622	Edmund Gunter	6.0° E
16 June 1634	Henry Gellibrand	4.1° E

William Borough was Comptroller of the Navy and Edmund Gunter was Gellibrand's predecessor as Professor of Astronomy. Although Gunter noted the difference between his and Borough's measurements he failed to appreciate the significance, attributing it to possible inaccuracy in the earliest measurement.

1.1.3 The Experiments of William Gilbert

Gerhard Mercator, mathematician and geographer, first realized in 1546 from observations of magnetic declination that the point to which the needle seeks could not lie in the heavens, leading him to fix the magnetic pole firmly on the Earth. Norman and Borough subsequently consolidated the view that magnetic directivity was associated with the Earth and even shifted the focus from the pole to a region closer to the center of the Earth. The culmination of this period of magnetic investigation was the publication in 1600 by William Gilbert, physician to Queen Elizabeth I, of his results of experimental studies in magnetism in a treatise entitled *De Magnete*.

In an impressive series of experiments Gilbert investigated the variation in inclination over the surface of a piece of lodestone cut into the shape of a sphere. He thus made the first practical dip circle. Although Gilbert must have leaned heavily on the work of Petrus Peregrinus (Smith, 1970a), he had a much wider basis of general observation to draw on than did Peregrinus. The most important gap in Peregrinus' knowledge was ignorance of magnetic inclination, and it was this extra information that enabled Gilbert to realize "*magnus magnes ipse est globus terrestris*" (the Earth globe itself is a great magnet). Had Peregrinus known that his magnets behaved like compass needles he might well have been able to generalize from his spherical piece of lodestone to the Earth as Gilbert did. Indeed, Mitchell (1939) has suggested that without Norman's discovery of inclination, *De Magnete* would never have been written. Gilbert's work was essentially the culmination of centuries of thought and experimentation in geomagnetism. His conclusions put a stop to the wild speculations that were then current concerning magnetism and the magnetic needle. Apart from the roundness of the Earth, magnetism was the first property to be attributed to the body of the Earth as a whole. Newton's gravitation came 87 years later with the publication of his *Principia*.

1.1.4 Magnetic Charts and the Search for the Poles

The magnetic compass was clearly useful as an aid in navigation with the result that methods for determining declination at sea were developed by Portuguese navigators. The most remarkable of these navigators was João de Castro who commanded one of the 11 ships that sailed to the East Indies in 1538. Using an instrument with an arrangement like a sundial with magnetic needle, he determined the magnetic azimuth of the Sun at equal altitudes before and after noon. The half-difference of these azimuths, measured clockwise and anticlockwise, respectively, was the magnetic declination. João de Castro also carried out observations during his voyage along the west coast of India and in the Red Sea, obtaining 43 determinations of declination between the years 1538 and 1541. These observations represent the first and most significant attempts to chart the variation in declination worldwide. The method was soon universally introduced on ships.

The first geomagnetic chart was drawn up by Edmund Halley following two voyages between 1698 and 1700 in the North and South Atlantic Oceans. These were the first sea voyages made for purely scientific purposes and led to the first declination chart of the Earth, published in 1702. These voyages also led Halley to discover the westward drift of some geomagnetic features. The first chart of inclination for the whole Earth was published by Johan Carl Wilcke in Stockholm in 1768.

The first surveys of global magnetic intensity, and the demonstration of its variation with latitude, are commonly attributed to Alexander von Humboldt, the German explorer. However, in 1838 Edward Sabine pointed out that the earliest surviving survey of global magnetic intensity, showing it to strengthen away from the equator, was made by De Rossel during the 1791–1794 expedition of D'Entrecasteaux. Even earlier measurements seem certain to have been made by the scientist Robert de Paul, chevalier de Lamanon, of the La Pérouse expedition, but any records were evidently lost when the expedition ships were wrecked (see Lilley and Day, 1993).

The D'Entrecasteaux expedition was sent to search for the missing La Pérouse expedition and reported six magnetic intensity measurements based on the period of oscillation (referred to as time of vibration by Sabine) of a vertical dip needle disturbed from rest in the plane of the magnetic meridian. The period of oscillation T of a magnet with moment of inertia $\Im$ and magnetic moment p in a magnetic field B (tesla) is given by

$$T = 2\pi \sqrt{\frac{\Im}{pB}} \ .$$

$$(1.1.1)$$

The measurements, as reported by Sabine (Lilley and Day, 1993), are listed below and clearly indicate the fact that the magnetic field intensity becomes greater at higher latitudes.

Station	*Date*	*Lat.*	*Long.*	*Magnetic dip*	*Period (s)*
Brest, France	20 Sept. 1791	48°24' N	4°26' W	71°30' N	2.020
Teneriffe	21 Oct. 1791	28°28' N	16°18' W	62°25' N	2.081
Tasmania[*]	11 May 1792	43°32' S	146°57' E	70°50' S	1.869
Ambon, Moluccas[*]	9 Oct. 1792	3°42' S	128°08' E	20°37' S	2.403
Tasmania[*]	7 Feb. 1793	43°34' S	146°57' E	72°22' S	1.850
Surabaya, Java	9 May 1794	7°14' S	112°42' E	25°20' S	2.429

[*]Present day names are given. Tasmania was referred to as Van Diemen's Land and Ambon as Amboyna.

Because the measurements made by De Rossel were not published until 1808, after those of Alexander von Humboldt in 1798–1803, this important observation in geomagnetism has frequently been credited to the wrong person. These determinations clearly entitle Admiral De Rossel with the distinction of having been the first to ascertain that the magnetic intensity is greater near the poles than at the equator. It was later that von Humboldt used the observation of the time of oscillation of a compass needle in the horizontal plane to determine relative measurements of horizontal intensity. This simpler method was generally adopted on scientific journeys. The first isodynamic charts for the horizontal intensity H and total intensity F were published by Christopher Hansteen, Professor of Applied Mathematics at the Norwegian University in 1825 and 1826.

The first representation of the geomagnetic field in mathematical form was made by the German mathematician Carl Frederich Gauss in his treatise *Allgemeine Theorie des Erdmagnitismus* of 1838. He calculated the coefficients in the spherical harmonic expression for the potential of the geomagnetic field from the values X, Y, Z derived from isomagnetic charts at 84 points, spaced at 30° of longitude along seven circles of latitude. For this purpose he used the three isomagnetic charts (Barlow's for declination, 1833; Horner's for inclination, 1836; and Sabine's for total intensity, 1837).

Gauss was able to forecast the positions of the geomagnetic poles, the points where the best fitting dipole axis cuts the surface of the Earth. However, the positions of the North and South Magnetic Poles (or dip poles where the inclination is vertical) are not so readily calculated and depend on extrapolation from existing maps. Their location was thus of considerable interest to early explorers of the polar regions. James Clark Ross discovered the position of the North Magnetic Pole on 1 June 1831 at 70°05' N, 96°46' W. He later attempted, unsuccessfully, to reach the southern counterpart and at least 78 yr were to pass before another attempt at location occurred. During Shackleton's 1907–1909 expedition to the South Pole a party of Australians, including two geologists — Professors T.W. Edgeworth David and Douglas Mawson — set out from Cape Royds and thought they had located the position of the South Magnetic Pole on 16 January 1909. However, it now appears that because of imprecise measurements, of necessity made in hurry, their location was incorrect. The most probable location for 1909 was subsequently calculated to have been at 71°36' S, 152°0' E, about 130 km northwest of David and Mawson's extreme station (Webb and Chree, 1925).

1.2 Fossil Magnetism and the Magnetic Field in the Past

1.2.1 Early Observations

The first observations that certain rocks were magnetized parallel to the Earth's magnetic field were made by Delesse in 1849 and Melloni in 1853. During the late eighteenth century it had already been observed, because of their effect on the compass needle, that some rocks possessed extremely strong remanent magnetization. These effects were attributed to lightning strikes by von Humboldt in 1797. Folgerhaiter (1899) extended the work of Delesse and Melloni, but also studied the magnetization of bricks and pottery. He argued that if the position of the brick or pot in the firing kiln were known, then the remanent magnetization it acquired on cooling should provide a record of the direction of the Earth's magnetic field.

1.2.2 Reversals of the Magnetic Field

Following on the work of Folgerhaiter, David (1904) and Brunhes (1906) investigated material baked by lava flows rather than those baked by man, comparing the direction of magnetization of the flows with that of the underlying baked clay. They reported the first discovery of natural remanent magnetization roughly opposed to that of the present field. The fact that the baked clays were also reversely magnetized led to the first speculation that the Earth's magnetic field had reversed itself in the past. Mercanton (1926) pointed out that if the Earth's magnetic field had reversed itself in the past then reversely magnetized rocks should be found in all parts of the world. He obtained samples from Spitsbergen, Greenland, Iceland, the Faroe Islands, Mull, Jan Mayen Land, and Australia, and found that some were magnetized in the same sense as the present field and others were roughly reversed from it. At the same time Matuyama (1929) observed similar effects in lavas covering the past one or two million years from Japan and Manchuria. He noticed, however, that the reverse lavas were always older than those in the same sense as the present field (normal). This was the first suggestion of a time sequence associated with reversely magnetized rocks. This time sequence was noted in detail from alternating polarities observed in the Massif Central, France, by Roche (1951, 1953, 1956, 1958), who placed the time of the last reversal as the middle of the Lower Pleistocene. Similar observations were made by Hospers (1953, 1954a) in lava sequences in Iceland, and by Khramov (1955, 1957, 1958) in sedimentary sequences in western Turkmenia.

Although Hospers (1955) appears to have been the first to suggest the use of reversals as a means of stratigraphic correlation, Khramov (1955, 1957) undertook the first application. In his book, Khramov (1958; English translation, 1960) suggested the possibility of a strict worldwide correlation of volcanic and sedimentary rocks and the creation of a single geochronological paleomagnetic time scale valid for the whole Earth. At this early stage the data indicated rhythmic variations of polarity. Glen (1982), in an historical account of the development of the polarity time scale, notes that Khramov's work clearly influenced the early work on polarity time scales of Cox, Doell and Dalrymple in the United States, and of McDougall and Tarling in Australia in the early 1960s.

1.2.3 Secular Variation

Studies of secular variation in the past using fossil magnetizations were first attempted by Chevalier (1925), using lava flows from Mount Etna (Fig. 1.4). The use of archaeomagnetic techniques on baked hearths and pottery was detailed by Thellier and Thellier (1951, 1952) in France and by Cook and Belshé (1958) in England. The first attempt at determining recent secular variation from

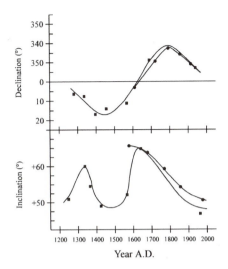

Fig. 1.4. Secular variation in Sicily determined by Chevalier (1925) from the magnetization of historic lavas of Mount Etna (shown as squares). The circles represent direct observations of the Earth's magnetic field.

sediments was made by Johnson *et al.* (1948) in the United States on some New England varved sediments covering a time span of about 5000 yr more than 10,000 years ago. A similar attempt using Swedish varved clays was made by Griffiths (1953). It was the development of a portable core sampler for lake deposits (Mackereth, 1958) that saw the start of extensive studies of lake sediments for obtaining prehistoric records of secular variation (Mackereth, 1971; Creer *et al.*, 1972). Studies of intensity variations in the past using archaeological material were pioneered by Thellier (1937a,b). A summary of results and details of the Thellier technique was later published by Thellier and Thellier (1959a,b).

1.2.4 Continental Drift

The potential use of fossil magnetism in rocks was well established by 1930, the most astute suggestion having been made by Mercanton (1926). He proposed that, because of the approximate correlation of the geomagnetic and rotational axis at the present, it might be possible to test the hypothesis of polar wandering and continental drift. Coming at the time of Wegener's (1924) continental drift hypothesis, it might then have served as a simple way of settling the dispute. However, paleomagnetism was a new technique and regarded by most as unpromising, and in any case Wegener's hypothesis was in general disrepute in both Europe and North America. Twenty years was to lapse until paleomagnetism was again being studied to test the polar wandering and continental drift hypothesis. Graham (1949) in the United States paved the way with a classic paper describing some of the field methods still in use today, and Blackett (1952) in England developed a sensitive astatic magnetometer that could be used for the measurement of the weak magnetization of sedimentary rocks.

Work carried out by groups in England led by Blackett and Runcorn during the early 1950s was crucial in placing directional studies from paleomagnetism on a firm footing. Of particular importance was the development by Fisher (1953) of a statistical method for the analysis of directions on a sphere. Using this method Hospers (1953, 1954a) was able to show that when averaged over tens of thousands of years the direction of the geomagnetic field approximated closely to that of an axial geocentric dipole. This enabled him to calculate the first paleomagnetic poles and associated errors. Runcorn (1954) provided theoretical arguments that the Coriolis force dominated convective motions in the Earth's core and that this would on average produce an axial (but not necessarily dipole) field. Later Runcorn (1959) showed that the observed pre-Tertiary paleomagnetic inclinations did not fit those expected from either an axial dipole, quadrupole or octupole. Thus the paleomagnetic observations could not readily be explained by geomagnetic field variations. Using the technique of calculating paleomagnetic poles, Creer *et al.* (1954) drew the first apparent polar wander path from which the method for making the paleomagnetic test for continental drift became obvious.

The data acquired by the groups in England led by Blackett and Runcorn appeared to indicate polar differences of about 25° between Europe and North America. This suggested that continental drift had occurred and that these continents had previously been very much closer (Runcorn, 1956; Irving, 1956). More dramatic differences emerged in data from India and Australia (Irving, 1956), South Africa (Graham and Hales, 1957), and South America (Creer, 1958). However, reviews of the data by Irving (1959), Blackett *et al.* (1960), and Cox and Doell (1960) arrived at differing conclusions. Irving (1959) and Blackett *et al.* (1960), like the majority of English paleomagnetists at that time, concluded that the data favored continental drift. Cox and Doell (1960), like most North Americans, took a more cautious approach and concluded that the data might be explained more plausibly by a relatively rapidly changing magnetic field with or without polar wandering. Khramov (1958) had already identified himself as an outspoken member of the very small community of Russian proponents of continental drift based on paleomagnetic data. Improved techniques in paleomagnetism involving various forms of "magnetic cleaning" considerably improved the database in the 1960s and showed more and more the strong case for continental drift. However, it was the interpretation by Vine and Matthews (1963) and Morley and Larochelle (1964) of the linear magnetic anomalies observed at sea as representing alternating blocks of normal and reversely magnetized oceanic crust that finally convinced geologists and geophysicists of the reality of continental drift (see Chapter 5).

1.3 Investigations of the External Magnetic Field

1.3.1 Transient Magnetic Variations

Careful observation of a compass needle with a microscope in 1722 led George Graham, an instrument maker in London, to discover transient magnetic variations that are more rapid than the slow secular variation. On some days he noted slow and regular changes of declination and on other days irregular changes sometimes larger and more rapid. The larger changes were later referred to as *magnetic storms*. Graham thus made the important distinction between days that are magnetically quiet and those that are disturbed. In 1759 Canton noted in London that the quiet daily variation was greater in summer than in winter, the first mention of an annual variation. In 1782 in Paris, Dominique Cassini showed that the daily variation could not be due to a daily variation in temperature. In 1777 Charles Coulomb greatly increased the sensitivity of magnetic measurements by suspending a magnetic needle from a fine string. These instruments could be made even more sensitive by attaching a small mirror that moved a light beam. This type of instrument dominated geomagnetism for nearly 200 yr.

Gauss and his successors from 1834 onward discovered the daily variation in the other geomagnetic elements. Many speculations were aroused but gradually the morphology of the quiet-day solar variation (S_q) became clear. Edward Sabine in 1852 found that S_q varied in intensity by 50% or more in correlation with rises and falls in the annual mean sunspot number. He calculated the lunar daily variation that had been discovered in 1850 by Kreil in Prague. These observations led Stewart in 1882 to infer that S_q must have its origin in electric currents in the upper atmosphere. Schuster in 1889 applied spherical harmonic analysis to S_q and confirmed Stewart's inference that S_q arises from an external source, but also found a smaller part of internal origin. He correctly ascribed this to electric currents induced in the Earth by the external source.

The proposition that there is an electrically conductive upper atmospheric layer was the first suggestion of the existence of the ionosphere. Final proof of this had to await the first soundings with radio waves by Breit and Tuve (1926). In 1922 the Department of Terrestrial Magnetism of the Carnegie Institution of

Washington set up a geomagnetic observatory at Huancayo in Peru, very close to the magnetic equator. This revealed a hitherto unsuspected enhancement of the horizontal component intensity of S_q by a factor of about two. Subsequently this enhancement was found to occur all along the magnetic equator within a narrow band of about 2° latitude on either side. The narrow equatorial current system implied from these observations was named the *equatorial electrojet* by Chapman (1951).

1.3.2 Early Theories of Magnetic Storms and Auroras

In addition to the quiet solar daily variation S_q, there is the disturbance daily variation S_D. Both S_D and the additional fields that arise during magnetic storms have been intensely studied over the past century, especially the relation of magnetic storms to auroras. Celsius found that the large magnetic disturbance of 5 April, 1741 was detected simultaneously by himself in Uppsala, Sweden and by Graham in London, thus demonstrating that magnetic storms were not just local phenomena. The magnetic network started by Gauss and Weber later showed the storms to be worldwide phenomena. Celsius also noted that the magnetic needle was disturbed during auroral displays. The British astronomer Richard Carrington first noted the occurrence of solar flares in 1859, noting also that an unusually intense magnetic storm followed 17 hr later, accompanied by polar auroras that could be seen far from the polar regions. He noted the coincidence but it was not until 1892 that George Hale devised the spectroheliograph, with which solar flares could be readily observed through filters that isolate the red $H\alpha$ brightenings near sunspots. Hale produced a series of photographs documenting the evolution of a large solar flare that was followed 19 hr later by an intense magnetic storm. Many such correlations soon followed, leaving no doubt that something was propagating from the Sun to the Earth at about 1000 km s^{-1} or faster, causing a magnetic disturbance on arrival.

   Many theories of magnetic storms and auroras have been proposed, and just as many have fallen by the wayside. Elias Loomis of Yale published a map in 1860 showing contours of equal auroral frequency in the northern hemisphere. The contours centered around the magnetic pole rather than the geographic one. In 1896 Birkeland showed that when a stream of electrons is projected toward a magnetized sphere they are deflected toward the poles by the magnetic field. The pieces suddenly seemed to fall into place. Solar flares or sunspots apparently emitted electron streams that were steered by the Earth's magnetic field toward the auroral zones. Since a stream of electrons carried an electric current, a magnetic disturbance would be produced. Birkeland asked his colleague Carl Störmer to calculate the motion of electrons in a dipole field. Unfortunately the problem has no analytical solution, but Störmer (1907–13) showed that a wide class of orbits existed in the dipole field that were trapped and did not extend to infinity. He also showed that for sufficiently low particle energies all orbits hitting the Earth at low and middle latitudes were trapped, so that particles arriving from a distant source never reached those latitudes but were always steered to the polar regions or turned away in accord with Birkeland's experiments.

Such a stream ejected from the Sun would require a very high density of charged particles to produce auroras and magnetic storms. Such high densities of charged particles of one sign would, however, be dispersed by their mutual electrostatic repulsion long before the stream reached the Earth. Lindemann (1919) therefore suggested that the solar corpuscular stream might be electrostatically neutral but ionized. It took more than 10 yr before Sydney Chapman together with newly graduated Vincent Ferraro worked out how a neutral beam could cause magnetic disturbances. They realized that an electrically neutral mixture of ions and electrons (what today would be called a plasma) would be a very good conductor of electricity. Electric currents would be induced in a cloud of such matter as it approached the Earth, thereby creating a magnetic disturbance. As a large plasma cloud nears the Earth its front boundary appears like an approaching "wall." Furthermore, if the cloud is a perfect electrical conductor, all induced currents flow on the surface of the "wall." Following some original work of Maxwell's, Chapman and Ferraro (1931, 1932, 1933) showed that the initial magnetic disturbance caused by the cloud should resemble the field of an image dipole at twice the distance of the cloud, rushing toward the Earth at twice the cloud's speed. This was how they explained the *sudden commencement* (§2.5.3), a rapid increase in the magnetic field, that occurs at the onset of many (but not all) magnetic storms.

1.3.3 The Magnetosphere

The Earth's magnetic field also exerts a force on the induced currents, a force that grows stronger as the plasma cloud draws nearer. Chapman and Ferraro argued that the force would become strong enough to stop any further frontal advance of the cloud toward Earth. However, the flanks would continue to advance so that soon a hollow, or magnetic cavity, would be formed enveloping the Earth (Fig. 1.5). This was known for many years as the *Chapman–Ferraro cavity*, the region from which the plasma of the cloud was excluded and within which the geomagnetic field would be confined. This region was later labeled by Gold (1959) as the *magnetosphere*.

The Chapman–Ferraro cavity would, however, be fully formed within a few minutes, but a typical magnetic storm lasts much longer with the *main phase* of gradual weakening of the field lasting 6–12 hr and the slow recovery lasting 1–3 days. Following Störmer's (1907–13) work on such ring currents, Schmidt (1924) suggested that a ring current was the cause of the main phase of magnetic

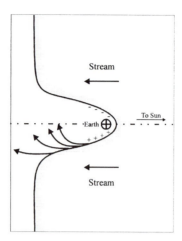

Fig. 1.5. The formation of the Chapman-Ferraro cavity, showing an equatorial section with the sign of the electric charges on the surface of the cavity carved out by the advancing neutral ionized stream. The curved arrows trace the paths of ions and electrons that Chapman and Ferraro proposed to account for ring current effects. Constructed from Chapman and Bartels (1940).

storms. As part of their own theory Chapman and Ferraro (1933) also produced a version of the ring current, set up (somehow) inside the Chapman–Ferraro cavity. The curved arrows of Fig. 1.5 are related to this.

Hannes Alfvén in Sweden did not, however, believe in the Chapman–Ferraro theory and viewed the plasma cloud as a collection of individually moving particles (Alfvén, 1939) rather than as a continuous field. He proposed that ring current effects as well as auroras were due to the entry into the Earth's magnetic field of particles from the plasma cloud, convected there by an electric field due to the cloud's motion. In this he used his "guiding-center" approximation to show that a cavity would be created around the Earth leading to the ring current field. Alfvén's (1939) paper has widely influenced contemporary thinking on plasma dynamics in a planetary magnetosphere. However, this important paper was rejected for publication by a leading journal on the grounds that it disagreed with the theories of Chapman and his colleagues, and so was published in an obscure Swedish journal. It was still poorly known and appreciated (ignored?) outside Scandinavia when the first satellite was launched in 1957 (Stern, 1989).

In the 1930s the existence of a plasma around the Earth, extending to several tens of kilometres, was not known, nor was it known that there exists a steady plasma flow out of the Sun. Biermann (1951) inferred the existence of such an outflow from comet observations, and Parker (1958) concluded that the source might be a hydrodynamic expansion of the ionized gas in the solar corona. This *solar wind* would of course be much enhanced during periods of solar activity leading to the disturbance observed during the initial phase of a magnetic storm as originally proposed by Chapman and Ferraro. Finally, exploration of the upper atmosphere with rockets and satellites led to the discovery of the existence of geomagnetically trapped radiation by Van Allen *et al.* (1958). Within the year following that discovery the gross features of the trapped radiation in the so-called *Van Allen belts* had been determined. Stern (1989) has given a brief history of magnetospheric physics before the era of spaceflight.

1.4 Origin of the Earth's Magnetic Field

William Gilbert, in his *De Magnete*, first identified the Earth's magnetic field as having its origin inside the Earth. He believed that its origin lay in permanent magnetization (lodestone) at the center of the Earth. Later in the first half of the seventeenth century René Descartes developed a large following for his ideas on the origin of the Earth's magnetic field (Mattis, 1965). Descartes believed that the Earth's magnetism was associated with "threaded parts" that were channeled in one-way ducts through the Earth whose main entrances and exits were the North and South Poles. There were two types of threaded parts, one entered the North Pole and exited the South Pole and the other was the inverse of this. In both cases the threaded parts had what in modern terms would be called "closed field lines." The threaded parts traveled through air to connect with the parts that traveled in the Earth. If the threaded parts chanced upon some lodestone during the course of their voyage, they would abandon their trip and would pass through the lodestone, giving rise to complex magnetic vortices. This is apparently the first explanation for the nondipole field.

In more modern times Einstein, shortly after writing his special relativity paper in 1905, described the problem of the origin of the Earth's magnetic field as being one of the most important unsolved problems in physics. Blackett (1947) proposed that large astronomical bodies might, because of some new as-yet-unexplained law of physics, have dipole moments that are directly proportional to their angular momentum. To test this hypothesis Blackett (1952) attempted to measure the weak magnetic field predicted from a pure gold sphere as it rotated with the Earth. In spite of using a sensitive astatic magnetometer specially developed for this purpose, he obtained negative results and he concluded his hypothesis was therefore in error. The sensitive astatic magnetometer he produced was subsequently used to measure the weak magnetizations of sedimentary rocks and was an important instrument in the early days of paleomagnetism (Chapter 3). Rotation of a body in which there was charge separation would produce a magnetic field external to the body and this could provide a possible explanation for the Blackett hypothesis. However, Inglis

(1955) discounted the possibility that large magnetic fields in the Earth might originate from charge separation due to temperature and/or pressure gradients. So although this process is probably unimportant in the Earth, it may still play an important role in some other large bodies.

Sir Joseph Larmor (1919a,b) was the first to suggest that large astronomical bodies, such as the Sun, might have magnetic fields caused by a self-exciting dynamo process. However, enthusiasm for this suggestion was dampened by the theoretical finding of Cowling (1934) who concluded that "The theory proposed by Sir Joseph Larmor, that the magnetic field of a sunspot is sustained by the currents it induces in moving matter, is examined and shown to be faulty; the same result also applies for the similar theory of the maintenance of the general field of Earth and Sun." What Cowling's analysis actually showed was that a magnetic field symmetric about an axis cannot be maintained by a symmetric motion, and so was much less general than suggested by his conclusion. It was, however, the first of the antidynamo theorems and was a prelude to extensive theoretical work to produce a more general antidynamo theorem. This continued until some important theoretical breakthroughs by Childress (1970) and G.O. Roberts (1970) demonstrated that no such general antidynamo theorem existed (see §9.3).

The first important mathematical contributions to dynamo theory were made by Elsasser (1946a,b, 1947) and Bullard (1949a,b). They were the first to discuss dynamos in a more modern sense. They used magnetohydrodynamic (MHD) dynamo theory in a liquid core to produce models for a self-sustaining dynamo, rather than the homopolar dynamo approach that relied on wires, rigid disks, etc. Alfvén (1942b, 1950) first showed that the well-known Kelvin–Helmholtz theorems, developed before the turn of the twentieth century, were applicable to a perfectly conducting MHD fluid. In doing so he demonstrated that the magnetic flux through any closed contour moving with the fluid is constant (the frozen-in-field theorem), a contribution that played a central role in the theories that followed for the Earth's dynamo. It follows that a conducting fluid permeated by a magnetic field will support wave motion, with the magnetic field lines behaving rather like elastic strings. The resulting waves are referred to as Alfvén waves.

Chapter Two

The Present Geomagnetic Field: Analysis and Description from Historical Observations

2.1 Magnetic Elements and Charts

Direct measurements of the Earth's magnetic field (usually quoted in the magnetic induction units of nanotesla) are made continuously at magnetic observatories, and are obtained from various oceanographic, land, aircraft, and satellite surveys. In addition, Bloxham and Gubbins (1985) analyzed the data of ancient mariners to achieve an extremely valuable extension of the historical directional analysis back to the early 17th century (and at the same time upgraded the navigational information associated with those data). A common way of describing the magnetic field is simply to plot different *magnetic elements*, such as total intensity or inclination. The various elements used in such plots are defined in Fig. 2.1. Although the symbols for the magnetic elements given in Fig. 2.1 are widely referred to in describing the magnetic field, somewhat different symbols are used in other branches of geomagnetism. In particular, **H** is generally used to describe the magnetic field and *H* its magnitude, a notation that will be used throughout this text. The most common display format is an *isomagnetic chart*, i.e. a contour map of equal values of a particular magnetic element. These charts are termed *isogonic* for declination, *isoclinic* for inclination, and *isodynamic* for equal intensity (or for intensity of a

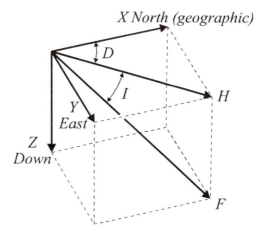

Fig. 2.1. The main elements of the geomagnetic field. The deviation, D, of a compass needle from true north is referred to as the *declination* (reckoned positive eastward). The compass needle lies in the magnetic meridian containing the total field F, which is at an angle I, termed the *inclination* (or dip), to the horizontal. The inclination is reckoned positive downward (as in the northern hemisphere) and negative upward (as in the southern hemisphere). The horizontal (H) and vertical (Z) components of F are given by $H = F\cos I$ and $Z = F\sin I$, respectively. Z is reckoned positive downward, as for I. The horizontal component can be resolved into two components: X (northward) $= H\cos D$ and Y (eastward) $= H\sin D$. Then $\tan D = Y/X$ and $\tan I = Z/H$.

particular field component, say, the horizontal field). *Isoporic* charts refer to isomagnetic charts of secular variation.

Following the work of Gilbert in the seventeenth century, geomagnetists have recognized that the magnetic field at the surface of the Earth can be well approximated by a magnetic dipole placed at the Earth's center and tilted about $11°$ with respect to the axis of rotation. Such a dipole accounts for roughly 80% of the Earth's magnetic field at the surface. The magnetic field inclination is downward throughout most of the northern hemisphere and upward throughout most of the southern hemisphere. A line that passes through the center of the Earth along the dipole axis intersects the Earth's surface at two points, referred to as the *geomagnetic poles*. These poles differ from the *magnetic poles*, which are the two points on the Earth's surface close to the geomagnetic poles where the field is vertical. The geomagnetic poles and magnetic poles would of course coincide if the field at the Earth's surface were perfectly described by a geocentric dipole, but this is not the case because about 20% of the field at the Earth's surface remains after the best fitting geocentric dipole field is removed. This remaining part of the field is referred to as the *nondipole field*. Both the dipole and the nondipole parts of the Earth's magnetic field vary with time.

The most important and commonly used method of describing the Earth's magnetic field quantitatively is through spherical harmonic analysis, although other quantitative ways of representing the Earth's magnetic field will be discussed later in this chapter. Various combinations of global observatory, satellite, and magnetic field survey data are used to obtain the coefficients of truncated spherical harmonic series descriptions of the Earth's magnetic field and its secular variation. Models of the field itself are determined essentially from land, sea, airborne, and satellite data. The secular variation models (which are usually truncated at a lower degree) are based mainly on data from continuously recording observatories and to some extent on repeat-station data. In the absence of new global coverage of vector satellite data the need for well-distributed observatory data is growing. The distribution of permanent magnetic observatories around the world, as shown in Fig. 2.2, is far from ideal. There are adequate concentrations only in Europe and parts of North America, and poor coverage in much of the southern hemisphere and in underdeveloped countries. Many oceanic regions contain no observatories.

The uneven distribution of magnetic observatories has a considerable effect on the accuracy of both the main field and the secular variation models over much of the globe. The number of observatories has grown with time without any overall plan. Many of them were established before there was any clear idea of the processes and scales of geomagnetic field phenomena and long before the era of satellites. Also, several observatories in key locations, especially those in underdeveloped countries, have closed due to lack of funding. Langel *et al.* (1995) have discussed this problem in detail and have presented the design for a permanent global network that covers geomagnetic phenomena with scales of about 1000 km in Europe and 4000 km and greater globally. To derive adequate spherical harmonic models of the main field and its secular variation observatories would need to be spaced no more than about 2000 km apart. To achieve this Langel *et al.* (1995) proposed an ideal distribution of 92 observatories worldwide that is economically realistic and would optimize the determination of global field models. Their scheme includes the establishment of eight sea-bottom observatories to cover those oceanic regions where magnetic field measurements cannot be made in any other way.

A considerable amount of information concerning the Earth's magnetic field and its variation in time is contained in the spherical harmonic coefficients. Often this is not well appreciated and, just as often, incorrect "physical" interpretations are drawn from the use of spherical harmonic representations of the field. Consequently, §2.2 and Appendix B are devoted to spherical harmonic analysis and an understanding of its use in describing the Earth's magnetic field, especially during the past 300 yr or so. Examples of various magnetic charts that can be derived from use of the spherical harmonic coefficients, giving a pictorial representation of the Earth's magnetic field and its secular variation, are shown in Fig. 2.3.

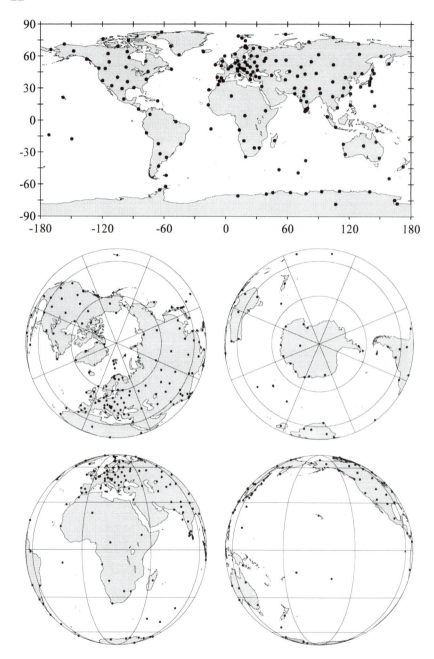

Fig. 2.2. The distribution of magnetic observatories on the Earth's surface that are used in determining the spherical harmonic description of the geomagnetic field.

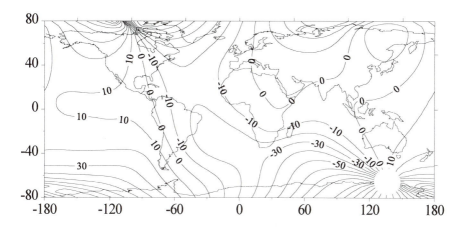

Fig. 2.3a. Isogonic chart for 1990 showing the variation in declination in degrees over the Earth's surface.

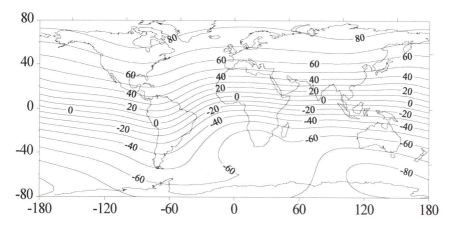

Fig. 2.3b. Isoclinic chart for 1990 showing the variation of inclination in degrees over the Earth's surface.

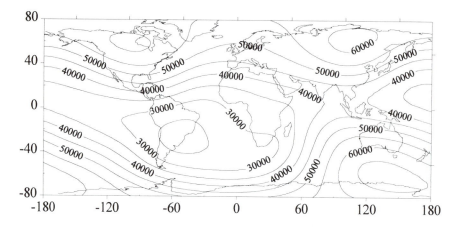

Fig. 2.3c. Isodynamic chart for 1990 showing the variation of total intensity over the Earth's surface. Contours are labeled in nT.

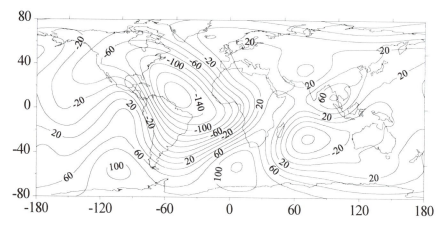

Fig. 2.3d. Isoporic chart for 1990 showing the rate of secular change in intensity of the vertical component. Values are in nT yr^{-1}.

2.2 Spherical Harmonic Description of the Earth's Magnetic Field

2.2.1 Scalar Potential for the Magnetic Field

The two Maxwell equations relating to magnetic field are

$$\nabla \times \mathbf{H} = \mathbf{J} + \frac{\partial \mathbf{D}}{\partial t} \qquad (2.2.1)$$

and

$$\nabla \cdot \mathbf{B} = 0 \ , \qquad (2.2.2)$$

where **H** is the magnetic field, **B** is the magnetic induction, **J** is the electric current density, and $\partial \mathbf{D}/\partial t$ is the electric displacement current density. As noted in §2.5.2, with the exception of thunderstorms the region from the Earth's surface up to about 50 km can be regarded as an electromagnetic vacuum. Thus for the region of interest it is reasonable to assume that **J**=**0** and that $\partial \mathbf{D}/\partial t$=**0**, so from (2.2.1) we have $\nabla \times \mathbf{H}$=**0**. This means that **H** is a conservative vector field in the region of interest so there exists a scalar potential ψ such that

$$\mathbf{H} = -\nabla \psi \ . \qquad (2.2.3)$$

Because **B** = μ_0**H** above the Earth's surface, where μ_0 (=$4\pi \times 10^{-7}$ Hm^{-1}) is the permeability of free space, if follows from (2.2.2) that $\nabla \cdot$H=0 also. Combining this with (2.2.3) shows that the potential ψ must satisfy Laplace's equation:

$$\nabla^2 \psi = 0 \qquad (2.2.4)$$

In spherical coordinates (r,θ,ϕ), where r is the distance from the center of the Earth, θ is the longitude and ϕ is the colatitude (90° minus the latitude) as shown in Fig. 2.4, Laplace's equation takes the form

$$\frac{1}{r}\frac{\partial^2}{\partial r^2}(r\psi) + \frac{1}{r^2 \sin\theta}\frac{\partial}{\partial\theta}\left(\sin\theta \frac{\partial\psi}{\partial\theta}\right) + \frac{1}{r^2 \sin^2\theta}\frac{\partial^2\psi}{\partial\phi^2} = 0 \ . \qquad (2.2.5)$$

This equation can, very conveniently, be solved through separation of variables as

$$\psi = \frac{U(r)}{r} P(\theta) Q(\phi) \ . \qquad (2.2.6)$$

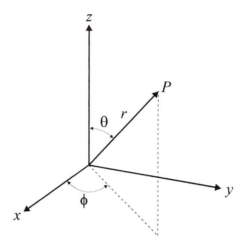

Fig. 2.4. Spherical polar coordinates.

The process for solution is well known and appears in many textbooks. The important point here is that *spherical harmonics* provide a completely general solution to (2.2.5), so §2.2.2 provides a brief review of this general solution.

2.2.2 Basics of Spherical Harmonics

For the special case in which there is no azimuthal (longitudinal) variation, simple substitution shows that the solution is

$$\psi(r,\theta) = \sum_{l=0}^{\infty}\left(A_l r^l + \frac{B_l}{r^{l+1}}\right)P_l(\cos\theta) \quad , \tag{2.2.7}$$

where A_l and B_l are constants and P_l are the Legendre polynomials given by Rodriques' formula:

$$P_l(\cos\theta) = P_l(\chi) = \frac{1}{2^l l!}\frac{d^l}{d\chi^l}\left(\chi^2 - 1\right)^l . \tag{2.2.8}$$

Appendix B gives the first 11 Legendre polynomials and Fig. 2.5 shows the first four as functions of θ up to 180°.

Substitution shows that a general solution to Laplace's equation is

$$\psi(r,\theta,\phi) = \sum_{l=0}^{\infty}\sum_{m=0}^{l}\left[A_{lm}r^l + B_{lm}r^{-(l+1)}\right]Y_l^m(\theta,\phi) \quad , \tag{2.2.9}$$

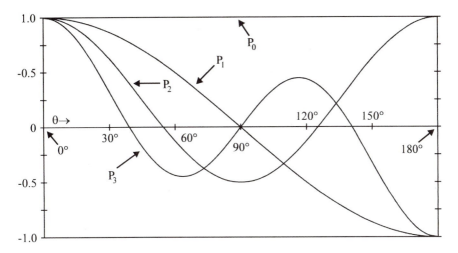

Fig. 2.5. The Legendre polynomials $P_l(\cos\theta)$ up to $l = 3$ as functions of θ.

where A_{lm} and B_{lm} are constants and the $Y_l^m(\theta,\phi)$ are *surface harmonics* of degree l and order m, given by

$$Y_l^m(\theta,\phi) = \left[\frac{(2l+1)(l-m)!}{4\pi(l+m)!}\right]^{\frac{1}{2}} P_{l,m}(\cos\theta)\, e^{im\phi} \quad , \qquad (2.2.10)$$

where $P_{l,m}(\cos\theta)=P_{l,m}(\chi)$ are called the *associated Legendre polynomials* and are given by the recursion formula

$$P_{l,m}(\chi) = \frac{1}{2^l\, l!}(1-\chi^2)^{\frac{m}{2}} \frac{d^{l+m}}{d\chi^{l+m}}(\chi^2-1)^l \quad . \qquad (2.2.11)$$

Unfortunately, the surface harmonics $Y_l^m(\theta,\phi)$ are often referred to as spherical harmonics: the term spherical harmonics will be used here to include the radial dependence r as well as the (θ,ϕ) dependence. These surface harmonics are orthonormal, the orthogonalization condition being

$$\int_0^{2\pi}d\phi \int_0^{\pi} \left[Y_l^m(\theta,\phi)\right]^* Y_l^m(\theta,\phi)\sin\theta\, d\theta = \delta_{ll'}\delta_{mm'} \quad , \qquad (2.2.12)$$

where the asterisk refers to the complex conjugate, and $\delta_{ll'}$ and $\delta_{mm'}$ are Kronecker delta functions ($\delta_{ll'} = 0$ for $l{\neq}l'$ and $\delta_{ll'} = +1$ for $l{=}l'$).

A particularly important property of spherical harmonic analysis for geomagnetism is that it allows for separation into internal (smaller r) and external (larger r) sources relative to the point of reference. If there are no external sources the radial component of the field, $-\partial\psi/\partial r$, must vanish at infinity and so no positive powers of r can occur in ψ. Thus the B_{lm} coefficients in (2.2.9) must represent internal sources. Conversely, if there are no internal sources then $-\partial\psi/\partial r$ must be finite throughout the interior and no negative powers of r can appear in ψ. Thus the A_{lm} coefficients in (2.2.9) must represent external sources.

In geomagnetism one uses *partially normalized Schmidt functions* P_l^m (see Schmidt, 1935; Chapman and Bartels, 1940) related to the associated Legendre polynomials, $P_{l,m}$, by

$$P_l^m = P_{l,m} \qquad\qquad \text{for } m = 0$$

$$P_l^m = \left[\frac{2(l-m)!}{(l+m)!}\right]^{\frac{1}{2}} P_{l,m} \qquad \text{for } m > 0 \quad . \qquad (2.2.13)$$

Appendix B provides further information on this partial normalization, gives $P_{l,m}$ and P_l^m for $l = 1$ to 4, and presents a polynomial form for $P_{l,m}$ that is convenient for computer coding algorithms.

The associated Legendre polynomials $P_{l,m}(\cos\theta)$ have $(l-m)$ zeros in $0 \leq \theta \leq 180$, dividing a meridian (or longitudinal line) into $(l-m+1)$ zones of alternating sign. Similarly, $\sin m\phi$ (or $\cos m\phi$) has $2m$ zeros, dividing a line of latitude into $2m$ meridional (or longitudinal) sectors of alternating sign at equal intervals of π/m. Thus the surface harmonics $P_l^m(\cos\theta)\sin m\phi$ (or $\cos m\phi$) divide up the surface of a sphere into regions (or tessera) created by the intersecting latitudinal zones and longitudinal sectors. In each region the sign of the surface harmonic is constant, but changes on crossing a zero line into an adjacent region. When $m=0$, the surface harmonics are described by Legendre polynomials (see Fig. 2.5) and are referred to as *zonal harmonics*, as illustrated in Fig. 2.6 for $P_7^0(\cos\theta)$. When $m=l$ (note that $0!=1$ by definition) the surface harmonics are referred to as *sectoral harmonics*, as illustrated in Fig. 2.6 for $P_7^7(\cos\theta)\cos 7\phi$. For the general case of $0<m<l$, the surface is divided into $2m(l-m+1)$ tessera and the harmonics are called *tesseral harmonics*, as illustrated in Fig. 2.6 for $P_7^4(\cos\theta)\cos 4\phi$.

2.2.3 Application of Spherical Harmonics to the Earth's Magnetic Field

The fact that $\nabla\cdot\mathbf{B}=0$ denies the existence of magnetic monopoles means that for a magnetic field there is no degree zero ($l=0$) term. Furthermore, it is convenient to normalize r to the Earth's mean radius a (6371 km; the radius of a sphere with the same volume as the Earth) so that the coefficients have the same dimensions. As already noted, in geomagnetism it is conventional to use the partially normalized Schmidt functions P_l^m, so the magnetic scalar potential ψ is written (Chapman and Bartels, 1940, 1962)

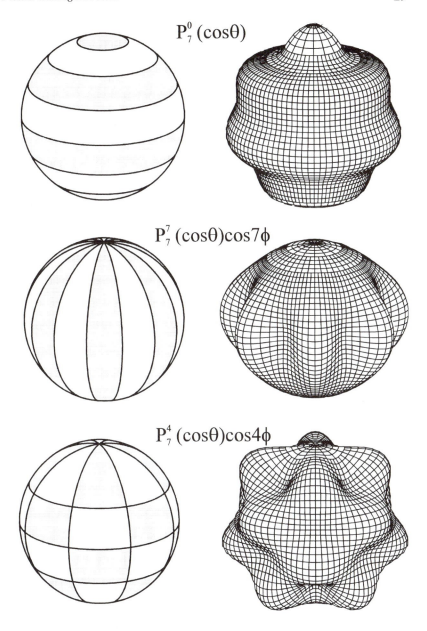

Fig. 2.6. Surface harmonics, showing the special cases of a zonal and a sectoral harmonic, and the general case of a tesseral harmonic.

$$\psi = \frac{a}{\mu_0} \sum_{l=1}^{\infty} \sum_{m=0}^{l} P_l^m(\cos\theta) \left\{ \left[C_l^m \left(\frac{r}{a}\right)^l + (1 - C_l^m)\left(\frac{a}{r}\right)^{l+1} \right] g_l^m \cos m\phi \right.$$

$$\left. + \left[S_l^m \left(\frac{r}{a}\right)^l + (1 - S_l^m)\left(\frac{a}{r}\right)^{l+1} \right] h_l^m \sin m\phi \right\} . \tag{2.2.14}$$

C_l^m and S_l^m are positive numbers between 0 and 1 that indicate the fraction of the potential associated with sources of external origin ($r>a$). The coefficients $(1 - C_l^m)$ and $(1 - S_l^m)$ indicate the fraction of the potential associated with sources of internal origin ($r<a$).

The coefficients g_l^m and h_l^m are to be evaluated from observations. In SI units these are traditionally given in units of magnetic induction (commonly quoted in nanotesla, see Table 2.1). In this way the coefficients in SI units have the same numerical values as in the cgs emu system of units where they were previously normally quoted in gamma. Hence in (2.2.14) the factor μ_0 is included to correct the dimensions on the right-hand side so the scalar potential has units of ampère as required for the magnetic field **H** (see Appendix A). It should be noted that above the Earth's surface there is no permanent magnetization so **B** is also a conservative field (i.e., $\nabla \times \mathbf{B} = 0$). Thus one can define a scalar potential ψ_B for the magnetic induction such that $\mathbf{B} = -\nabla \psi_B$. Naturally it follows that $\nabla^2 \psi_B = 0$ so ψ_B can be expanded exactly as in (2.2.14), but without the μ_0 when (g_l^m, h_l^m) are in nanotesla (this is the common practice for IAGA publications relating to the International Geomagnetic Reference Field (IGRF); see also Langel, 1987).

Schmidt (1939) also developed general formulas for a spheroidal body; however, for the Earth, with polar radius some 21 km smaller than the equatorial radius, the coefficients are only slightly modified by Schmidt's refinement (see Vestine, 1967).

Recalling from Fig. 2.1 the definition of the components X, Y, and Z of the magnetic field and the definition of the spherical polar coordinates from Fig. 2.4, it is apparent that X is the negative of the θ component, Y is the ϕ component, and Z is the negative of the radial component. Thus the components

$$X = \frac{1}{r} \frac{\partial \psi}{\partial \theta}, \qquad Y = -\frac{1}{r \sin\theta} \frac{\partial \psi}{\partial \phi}, \qquad Z = \frac{\partial \psi}{\partial r} \tag{2.2.15}$$

can easily be obtained in terms of the Schmidt polynomials. Indeed, the coefficients in (2.2.14) are typically obtained by using many sets of observations of (X, Y, Z) and then solving for the coefficients. By truncating the series at some value l, the relative importance of the internal $(1 - S_l^m, 1 - C_l^m)$ versus external $(S_l^m$ and $C_l^m)$ sources can be determined. The coefficients determined this way vary with time, as changes in ionospheric currents, associated changes in induced electric currents in the crust and upper mantle, and currents in the Earth's core lead to changes in the magnetic field at the Earth's surface. However, if mean values of the field as determined over several years are used, the above method of analysis shows that S_l^m and C_l^m do not differ significantly from zero (see Vestine, 1967). In this case (2.2.14) reduces to

$$\psi = \frac{a}{\mu_0} \sum_{l=1}^{\infty} \sum_{m=0}^{l} \left(\frac{a}{r}\right)^{l+1} P_l^m(\cos\theta)(g_l^m \cos m\phi + h_l^m \sin m\phi) . \qquad (2.2.16)$$

The coefficients g_l^m and h_l^m are called the Gauss coefficients (Chapman and Bartels, 1940, 1962) appropriate to the Schmidt polynomials, P_l^m. Discussion of the external part of the Earth's magnetic field will be taken up again in §2.5.

When written in the form of (2.2.16), the separation of the r, θ, and ϕ components of the spherical harmonics is particularly apparent. It is immediately obvious that the surface harmonic for a given r is simply a Fourier series for a given co-latitude and an associated Legendre series for a given longitude.

Table 2.1
IGRF 1995 Epoch Model Coefficients

| | | Main field (nT) | | Secular change (nT yr^{-1}) | |
l	*m*	*g*	*h*	*g*	*h*
1	0	-29,682.0	0.0	17.60	0.00
1	1	-1,789.0	5,318.0	13.00	-18.30
2	0	-2,197.0	0.0	-13.20	0.00
2	1	3,074.0	-2,356.0	3.70	-15.00
2	2	1,685.0	-425.0	-0.80	-8.80
3	0	1,329.0	0.0	1.50	0.00
3	1	-2,268.0	-263.0	-6.40	4.10
3	2	1,249.0	302.0	-0.20	2.20
3	3	769.0	-406.0	-8.10	-12.10
4	0	941.0	0.0	0.80	0.00
4	1	782.0	262.0	0.90	1.80
4	2	291.0	-232.0	-6.90	1.20
4	3	-421.0	98.0	0.50	2.70
4	4	116.0	-301.0	-4.60	-1.00
5	0	-210.0	0.0	0.80	0.00
5	1	352.0	44.0	0.10	0.20
5	2	237.0	157.0	-1.50	1.20
5	3	-122.0	-152.0	-2.00	0.30
5	4	-167.0	-64.0	-0.10	1.80
5	5	-26.0	99.0	2.30	0.90

Table 2.1 — *continued*

l	m	Main field (nT)		Secular change (nT yr^{-1})	
		g	h	g	h
6	0	66.0	0.0	0.50	0.00
6	1	64.0	-16.0	-0.40	0.30
6	2	65.0	77.0	0.60	-1.60
6	3	-172.0	67.0	1.90	-0.20
6	4	2.0	-57.0	-0.20	-0.90
6	5	17.0	4.0	-0.20	1.00
6	6	-94.0	28.0	0.00	2.20
7	0	78.0	0.0	-0.20	0.00
7	1	-67.0	-77.0	-0.80	0.80
7	2	1.0	-25.0	-0.60	0.20
7	3	29.0	3.0	0.60	0.60
7	4	4.0	22.0	1.20	-0.40
7	5	8.0	16.0	0.10	0.00
7	6	10.0	-23.0	0.20	-0.30
7	7	-2.0	-3.0	-0.60	0.00
8	0	24.0	0.0	0.30	0.00
8	1	4.0	12.0	-0.20	0.40
8	2	-1.0	-20.0	0.10	-0.20
8	3	-9.0	7.0	0.40	0.20
8	4	-14.0	-21.0	-1.10	0.70
8	5	4.0	12.0	0.30	0.00
8	6	5.0	10.0	0.20	-1.20
8	7	0.0	-17.0	-0.90	-0.70
8	8	-7.0	-10.0	-0.30	-0.60
9	0	4.0	0.0	0.00	0.00
9	1	9.0	-19.0	0.00	0.00
9	2	1.0	15.0	0.00	0.00
9	3	-12.0	11.0	0.00	0.00
9	4	9.0	-7.0	0.00	0.00
9	5	-4.0	-7.0	0.00	0.00
9	6	-2.0	9.0	0.00	0.00
9	7	7.0	7.0	0.00	0.00
9	8	0.0	-8.0	0.00	0.00
9	9	-6.0	1.0	0.00	0.00
10	0	-3.0	0.0	0.00	0.00
10	1	-4.0	2.0	0.00	0.00
10	2	2.0	1.0	0.00	0.00
10	3	-5.0	3.0	0.00	0.00
10	4	-2.0	6.0	0.00	0.00
10	5	4.0	-4.0	0.00	0.00
10	6	3.0	0.0	0.00	0.00
10	7	1.0	-2.0	0.00	0.00
10	8	3.0	3.0	0.00	0.00
10	9	3.0	-1.0	0.00	0.00
10	10	0.0	-6.0	0.00	0.00

2.2.4 Determination of the Gauss Coefficients

In principle, the number of measurements on the Earth's surface required to obtain the Gauss coefficients up to $l = N$ is straightforward. As already noted, g_0^0 is zero because there are no magnetic monopoles. Then for $l = 1$ there are two g_l^m terms, g_1^0 and g_1^1; for $l = 2$ there are 3 terms, $g_2^0, g_2^1,$ and g_2^2; and for $l = N$ there are $N + 1$ terms. Therefore, the total number of terms g_l^m to degree N is

$$\left[\frac{2 + (N + 1)}{2} \right] N = \frac{N^2 + 3N}{2} \ . \tag{2.2.17}$$

Following a similar procedure for the h_l^m terms (note that all h_l^0 terms vanish), one finds that the total number of terms h_l^m to degree N is

$$\left[\frac{1 + N}{2} \right] N \ . \tag{2.2.18}$$

By summing (2.2.17) and (2.2.18) the total number of measurements required to determine the potential ψ given by (2.2.16) up to degree N is

$$(N + 1)^2 - 1 \ . \tag{2.2.19}$$

Gauss (1839) calculated coefficients up to $l = 4$, while subsequently the series has often been computed to $l = 6$ to 10 (Vestine, 1967). Since 1965 there has been a tendency to calculate more terms, even beyond $l = 25$, although the interpretation of terms greater than $l = 14$ is complex (Kolesova and Kropachev, 1973; Alldredge and Stearns, 1974; Cain, 1975; Harrison and Carle, 1981; Shure and Parker, 1981; Langel, 1987). Note that for $l = 6$ at least 48 independent measurements of the Earth's field are required. These measurements come from a variety of sources, including continuously recording magnetic observatories and repeat stations, and more recently from ship, aircraft, and satellite surveys (e.g., Vestine *et al.*, 1947a,b; Alldredge and Stearns, 1974; Langel, 1987).

In practice, many more than 48 independent measurements are needed to determine ψ up to degree 6 reliably, and the measurements should ideally be uniformly distributed over the Earth's surface. Usually a method of weighted least squares is used to determine these coefficients from the measurements. The reliability with which these coefficients have actually been determined is difficult to assess, as has been pointed out explicitly in the analyses of the statistical problem by Wells (1973) and Langel (1987). Table 2.1 gives the Gauss coefficients for the 1995 epoch IGRF model up to degree 10 for the field itself and up to degree 8 for the secular variation. The main-field model is truncated at degree 10 (120 coefficients) as a practical compromise that has been adopted to produce a well-determined main-field model. The main-field coefficients are rounded to the nearest nanotesla to reflect the limit of resolution of the observational data. The prospective secular variation model is truncated at degree 8 (80 coefficients) with coefficients rounded to the nearest 0.1 nT yr⁻¹, which reflects the effective limit of the resolution of the data.

2.2.5 Interpretation of Spherical Harmonic Terms

Individual terms in (2.2.16) can be interpreted in terms of equivalent "sources" at the Earth's center. That this is unrealistic is obvious from the fact that the field is generated in the outer core, so the source cannot be at the Earth's center, and that it is not practically possible to create a current source that would generate precisely the field described by a single harmonic term.

The g_1^0 term is given by

$$\frac{g_1^0}{\mu_0} P_1^0(\cos\theta)\frac{a^3}{r^2} = \left(\frac{g_1^0 4\pi a^3}{\mu_0}\right)\frac{\cos\theta}{4\pi r^2} \ . \tag{2.2.20}$$

This term is the potential associated with a geocentric dipole with strength $(g_1^0 4\pi a^3 / \mu_0)$ oriented along the z axis (in the $+z$ direction; Fig. 2.4 gives the relationship between the axes). Thus the magnitude of g_1^0 in Table 2.1 is linearly related to the strength of the *axial geocentric dipole*, which points downward because the sign of g_1^0 is negative. The g_1^1 term is

$$\left(\frac{g_1^1 4\pi a^3}{\mu_0}\right)\left(\frac{\cos\phi\sin\theta}{4\pi r^2}\right) \ . \tag{2.2.21}$$

If γ is the angle between $\mathbf{r}$ and the x axis in Fig. 2.4, then it is easy to show that

$$\cos\gamma = \cos\phi\sin\theta \ . \tag{2.2.22}$$

Therefore, the g_1^1 term given above is that of a geocentric dipole oriented in the $+x$ direction. Similarly, the term $(h_1^1 4\pi a^3/\mu_0)(\sin\phi \sin\theta/4\pi r^2)$ corresponds to a geocentric dipole oriented along the $+y$ axis. The magnitude and direction of the geocentric dipole, p, is ,therefore,

$$p = \frac{4\pi a^3}{\mu_0}\sqrt{(g_1^0)^2 + (g_1^1)^2 + (h_1^1)^2} \tag{2.2.23}$$

and is found to be tilted at roughly 11° to the rotation axis. Equations (2.2.20) and (2.2.21) clearly show that the *potential* for a dipole falls off as r^{-2}. Thus, from (2.2.15), it is evident that the strength of the field itself falls of as r^{-3} for a dipole. Similarly, the $l = 2$ terms represent the geocentric quadrupole (potential falls off as r^{-3} and field strength falls off as r^{-4}), the $l = 3$ terms represent the geocentric octupole (potential falls off as r^{-4} and field strength falls off as r^{-5}), and so forth.

The g_1^0, g_1^1, and h_1^1 terms collectively represent the *dipole field*, while the remaining terms collectively represent the *nondipole field* (however, the dipole and quadrupole terms are sometimes combined to give an estimate of the "best" eccentric dipole field; see §2.2.3). The majority of the geomagnetic field at the Earth's surface can certainly be described by this tilted geocentric dipole (Table 2.1; see also Vestine (1967)). Strangely enough though, it is not immediately obvious how to determine precisely what proportion of the surface field originates from the geocentric dipole, nor is it always clear what is actually implied by the result. One answer has been provided by Stacey (1992), who subtracts the square of the nondipole terms in a spherical harmonic analysis from the square of the dipole field terms. He then takes the square root of this to arrive at the result that somewhat more than 20% of the present field at the Earth's surface is dipolar. This is a legitimate and straightforward way to handle the problem. However, all zonal harmonics give rise to vertical fields at the geographic poles, but not all of the same sign. Hence by using the root-mean-square approach, Stacey obtains a result that maximizes the implied nondipole contribution. Other methods include determining the best fitting geocentric (or alternatively nongeocentric) dipole field to the measurements. Such procedures lead to lower estimates for the nondipole field contribution and this accounts largely for the range of published estimates of the contribution of the nondipole at the Earth's surface to the dipole field (10 to 25%). Stacey's estimate is more important with regard to energy considerations, while the other methods appear more relevant to paleomagnetic directional analyses.

The coefficients in Table 2.1 indicate that the contribution from higher degree terms decreases (Fig. 2.7). The contribution decreases out to degree $l = 12$ to 14, beyond which the errors in the determination of the coefficients are typically comparable in size to the coefficients themselves (e.g., Kolesova and Kropachev, 1973) and significant contributions from noncore sources also seem likely. The lowest degree terms correspond to the largest wavelength features of the field, as can be appreciated by considering the zonal harmonics. The g_1^0 term has a cosθ dependence, so at a distance r the "wavelength" of the associated magnetic feature is $2\pi r$, the associated wavelength increasing linearly with distance from the source. The g_2^0 term has a wavelength of πr, the g_3^0 term a wavelength of $(2\pi r/3)$, and the g_l^0 term a wavelength of $(2\pi r/l)$. The fact that the longest wavelength associated with a source is $2\pi r$ means that an observer's distance from a source is less than the maximum wavelength associated with the source. Combining this with the fact that the intensity falls off rapidly (at least as r^{-3}, and more rapidly for the higher degree terms), as a general rule-of-thumb the depth of the source of an easily observable magnetic field feature is expected to be less

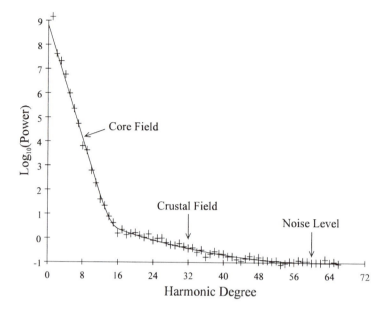

Fig. 2.7. Normalized power in spherical harmonics of the internal geomagnetic field as computed by Cain (pers. comm., see also Cain *et al.* 1974). The power is computed from the equation $W_l = (1/(2l+1))\Sigma_m\{(g_l^m)^2 + (h_l^m)^2\}$. Over several values of l the power varies according to $W_l \approx W_o K^{-l}$. For l from about 2 to about 14, $K \approx 3.9$, and for l from about 17 to about 34, $K \approx 1.1$.

than the wavelength of that feature. This nonunique physical argument is used as the basis for a rough separation of the field of internal origin into crustal sources and core-field sources.

The wavelengths for the field of internal origin that have harmonics of degree 14 or less are usually attributed to sources within the Earth's core. Crustal magnetic sources (associated mostly with remanent magnetization, but some is due to induced magnetization) generally have maximum associated wavelengths of a few tens, or occasionally a few hundreds, of kilometres. Away from the poles, such wavelengths are shorter than the shortest wavelengths in a degree 14 model. It is conceivable, however, that some long-wavelength features could be associated with crustal remanence, as might occur say from continental-sized magnetic anomalies. A review of the observed long-wavelength crustal magnetic anomalies has been given by Cain (1971, 1975) and Langel (1987, 1990a). The interpretation of terms with l = 12 to l = 22 is controversial (Cain, 1975; Harrison and Carle, 1981; Shure and Parker, 1981). That these intermediate wavelength features have changed with time between recent magnetic epochs suggests either that they are partly of core origin or that there has not been a complete removal of external magnetic field effects in the analysis.

The decreased power in intermediate wavelength features, less than 4000 km or so ($l > 10$) and larger than a few hundred kilometres, probably reflects the lack of magnetic sources in most of the mantle. This is to be expected because only the uppermost mantle can have temperatures low enough to permit permanently magnetized material. There are induced electric currents in the mantle associated with variations in the external magnetic field, but these currents are not expected to make substantial contributions to the apparent "main" field when averaged over long times (for further discussion on potential erroneous effects stemming from external sources see Alldredge and Stearns (1974), Lowes (1975, 1976), Langel (1987, 1993), and Barton (1989)).

There is controversy as to whether the coefficients of an accurate surface harmonic model of the field at the core–mantle interface decrease with increasing degree (e.g., Lowes, 1974). As is well known, upward continuation (i.e., increasing the value of r in Eq. (2.2.16)) is stable because the higher the degree of a harmonic the more it is attenuated. Conversely, downward continuation (i.e., decreasing r) is unstable because the higher the degree of a harmonic the more it is magnified. Thus any small errors in estimation of coefficients at the Earth's surface for the higher degree terms are also greatly magnified. In addition, downward continuation from a radius of r_1 to a smaller radius r_2 is valid only if all the sources internal to r_1 are also internal to r_2. Thus, to be strictly valid, all contributions from crustal and mantle sources must have been removed from the field that is downward continued to the core–mantle interface. Using magnetometer data obtained from the satellite MAGSAT, Benton *et al.* (1982) have argued from the sensitivity of locations and intersections of key geomagnetic indicators (specifically curves where $H_R=0$, $\partial H_R/\partial\theta=0$, $\partial H_R/\partial\phi=0$; see §9.1) that spherical harmonic expansions of the geomagnetic field should be truncated at or below degree 8 if they are to be used to infer sources in the core. For a discussion of the use of satellite data for modeling core and crustal fields, readers are referred to Langel (1993).

2.3 Uniqueness and Other Mathematical Problems

2.3.1 Application of Laplace's Equation to the Earth

In §2.2 it was assumed that $\nabla\times\mathbf{H} = \mathbf{0}$, from which it followed that $\mathbf{H}$ could be derived from a scalar potential ψ as $\mathbf{H}=-\nabla\psi$ and that ψ satisfied Laplace's equation, $\nabla^2\psi=0$. The basic requirement that $\nabla\times\mathbf{H} = \mathbf{0}$ is clearly not satisfied throughout the Earth's interior, because there are strong internal sources. The question arises as to whether this condition is even satisfied at the Earth's surface, because electric currents (such as lightning) do sometimes cross the Earth's surface.

If $\nabla \times \mathbf{H} = \mathbf{0}$ then the line integral of the horizontal component, $\mathbf{h}$, of the field around any closed curve, L, on the Earth's surface must satisfy

$$\oint \mathbf{h} \cdot d L = 0 \ . \tag{2.3.1}$$

Analysis by several workers (e.g., Schmidt, 1939; Chapman, 1942; Vestine *et al.,* 1947a,b; Vestine, 1967) of charted magnetic data indicates that the above condition is "reasonably well satisfied" for the Earth. By "reasonably well" is meant that although there are some small electric currents crossing the Earth's surface even in fair weather conditions, the lowest degree coefficients (such as those given in the model of Table 2.1) are not affected much by these currents. That is, for most magnetic epochs the contoured magnetic data satisfy (2.3.1) for almost all closed curves L on the Earth's surface. Departures are so small that they could be due to various errors involved in the construction of the charts. Therefore, the assumption of a scalar potential satisfying Laplace's equation at the Earth's surface in §2.2 is regarded as valid.

2.3.2 Approximation with a Truncated Series

Errors can of course arise out of using a truncated series of spherical harmonic terms to represent the geopotential. A central concept associated with orthogonal functions is the question of a least-squares fit to the actual function, and is an important aspect of *Parseval's approximation in the mean.*

Consider a complete set of functions $u_l(\chi)$ that are orthonormal over an interval $\Im$, so that

$$\int_{\Im} u_l(\chi) u_{l'}(\chi) d\chi = \delta_{ll'} \ . \tag{2.3.2}$$

As a specific example, one could consider the $u_l(\chi)$ to be the fully normalized Legendre polynomials (see Appendix B), so that

$$u_l(\chi) = \sqrt{\frac{2l+1}{2}} \ P_l(\chi) \ , \tag{2.3.3}$$

where $\chi = \cos \theta$ as before, and the interval $\Im$ would then be from -1 to +1. However, all that matters is that the $u_l(\chi)$ be orthonormal over the relevant interval (or relevant space for a multidimensional function).

Suppose there is a function $\psi(\chi)$, such as a scalar potential, that can be precisely represented by the infinite series

$$\psi(\chi) = \sum_{l=0}^{\infty} a_l \, u_l(\chi) \, , \qquad (2.3.4)$$

and consider the function $S_\aleph(\chi)$ that is the sum of an arbitrary, finite subset of the $u_l(\chi)$, i.e., l is contained a particular finite set $\aleph$ of integers, so that

$$S_\aleph(\chi) = \sum_{l \subset \aleph} A_l \, u_l(\chi) \, . \qquad (2.3.5)$$

Now $S_\aleph(\chi)$ affords an approximation to the function $\psi(\chi)$ and we wish to know how to choose the A_l so that $S_\aleph(\chi)$ provides the best approximation in a least-squares sense for that particular set of $u_l(\chi)$.

The quantity

$$\varepsilon_m \equiv \int_{\Im} \{\psi(\chi) - S_\aleph(\chi)\}^2 d\chi \qquad (2.3.6)$$

is a positive number that measures the departure of $S_\aleph(\chi)$ from $\psi(\chi)$ on the interval and we wish to choose the A_l to minimize ε_m. Substitution from (2.3.4) and (2.3.5) into (2.3.6) and using the orthonormalisation condition (2.3.2) gives

$$\varepsilon_m = \int_{\Im} \psi(\chi)^2 dx - \sum_{l \subset \aleph} (a_l)^2 + \sum_{l \subset \aleph} (A_l - a_l)^2 \, , \qquad (2.3.7)$$

which clearly has its smallest value for $A_l = a_l$. Since any set of orthogonal functions can be made orthonormal simply by multiplying by the right set of scalars (as done, for example, for the $P_l(\chi)$ in Eq. (2.3.3)) and the result of (2.3.7) is dependent only on the orthonormality of the functions, this conclusion is general for any complete set of orthogonal functions. This is extremely valuable because the a_l are independent of each other and may be obtained easily by exploiting the orthogonality, as is done when estimating the coefficients in a Fourier series. Unfortunately, as shown below, when attemtping to model with a finite number of poorly distributed observations, the orthogonality is lost and the A_l are no longer independent.

Of particular interest here is the case where we want to approximate the scalar potential of the magnetic field with a model that only uses harmonic degrees 1 to N (excluding 0 because there are no monopoles), such as the model represented by the coefficients in Table 2.1. Here the set $\aleph$ is simply the contiguous set of integers $l = 1$ to N, and $S_\aleph(\chi)$ may be denoted just by S_N. The above analysis tells us that the S_N which is the best approximation in a least-squares sense to the actual ψ is obtained simply by truncating the infinite series.

In reality, we are usually trying to model a continuous function on the basis of a finite set of observations. The integral of (2.3.2) has a uniform, and continuous, distribution of χ. If the observation points χ_1, χ_2, χ_3,..., are not similarly distributed then the $u_i(\chi)$ will not form an orthogonal set over the space defined by the data points. The greater the nonuniformity of the data distribution, the greater the divergence of the $u_i(\chi)$ from an orthogonal set. This results in an interdependence in the estimates of the coefficients. This is demonstrated very simply using the orthogonal functions $\sin(n\pi x)$ and $\cos(n\pi x$). The first of these is an odd function $[\sin(n\pi x) = -\sin(-n\pi x)]$ and the second is an even function. Suppose we sample the function $\cos \pi x$ at N random sites in $0-\varepsilon < x < 1-\varepsilon$, where ε is a small positive number. We can then obtain a perfect fit to the N observations with the function

$$F(x) = \sum_{n=1}^{N} A_n \sin(n\pi x) \qquad (2.3.8)$$

even though the generating function was an even function. Quite obviously, though, if we include a $\cos \pi x$ term in the series and then fit to the observations, all of the sine terms will vanish.

Even with perfect, uniformly spaced observations there must be at least two observations per wavelength to define a given harmonic. The Nyquist wavelength is twice the distance between uniformly spaced observations and any wavelengths shorter than this appear in the observations as if they had been generated by a longer wavelength harmonic (i.e., they are aliased down into wavelengths longer than the Nyquist wavelength). If the observations are noisy (as is inevitably the case) and nonuniformly spaced, then the more observations per "wavelength" of any given harmonic, the better the estimation of the coefficient for that harmonic. If in a model we then include short-wavelength harmonics that are poorly constrained by the data, the effective nonorthogonality of the harmonics means that poor estimation of the coefficient for this harmonic will affect the estimation of coefficients for the lower degree harmonics. The expansion to too high a degree produces an unstable model. Thus, in general, it is better to fit a model in which the harmonic series is simply truncated above some degree, rather than including some high-degree harmonics and excluding some low-degree harmonics. It has already been noted in §2.1 that ground-based magnetic observations are poorly distributed, so these problems occur in spherical harmonic modeling of the Earth's magnetic field.

Clearly it is desirable to use those orthogonal functions that are most efficient for the problem at hand. The Earth's mean equatorial radius is about 21 km greater than its polar radius, so elliptical harmonics are sometimes used to model the geomagnetic field. However, compared with the equatorial radius of 6378 km, the difference in radii is small and spherical harmonics remain the preferred choice by the vast majority of those who model the Earth's magnetic field.

Because of the problems discussed above, the typical spherical harmonic model (e.g., the IGRF given in Table 2.1) is simply a truncated series. An extreme cutoff like this is of course unrealistic and, as already mentioned (§2.2.5), simple downward continuation to the core–mantle boundary of the spherical harmonic model at the Earth's surface is unstable. Other methods (including so-called stochastic inversion) have therefore been devised in an attempt to produce reasonable models of the field at the core–mantle boundary (e.g., see Bloxham and Gubbins, 1985, 1986; Benton *et al.*, 1987). An interesting approach is that of "smooth" models, such as the cubic spline model of Shure *et al.* (1982) and the spherical triangle tesselation model of Constable *et al.* (1993). These models are smooth in the sense that some norm (e.g., $\int_S (B_r)^2 \, dS$, where B_r is the radial component of **B** and S is the core–mantle boundary) is minimized, and one seeks the smoothest model at the core–mantle boundary consistent with the available observations. The advantage of such a "smoothest" approach is that any structure in the model is required by the data, and so is likely real. The complementary disadvantage is that it is an extreme model and it is likely that the real Earth has more structure than the resulting models.

2.3.3 Uniqueness of Source

It has been argued that magnetic sources represented by harmonics of degree $l = 14$ and less probably originate primarily in the core. However, it is important to realize that the spherical harmonic procedure represents the field by *hypothetical* sources placed precisely at the Earth's center. It is highly unlikely that the actual sources of the Earth's magnetic field lie in the solid inner core of the Earth (see Chapter 8), so a spherical harmonic analysis does not give the sources uniquely. As a corollary, the individual terms in a spherical harmonic analysis should not be considered as representing separate "real" physical entities.

The ambiguity in a spherical harmonic analysis can be illustrated by considering a simple example of a dipole source displaced from the Earth's center a distance d along the z axis. If the dipole were at the center, it would be represented by the g_1^0 term,

$$\frac{p\cos\theta}{4\pi r^2}\ ,\qquad\qquad\qquad (2.3.9)$$

in the spherical harmonic expansion, where $p = (g_1^0 4\pi a^3 / \mu_0)$ has been substituted for the dipole strength. The potential for the offset dipole is

$$\frac{p\cos\theta}{4\pi(r^2 + d^2 - 2rd\cos\theta)} = \left(\frac{p\cos\theta}{4\pi r^2}\right)\left(\frac{1}{1 + \frac{d^2}{r^2} - 2\frac{d}{r}\cos\theta}\right)\ .\qquad (2.3.10)$$

The second term on the right side can be expanded in a Taylor series for $d < r$ to obtain the potential, ψ_d,

$$\psi_d = \frac{p}{4\pi r^2}\cos\theta + \frac{2pd}{4\pi r^3}\left(p_2^0\right) + \ \cdots\cdots\ .\qquad (2.3.11)$$

That is, an offset dipole will be represented in a spherical harmonic expansion by an axial geocentric dipole of the same strength plus an axial geocentric quadrupole of strength $(2pd)$, and so forth. The larger d is, the larger the number of zonal harmonics required to approximate that offset dipole. Thus, an offset dipole in the core is mathematically indistinguishable from a dipole plus a series of higher degree multipole fields at the Earth's center. This not only indicates the nonuniqueness of spherical harmonic analyses (*or any other analyses*), but it also emphasizes that separate terms in a spherical harmonic analysis do not represent separate "real" magnetic sources. It is experimentally impossible to determine the magnetic sources uniquely in the core no matter how many measurements of the field are made at the Earth's surface, even in the ideal case when those measurements are perfect estimates of the mean magnetic field! Of course one can add sufficient "physical constraints" to improve and constrain the variability of allowed solutions, but this increased resolution is directly dependent on the applicability of the "physical constraints".

2.3.4 Nonspherical Harmonic Representation of the Earth's Magnetic Field

Because of their mathematical advantages (as is exemplified, for example, by their use in dynamo theory; see Chapters 8 and 9), it is frequently desirable (and often necessary) to express magnetic field models in terms of spherical harmonic coefficients. However, as discussed in §2.3.3, such representations do not have direct physical significance and so, not surprisingly, many people prefer models that have more physical significance to them, even though these models are themselves highly idealized. Ultimately, one hopes to be able to construct models that do in fact have more physical significance than spherical harmonic models.

Most alternative descriptions have used magnetic dipoles as elementary sources of the internal magnetic field, as pioneered by McNish (1940) and Lowes and Runcorn (1951; also see Lowes, 1955). Several models consisting of one centered axial dipole with several eccentric radial dipoles have subsequently been advocated (Alldredge and Hurwitz, 1964; Pudovkin and Kolesova, 1968; Pudovkin *et al.*, 1968; Alldredge and Stearns, 1969; Zidarov and Bochev, 1965, 1969). Most of the above workers attempted to explain the observed geomagnetic secular variation in terms of their models. Much of the earlier work and some of the later work constrains the eccentric dipoles to lie close to the core–mantle boundary, based on the argument that the sources giving rise to the observed secular variation cannot lie too deep in the core because of the electromagnetic screening by the electrically conducting core (see Chapter 9 for further discussion). Similar arguments have been advanced by Cox (1968) and Harrison and Ramirez (1975) who use dipoles to model some changes in the magnetic field as observed in the paleomagnetic record. Some of the later models have tended to relax the condition that all the dipoles be radial and even in some cases to allow the magnitude of the dipoles to change with time (Bochev, 1965, 1969, 1975; see also Zidarov, 1969, 1970, 1974).

A few workers have adopted circular current loops as more physical representations of the Earth's magnetic field, as pioneered by Zidarov and Petrova (1974). Peddie (1979), for example, considers models that constrain circular current loops to the core–mantle boundary (up to seven loops) as well as models for unconstrained current loops (up to four loops).

None of the models given above is very satisfactory in providing a physical picture of the internal sources. This is partly manifested in the work by Peddie (1979), which constrains the current loops to lie at the core–mantle interface, a constraint that results in intersecting current loops. Moreover, because of uniqueness problems already mentioned, it is questionable whether any given model will ever be completely satisfactory. In addition, modern dynamo theory (Chapters 8 and 9) suggests that "real" sources are probably far more numerous and complex than suggested by any of the above modeling schemes.

In view of these difficulties and because of the mathematical advantages already noted, spherical harmonic representations are used predominantly in this book. At present, there is no evidence that alternative models to spherical harmonic analyses are as accurate or as effective in modeling the Earth's magnetic field of core origin and its secular variation. Nevertheless, this should not be misconstrued as a reason for suspending research to construct more viable physical models such as undertaken by Peddie (1979).

The use of spherical harmonics to represent crustal magnetic sources is a different matter. Crustal sources are close to the observer, even when satellites such as POGO and MAGSAT are used. Thus a very large number of spherical harmonic terms are needed to represent the crustal fields. These fields typically vary significantly over short distances and rapid convergence of the spherical harmonic series should not be expected. Thus alternative representations of the fields of crustal origin are probably desirable. Discussion of such problems with respect to MAGSAT data can be found in Benton *et al.* (1982), Galliher and Mayhew (1982), Langel and Estes (1982), Schmitz *et al.* (1982), and Langel (1987, 1990b, 1993).

2.3.5 Magnetic Annihilator

Backus and Gilbert (1967, 1968, 1970) have developed a method to optimize the information that can be obtained from nonunique problems in geophysics. The method requires that there exists a viable solution to *the forward problem*. In the case of geomagnetism this would require that the hydromagnetic dynamo problem could be solved to the extent that the observed properties of the Earth's magnetic field could be obtained from the theory. Because this has not yet been done, it might be supposed that the application of techniques like those developed by Backus and Gilbert would be inappropriate to geomagnetism. Although such is sometimes the case, there are several situations in which the theory is applicable and still others waiting to be tried. For example, inverse theory might be applicable to the source problem of the last subsection when (say) dipole sources are assumed to exist in the outer core and to give rise to the observed instantaneous field. A particular distribution of such dipole sources that produce the observed field would then be taken as a solution of the forward problem. One would then use the inverse theory methods to investigate the statistical variations associated with the set of all allowable solutions, commonly referred to as the *inverse* problem. Such an approach has not yet been tried on this problem, but it has been attempted on the related problem that deals with the inversion of short-wavelength magnetic anomaly data to obtain crustal magnetic sources (Parker and Huestis, 1974).

Direct application of the Backus–Gilbert inverse technique to the magnetic anomaly problem resulted in such wide limits for acceptable models that additional modifications of the techniques and certain geophysically reasonable constraints were needed to obtain useful results (Parker and Huestis, 1974). The Parker–Huestis approach will not be extensively discussed here other than to introduce the concept of an annihilator, which is useful in dealing with nonuniqueness problems. Following Parker (1977) it is convenient to assume for the purposes of illustration that the problem is linear, although most inverse problems in geomagnetism are actually nonlinear. Methods for treating nonlinear problems are beyond the scope of this section, but they have been discussed by Sabatier (1977) and Parker (1977).

Suppose that F(z) represents the unknown source function to be determined. Information concerning this source function is represented by data obtained at the surface of the Earth that can (say) be represented by the function $D(x)$. We assume that $D(x)$ and $F(z)$ are linearly related, so that

$$D(x) = \int G(x,z)F(z)\mathrm{d}z \ , \qquad (2.3.12)$$

where $G(x,z)$ is a kernel that describes the mapping of $F(z)$ into $D(x)$. If there is no function $f(z)$, such that

$$\int G(x,z)\,f(z)\mathrm{d}z = 0 \ , \qquad (2.3.13)$$

then $F(z)$ in (2.3.12) can be shown to be unique. A solution of the forward problem in such a case is sufficient. Usually, however, there are nontrivial solutions of (2.3.13). There is no way to learn about $f(z)$ from the data, because the substitution of $[F(z) + af(z)]$, where a is any constant, for $F(z)$ in (2.3.12) does not change $D(x)$. The set of all functions $f(z)$ that satisfies (2.3.13) is called the *annihilator* (Parker, 1977). The annihilator is a manifestation of the nonuniqueness of the problem that cannot be removed, no matter how perfect and complete the data may be. A simple example of a magnetic annihilator in the flat-Earth model often used in magnetic anomaly studies is a uniformly magnetized horizontal layer (Parker and Huestis, 1974), because no magnetic field exists external to such a layer.

Equation (2.3.12) can also be used to illustrate the concept of mathematical stability. Parker (1977) points out that if $G(x,z)$ greatly smoothes $F(z)$, then the inversion of $D(x)$ to obtain $F(z)$ will be unstable. This means that small changes in $D(x)$ can result in large changes in $F(z)$. Unfortunately, one of the best known examples of an unstable problem is the inversion (in this instance, downward continuation) of a potential field that satisfies Laplace's equation (Bullard and Cooper, 1948). That is, small changes in the magnetic scalar potential at the Earth's surface can lead to very large changes in potential at the core–mantle interface (as previously noted in §2.2.3).

The conclusion is that there is a large degree of nonuniqueness in the magnetic source problem that can never be resolved, no matter how perfect and complete the data. In addition, small errors in measurements at the Earth's surface will be greatly magnified by downward continuation to the core–mantle interface. These facts only serve to strengthen the view that spherical harmonic expansions are to be preferred, when possible, for representation of the magnetic data. This all serves to illustrate that it will probably never be possible to define the magnetic sources and core source regions very well. In particular, it is doubtful if the process of downward continuation of the magnetic potential to the core–mantle boundary will ever show convincingly whether the core is turbulent or not, or show whether the sources of the nondipole field lie at the core–mantle boundary or deeper in the core (see Chapters 8 and 9). This conclusion does not, of course, mean that there are not other ways of answering such questions.

2.4 Geomagnetic Secular Variation

2.4.1 Overview

It has been known for over 300 yr, since the discovery by Gellibrand in 1635, that the geomagnetic field changes with time. For example, the declination at London, England, is known to have changed gradually from 11½°E in 1576 to 24°W in 1823, before turning eastward again (Fig. 2.8).

Variations in the geomagnetic field observed at the Earth's surface occur on time scales ranging from milliseconds to millions of years. The short-term variations arise mostly from currents flowing in the ionosphere and magnetosphere giving rise to magnetic storms and the more or less regular daily variation (see §2.5.3). These rapid fluctuations are superimposed upon much slower changes with periods of a year to millions of years. These slower changes

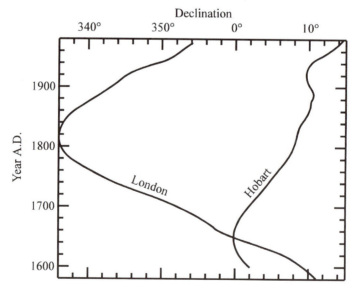

Fig. 2.8. Variation in declination at London, England (51.5°N), and at Hobart, Tasmania (42.9°S), from observatory measurements. The earliest measurement in the Tasmanian region was made by Abel Tasman at sea in 1642 in the vicinity of the present location of Hobart. Preobservatory data have been derived also by interpolation from isogonic charts.

are generally referred to as the *geomagnetic secular variation.*

Although the cutoff between internal and external sources in terms of period is probably not as distinct as is sometimes argued, it appears that periods of a few tens of years or longer are of internal origin while periods less than a year or so are of external origin. Some periods of 11 and 21½ yr are sufficiently close to fundamental solar periods (see Chapter 10) that they may be of external origin, but these signals are relatively small (e.g., Barton, 1989). Periods of internal origin shorter than a year or so are not seen at the Earth's surface because of the screening effect of the lowermost electrically conducting mantle.

2.4.2 The Magnetic Jerk and Screening by the Mantle

Temporal variations in the core's magnetic field have to pass through the electrically conducting mantle and so the variations as seen at the Earth's surface are a filtered version. To predict the effects of this filter, it is necessary to know the distribution of electrical conductivity through the mantle. Magnetic fields of external origin are used to obtain estimates of the conductivity of the upper mantle, while fields that originate within the core are used to obtain estimates of the conductivity for the lower mantle. In addition, there are estimates that come from mineral physics (see §7.2).

Variations in the external magnetic field (caused, for example, by changes in ionospheric currents) will induce eddy currents within the Earth according to Lenz's law. The change in the magnetic field measured at the Earth's surface is a combination of the external field variations and the associated induced fields. In practice two time series can be used (one for the external variations seen at the Earth's surface and the other for the fields induced in the Earth) and these series can be separated using cross-correlation techniques. Sometimes only the electric or only the magnetic field variations are used. The so-called magnetotelluric method, which has been widely used to provide estimates of the upper mantle electrical conductivity distribution, uses both the electric and the magnetic field.

In a stationary conductor the magnetic induction equation (derived in §8.2) reduces to the vector diffusion equation

$$\frac{\partial \mathbf{H}}{\partial t} = \frac{1}{\sigma \mu} \nabla^2 \mathbf{H} \ , \qquad (2.4.1)$$

where $\mathbf{H}$ is the magnetic field, σ is the electrical conductivity, and μ is the magnetic permeability (which is typically, and reasonably, assumed to be the free-air permeability μ_0). This shows that the magnetic field will diffuse into a conductor, in this case the crust and mantle. A similar equation can be derived from Maxwell's equations for the electric field. For illustration purposes consider a magnetic field varying with a single frequency ω, diffusing into a half-space with a constant conductivity. The solution of (2.4.1) in this case is easily shown to be

$$\mathbf{H} = \mathbf{H}_0 \exp(-\tfrac{z}{\delta}) \exp[i(\omega t - \tfrac{z}{\delta})]$$

$$\delta = \left(\frac{2}{\omega\mu\sigma}\right)^{1/2} \quad , \tag{2.4.2}$$

where z is depth and δ is the skin depth, the depth at which the amplitude of the field has fallen to $1/e$ of its original value. Note that the skin depth is a function of both conductivity and frequency, increasing for lower frequency signals and for less conductive materials. In practice there will be a range of frequencies examined and this allows determination of the conductivity as a function of depth.

Induction models were pioneered and first applied to the mantle by Lahiri and Price (1939). Over the next three decades several models emerged showing a smoothly varying conductivity with depth (e.g., McDonald, 1957; Banks, 1969). Layered models with considerably increased variation with depth subsequently emerged (e.g., Banks, 1972; Larsen, 1975; Filloux, 1980b). A puzzling aspect of these models was that as more layers were included in the model there was a general tendency for larger fluctuations in conductivity with depth; i.e., the conductivity structure in the models increased dramatically with increasing numbers of layers. This puzzle was explained by Parker and Whaler (1981; also see Parker, 1994). They showed that in the one-dimensional case without constraints on conductivity or on layer thickness, the model that minimizes errors (in a least squares sense) contains layers with no conductivity separated by infinitely thin layers with finite conductivity, now commonly referred to as the D^+ model. The model is clearly not very physical, but it emphasizes the nonuniqueness problem and illustrates that if one increasingly overfits the data, the resulting model will increasingly (and inappropriately) become more complex, and in the limit converge to the D^+ model. To avoid this problem several investigators have specified a level of error tolerance and proceeded to determine the smoothest model compatible with the data (e.g., Constable *et al.*, 1987; Egbert and Booker, 1992). As noted in §2.3.2, smoothest models probably provide a good picture of large-scale trends but fail to resolve large variations that may occur on small spatial scales. This means that partial melt or material such as graphite, which are both likely to be present in the mantle and can lead to changes in the electrical conductivity by orders of magnitude over small spatial scales, may be missed. Two- and three-dimensional inversion models have been applied to determine the Earth's outermost electrical conductivity structure and it is only a matter of time before they are applied to the entire upper mantle.

Interestingly, essentially all published models for electrical conductivity give a value of conductivity near 1 Sm^{-1} at the 670-km seismic discontinuity. This represents a substantial increase from crustal conductivities, which are typically more than two orders of magnitude smaller. Although some of the models provide values even deeper in the mantle (to depths around 1000 km), resolution below 670 km is very poor. Usually the conductivity of the lower mantle is estimated by other means, such as using information associated with variations in the magnetic field of core origin.

The problem in using the above theory is that either one has to know the mantle conductivity to analyze the geomagnetic spectrum or one has to know the geomagnetic spectrum to obtain the mantle conductivity. Runcorn (1955) noted that for a flat-earth model, in which the mantle is approximated by a plane of constant conductivity, σ_0, and thickness, L, the amplitude of the spectrum at the surface has the form

$$\mathbf{H} = \mathbf{H}_0 \exp\left(-\frac{L}{\sqrt{\mu_0 \sigma_0 \omega}} \right) \quad . \tag{2.4.3}$$

Therefore, by assuming that the secular variation at the core–mantle interface could be described by a white spectrum, Runcorn could estimate the mean mantle conductivity.

A quick estimate of the minimum period that will not be significantly screened by the lower mantle can be obtained by assuming Runcorn's planar screening model. The skin depth in kilometres is

$$\delta = \frac{1}{2}\left(\frac{T_c}{\sigma_0} \right)^{\frac{1}{2}} \quad , \tag{2.4.4}$$

where T_c is the period. For this estimate we take σ_0 in the lower mantle equal to 1 Sm^{-1} (as we shall see this is a minimum estimate) and δ equal to 2000 km. Then $T_c \approx 16 \times 10^6$ s, roughly half of a year. This represents a very good first-order estimate of the minimum period for a magnetic field of core origin that can be seen at the Earth's surface. Periods much less than this can safely be attributed to sources above the Earth's core. McDonald (1957) further improved on this procedure by considering a spherical model and by allowing for geometrical spreading in a mantle in which the conductivity was allowed to vary according to a power law of the radius.

There is a constraint on the lower mantle conductivity; it cannot be too high or much of the short-period secular variation observed at the Earth's surface would be screened out. The shortest period variation widely believed to be of core origin is the *secular variation impulse* (*saut de variation séculaire*) of 1969 as named by Courtillot *et al.* (1978). Its date was established as 1969.5 by Ducruix *et al.* (1980) with an uncertainty of only a few months. Following Malin *et al.* (1983), it should be called an *impulsive jerk*, not a very appealing name as Courtillot and Le Mouël (1988) have pointed out. The term *jerk* (or *magnetic jerk*) has become the most popular and widely used. The phenomenon is best seen in the third derivative of the eastward component of the magnetic field (i.e., in the second derivative of the secular variation) at European observatories (Fig. 2.9). It is also present in observatory data worldwide (Le Mouël *et al.*, 1982) and is widely believed to reflect a major magnetic perturbation in the core at some time before 1970 (e.g., Malin and Hodder, 1982). Although Alldredge (1984) prefers a time scale of one or two decades, the consensus view is that the 1969 jerk occurred over a time interval of less than a few years, and quite probably on the order of a year (e.g., Courtillot and Le Mouël, 1984; Backus *et al.*, 1987). This originally suggested to many that the estimate of the conductivity of the lowermost mantle should be lowered from the most commonly accepted value of about 3×10^2 Sm^{-1}. Previously a 3.7-yr cutoff period was indicated from spectral analysis of observatory data (Currie, 1968).

Backus (1983) first carried out an analysis of the screening problem related to the observation of magnetic jerks. He argued that the short duration of the 1969 jerk as observed in Europe may be due to the mixing of harmonic modes. Courtillot *et al.* (1984) commented on this, arguing that some of the observations

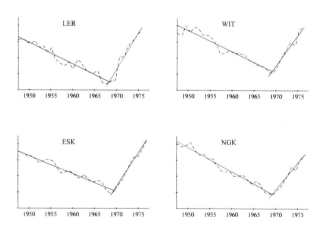

Fig. 2.9. Secular variation (first differences of annual means) of the eastward component (Y) of the geomagnetic field at four European observatories (intervals of 10 nT yr^{-1}). Solid lines are least-squares bilinear fits to the data, showing a jerk at 1969. After Courtillot *et al.* (1979).

relating to the jerk may be contaminated by a remaining solar cycle effect. Backus (1983) had acknowledged that the presence of external sources was an alternative explanation, but did not develop this point. Later Backus (1991) produced a rigorous analysis that can be summarized as follows. If the mantle conductivity cannot be neglected, one can no longer assume $\mathbf{H} = -\nabla\psi$. Instead the field must be divided into its poloidal and toroidal parts (see Chapter 9),

$$\mathbf{H} = \nabla \times \Lambda p + \Lambda q \ , \tag{2.4.5}$$

where the angular operator Λ is defined by

$$\Lambda = \mathbf{r} \times \nabla \ , \tag{2.4.6}$$

and the scalar function S that gives rise to the poloidal field can be obtained from

$$S = \Lambda^{-2}(\mathbf{r} \cdot \mathbf{H}) \ . \tag{2.4.7}$$

Backus was able to show that at the Earth's surface, radius R_s,

$$S_l^m(R_s, t) = \left(\frac{R_c}{R_s}\right)^{l+1} \int_0^\infty F_l(\tau) P_l^m(R_c, t - \tau)d\tau \ , \tag{2.4.8}$$

where R_c is the radius of the core, t is the time, and S_l^m is the scalar function (which depends on the order and degree of the spherical harmonic) from which the poloidal magnetic field can be obtained. The poloidal field is the only field of core origin that is observable at the Earth's surface. $F_l(t)$ is the impulse response function and it depends on l but not on m (Backus, 1983, 1991). It vanishes as τ approaches zero or infinity, it is positive, and it integrates to one. Therefore it can be thought of as a probability density function. Let τ_1 be the mean and τ_2 the standard deviation of this function. Then $S_l^m(R_s, t)$ can be obtained by attenuating P_l^m by $(R_c/R_s)^{l+1}$, averaging over the time interval of length $2\tau_2$ and delaying it by τ_1. Backus (1983) showed that the value of τ_1 is roughly twice that of τ_2 and that τ_1 is probably between 1 and 10 yr. In short Backus (1983) showed that the 1969 jerk (and other similar impulses) may actually have occurred 1 to 10 yr earlier, and the claimed 1-yr duration probably reflects a more gradual change that occurred over a few years. However, Courtillot et al. (1984) argued that the presence of a sunspot-related signal in the data and the fact that the jerk is both worldwide and synchronous (Le Mouël et al., 1982) can change this conclusion. In this case the application of the Backus (1983, 1991) filter theory would indicate that τ_1 is likely to be less than a few years.

The whole question of jerks is central not only to our understanding of mantle conductivity, but also to dynamo theory, for if they are of internal origin then they cannot be considered independently of the mechanism for generating the Earth's main magnetic field. For example, Hulot *et al.* (1993) indicated that jerks could be related to major changes within the large-scale flow driving the core convection, and Jault and Le Mouël (1994) have shown that jerks cannot be created by superficial flows in the core. Hulot *et al.* (1993) showed evidence that jerks also occurred around 1914 and 1978. Alexandrescu *et al.* (1995) have recently developed a wavelet analysis for the detection of geomagnetic jerks, and they conclude that in the *Y* component of European observatory records there are five, and only five, such events, occurring in 1901, 1913, 1925, 1969, and 1978. Although each of these events had previously been suggested, Alexandrescu *et al.* (1995) claim their approach to be much more reliable and, perhaps more importantly, that their analysis rejects the subtle possibility of sharpening the jerk with some additional external signal. Clearly this reinforces the concept that jerks are of internal origin.

The above results do not appear to require major changes in the estimates of the electrical conductivity of the lowermost mantle of 3×10^2 Sm^{-1}, a value that is about three orders of magnitude below that for the core. Note, however, that local (discontinuous) regions of very high conductivity in the lowermost mantle are possible; indeed such regions might be a cause for some of the geographic variations observed for the 1969 jerk (Le Mouël *et al.*, 1982). One can combine all the data to estimate that the mantle cutoff period (at which the signal is damped by 1/e) is between a few months (no increase in conductivity below 1500 km) and 1 yr (magnetic jerk). This range is significantly lower than the older, but widely cited, 4-yr cutoff period estimated by Currie (1968). Discussion of the variation of electrical conductivity with depth is given in Chapter 7.

2.4.3 Methods Used to Determine the Secular Variation

Temporal variations in the internal field are modeled by expanding the Gauss coefficients in a Taylor series in time about some epoch t_e, e.g.,

$$g_e^m(t) = g_e^m(t_e) + \left(\frac{dg_e^m}{dt}\right)_{t_e}(t - t_e) + \left(\frac{d^2 g}{dt^2}\right)\frac{(t - t_e)^2}{2!} + \cdots \quad . \qquad (2.4.9)$$

Most models include only the first two terms on the right, although some have incorporated the secular acceleration term and even the third derivative term.

There are three general methods used to obtain the coefficients in (2.4.9). The first is to estimate the temporal changes in the measured components directly and then determine a spherical harmonic fit to those data. The second is to estimate the first derivatives by using the differences between annual means for different epochs (e.g., $dX/dt \approx \Delta X/\Delta t$), i.e., by using the mean slope of the curve for two annual means. The final method is to substitute an equation, such as (2.4.1), into the appropriate spherical harmonic expansion and solve directly for terms such as g_e^m and $d g_e^m / dt$ (which in this case doubles the data requirement). Detailed procedures for doing this for all methods, including error analyses, have been well described by Langel (1987). An example of the secular variation for the 1990 epoch is given in Fig. 2.3d.

2.4.4 Drift of the Nondipole Field

The largest changes in the direction of the Earth's magnetic field during the past 150 years are associated with the nondipole part of the field. This can readily be seen as a westward shift of various isoporic foci (maxima or minima) on isomagnetic charts between different magnetic epochs. Figure 2.10 compares the vertical component of the nondipole field for 1930 and 1990. Bullard *et al.* (1950) made use of data tabulated by Vestine *et al.* (1947a,b) for the time period spanned by the 1907 epoch to the 1945 epoch and determined the average velocity of the nondipole field to be 0.18° per year westward, the so-called *westward drift of the nondipole field.* Thus, a whole circuit of the Earth would take 2000 yr, a period sometimes considered, erroneously, to be "sufficient time" to average out secular variation in paleomagnetism.

Prior to the mid-1980s the historical record of the magnetic field extended from the present back to the first analyses of Gauss in the mid 19th century. One of the most important contributions to geomagnetism in the past decade has been the use of ancient mariners' data to extend this record back to 1715 A.D. (Bloxham and Gubbins, 1985, 1986). They calculated the field values at the core–mantle boundary by using what they referred to as a stochastic inversion technique. There has been some controversy concerning how well the field is resolved there and concerning some interpretations of the data. For example, the magnetic field can change because it diffuses from one spot to another or because it is advected by fluid motions (when diffusion is negligible, the so-called frozen flux approximation is applicable; see Chapter 8). Bloxham and Gubbins suggested that both processes sometimes occurred and, if so, this could cause problems in using the data to estimate core velocities (see §11.1). However, Constable *et al.* (1993) claim the data are compatible with a frozen flux hypothesis. Of course this does not rule out the possibility that diffusion occurs on the 350-yr time scale of the historical data, but the present consensus is that diffusion appears to be relatively unimportant for the lowest degree harmonics over the interval of the historical data.

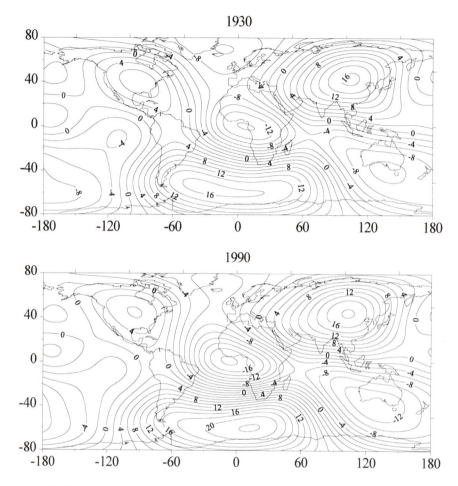

Fig. 2.10. The vertical component of the nondipole field for 1930 and 1990. Contours are labeled in units of 1000 nT.

Such controversies should not detract from the major contributions of Bloxham and Gubbins, particularly that of more than doubling the historical record. As an interesting bonus for historians, they were also able to upgrade the navigational information associated with those data. Unfortunately, the extension of the data record is only for the directional data and does not include intensity data. Their analyses of these data indicate that the westward drift of the nondipole field is primarily confined to the Atlantic hemisphere. Indeed the nondipole field is quite small beneath the Pacific Ocean and this is referred to as the *historical dipole window*. Additional analyses and interpretation of these data by Bloxham and Gubbins will be given in later chapters.

Yukutake and Tachinaka (1968) noted that there were two main types of nondipole anomalies: those that clearly drifted westward (such as the central African anomaly in Fig. 2.10) and those that have remained stationary and just increased or decreased in magnitude (such as the Mongolian anomaly, which increased by 50% over the 150 yr). This is commonly referred to as the drifting and standing model for secular variation. Yukutake (1979) subsequently divided the field into westward and eastward drifting components; however, he suggests that much of the drifting part of the field is due to the *equatorial* dipole field, which he separated into two components drifting in opposite directions. Braginsky (1972) made a similar suggestion but the amplitudes and velocities of the westward and eastward drifting components differ considerably in the two analyses.

To complicate the matter further, James (1971, 1974) points out that observed changes in the rotational axisymmetric part of the nondipole field, as manifested in changes in the zonal harmonics, cannot be explained by westward drift and he suggests significant meridional drift. This may be important for theoretical reasons as is further discussed in §9.5.

In summary, there are several extreme models put forward to explain essentially the same secular variation data: westward drift of the nondipole field; the standing and drifting models and their variants; poleward drift of the nondipole field; and random changes in the nondipole field. Part of the problem is that, despite Bloxham and Gubbins doubling its length, the historical record is still only a few hundred years long. That is, it is approximately an order of magnitude shorter than the westward drift period proposed by Bullard *et al.* (1950). Moreover, there are other evident changes (e.g., in intensity) that significantly affect the nondipole field during the historical time that must be removed to discern longer term trends. The best way to resolve some of these uncertainties seems to be through the use of paleomagnetic data to extend the time window and this subject will be returned to in Chapter 4.

2.4.5 Variations of the Dipole Field with Time

The dipole field is also changing with time. The intensity of the dipole field has decreased at the rate of about 5% per century since the time of Gauss' analysis in 1835 (Leaton and Malin, 1967; Vestine, 1967; McDonald and Gunst, 1968; Langel 1987; Fraser-Smith, 1987) (Fig. 2.11a). Indeed, Leaton and Malin (1967) and McDonald and Gunst (1968) have speculated on the demise of the main dipole around A.D. 3700 to 4000 if the present trends continue. In contrast, the dipole axis, as represented by the position of the North Geomagnetic Pole, has hardly changed its position since the analysis of Gauss (Bullard *et al.*, 1950). Over the past 150 yr there appears to have been a slow westward change of about 0.05° to 0.1° per year in azimuth angle, but no progressive motion in polar angle (Nagata, 1965; McDonald and Gunst, 1968; Barton, 1989; Fraser-Smith, 1987).

When there are only declination values available the method due to Bauer (1894) can be used and has been discussed in detail by Benkova *et al.* (1970). For any point on the Earth's surface, the relation

$$X \sin D = Y \cos D \qquad (2.4.10)$$

can be rewritten by expanding X and Y in spherical harmonics. If there are N values of D, a series of N inhomogeneous equations result and may be solved to

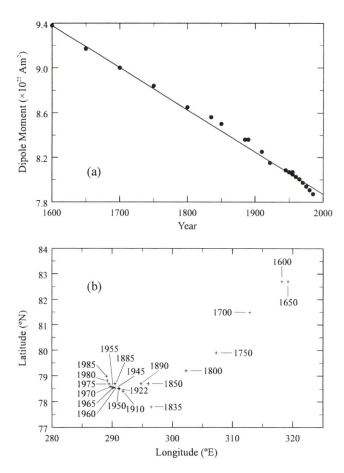

Fig. 2.11. Variations of the dipole field with time since A.D. 1600. (a) Variation of the dipole moment from successive spherical harmonic analyses. (b) Variation of the dipole axis as represented by the change in position of North Geomagnetic Pole. After Fraser-Smith (1987).

obtain relative values of the coefficients (see Barraclough, 1974). If values of inclination are available at points for which declination data also exist, then Bauer (1894) proposed the relation

$$Z \cos I = H \sin I = \frac{X}{\cos D} \sin I \qquad (2.4.11)$$

be used in addition. These methods, and values obtained from them, are now largely of historical interest, having been succeeded by those of Bloxham and Gubbins (1985). However, the latter could not obtain reliable paleointensity estimates. A method used by Braginsky and Kulanin (1971), Yukutake (1971), Braginsky (1972), Benkova *et al.* (1974), and Barraclough (1974) determines values of g_1^0 by extrapolating back in time from values determined since the time of Gauss. Barraclough (1974) fitted a straight line to values of g_1^0 from 170 spherical harmonic models of the field between 1829 and 1970 to derive the relation

$$g_1^0(t) = -31110.3 + 15.46(t - 1914) \; , \qquad (2.4.12)$$

where t is the epoch in years A.D.

Barraclough (1974) has analyzed the field for epochs since 1600. His estimates of the positions of the North Geomagnetic Pole since 1600 are plotted in Fig. 2.11b as summarized by Fraser-Smith (1987). They show that the dipole field has drifted westward at about 0.08° per year since 1600 and has changed latitude at the very much slower rate of 0.01° per year.

2.5 The External Magnetic Field

2.5.1 The Magnetosphere

The Earth is immersed in a rapidly expanding solar atmosphere consisting mainly of H^+ and $^4He^{2+}$ and electrons that are being transported away from the Sun. The movement of these particles is known as the *solar wind* (see also Chapter 10), the speed of which relative to the Earth varies considerably over a 27-day period (one solar rotation as seen from the Earth), ranging from 270 km s^{-1} in relatively cool plasma to 650 km s^{-1} in hotter more tenuous plasma (e.g., Isenberg, 1991). The average speed is about 10 times the local Alfvén speed (defined in Chapter 10) and so the flow is termed supermagnetosonic.

Because the solar wind consists of charged particles, it interacts with the geomagnetic field. Far from the Earth on the Sun side, the field is weak and is compressed by the solar wind. Nearer the Earth the field strength is sufficient that the magnetic pressure acts as a barrier and forces the solar wind to separate and flow around the Earth, producing a cavity into which there is little penetration of the solar wind and within which the geomagnetic field is confined. Birkeland (1908) anticipated the existence of such a cavity and Gold (1959) first described it in its modern form and termed it the *magnetosphere* (Fig. 2.12). The solar wind effectively drags the geomagnetic field along with it and an elongated comet-like tail is formed on the dark side, first discovered by Ness (1965) and known as the *geomagnetic tail*, or *magnetotail*.

The boundary of the magnetosphere is known as the *magnetopause*, effectively a current sheet with no large-scale flow through it. Its position along the Sun–Earth line on the sun side (sometimes called the *stagnation point*) may be estimated by balancing the dynamic pressure of the solar wind ($2\rho v^2$, where ρ is mass density of the solar wind and v its speed) against the magnetic pressure there ($B^2/2\mu_0$; see §8.4). Substitution of appropriate values leads to an estimate for B of around 70 nT (e.g., Schultz, 1991), which occurs at a distance of about 8 Earth radii on the Sun–Earth line (e.g., Paschmann, 1991).

The motion of the Earth in the solar wind creates a shock front referred to as the *bow shock*, which is separated from the magnetopause by a region referred to as the *magnetosheath*. It is within this region that the solar wind particles slow to sub-magnetosonic speeds relative to the Earth. Because the high-latitude

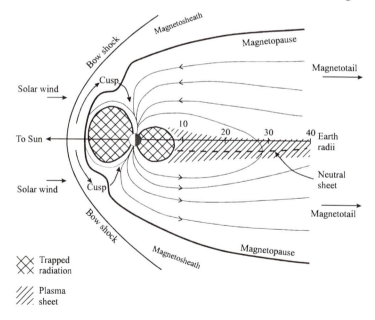

Fig. 2.12. The different regions of the magnetosphere in the noon–midnight meridian plane.

geomagnetic lines of force are mostly dipolar and are either compressed on the Sun side of the Earth by the solar wind or swept away by the solar wind on the dark side, a *cusp* geometry will occur. This cusp (or cleft) is the region of the last line of force that crosses the front side of the Earth and still goes to the geomagnetic tail. The low magnetic field there allows some penetration of the solar wind into the ionosphere, causing the kink in the magnetopause.

As the magnetic field lines are stretched in the magnetotail, field lines of opposite senses along the center of the tail are brought very close to each other, effectively canceling. There is thus a narrow region, known as the *neutral sheet*, lying close to the ecliptic on the dark side, in which the field is very weak. This sheet, discovered by Ness (1965), is important in magnetospheric electrodynamics (e.g., Speiser and Ness, 1967).

Other planets with significant magnetic fields of internal origin also have magnetospheres, but with widely varying shapes. Jupiter and Saturn, for example, have rapid rotation rates (Chapter 10) and so their magnetospheres are more extended in their equatorial planes. Uranus has one of the most complex and varying magnetospheres. This is because its rotational axis lies in the ecliptic plane and its magnetic field of internal origin is highly nondipolar at the magnetopause (e.g., Isenberg, 1991).

It is commonly argued that the magnetosphere is probably "open" (e.g., Stern, 1977), following the pioneering work of Dungey (1961, 1963) and Axford *et al.* (1965). The term "open" means that some of the geomagnetic field lines connect with those of the interplanetary magnetic field. A schematic view of an open magnetosphere is shown in Fig. 2.13. Note that the topology of the magnetic field lines is such that there are always at least two points (A and B in Fig. 2.13) at which the intensity of the Earth's magnetic field is zero. At such points, sometimes referred to as *neutral points,* the very concept of magnetic field lines becomes vague; magnetic "field lines" can open up and form new connections, a process referred to as *magnetic merging* or *magnetic reconnection*. This process is also important in many dynamo models discussed in Chapters 8 and 9. Point A (Fig. 2.13) provides an opening for charged particles spiralling along a magnetic field line from the Sun to find their way into the magnetosphere. Axford (1969) and others suggest that the neutral point B (Fig. 2.13) is located within the neutral sheet.

The solar wind is highly variable in time. Therefore its pressure on the geomagnetic field will vary with time and will cause the magnetopause to change shape. This invariably leads to changes in the magnetic field at the Earth's surface. However, tracing the sources of disturbance of the magnetic field at the Earth's surface is extremely difficult. This is because processes that occur in distant regions can induce other major disturbances within the ionosphere. Conversely, changes in ionospheric currents could lead to changes far above the ionosphere.

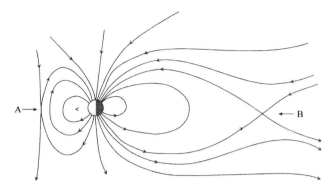

Fig. 2.13. Schematic view of an open magnetosphere with an extended tail. Points A and B are neutral points at which the Earth's magnetic field is zero.

Readers are referred to Parks (1991) for an introduction to the underlying physics of magnetospheric processes, Jacobs (1991) for a discussion of many topics at the frontiers of the field, and Moore and Delcourt (1995) for an outstanding review.

2.5.2 The Ionosphere

With the exception of thunderstorms, the region from the Earth's surface up to about 50 km can be regarded as an electromagnetic vacuum. The movement of cumulonimbus clouds overhead can change the magnetic field at the Earth's surface by 20 nT or so and lightning discharges can have even much larger effects. The ionosphere, roughly between 50 and 1500 km above the Earth's surface, is subdivided into the *D region* (50–90 km), the *E region* (90–120 km), and the *F region* (120–1500 km). The ionization originates from the interaction of solar ultraviolet radiation with various atmospheric constituents. The electron density increases from the D region to the F region, where it will typically reach 10^5–10^6 cm^{-3} at midday. Electron energies are around a few eV.

Particles in the ionosphere are affected by the rotation of the Earth and by tidal forces from the Moon and Sun. Thermal effects are important in the first-order analysis only in the lowermost regions of the ionosphere. Neutral particles in the D region and below rotate in unison with the Earth. The degree of ionization is so small in the D region that regional collisions with neutral particles result in synchronous rotation. The D region disappears at night. In the E region collisions are less frequent than at lower altitudes and the degree of ionization is larger so that there is a partial decoupling between the ionized particles and neutral particles. In the F region it is usually assumed that, to a first approximation, there is complete decoupling between the neutral and ionized particles. That is, electromagnetic forces are much stronger than collisional forces and hence the motions of charged particles are controlled by processes that occur in the magnetosphere and solar winds.

2.5.3 Transient Magnetic Variations, Storms, and Substorms

Electromagnetic forces in the ionosphere can lead to large electric currents. Some of these currents have associated magnetic fields of magnitude up to 1000 nT at the Earth's surface (see below). There are numerous sources for large-scale electric fields in the Earth's magnetosphere. They are caused by temporal variations in the magnetic field, rotation of the Earth, and variations in the solar wind. Associated with these fields is a host of electrical currents. One needs to understand the so-called $\mathbf{E} \times \mathbf{B}$ drift to understand many magnetospheric processes. This drift occurs when charged particles have a velocity perpendicular to the magnetic field. Equating the electrical field to the Lorentz field yields

$$\mathbf{E} = \mathbf{v} \times \mathbf{B} \ . \qquad (2.5.1)$$

To obtain the drift velocity, take the cross product of (2.5.1) with $\mathbf{B}$, expand the right-hand side, and divide by B^2 to obtain

$$\frac{\mathbf{E} \times \mathbf{B}}{B^2} = \mathbf{v} \ , \qquad (2.5.2)$$

where we have used $\nabla \cdot \mathbf{v} = \nabla \cdot \mathbf{B} = 0$ ($\mathbf{v}$ is assumed perpendicular to $\mathbf{B}$). There are also currents directed along the $\mathbf{B}$ field and these are referred to as *Birkeland currents*, whose origins are complex and still not fully understood (e.g., Schultz, 1991). The α-effect (see Chapters 8 and 9) gives rise to field-aligned currents.

Plasma (ionized gas) within the magnetosphere can cause *ring currents* associated with the drift of ions (positive and negative ions drifting in the opposite directions) along closed field lines. In particular, at radial distances of 4-6 Earth radii, ions in the *Van Allen radiation belts* (Fig. 2.14) drift because of gradients in the main field. The growth (growth phase) and decay (recovery phase) of these currents over several days is called a *magnetic storm* (e.g., McPherron, 1991). Variations in the horizontal component of the magnetic field of 100 nT are typically observed for magnetic storms that are effectively extreme disturbances in the magnetic field. Very great magnetic storms occur about two or three times per solar cycle and can have associated changes in $\mathbf{B}$ that exceed 500 nT. There are also *magnetic substorms*, first defined by Akasofu and Chapman (1961). According to McPherron (1979), magnetic substorms are defined as "a transient process initiated on the night side of the Earth in which a significant amount of energy derived from the solar wind–magnetosphere interaction is deposited in the aurora, ionosphere and magnetosphere." Typical variations in $\mathbf{B}$ for substorms are on the order of 40 nT.

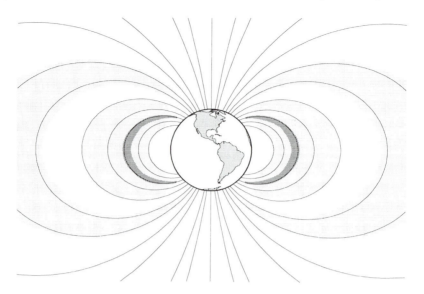

Fig. 2.14. The Van Allen radiation belts (light shading) consist of an inner belt of protons and an outer belt of electrons. The darkly shaded region is a recently discovered belt, within the inner belt, where cosmic rays collect. All these belts approach closest to the Earth in the South Atlantic region because of the offset of the Earth's magnetic dipole. After Mewaldt *et al.* (1994).

As an example of substorms consider the effects of auroral phenomena that occur in the higher latitude regions in the ionosphere; the maximum variation occurs in a geomagnetic latitude belt of 65° to 70°. The soft green light typically associated with auroras comes from the forbidden emission (5577 Å) of oxygen precipitated by the bombardment of electrons from the outer radiation belts. Auroral events are usually accompanied by large-scale electric fields. These fields move charges around in a manner to produce large ionospheric currents. The strongest of these currents runs from dawn to dusk, peaks around midnight, and is referred to as the *auroral electrojet* (Stern, 1977). The current can be so strong that magnetic fields at the Earth's surface underneath it can change as much as 1000 nT, although typical variations are more like 200 and 300 nT. The deviations in the magnetic field observed at the Earth's surface from auroral substorms, when plotted as a function of time, often resemble bays on a geographic map and consequently are frequently referred to as *magnetic bays*.

Even without variations in the solar wind, there would be variations in the magnetic field due to tidal effects on the magnetosphere from the Sun and Moon and due to the fact that the dipole is tilted with respect to the general direction of the solar wind as the Earth rotates. The degree of variation in the magnetic elements thus varies widely from day to day and from season to season. Furthermore there are considerable variations due to changes in sunspot activity, either due to the sudden occurrence of solar flares or else corresponding to the sunspot cycle (Chapter 10). At any station the days when the three principal geomagnetic elements (*H, D* and *Z*) undergo smooth and regular variations are termed *magnetically quiet days,* and when these variations are more or less irregular the days are termed *magnetically disturbed days.* When extreme, a magnetic disturbance is identified as a magnetic storm. Except during magnetic storms it is evident from magnetic observatory records that there is a regular daily variation in the magnetic elements. The main part is the *solar daily variation,* **S**, with a period of 24 h together with the much smaller *lunar daily variation,* **L**, with a period near 25 h. The *disturbance variation,* **D**, refers to the additional field(s) present during disturbed and extremely disturbed days. **S** and **D** are very easily recognized in the magnetic records but **L** can only be determined after careful analysis of a large data set.

The solar daily variation is seen in its pure form on extremely quiet days, and the mean taken over these days is referred to as the *solar quiet day variation,* S_q. On normal days, or days with only minor disturbance, there is in addition the *solar disturbance daily variation,* S_D. The variation S_D is part of the **D** field generally and only in the absence of storms and substorms will **D** $\approx S_D$. The disturbance field **D** is defined as the difference ΔF between the total field at any instant and the average field (over a month or more) after allowing for the S_q and **L** fields, so that

$$\mathbf{D} = \Delta \mathbf{F} - \mathbf{S}_q - \mathbf{L} \qquad (2.5.3)$$

The intensity of S_D varies with the intensity of the general disturbance. Both the S_q and the S_D fields vary with latitude and local time (that is, their time variation is roughly constant relative to the meridian plane containing the Sun). However, both the latitude and the time variations for the S_q and S_D fields are themselves very different (Fig. 2.15). In particular, whereas the horizontal component of the S_q field changes sign at about 30° latitude, that of the S_D field changes sign at about 55° latitude. Also the time variation in S_D is independent of the intensity of the disturbance, ranging from the very slight disturbances present on average days to the very intense ones present during magnetic storms (Fig. 2.15). For a general discussion of S_q, S_D, and **L** see Chapman and Bartels (1940, 1962) and Matsushita (1967). Both S_q and **L** most likely arise from tidal motions of the air in the ionosphere in a horizontal direction across the lines of force of the Earth's magnetic field so inducing electric currents. The solar atmospheric tide is probably caused by thermal rather than gravitational action of the Sun on the atmosphere. For this reason S_q is very much larger than **L**, which is probably due to gravitational tides.

Along the magnetic equator within a narrow band of about 2° latitude there is an enhancement of the horizontal component of S_q by about a factor of two. The

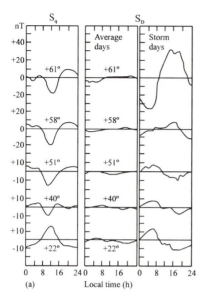 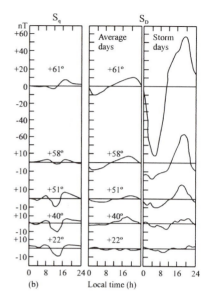

Fig. 2.15. Average S_q and S_D variations for various northern latitudes. In the southern hemisphere the vertical component variations will be reversed in sign. (a) Horizontal component; (b) vertical component. From Chapman and Bartels (1962).

narrow equatorial current system inferred from this observation flowing from east to west on the sunward hemisphere was named the *equatorial electrojet* by Chapman (1951). Details of this effect and its theory are given by Onwumechilli (1967). The cause of the equatorial electrojet lies in a special equatorial feature of the electrical conductivity of the ionosphere. When there are crossed electric and magnetic fields a Hall current usually flows perpendicular to both the electric and the magnetic fields. When the medium is bounded in the direction of the Hall current, polarization results and opposes the flow. The result is enhanced conductivity of the medium. In the ionosphere the Hall current leads to polarization but this easily leaks away along the lines of force of the geomagnetic field, except in the region of the magnetic equator where the field is entirely horizontal. The resulting increased ionospheric conductivity is sufficient to account for the observed enhancement of S_q. The equatorial electrojet and associated currents as seen in MAGSAT data are discussed by Langel *et al.* (1993).

Magnetic storms typically can be divided into three phases termed the initial, main, and recovery phases. The initial phase may be gradual or be represented by an abrupt change called a *sudden commencement*. In addition to the S_D component (related to local time) there is a component related in time to the beginning of the storm termed the *storm-time variation*, $\mathbf{D}_{st}$, so that during magnetic storms

$$\mathbf{D} = \mathbf{D}_{st} + \mathbf{S}_D \; . \tag{2.5.4}$$

There is also an irregular part, $\mathbf{D}_i$, that only becomes significant at high latitudes. The main characteristic of magnetic storms is the reduction in the horizontal component of the geomagnetic field during the main phase (Fig. 2.16). The main phase of a storm is defined as beginning when the horizontal component of the $\mathbf{D}_{st}$ field decreases below its pre-sudden commencement value and ends when this reaches its maximum decrease. The initial and recovery phases are simply before and after the main phase. The magnitude of the reduction in the horizontal component of $\mathbf{D}_{st}$ during the main phase has a maximum at the equator and decreases to a minimum at about 60° latitude, thereafter increasing rapidly toward the auroral zone. The actual observed changes in the horizontal component H of the geomagnetic field during a magnetic storm will depend very much on local time. The component $\mathbf{D}_{st}$ will have the form shown in Fig. 2.16, but the resulting geomagnetic field variation will depend on the phase of $\mathbf{S}_D$ at the beginning of the storm, in the limits either greatly enhancing the reduction in H or greatly diminishing it (see Chapman and Bartels, 1940, 1962; McPherron,

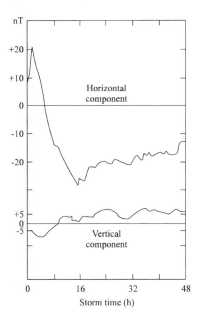

Fig. 2.16. Typical magnetic storm effects showing the average horizontal and vertical $\mathbf{D}_{st}$ variation at 40° N. In the southern hemisphere the vertical component variations would be reversed in sign. From Chapman and Bartels (1962).

TABLE 2.2
Periods and Average Amplitudes of Various Geomagnetic Phenomena as Observed in Midlatitudes

Phenomenon	Period (amplitude in nT)	
Micropulsations	1 ms-3 min	(~<1)
Magnetospheric substorms and bays	1-2 h	(~10)
Solar quiet day variation, S_q	24 h	(~20)
Solar disturbance daily variation, S_D	24 h	(~5-20)
Lunar daily variation, **L**	25 h	(~1)
Storm-time variation, $\mathbf{D}_{st}$		
(a) initial phase	~ 4 h	(~15)
(b) main phase	~8 h	(~35)
(c) recovery phase	~ 60 h	(~35)
External magnetic field	~<4 yr	
Internal magnetic field	~>4 yr	
Sunspot cycle variation	11 and 22 yr	

1991).

Examples of some of the various geomagnetic phenomena produced externally and seen at the Earth's surface are given in Table 2.2.

2.5.4 Magnetic Indices

The magnetic activity associated with the external field and recorded by magnetometers at the Earth's surface can clearly be very complex and its characterization is not an easy task. Traditionally this is done through the use of magnetic indices. Since continuous recordings of variations in the Earth's magnetic field began, many indices have been derived to provide information that summarizes the geomagnetic activity level. Mayaud (1980) has given an extensive review of these indices and Menvielle and Berthelier (1991) have provided an easily followed short review of the subject.

The first attempt to characterize geomagnetic activity on a daily basis was the determination of a daily range of some selected geomagnetic component. However, such an index was sensitive to both the regular and the irregular variations. The characterization of the irregular activity has been made on a daily basis since 1884 using the so-called C numbers. For each Greenwich day the observer in charge of an observatory assigned a number 0, 1, or 2, describing the relative degree of disturbance present in the magnetograms (quiet, moderately disturbed, or disturbed). The planetary or international index, C_i, was then defined as the mean of the C numbers for all the cooperating observatories. This C_i index was calculated during the years 1884–1975 inclusive and was used to determine the five international quietest days of each month until 1942. Because of the subjective nature of this index, Bartels *et al.* (1939) introduced the now well-known K index (the letter K coming from the German *kennziffer*, meaning literally *range index*) to allow a more objective monitoring of the irregular variations.

The K indices have been regularly calculated at almost all the geomagnetic observatories (about 200 in 1988) since their introduction by Bartels *et al.* (1939). The 3-hr K index was chosen because this time interval appeared to be long enough to give correct indications for such details as magnetic bays and other perturbations of only 1 to 2 hr duration. Ten ranges of geomagnetic activity were formulated following a quasi-logarithmic scale giving a single digit 0 to 9. Details of their determination are summarized by Menvielle and Berthelier (1991). Bartels *et al.* (1940) attempted to derive an index of worldwide activity, Kw, following the same scheme as C_i but eliminating the local time and seasonal variations at each observatory that appeared in the K values. The K value at each observatory is replaced by a standardized value, K_s, obtained using conversion tables in such a way that the frequency distribution no longer depends on the local time or season.

Bartels (1949) later defined the planetary index, Kp, that is still used and has remained the same since its introduction. The Kp index is calculated in the same way as the original Kw index, except that a more carefully built set of conversion tables is used with a modified network of stations. The Kp network is now made up of 13 stations, mostly located in western Europe and North America with only 2 stations is the southern hemisphere (Canberra in Australia and Eyrewell in New Zealand). However, the Kp network includes auroral stations for which the standardization procedure is the most questionable, so by the end of the 1950s with the increase in the number of magnetic observatories it became possible to build a network made up of mainly subauroral stations. The K indices calculated from these stations would be most relevant to give a global characterization of geomagnetic activity representing all longitudes in both hemispheres. A new index was thus introduced by Mayaud (1968) known as the *am* planetary (mondial) index with subsets *an* and *as* being the hemispheric indices. In addition Mayaud was able to define an antipodal activity index, *aa*, based upon the availability of records since 1868 from two old observatories, Melbourne and Greenwich, that are almost antipodal (today, the Canberra and Hartwell observatories are used). The *am* index is based upon a network of 22 mainly subauroral stations that are divided into groups according to their longitude, with five longitude sectors in the northern hemisphere and four in the southern hemisphere. The corresponding equivalent K indices (Km, Kn, Ks) can be calculated using conversion tables.

The best characterization of worldwide geomagnetic activity is given by the *am* or Km indices, although Kp is generally used from force of habit (Menvielle and Berthelier, 1991). Km and Kp both indicate the level of activity and it is not surprising that they have close and sometimes identical values. Provisional values of K-derived indices are circulated on a monthly basis. Values of Kp and the related series *an*, *Kn*, *as*, *Ks*, *am*, and *Km* are published regularly by the International Association of Geomagnetism and Aeronomy (IAGA) or are available from World Data Centers.

Chapter 3

Foundations of Paleomagnetism

3.1 Rock Magnetism

3.1.1 Types of Magnetization Acquired by Rocks

When a rock forms it usually acquires a magnetization parallel to the ambient magnetic field (usually presumed to be the Earth's magnetic field) and this is referred to as a *primary magnetization*. This primary magnetization then provides information about the direction and intensity of the magnetic field in which the rock formed. However, there are numerous pitfalls that await the unwary: first, in sorting out the primary magnetization from *secondary magnetizations* (acquired subsequent to formation), and second, in extrapolating the properties of the primary magnetization to those of the Earth's magnetic field. For more detailed reviews of rock magnetism readers are referred to Fuller (1970), Stacey and Banerjee (1974), O'Reilly (1984), and Dunlop (1990).

Magnetism in minerals occurs because the negative exchange energy from the interaction of the spin angular momentum of electrons dominates the magnetic ordering. If adjacent magnetic moments are parallel, as in iron, then the material is *ferromagnetic*. If adjacent magnetic moments are antiparallel then the material is either *antiferromagnetic* (the magnitudes of the opposing moments are equal) or *ferrimagnetic* (the magnitudes of the opposing moments are unequal). The most common magnetic minerals in rocks are either antiferromagnetic or ferrimagnetic. A ferrimagnetic substance, such as magnetite, exhibits spontaneous magnetization because of the inequality of opposing magnetic moments, and may therefore have remanence. As the temperature is increased the disordering effect of thermal energy will eventually overcome the ordering force of the exchange energy. The temperature at which the magnetic ordering is destroyed is known as the *Curie temperature* for ferromagnetic materials (e.g., iron). For an antiferromagnetic (e.g., ulvospinel) or a ferrimagnetic material (e.g., magnetite) the critical point temperature is called the *Néel temperature*. However, following common practice in paleomagnetism, we will use "Curie temperature" in place of "Néel temperature" for ferrimagnetic materials.

When an igneous rock cools from a temperature above the Curie temperature of its magnetic minerals in an external magnetic field (such as the Earth's magnetic field), it acquires a remanent magnetization referred to as *thermoremanent magnetization* or TRM. Technically speaking, a *remanent magnetization* or RM is the net magnetization present in the material in zero external magnetic field. In a magnetically isotropic rock the TRM is aligned parallel (or in very rare cases antiparallel) to the external field in which the rock cooled. Although essentially all magnetic minerals exhibit magnetic anisotropy (that is, there are preferred directions of magnetization in any magnetic mineral grain), the majority of igneous rocks are magnetically isotropic because the magnetic minerals they contain are randomly oriented. When the TRM is produced antiparallel to the external field the rock is said to have *self-reversed*. Self-reversal was predicted on theoretical grounds by Néel (1955), but in practice appears to be rare in nature. The first discovered self-reversing rock was studied extensively by Uyeda (1958). The numerous occurrences of rocks with magnetization almost exactly opposed to the direction of the present Earth's magnetic field are now known to be due almost entirely to the fact that the field has reversed its polarity many times in the past. The evidence relating to this and a full discussion of self-reversal mechanisms in rocks is given in Chapter 5.

The magnetization first measured in the laboratory is referred to as the *natural remanent magnetization* or NRM, which typically consists of a primary and secondary magnetizations. Primary magnetizations are usually a TRM in igneous rocks and *depositional remanent magnetization* (DRM), *postdepositional RM* (post-DRM) or occasionally *chemical remanent magnetization* (CRM) in sedimentary rocks. The secondary magnetizations have many possible origins, such as subsequent chemical alteration and relaxation effects, and well over 20 different kinds of remanent magnetization have been proposed. The definitions of the most common forms of RM are given in Table 3.1. In recent years there has been a growing recognition of the need for identification, isolation, and dating of secondary magnetizations, and this has required far greater rock magnetic input to paleomagnetic studies.

TABLE 3.1
Some Common Types of Remanent Magnetizations

Thermoremanent Magnetization (TRM): That RM acquired by a sample during cooling from a temperature above the Curie temperature in an external magnetic field (usually in a weak field such as that of the Earth).

Chemical Remanent Magnetization (CRM): That RM acquired by a sample during a chemical change in an external magnetic field.

Viscous Remanent Magnetization (VRM): That RM acquired over a long time in an external magnetic field.

Isothermal Remanent Magnetization (IRM): That RM acquired in a very short time at one temperature (usually room temperature) in an external (usually strong) magnetic field.

Anhysteretic Remanent Magnetization (ARM): That RM acquired when an alternating magnetic field is decreased from some large value to zero in the presence of a weak steady magnetic field (this definition differs from that used in magnetic tape research in which the intensity of the steady field is also decreased to zero with the alternating field).

Natural Remanent Magnetization (NRM) : That RM acquired by a sample under natural conditions.

Depositional Remanent Magnetization (DRM): That RM acquired by sediments when grains settle out of water in the presence of an external magnetic field.

Postdepositional Remanent Magnetization (post-DRM): That RM acquired by physical processes that cause translation or rotation of sedimentary grains after deposition.

3.1.2 Magnetic Hysteresis

Some idea of the origin and properties of remanent magnetization in rocks may be obtained by considering magnetic hysteresis. Consider a highly simplified example of uniformly magnetized (single-domain) grains that are identical and noninteracting (the magnetic field produced by the magnetization of any grain is negligibly small at the site of any other grain). Assume that the grain possesses some uniaxial magnetic *anisotropy energy*, E_{an}, per unit volume where

$$E_{an} = K \sin^2 \theta \ . \tag{3.1.1}$$

K is defined as the uniaxial anisotropy constant and θ is the angle the magnetization makes with the z axis in a Cartesian coordinate system (x,y,z). The anisotropy energy is lowest for $\theta = 0$ and largest for $\theta = \pi/2$, so z is an *easy axis of magnetization* and it is harder to acquire a magnetization in the x–y plane. Magnetic anisotropy of some type is a necessary, but not sufficient, condition for a grain to possess a remanent magnetization. Its physical origins are discussed briefly in §3.1.3.

Easy axis

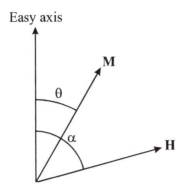

Fig. 3.1. The geometry of the easy axis of magnetization (vertical axis), the magnetization direction (M), and the magnetic field (H).

There will be an extra energy associated with a grain having magnetization **M** in an external field **H** when **M** and **H** are not parallel. From Fig. 3.1 the total energy E per unit volume originating from the anisotropy and the interaction of the grain's magnetization with the external field is given by

$$E = K \sin^2 \theta - \mu_0 M_s H \cos(\theta - \alpha) , \qquad (3.1.2)$$

where M_s is the saturation magnetization (the largest magnetization the grain can have) and is used because it is assumed that all the grains are uniformly magnetized. The grain's magnetic moment is $m = M_s V$, where V is the grain volume. Equilibrium occurs where $dE/d\theta = 0$, with a minimum energy state (stable equilibrium) if $d^2E/d\theta^2 > 0$. With **H** along $+z$ (i.e., $\alpha = 0$) it is easily seen that stable equilibrium occurs at $\theta = 0$. Suppose all grains are magnetized

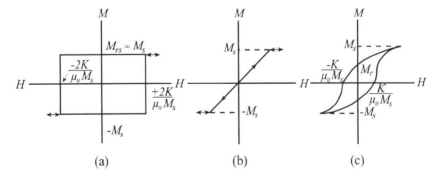

Fig. 3.2. Hysteresis loops of single-domain particles. (a) Square loop produced when H and M are confined to the same axis. (b) No hysteresis produced if H and M are perpendicular. (c) Hysteresis loop produced by a set of randomly oriented grains.

parallel to $+z$ near absolute zero temperature. This can be achieved by increasing **H** enough so that the magnetization is "saturated." If **H** is now reduced to zero, the magnetization will retain its saturation value M_s. That is, the saturation isothermal remanent magnetization, M_{rs}, will equal M_s (Fig. 3.2). If **H** is now increased in the opposite direction ($-z$; $\alpha = 180°$ in (3.1.2)) then it is easily shown that an energy minimum still exists at $\theta = 0$ for $H < 2K/\mu_0 M_s$. However, for $H > 2K/\mu_0 M_s$, the $\theta = 0$ direction becomes a maximum energy direction and the magnetization will change to $\theta = 180°$. This produces half of the rectangular hysteresis loop shown in Fig. 3.2a, the other half being obvious from symmetry. However, if **H** is applied at right angles to z ($\alpha = 90°$ in Eq. (3.2)) the conditions for an energy minimum are

$$\sin\theta = \frac{\mu_0 M_s H}{2K} \text{ for } |H| \leq \frac{2K}{\mu_0 M_s}$$

$$\theta = \frac{\pi}{2} \text{ for } |H| \geq \frac{2K}{\mu_0 M_s} \quad . \tag{3.1.3}$$

The magnetization along the z axis is

$$M = M_s \sin\theta = \frac{\mu_0 M_s^2 H}{2K} \quad \text{for} \quad |H| \leq \frac{2K}{\mu_0 M_s} \quad . \tag{3.1.4}$$

Therefore, there is no hysteresis and no remanence, as shown in Fig. 3.2b. Suppose one is dealing with an assembly of uniaxial single-domain grains that have isotropic random orientations of their easy axes. Integration over all possible angles **H** can make with the easy axis is then required, which produces the hysteresis loop shown in Fig. 3.2c. The M_{rs}/M_s value is around 0.5 and the bulk coercivity H_c (the value of H that reduces the magnetization to zero) is $K/\mu_0 M_s$. The saturation remanent magnetization is M_{rs}.

TABLE 3.2
Some Hysteresis Loop Parameters that are Useful in Distinguishing Single-Domain (SD) from Multidomain (MD) Grains.

	SD	MD
$\dfrac{M_{rs}}{M_s}$	Ideally 0.5 in an isotropic sample but typically between 0.3 and 0.5	Less that 0.1 (and often much less)
$\dfrac{H_{rc}}{H_c}$	Between 1 and 1.2	Variable but typically much higher (~5) than SD grains

Note. M_s is the saturation magnetization. M_{rs} is the saturation remanent magnetization (that remaining after the sample has been exposed to a saturation field). H_c is the bulk coercivity (the magnetic field required to reduce an initially magnetized sample to zero net magnetization in the presence of the external field). H_{rc} is the remanent coercive force (the field required to reduce the remanence to zero, in zero external field).

However, only small grains are uniformly magnetized. Larger grains are typically multidomain (see §3.1.5), so are nonuniformly magnetized, and parameters of their hysteresis loops often differ substantially from those of single-domain grains. Thus hysteresis parameters are often useful for determining the magnetic character of a sample and for distinguishing grains with single-domain assemblages from grains with multidomain assemblages (see Table 3.2), as was pioneered by Stoner and Wohlfarth (1948) and extended to geophysics by Wasilewski (1973), Fuller (1974), Day *et al.* (1976, 1977), and Day (1977). The method is not without its shortcomings, however, as is discussed more fully by Day (1977) and Levi and Merrill (1978). Other methods are therefore often used to supplement the hysteresis loop analysis for characterizing the state of remanence (see Lowrie and Fuller, 1971; Halgedahl, 1989; Dunlop, 1990; Moskowitz, 1993).

3.1.3 The Demagnetizing Field and Magnetic Anisotropy

Once a sample has been magnetized it produces its own internal magnetic field that tends to oppose the magnetization. This arises from the shape of the body because it is more easily magnetized in some directions than others. Because the demagnetizing field of nonuniformly magnetized grains is complicated, the discussion will be restricted to consideration of uniformly magnetized grains (Merrill, 1977, 1981; Williams and Dunlop, 1989). The demagnetizing field H_d is uniform only in simple shapes such as ellipsoids and is usually expressed in the form

$$\mathbf{H}_d = \overline{\mathbf{N}} \cdot \mathbf{M}_s \ , \qquad (3.1.5)$$

where $\overline{\mathbf{N}}$ is the demagnetization tensor of second order. In many cases, such as when the magnetization is along one of the major axes of an ellipsoid, $\overline{\mathbf{N}}$ reduces to a scalar N, referred to as the *demagnetizing factor*. Consider the case when rectangular coordinates are aligned along the axes of an ellipsoidal grain that is uniformly magnetized. If N_x, N_y and N_z are the appropriate demagnetizing factors for the ellipsoid then it can be shown (Chikazumi, 1964) that

$$N_x + N_y + N_z = 1 \ . \qquad (3.1.6)$$

Thus for a sphere $N_x = N_y = N_z = \frac{1}{3}$, for an infinite cylinder $N_z = 0$ and $N_x = N_y = \frac{1}{2}$ (where z lies along the cylinder axis), and for a thin disc $N_z = 1$ and $N_x = N_y = 0$ (where z lies perpendicular to the disc).

Because the demagnetizing field depends on the direction of magnetization for all nonspheres, it is easiest to magnetize samples along those directions in which the demagnetizing field is smallest. For example, it is easier to magnetize the infinite cylinder above along the z axis rather than perpendicular to it. This type of magnetic anisotropy is referred to as *shape anisotropy*. Consider a grain in the shape of a prolate spheroid with major axis in the z direction and let it have magnetization M_s at an angle θ to z. The energy associated with the demagnetizing field is called the *magnetostatic energy* and per unit volume for the spheroid example is given by (Cullity, 1972)

$$E_{ms} = \tfrac{1}{2}\mu_0 M_s^2 N_z + \tfrac{1}{2}(N_x - N_z)\mu_0 M_s^2 \sin^2\theta \ . \qquad (3.1.7)$$

The angle-dependent term has the same form as (3.1.1) and this defines the shape anisotropy constant K_s as

$$K_s = \tfrac{1}{2}(N_x - N_z)\mu_0 M_s^2 \ . \qquad (3.1.8)$$

For a spherical grain $N_x = N_z$ and $K_s = 0$. The shape anisotropy constant is thus determined by the axial ratio of the grain and its magnetization. Recall that the coercivity of a uniaxial grain is given by

$$H_c = \frac{2K}{\mu_0 M_s} \qquad (3.1.9)$$

so that the coercivity of a grain due to its shape anisotropy is given by

$$H_c(\text{shape}) = (N_x - N_z) M_s \ . \qquad (3.1.10)$$

Besides shape anisotropy there are other forms including *magnetostrictive* (or *stress*) *anisotropy, magnetocrystalline anisotropy*, and occasionally *exchange anisotropy*. Exchange anisotropy exists when there is exchange coupling between two different magnetic phases and will not be discussed further. Readers interested in learning more about anisotropy are referred to books such as Chikazumi (1964), Morrish (1965), Cullity (1972), Stacey and Banerjee (1974), and O'Reilly (1984).

Magnetocrystalline anisotropy is the only anisotropy that exists in an idealized infinitely large crystal free of defects and so it is sometimes referred to as an intrinsic anisotropy. Although in principle its origin is understood theoretically (it originates from three sources; spin–spin interaction, relativistic spin–orbit coupling, and ionic or atomic distortions), in practice it must be determined experimentally. It is described by a magnetocrystalline anisotropy tensor of second order, $\overline{\mathbf{K}}$. For a uniaxial crystal the coercivity is simply related to the first magnetocrystalline anisotropy constant K_1 and from (3.1.9)

$$H_c(\text{magnetocrystalline}) = \frac{2K_1}{\mu_0 M_s} \ . \qquad (3.1.11)$$

Moskowitz (1993) has recently determined the various anisotropy energies for titanomagnetites as a function of temperature.

Magnetostrictive (or stress) anisotropy originates from internal stress, such as would exist around dislocations. Like magnetocrystalline anisotropy its origin is understood in principle (see Mattis, 1965) but in practice experimental values must be used. Even an ideal crystal undergoes some strain when it is magnetized. Similarly any internal stress $\bar{\alpha}$ can be resolved into a strain $\bar{\varepsilon}$ not associated with the magnetization and a strain $\bar{\lambda}$ directly associated with the magnetization. $\bar{\lambda}$ is referred to as the magnetostrictive anisotropy tensor and is determined experimentally (Chikazumi, 1964). When the magnetostriction is isotropic and there is an angle θ between M_s and the internal stress α, there is a *magnetoelastic energy*, E_{me}, given by (Cullity, 1972)

$$E_{me} = \tfrac{3}{2} \lambda \alpha \sin^2 \theta \quad . \tag{3.1.12}$$

From (3.1.1) and (3.1.9) the corresponding coercivity due to magnetostriction is

$$H_c(\text{stress}) = \frac{3\lambda\alpha}{\mu_0 M_s} \quad . \tag{3.1.13}$$

In the above no attempt has been made to distinguish between *microscopic* and *macroscopic* coercivity. Microscopic coercivity refers to the field strength required to make irreversible changes within part of a grain, such as the passage of a domain wall over a local energy barrier (e.g., a dislocation). Macroscopic coercivity is the bulk coercivity, i.e., the coercivity of the entire sample. The two coercivities have always been assumed to be linearly related until recently when Xu and Merrill (1989) showed that this was not always the case.

3.1.4 Single-domain Theory for TRM

Néel (1949) presented a theory for an assembly of identical noninteracting uniformly magnetized grains with uniaxial anisotropy. Although the theory has some defects, it does provide valuable insight into many of the observed phenomena in rock magnetism, and remains at the foundations of rock magnetism.

For simplicity consider the case when the applied external field **H** is along the easy anisotropy axis of the grains. At all times it is assumed that the grains are uniformly magnetized (including during changes in direction). Under these assumptions there are two equilibrium magnetic states for the magnetization in the grain. The lowest state has energy $-V\mu_0 M_s H$ and the highest state has energy $+V\mu_0 M_s H$. Using Boltzmann statistics for n identical particles the number of grains in the upper, n_u, and lower, n_l, states is

$$n_u = \frac{ne^{-(V\mu_0 M_s H)/kT}}{e^{+(V\mu_0 M_s H)/kT} + e^{-(V\mu_0 M_s H)/kT}} \qquad (3.1.14)$$

$$n_l = \frac{ne^{+(V\mu_0 M_s H)/kT}}{e^{+(V\mu_0 M_s H)/kT} + e^{-(V\mu_0 M_s H)/kT}} \quad , \qquad (3.1.15)$$

where k is the Boltzmann constant and T is temperature. The denominator of (3.1.14) or (3.1.15) is referred to as the *partition function*. The magnetization is given by

$$M = n_l V M_s - n_u V M_s = nV M_s \tanh\left(\frac{V\mu_0 M_s H}{kT}\right) \quad , \qquad (3.1.16)$$

where (3.1.14) and (3.1.15) have been used for n_l and n_u. Equilibrium conditions have been invoked by using Boltzmann statistics (small relaxation times) and, in analogy with paramagnetism (which considers magnetic moments associated with individual atoms, not grains), the grains are called *superparamagnetic.*

At this point in the theory Néel (1949) made a crucial assumption; he assumed that there is a *blocking temperature* above which the grain assembly is in equilibrium, but below which the magnetization is locked into the sample. Therefore the room-temperature value for TRM, M_T, can be obtained from (3.1.16) and is:

$$M_T = nV M_s \tanh\left(\frac{V\mu_0 M_s H}{kT}\right)_{\text{blocking}} \quad , \qquad (3.1.17)$$

where "blocking" refers to the fact that the parameters in the argument of the hyperbolic tangent are to be evaluated at the blocking temperature.

An important second contribution from Néel was the explanation as to why the TRM might sometimes be stable over long time periods. He pointed out that dn_l/dt would be negative once the external field was removed, due to thermal fluctuations causing some of the magnetic moments to pass over the anisotropy barrier which has an energy height of $V\mu_0 M_s H_c/2$. dn_l/dt is proportional to $(n_u - n_l)$, because it will increase for moments traversing the barrier from the initially higher energy state and will decrease in the opposite case, so that

$$\frac{dn_l}{dt} = \frac{1}{2\tau}(n_u - n_l)$$

$$\frac{dn_u}{dt} = \frac{1}{2\tau}(n_l - n_u) \quad , \qquad (3.1.18)$$

where 2τ is the constant of proportionality. Subtracting the second equation from the first, using (3.1.16), and integrating,

$$M = M_0 e^{-t/\tau} \quad , \tag{3.1.19}$$

where M_0 is the initial remanence and τ is now seen to be a relaxation time. To obtain information on the geomagnetic field at some time t_p before present, it is necessary to have rocks that contain some magnetic grains with relaxation times greater than t_p.

From Boltzmann statistics, $1/\tau$ must be proportional to the probability of traversing the anisotropy barrier, so

$$\tau = \frac{1}{f}\exp\frac{V\mu_0 M_s H_c}{2kT} \quad , \tag{3.1.20}$$

where f has the dimensions of frequency. Néel (1949) and Brown (1959) derived different relationships for f, but they are similar in that f only varies slowly as a function of temperature. For most magnetic minerals in rocks, f can be taken to be on the order of 10^8 to 10^{10} s. In the case when a magnetic field H is present and along the easy axis, then the relaxation times of remanence toward and away from that field are respectively (Néel, 1949)

$$\tau_- = \frac{1}{f}\exp\frac{V\mu_0 M_s(H_c - H)^2}{2kTH_c}$$
$$\tau_+ = \frac{1}{f}\exp\frac{V\mu_0 M_s(H_c + H)^2}{2kTH_c} \quad . \tag{3.1.21}$$

It is readily seen by substituting reasonable values for elongated single-domain magnetite into (3.1.20) and using the above value for f, that a sample with a relaxation time of 1 s at 550°C (30° below magnetite's Curie temperature) will have a relaxation time longer than the age of the Earth at room temperature. These results indicate the reasonableness of the blocking temperature concept at the first-order approximation level and indicate why it is possible for rocks to retain a record of the ancient field directions for millions and even billions of years.

Taking f to be constant and selecting some relaxation time, τ_0 (say 1 yr $\approx$ 3×10^7 s), as indicating blocking, then the blocking temperature T_b can be obtained from (3.1.20) as

$$T_b = \frac{V\mu_0 M_s H_c}{2K \log_e(f\tau_0)} \quad , \tag{3.1.22}$$

where V, M_s, and H_c are to be evaluated at T_b. Because H_c, V, and M_s can vary considerably between individual grains in an actual rock sample, one typically finds a spectrum of blocking temperatures present. There is, therefore, a range of temperatures over which the TRM is "locked into" the rock. Roughly speaking, for a mineral such as magnetite, half of the TRM would be stabilized during cooling in the first 150° or so below its 580° Curie point. Many of the oversimplified assumptions in the single-domain theory given above have been modified in more recent formulations (e.g., see O'Reilly, 1984; Moon and Merrill, 1988; Williams and Dunlop, 1989).

3.1.5 Classical Magnetic Domains

Magnetic structures in minerals can be viewed by a variety of techniques including bittern pattern imaging, optical techniques, neutron scattering, and atomic force microscopy. Using such techniques one finds that only rarely are magnetic minerals in rocks approximately uniformly magnetized. Indeed, classical domain theory was developed to explain these nonuniform structures (e.g., Kittel, 1949). The key ingredient of this theory is that the magnetic structure could be determined by minimizing the appropriate magnetic energy. For example, a uniformly magnetized grain in zero external magnetic field has an energy of self-demagnetization of $\frac{1}{2}V(\mu_0\mathbf{M}_s\cdot\mathbf{H}_d)$, where V is the volume of the grain. This energy can be reduced by decreasing the demagnetization field (H_d) associated with the surface-bound magnetic poles. Two structures that have a lower demagnetization energy than a single-domain grain are shown in Fig 3.3. In Fig. 3.3a one has a two-domain grain; in classical domain theory the

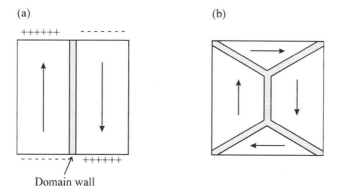

Fig. 3.3. Two hypothetical multidomain states. (a) A two-domain grain separated by a 180° domain wall. The magnetostatic energy of this grain is roughly one-half that of a uniformly magnetized grain of the same size and shape, but there is an additional domain wall energy. (b) A four-domain grain with closure domains (four 90° walls and one 180° wall). The magnetostatic energy is now zero, but there is an increase in domain wall energy compared with that in (a).

magnetization in each of the domains is uniformly magnetized. There is a narrow transition region separating the two domains that is referred to as a *domain wall*. Figure 3.3b shows the case where H_d is zero and, from the above, it might seem that this structure would always be preferred to the single-domain structure. This is not so, however, because the domain walls have an additional energy associated with the fact that most of the individual magnetic moments in the walls are not oriented along easy anisotropic directions. This additional energy is referred to as *domain wall energy*.

Consideration of the various energies involved led to the well-accepted classical domain theory. In this theory only very small grains are expected to be uniformly magnetized. For example, in the case of the mineral magnetite (Fe_3O_4), the size of uniformly magnetized equant grains was predicted to be approximately one order of magnitude smaller than can be observed optically (e.g., Butler and Banerjee, 1975). On the average, as the grain size increased more and more domains were expected and this expectation was confirmed experimentally.

There is no net magnetization of either grain shown in Fig. 3.3, and the grains are said to be *demagnetized*. However, multidomain grains can still have a remanent magnetization. For example, if the domain wall in the grain shown in Fig. 3.3a is displaced to the right, this produces an upward directed net magnetization. Such displacement can occur, for example, by applying an upward directed external magnetic field. If there are internal defects present in the crystal (e.g., dislocations), when the external field is removed the domain wall can be "pinned" by these defects and not return to the middle of the grain (e.g., Xu and Merrill, 1989). Hence the grain will exhibit a remanent magnetization. It was also predicted, and observed, that larger grains of a given mineralogy containing many domains were, on average, less stably magnetized than grains with fewer domains. For example, the remanent magnetization in large multidomain grains could usually be more easily altered by heating the sample or by applying external magnetic fields (e.g., Butler, 1992). Indeed, theory and observations suggest that the stability of a grain varies with grain size as illustrated in the curve shown in Fig. 3.4. A maximum in stability occurs either in the single-domain size range, as commonly supposed, or in the so-called *pseudo-single-domain* (PSD) size range as shown in Fig. 3.4. Although there are several definitions of PSD grains, essentially they are relatively small grains that exhibit hysteresis properties intermediate between single-domain (SD) and multidomain (MD) grains (see Table 3.2). The problem with defining such grains will become clearer in the next section. The classical SD and MD principles are well explained in such textbooks as Chikazumi (1964), Morrish (1965), Stacey and Banerjee (1974), O'Reilly (1984), and Butler (1992).

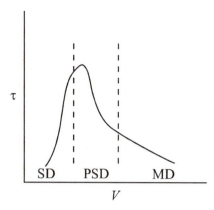

Fig. 3.4. Hypothetical change in relaxation time, τ, as a function of volume, V, all other factors held constant. The shape of the curve is not well known (see Banerjee, 1977; Levi and Merrill, 1978), but it appears to have a single maximum. SD, single-domain; PSD, pseudo-single-domain; MD, multidomain.

3.1.6 Modern Domain Concepts

Currently there is a revolution in our understanding of domain structure that will eventually affect almost all aspects of rock magnetism. Although it is difficult to pinpoint the precise start of this revolution, papers by Halgedahl and Fuller (1981, 1983) were very influential. Using bitter pattern techniques they showed that there were far fewer domains on average than was predicted by theory and, even more importantly, that grains could exhibit very different numbers of domains depending on their magnetic histories. Moon and Merrill (1984, 1985) used micromagnetic calculations to provide an explanation for these findings. Often a magnetic grain of a given size exhibited several local energy minima (LEM) states, rather than the one energy state implicitly assumed in classical domain theory.

A simple way to envisage the above is to consider a magnetic cube growing in size. In classical domain theory a small enough cube will be SD and when it grows to a critical size it will convert to a two-domain state. However, to become a two-domain grain it must nucleate a domain wall and in most instances this requires energy. That is, there is usually an energy barrier separating the SD grain from the two-domain grain state. Thus as the grain grows it may remain SD even though the two-domain state is a *global* or *absolute energy minimum state*. Energy can be supplied to this large SD grain, e.g., applying some appropriate external magnetic field, to provide the activation energy to surmount the energy barrier. A change of an SD grain to an MD grain in this process is an example of a *transdomain process* (Moon and Merrill, 1986). It also shows that the grain can exhibit more than one domain state depending on its history. In practice there are usually several LEM states that a grain can occupy.

Micromagnetic calculations have shown that domains are not regions of perfectly uniform magnetization as assumed in classical theory (Moon and Merrill, 1984), although they may be approximately so in larger grains (e.g., Ye and Merrill, 1991). Williams and Dunlop (1989) have shown that even SD grains often do not have the structure predicted by classical theory: three-dimensional micromagnetic calculations on a supercomputer for a small cube of magnetite indicate that SD grains are typically not uniformly magnetized. Although it seems preferable to retain the description "single-domain" for these grains, Williams and Dunlop (1989) have shown that both the domain structure of these grains and their reversal modes are somewhat different from that originally envisaged by Néel (1949). Moreover there is no simple SD to MD transition; instead, for example, in magnetite there is a transition from an SD-like state to a vortex state (e.g., Newell *et al.*, 1993; Enkin and Williams, 1994). Calculations indicate that classical MD-type magnetite grains occur only when the size is nearly an order of magnitude larger than that for the SD to vortex state transition.

Eventually the above will be incorporated into theories for remanence, susceptibility, etc. It is well known in statistical mechanics that one can determine the equilibrium properties of a system from knowledge of the partition function of the system (see §3.1.4). This partition function, which can be calculated once all the equilibrium energy states can be determined, is markedly different in modern domain theory where there are typically several LEM states available in addition to the single state used in classical theory. We can, therefore, safely predict that theories of remanence will change, even though it is premature to predict the precise nature of that change. Presumably our understanding of the dynamics of the system will also be dramatically altered since typically the dynamics are determined by using the so-called master equation of statistical mechanics, and this equation is significantly affected by the existence of LEM states (e.g., Moon and Merrill, 1988).

Much of the above is exemplified in the work of Halgedahl (1991) who heated and cooled titanomagnetite grains in a weak magnetic field and observed, at room temperature, the domain structure after each thermal cycle. The composition of the titanomagnetite grain was chosen so that it had a low Curie temperature to minimize the chance of chemical alteration on heating. The number of domains observed for one of her grains is shown in Fig. 3.5. This illustrates the multiplicity of LEM states available to a grain. Recently Ye and Merrill (1995) have applied renormalization group theory in an attempt to explain Halgedahl's observations, and suggest that the final room-temperature domain state is determined primarily by thermal fluctuations acting within a fraction of a degree of the Curie temperature. Moreover, Ye and Merrill predict that several forms of secondary magnetization (e.g., grain growth CRM) will

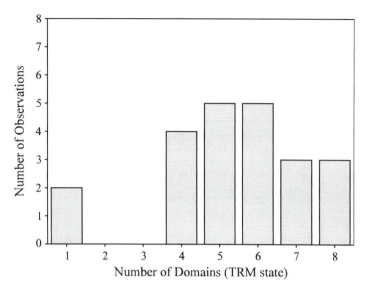

Fig. 3.5. Domain multiplicity distribution for a single grain given TRMs under "identical conditions." This experiment demonstrates the existence of LEM (local energy minima) domain states and that very small differences in initial conditions can lead to different domain structures. After Halgedahl (1991).

exhibit a much narrower range of domain states than shown in Fig. 3.5. If this is so, then domain imaging could become a very important tool in the future for distinguishing some forms of secondary magnetization from primary magnetization.

Recent reviews of this subject that provide some insight into the revolution taking place, and its effect on most aspects of theoretical and applied rock magnetism, have been given by Halgedahl (1989) and Dunlop (1990).

3.2 Magnetic Mineralogy

3.2.1 Properties of Magnetic Minerals

The magnetic minerals that can contribute significantly to the remanence in rocks include iron–titanium oxides, iron–manganese oxides and oxyhydroxides, iron sulfides, and iron and nickel–cobalt alloys. The purpose here is simply to give a few examples of these minerals and properties so that those not previously familiar with rock magnetism can gain some appreciation of the complexities

TABLE 3.3
Magnetic Properties of Some Common Minerals.

Mineral	Composition	Magnetic State	M_s $(10^3$ Am$^{-1})$	T_c (°C)
Magnetite	Fe_3O_4	Ferrimagnetic	476	580
Ulvospinel	Fe_2TiO_4	Antiferromagnetic		-153
Hematite	xFe_2O_3(hexagonal)	Antiferromagnetic with a parasitic ferromagnetism	~2.2	~680[a]
Ilmenite	$FeTiO_3$	Antiferromagnetic		-233
Maghemite	γFe_2O_3 (Cubic)	Ferrimagnetic	426	~600
Pyrrhotite	$Fe_{1-x}S(0<x<\frac{1}{7})$	Ferrimagnetic	90	~320
Trolite	FeS	Antiferromagnetic		~305
Jacobsite	$MnFe_2O_4$	Ferrimagnetic	424	~300
Goethite	$xFeOOH$	Antiferromagnetic with a parasitic ferromagnetism	~2(?)	~120
Iron	Fe	Ferromagnetic	1714	770
Cobalt	Co	Ferromagnetic	1422	1131
Nickel	Ni	Ferromagnetic	484	358
Awaruite	Ni_3Fe	Ferromagnetic	950	620
Wairauite	CoFe	Ferromagnetic	1936	986

Note. M_s, saturation magnetization; T_c, Curie (or Néel) temperature.
[a] Estimates vary from 675° to 725°C. More precise experiments done this decade suggest 696°C rather than 680°C.

and variations that arise from mineralogy. More detailed magnetic mineralogy can be found in reviews by Nagata (1961), Stacey and Banerjee (1974), Haggerty (1976), and O'Reilly (1984). A list of the properties of several magnetic minerals is given in Table 3.3.

For simplicity, discussion is restricted to the Fe–Ti oxides given in the ternary system shown in Fig. 3.6. The members within the solid solution series $xFe_2TiO_4(1-x)Fe_3O_4$ are referred to as *titanomagnetites* and those of the series $xFeTiO_3(1-x)Fe_2O_3$ are referred to as *titanohematites*. Titanomagnetite forms a *solid solution series*, which means that Fe_2TiO_4 and Fe_3O_4 are mutually soluble in each other. The Curie temperature of ferrimagnetic magnetite, an inverse spinel with a saturation magnetization value of 480×10^3 Am^{-1} (480 emu cm^{-3}), is 580°C, and to the first-order approximation, the Curie temperature decreases linearly with mole percent ulvospinel in solid solution to a value near -150°C for antiferromagnetic ulvospinel (Nagata, 1961). Similarly, the amount of ilmenite also decreases the Curie point nearly linearly in the titanohematite series from 680°C for hematite (the Curie temperature of hematite is still controversial; see Fuller (1970)) to near -223°C for ilmenite (Nagata, 1961). Although at high temperatures both the titanomagnetites and the titanohematites form solid solution series, at lower temperatures *exsolution* occurs (unmixing of phrases; so that in the titanomagnetites exsolution would result in an ulvospinel-rich phase

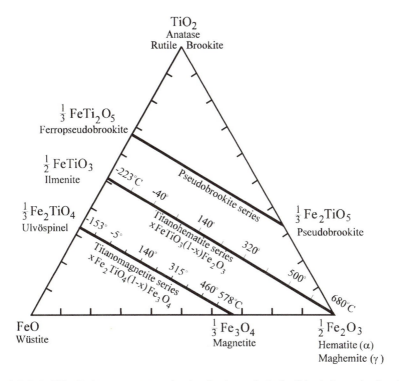

Fig. 3.6. FeO–TiO_2–Fe_2O_3 ternary system showing the three principal solid solution series found in igneous rocks. Members of the pseudobrookite series are all paramagnetic above liquid oxygen temperatures and therefore have little significance. Approximate Curie (or Néel) temperatures for various values of mole function x are indicated for the titanomagnetite and titanohematite series. In the titanomagnetite series M_s decreases with increasing x but in a complicated way depending on ordering of Fe^{2+} Fe^{3+} and Ti^{4+} on two possible lattice sites. In the titanohematite series there is complex magnetic variation with x. For $0 \le x < 0.5$ the Ti ions are disordered and the magnetization weak ($\sim 2 \times 10^3$ Am^{-1}); for $0.5 < x < 0.8$ the Ti ions are ordered and the material ferrimagnetic with a maximum magnetization near 150×10^3 Am^{-1}; for $1 > x > 0.8$ the material is antiferromagnetic with ferrimagnetic clusters.

and a magnetite-rich phase). The temperature of the *solvus* (the maximum temperature for exsolution) for both the titanohematites and the titanomagnetites are controversial. For further discussion see reviews by Merrill (1975), Haggerty (1976, 1978), and O'Reilly (1984).

In addition to exsolution, high-temperature (above the Curie point) oxidation commonly occurs after crystallization in terrestrial magnetic minerals. Fundamental work on the mineralogy associated with this oxidation was first carried out by Buddington and Lindsley (1964) and continued for many years (see Lindsley, 1976). This work assumes thermodynamic equilibrium conditions, and needs slight adjustments for the nonequilibrium conditions that seem to exist in nature (Sato and Wright, 1966; Grommé *et al.*, 1969). In addition, low-temperature oxidation (below 200°C or so) of titanomagnetites produces titanomaghemites (which are *cation-deficient spinels*, meaning that there are vacancies in some of the lattice sites that were occupied by cations before oxidation), with compositions lying in the field between titanomagnetites and titanohematites in Fig. 3.6. Titanomaghemites have been characterized magnetically by Readman and O'Reilly (1970, 1972), Ozima and Sakamoto (1971), Nishitani (1979), Keefer and Shive (1981), and Moskowitz (1993). Paleomagnetists often use a useful empirical classification scheme for high-temperature oxidation developed by Wilson and Haggerty (1966) and Ade-Hall *et al.* (1971) and a low-temperature scheme developed by Haggerty (1976) and Johnson and Hall (1978).

Reviews of the magnetic effects associated with chemical changes have been given by Merrill (1975), Henshaw and Merrill (1980), and Levi (1989) who show that sometimes chemical changes of magnetic minerals lead to pronounced magnetic changes and at other times little magnetic effects occur at all. Although many of these variations can be explained theoretically (Merrill, 1975; Henshaw and Merrill, 1980), often the variations depend on such subtle effects as minute exsolution and grain growth, that they are sometimes difficult to detect in practice, as is clearly shown by the work of Haigh (1958), Kobayashi (1961), Strangway *et al.* (1968), Larson *et al.* (1969), Evans and McElhinny (1969), Evans and Wayman (1974), Karlin *et al.* (1987), Özdemir and Dunlop (1988, 1989), and Pick and Tauxe (1991). There is little doubt that some remagnetization due to chemical changes occurs in nature, but the extent of this occurrence remains difficult to assess (see also §3.3.1).

The titanohematite system illustrates well the complexities of magnetic minerals in rocks. The higher temperature end member, hexagonal (rhombohedral) hematite (commonly found in red sandstones), is basically antiferromagnetic on which a weak or "parasitic" ferromagnetism is imposed. Above -15°C the magnetic moments lie in planes (basal planes) perpendicular to the C axis (Nagata, 1961). The coupling between moments in adjacent basal planes is essentially antiferromagnetic. However, a slight canting of the spins in the basal planes results in a small moment (weak ferromagnetism) that is perpendicular to the basal spins and is referred to as the spin-canted moment (Dzyaloshinski, 1958). In addition, there are defect moments parallel to the spin directions that probably originate from partial ordering of the defects on every other basal plane (a second source of weak ferromagnetism). The saturation magnetization is somewhat variable for hematite, but is approximately 2.2×10^3 Am^{-1} (2.2 emu cm^{-3}) at room temperature. Below -15°C (the Morin transition) the easy axis of magnetization becomes the C axis and the spin-canted moment vanishes, leaving only the defect moment, a fact exploited by Fuller

(1970) and Dunlop (1971) to study the origin and properties of hematite's weak ferromagnetism.

Both the defect and the spin-canted moments seem to persist when Ti is added to the system, although the Morin temperature decreases rapidly with small amounts of added Ti. At around $x = 0.5$ in the $x\text{FeTiO}_3(1\text{-}x)\text{Fe}_2\text{O}_3$ system there is an ordering phenomenon. For $x \approx 0.5$ the Ti ions are preferentially ordered on alternating basal planes (the symmetry group is $3\overline{\text{R}}\text{C}$), whereas in the range $0.5 < x < 0.8$ the Ti ions are disordered (symmetry group $3\overline{\text{R}}$). The room-temperature saturation magnetization increases more than 50-fold on this ordering, because the ordered phase is ferrimagnetic. FeTiO_3, at the end of the solid solution series, is antiferromagnetic. Thus, as the Ti content in the titanohematite is increased, the system varies from antiferromagnetism with a weak ferromagnetism to ferromagnetism and finally to antiferromagnetism (Ishikawa and Akimoto, 1958; Uyeda, 1958; Nagata, 1961; Ishikawa and Syono, 1963; Allan and Shive, 1974). The room-temperature saturation magnetization in titanohematites varies from zero to a value around 0.3 of that of magnetite. Some paleomagnetists have erroneously dismissed the magnetic role titanohematites can have in rocks because they considered only the end members of the solid solution series.

3.2.2 The Magnetic Record in Rocks

The following description of the magnetic records in rocks contains many generalities and oversimplifications, and is very incomplete. Its intent is solely to provide a broad perspective of some of the problems faced by paleomagnetists.

Igneous rocks are divided into extrusive rocks (e.g., lava flows) and intrusive igneous rocks (e.g., plutons). The majority of the extrusive rocks are basalts. The dominant magnetic mineral in most terrestrial basalts is titanomagnetite, with relatively high Curie temperatures (often within 100°C of the magnetite Curie point at 580°C). Basalts that cool very rapidly in the marine environment often contain skeletal titanomagnetite grains with much lower Curie temperatures (typically below 200°C), because they contain a higher fraction of ulvospinel in solid solution with magnetite. The titanomagnetites in marine basalts readily undergo low-temperature oxidation to titanomaghemites (which raises their Curie points). Lava flows typically represent "instantaneous" readings of the magnetic field because they often cool quickly relative to changes of the internal magnetic field. Indeed, the surfaces of these flows can cool so quickly that more than one geologist has found that the apparently solid lava flow they were standing on was being rotated by magma beneath it at about 1,100°C. The directions in the cooled outer surface in this case would no longer accurately reflect the Earth's magnetic field. Careful paleomagnetists are well aware of this and typically avoid *edge effects* by examining the consistency of magnetic directions over different regions of the flow. The NRM in fresh basalts is usually a TRM and has excellent rock magnetic properties. The main problem with using lava flows is that they usually provide a discontinuous record in time of the magnetic field. Moreover, even when large sequences of flows are present, the record is rarely even quasi-continuous; multiple lava flows are typically extruded over very short periods of time, with the result that there are groups of flows bunched together in time but separated from other such groups by discontinuities.

Intrusive igneous rocks cool much slower than extrusive rocks and so often have larger sized magnetic minerals with lower magnetic stabilities. Although in principle such rocks contain a more continuous record of the magnetic field than do lava flows, this record is often difficult to resolve. This is because the magnetic minerals in igneous rocks are rarely in equilibrium with each other (e.g., Sato and Wright, 1966; Grommé *et al.*, 1969) and because the rate of chemical alteration increases dramatically with temperature. Thus the slower cooling intrusive rocks often undergo chemical alteration below the Curie points of the titanomagnetites and titanohematites, the minerals that typically dominate their magnetic records.

Marine sediments receive contributions from terrestrial rocks, other marine rocks, extraterrestrial sources, chemical sources, and biological sources. As extreme examples, bacteria-produced single-domain magnetites dominate the magnetic mineralogy of some marine sediments (e.g., Kirschvink and Lowenstam, 1979), while others are dominated by aeolean contributions. Pelagic sediments are often oxidizing with depth (e.g., Henshaw and Merrill, 1980). Some pelagic deep-sea cores have provided excellent records of the reversal chronology but, because of their low sedimentation rates, not of reversal transitions. In contrast, more rapid sedimentation often occurs on the continental shelf environment, but with a whole host of other problems. Often the environment becomes more reducing with depth and chemical changes are significant, including changes facilitated by the bacterial mobilization of certain elements (e.g., Karlin and Levi, 1985; Channell, 1989).

The NRM in sediments is often a DRM or post-DRM, so secondary overprints are not uncommon (e.g., Verosub, 1977; Barton *et al.*, 1980). The magnetization is locked in over a depth that varies significantly from one sediment to another, depending on how the sediments are compacted and how they are mixed by benthic organisms (e.g., Channell, 1989). That is, sediments have typically filtered the magnetic signal, often in a manner that is difficult, or impossible, to unravel. Nevertheless, a few marine and terrestrial sedimentary cores (particularly for rapidly deposited sediments) have produced some of the best continuous records of reversal transitions available. In addition, certain lake sediments have proved to be very valuable in delineation of the secular variation of the Earth's magnetic field over a few thousand to a few tens of thousands of years.

The magnetic mineralogy of terrestrial sediments is highly variable, because the sediments originate in part from preexisting igneous, sedimentary and metamorphic rocks, and also because magnetic minerals sometimes form in sediments by authigenesis and diagenesis. Although the origin of the magnetization of so-called "red beds" (principally red sandstones and shales) is controversial, it is generally agreed that CRM in hematite typically contributes substantially to the remanence (Collinson, 1965; Roy and Park, 1972; Larson and Walker, 1975; Elston and Purucker, 1979; Butler, 1982). The controversy largely involves whether the CRM occurred rapidly soon after deposition, or whether it was graded, making the NRM unsuitable for paleomagnetic studies. Although no attempt is made to resolve this controversy here, no significant interpretations made in this book depend on the outcome.

Although magnetic properties have been generalized with respect to different rock types above, it needs to be emphasized that exceptions occur all too often. For example, it is quite possible in a given study to find that a particular granite is more strongly and stably magnetized than a particular basalt. Because subtle changes in factors such as grain size and oxidation state can have pronounced affects on the magnetic properties of rocks, it should not be expected that generalizations, such as given above, can be applied with much confidence to any given situation.

3.3 Paleomagnetic Directions and Poles

3.3.1 Demagnetization Procedures, Remagnetization, and Consistency Checks

Although all magnetic minerals exhibit anisotropy, the majority of igneous rocks and sedimentary rock used in paleomagnetism are not significantly magnetically anisotropic as a whole. On the other hand metamorphic rocks, which are rarely used in paleomagnetic studies, often do exhibit magnetic anisotropy. The absence of magnetic anisotropy in rocks occurs because the orientations of individual grains in a sample are random, or nearly random. Of course, because all rocks are not magnetically isotropic, paleomagnetists should determine in the laboratory whether the isotropic condition is satisfied in any given study.

The remanence acquired when a rock forms is called a primary magnetization and this will be parallel (or, very rarely, antiparallel) to the external field during its formation provided that the rock is magnetically isotropic. This follows since the external field is the only parameter with a preferred direction left in the problem. Subsequent to formation, secondary magnetization(s) (CRM, VRM, etc.) are often acquired. To distinguish between primary and secondary magnetizations paleomagnetists use various demagnetization procedures on the NRM. The two most commonly employed techniques are alternating field (AF) demagnetization and thermal demagnetization.

In *AF demagnetization* a sample is placed in an alternating field of amplitude H that is slowly (relative to the period of the alternating field) reduced to zero while the sample is tumbled as randomly as possible. This process randomizes the remanence due to those grains having relaxation times corresponding to coercivities $H_c \leq H_0$ (recall that τ is a function of H_c; see Eq. (3.1.20)). H_0 is then increased in steps, so that the directions of more and more stable (larger τ) remanence can be ascertained. If only one remanence contributes to the NRM, no change in direction will be observed on demagnetization. If two or more remanences are present, and have different directions, then a directional change on demagnetization is to be expected since it would be coincidental if the relaxation time spectra of the two were identical. The most stable component of remanence is often argued to be the primary magnetization, but a variety of more sophisticated (and more reliable) criteria have been used in recent years. AF demagnetization then measures the stability of remanence in alternating fields, from which stability in time is inferred.

Thermal demagnetization measures stability of remanence as a function of temperature, from which stability in time is also inferred. This is carried out by heating the sample to some temperature, T_0, and cooling in a nonmagnetic space. The remanence with blocking temperatures less than T_0 will be randomized (providing magnetic interactions between blocked remanence and unblocked remanence are negligible, as often seems to be the case). T_0 is stepwise increased up to the Curie temperature. Analogous with AF demagnetization, thermal demagnetization makes it possible to discern different remanences, if present. When multiple magnetizations are present the direction of the primary magnetization is often taken to be that of the remanence with the highest blocking temperature.

Other demagnetization procedures including chemical demagnetization, low-temperature demagnetization, and various combinations of the above are also used in actual paleomagnetic studies. Descriptions of these are given by O'Reilly (1984), Langereis *et al.* (1989), and Butler (1992).

A major problem in paleomagnetism involves remagnetization resulting in the acquisition of secondary magnetization. This acquisition is now accepted as a widespread phenomenon, particularly in sedimentary rocks (e.g., Elmore and McCabe, 1991; Van der Voo, 1993). On average the problem increases with age of the rock and is particularly serious in Mesozoic and older rocks (e.g., McCabe and Elmore, 1989). The recognition of remagnetization was greatly facilitated by the use of multicomponent analyses (e.g., Jones *et al.*, 1975; Halls, 1978; Hoffman and Day, 1978; Dunlop, 1979; Kirschvink, 1980; McFadden and McElhinny, 1988; Van der Voo, 1993). In paleomagnetism the change in the vector magnetization observed during demagnetization is usually illustrated by means of a Zijderveld diagram (Zijderveld, 1967) in which the projections of the observed vector onto two orthogonal planes are plotted at each demagnetization step. Figure 3.7 is an example of a Zijderveld diagram in which thermal demagnetization shows the presence of more than one component of magnetization. Only one component appears to be demagnetized between room temperature and 360°C. A component with a very different direction is demagnetized above 360°C. Most paleomagnetists would conclude that there are two components present in this example. However, this method is not foolproof. If two components have similar responses to AF or thermal demagnetization, then the Zijderveld diagram will give only one component, usually with an incorrect direction! To illustrate this further, suppose the actual sample has three components of magnetization, all with different directions. Further suppose that there is a 15% probability that the demagnetization curves for any two components will be very similar for *part* of the demagnetization process. Then the probability that one will obtain incorrect information on a multicomponent decomposition analysis is only slightly less than a 50%! Typically the larger the number of components, the larger the possibility for error. Such errors are not incorporated in the usual error analyses discussed in §3.3.3. Although multicomponent analyses became a very important tool during the 1980's, many paleomagnetists seem unaware of some of its pitfalls.

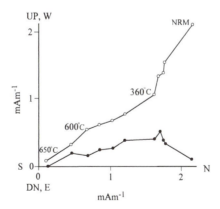

Fig. 3.7. Orthogonal Zijderveld plots of magnetization vectors during thermal demagnetization of a sedimentary rock. The x axis is the North–South direction that is common to both the horizontal and the vertical planes. The y axis is the West–East direction lying in the horizontal plane and also the up-down direction lying in the vertical plane. Solid (open) symbols are projections of the vectors on the horizontal (vertical) plane. After Li *et al.* (1991).

The sources of stable remagnetization appear to come primarily from authigenesis (grain growth CRM), diagenesis (the replacement of a precursor mineral by another), or from reheating associated with tectonics (e.g., Elmore and McCabe, 1991: Van der Voo, 1993). Often fluids are involved. VRM also can be a significant source of remagnetization but, although exceptions may occur (e.g., Moon and Merrill, 1986), it appears to be less stable than most forms of primary magnetization. There can be little doubt that paleomagnetists will on occasion end up with erroneous directions. However, substantive errors are significantly reduced by the availability of other criteria that provide tests of, and therefore confidence in, the paleomagnetic data. An example of this is when different rock types from the same general region and of the same relative age give the same magnetic directions. It would be rather surprising if the magnetization of a sediment (typically DRM or post-DRM) agreed with that of an igneous rock (typically TRM) unless their magnetizations were both primary. Consistency of directions from different rock formations in the same general locality and of the same age provide considerable confidence for the reliability of paleomagnetic directions. Also, there are several field tests, as illustrated in Fig. 3.8, that make it possible to place a minimum age on the magnetization, even supposing it may not be primary.

By using such field criteria, it is possible to obtain some idea of the reliability of most paleomagnetic data. The main problem is that each paleomagnetic study will involve different techniques and tests depending on the specific circumstances. Some will demonstrate firmly that the NRM is a primary

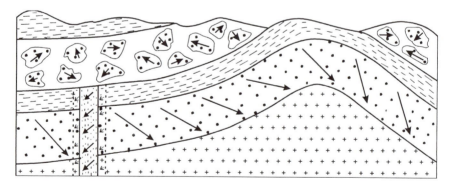

Fig. 3.8. Fold and Conglomerate test of Graham (1949) and the Baked Contact test for paleomagnetic directions. If the directions of magnetization of samples collected from different limbs of a fold converge after unfolding the beds, the magnetization pre-dates the folding and has been stable since. If directions of magnetization in conglomerate pebbles derived from the beds under investigation are randomly oriented, the magnetization of the parent beds has been stable since formation of the conglomerate. If directions of magnetization in the baked zone surrounding an intrusion are parallel to those observed in the intrusion, but differ from that of the unbaked country rock, the magnetization of the intrusion has been stable since formation.

remanence, while others will rely on less convincing arguments. There is no standard applicable to all studies because in many cases it might be impossible to do all that a given standard might require. Inevitably, therefore, a few erroneous directions arise, but analysis methods are now sufficiently sophisticated that substantive errors should constitute no more than a minute fraction of the whole. The analysis of a mixed set of data on a global scale, as is carried out in the next few chapters, thus relies on careful selection of the data according to some minimum criteria of acceptability. Examples of some modern methods of selection are discussed by Van der Voo (1993).

3.3.2 The Geocentric Axial Dipole Field Hypothesis

Suppose the direction of the primary magnetization of a rock at a particular location has been determined and that a reliable age for the time of formation of the rock is available. As discussed in Chapter 2, the instantaneous field is moderately complex at any time so the problem arises as to how a local measure of that field at some time in the past can be analyzed. The problem is complicated by the fact that not all rocks record the field in the same way. Imagine a series of lava flows, each with a thickness of a few metres, extruded successively at random times by a volcano. Each lava flow will cool to the ambient temperature over a time, typically in days to years, that depends on its thickness. Thus the paleomagnetic direction measured in each lava will be an average of the local variations over at most a few years. Over such time intervals the field is essentially unchanged and so each lava flow has effectively recorded the instantaneous field. Thicker lavas, up to several tens of metres thick, will cool more slowly, so that the field recorded in the flow will be more complex. At any given horizon the instantaneous field will be that recorded when it cooled through the blocking temperatures. A series of samples collected across a thick sequence of flows can thus record changes over a few hundred or even a thousand years. The average magnetization of such a thick sequence flows, or of large intrusions that cool much more slowly than lavas, will be a *time-averaged* magnetic field over several hundred or thousand years.

   Because the process of magnetization acquisition in sediments is usually either depositional or chemical, the time span covered by a typical specimen of thickness 2.5 cm, as is used in the laboratory, may already be more than a thousand years (deposition rates in the pelagic ocean environment are usually measured in millimetres per thousand years). Successive samples collected over great thicknesses could correspond to time spans of hundreds of thousands or millions of years. In these cases the field measured by just a thin specimen could be time averaged as much as a very thick lava flow or intrusion. Paleomagnetists therefore design their sampling schemes to attempt to measure the ancient field at as many different instances in time as they can. The average of these magnetizations, determined by finding the direction of the vector sum, is termed the *paleomagnetic field* for that rock formation.

The paleomagnetic field is thus a time-averaged field. From knowledge of the present field over historic times it is supposed that over periods of several thousands of years any locality would record the full variability available to the field. Thus, over such periods of time, the time-averaged field would to a first approximation correspond not to a dipole inclined to the axis of rotation but to a dipole aligned along that axis. This hypothesis is termed the *geocentric axial dipole field* or GAD hypothesis. The problem that arises is that one is never sure of the extent to which the time-averaging procedure has been carried out, so that each paleomagnetic direction will be different from every other such direction in the time factor. It might be supposed that over a sufficiently long enough time all paleomagnetic directions for a given age at any locality would coincide. The basic question of just how much time is sufficient remains, and will be posed in more detail in subsequent chapters.

For the moment we assume validity of the assumption and articulate its implications. For a dipole of strength p the radial and tangential components of the field at some colatitude θ at the Earth's surface (radius R) are respectively

$$F_R = -\frac{2\mu_0 p \cos\theta}{4\pi R^3} \tag{3.3.1}$$

and

$$F_\theta = -\frac{2\mu_0 p \sin\theta}{4\pi R^3} \quad , \tag{3.3.2}$$

where the field F is measured in tesla (see §2.2). Since the tangent of the magnetic inclination I is F_R/F_θ,

$$\tan I = 2\cot\theta \tag{3.3.3}$$

or

$$\tan I = 2\tan\lambda \quad , \tag{3.3.4}$$

where λ is the latitude. If the geocentric axial dipole field hypothesis is applicable then this magnetic latitude corresponds to the geographic latitude. There is no information about longitude.

Because the time-averaged inclination corresponds to the paleolatitude on the geocentric axial dipole field assumption, and the time-averaged declination indicates the direction of a meridian, the position of the corresponding geographic pole on the Earth's surface, relative to the observation site, can be calculated (Fig. 3.9). Given the site coordinates (λ_s, ϕ_s) and the paleomagnetic direction (D_m, I_m) at that site, the latitude of the pole is given by

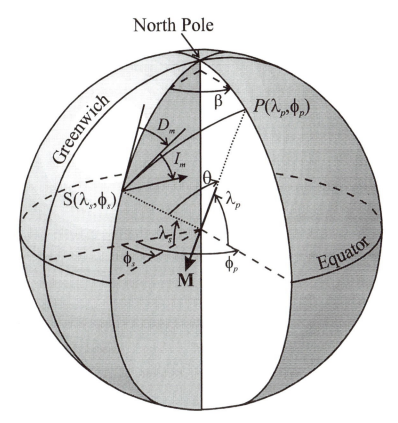

Fig. 3.9. Relationship to calculate the position (λ_p, ϕ_p) of the paleomagnetic pole P relative to the sampling site S at (λ_s, ϕ_s) with mean magnetic direction (D_m, I_m).

$$\sin \lambda_p = \sin \lambda_s \cos\theta + \cos\lambda_s \sin\theta \cos D_m \quad (-90° \le \lambda_p \le +90°) \quad . \quad (3.3.5)$$

The longitude of the pole is given by

$$\phi_p = \phi_s + \beta \qquad \text{when } \cos\theta \ge \sin \lambda_s \sin \lambda_p$$

or $\qquad\qquad\qquad\qquad\qquad\qquad\qquad\qquad\qquad\qquad\qquad\qquad\qquad$ (3.3.6)

$$\phi_p = \phi_s + 180 - \beta \quad \text{when } \cos\theta < \sin \lambda_s \sin \lambda_p \quad ,$$

where

$$\sin\beta = \sin\theta \sin D_m / \cos\lambda_p \quad (-90° \le \beta \le 90°) \quad . \quad (3.3.7)$$

The paleocolatitude θ is determined from (3.3.3). The pole position (λ_p, ϕ_p) calculated in this way is called the *paleomagnetic pole*. The use of the term

TABLE 3.4
Summary of Poles Used in Geomagnetism and Paleomagnetism

North (south) magnetic pole	Point on the Earth's surface where the magnetic inclination is observed to be +90° (-90°). The poles are not exactly opposite one another and for epoch 1990 lie at 78.1°N, 256.3°E and 64.9°S, 138.9°E.
Geomagnetic north (south) pole	Point where the axis of the calculated best fitting dipole cuts the surface of the Earth in the northern (southern) hemisphere, The poles lie opposite one another and for epoch 1990 are calculated to lie at 79.2°N, 288.9°E and 79.2°S, 108.9°E.
Virtual geomagnetic pole (VGP)	The position of the equivalent geomagnetic pole calculated from a spot reading of the paleomagnetic field. It represents only an instant in time, just as the present geomagnetic poles are only an instantaneous observation.
Paleomagnetic pole	The average paleomagnetic field over periods sufficiently long so as to give an estimate of the geographic pole. Averages over times of 10^4 to 10^5 years are estimated to be sufficient. The pole may be calculated from the average paleomagnetic, field or from the average of the corresponding VGPs.

implies that "sufficient" time averaging has been carried out. Alternatively, any instantaneous paleofield direction may be converted to a pole position using (3.3.5), (3.3.6) and (3.3.7), in which case the pole is termed a *virtual geomagnetic pole* (VGP). The VGP is the paleomagnetic analogue of the geomagnetic poles of the present field. The paleomagnetic pole may then be calculated alternatively by finding the average of many VGPs, corresponding to many paleodirections. Table 3.4 gives a summary of the various types of poles used in geomagnetism and paleomagnetism.

Conversely of course, given a paleomagnetic pole position (λ_p, ϕ_p) the corresponding expected mean direction of magnetization (D_m, I_m) may be calculated for any site location (λ_s, ϕ_s) (Fig. 3.9). The paleocolatitude θ is given by

$$\cos\theta = \sin\lambda_s \sin\lambda_p + \cos\lambda_s \cos\lambda_p \cos(\phi_p - \phi_s) \quad , \qquad (3.3.8)$$

and the inclination I_m may then be calculated from (3.3.3). The corresponding declination D_m is given by

$$\cos D_m = \frac{\sin\lambda_p - \sin\lambda_s \cos\theta}{\cos\lambda_s \sin\theta} \quad . \qquad (3.3.9)$$

It is geometrically obvious that

$$0° \le D_m \le 180° \quad \text{for} \quad 0° \le (\phi_p - \phi_s) \le 180°$$
$$180° < D_m < 360° \quad \text{for } 180° < (\phi_p - \phi_s) < 360° \quad .$$

The declination is indeterminate (so any value may be chosen) if the site and pole position coincide, and if $\lambda_s = \pm 90°$ then D_m is defined as being equal to λ_p, the longitude of the geomagnetic pole.

If the location of the rocks used for paleomagnetic study has moved since its formation relative to the present geographic pole, the paleomagnetic pole will depart from the present geographic pole. Indeed, the positions of the paleomagnetic poles from all continents depart more and more from the present geographic pole the further one goes back in time. The path traced by successive positions of the paleomagnetic poles with time plotted on the present longitude–latitude grid for a given continental block is termed the *apparent polar wander path* for that block. The comparison of such paths between different continental blocks provides the evidence that continental drift has occurred. The matching of two paths over any time span then provides a method for determination of the relative positions of two blocks in the past. An example is given in Fig. 3.10. The pole paths for Europe and North America diverge over the past 350 million years. If various segments of the apparent polar wander paths are made to overlap, it is possible to find the relative locations of Europe and North America in the past (Graham *et al.*, 1964). This shows that these two continents were previously joined together and have subsequently separated to form what is now

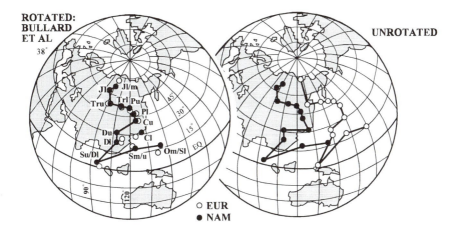

Fig. 3.10. Apparent polar wander paths for Europe (EUR) and North America (NAM) for the time interval Middle Ordovician (~470 Ma) through Early to Middle Jurassic (~175 Ma). Rotated: after closing the Atlantic Ocean according to the parameters of Bullard *et al.* (1965). Unrotated: with the continents in their present-day coordinates. After Van der Voo (1993).

the North Atlantic Ocean (see Fig. 3.10). Examples of maps showing possible positions of the continents over the past 600 Myr are given in Irving (1977, 1979), Morel and Irving (1978), Smith *et al.* (1981), and Van der Voo (1993). A full discussion of paleomagnetism and its relation to continental drift and plate tectonics is given by McElhinny (1973), Butler (1992), and Van der Voo (1993).

The close correspondence of paleomagnetic poles of a given age from widely separated regions of continental extent provides the key evidence that, to a first approximation, the geocentric dipole assumption is valid. In some instances a paleomeridian may pass lengthwise across a continent so that the latitude variation is maximized. In such a case the geocentric dipole assumption may be tested explicitly, as has been done for Africa by McElhinny and Brock (1975). However, these tests do not establish the *axial* nature of the paleomagnetic field. For this it is necessary to invoke paleoclimatic evidence. Past climate and temperature distribution can be expected to have an equator to pole variation dependent essentially on the axis of rotation. If the paleolatitudes of various paleoclimatic indicators determined from paleomagnetic data also show such an equator to pole distribution, then the *axial* nature of the paleomagnetic field is strongly supported (Irving and Gaskell, 1962; Irving and Briden, 1962; Briden and Irving, 1964; Drewry *et al.*, 1974; see also summary by McElhinny, 1973; Van der Voo, 1993). Further discussion of all these points is given in §6.1 and §6.2.

3.3.3 Standard Statistical Methods in Paleomagnetism

The scatter in paleomagnetic observations is typically large, so statistical analyses play a central role. At a given sampling site errors may be introduced by inaccurate orientation, local magnetic anomalies, etc., and these may be further compounded in the laboratory because noise may be introduced during procedures to remove secondary components. All of these errors produce a certain amount of scatter in the magnetizations between samples collected at a site and are referred to as the *within-site scatter*. Because sampling procedures are designed so that sites represent as far as is possible single points in time, the variation in magnetizations between sites will tend to represent variations produced by the secular variation. The *between-site scatter* is thus often used as a parameter to investigate changes in secular variation in the past. The statistical analysis of directional data is dealt with in several texts (e.g., Fisher *et al.*, 1987), so only a brief introduction to Fisher statistics, the most common used in paleomagnetism, is given here. It is assumed that readers are familiar with standard statistical tests such as chi-square and F tests.

Analogous to a normal distribution, Fisher (1953) suggested that the distribution of vectors on a unit sphere has a probability density $p(A)$ given by

$$p(A)\,d\,A = c\exp(\kappa\cos\vartheta)\,d\,A \qquad (3.3.10)$$

where dA is the element of area at (ϑ,ϕ), ϑ being the polar angle to the mean direction of the distribution and ϕ a uniformly distributed azimuthal angle about the mean. The requirement that the integral of $p(A)$ over the whole sphere must equal 1 gives $c = \kappa/(4\pi\sinh\kappa)$. κ is referred to as *Fisher's precision parameter* and describes the dispersion of the distribution. For small κ the distribution is highly dispersed, the extreme case being a uniform distribution over the sphere for $\kappa = 0$. For large κ the distribution is highly concentrated, the extreme case being no dispersion at all in the limit $\kappa\to\infty$.

For κ large, ϑ is dominated by small values, so $\cos\vartheta \approx 1-(\tfrac{1}{2})\vartheta^2$ (note that this approximation is in error by only 2% for $\vartheta = 45°$). Thus

$$p(A) \approx \frac{\kappa}{2\pi}\exp(-\frac{\kappa\vartheta^2}{2}) \quad, \qquad (3.1.11)$$

which is a bivariate normal distribution, with $\vartheta^2 = \alpha^2 + \beta^2$, where α and β are any two orthogonal polar angles from the mean. Thus it is seen that κ is in effect the *invariance*, or the reciprocal of variance. This approximation is why most of the hypothesis tests can use can use the chi-square of F distributions as good approximations (e.g., see McFadden and Lowes, 1981).

In paleomagnetism the direction of magnetization of a rock sample is specified by the declination, D, measured clockwise from true north, and the inclination, I, measured positively downward from the horizontal. This direction may be specified by its three direction cosines, as follows

North component	$l = \cos D \cos I$
East component	$m = \sin D \cos I$
Down component	$n = \sin I$

The direction cosines (X,Y,Z) of the resultant of N such directions of magnetization are proportional to the sum of the separate direction cosines, and are given by

$$X = \frac{1}{R}\sum_{i=1}^{N} l_i ; \quad Y = \frac{1}{R}\sum_{i=1}^{N} m_i ; \quad Z = \frac{1}{R}\sum_{i=1}^{N} n_i \quad, \qquad (3.3.12)$$

where R is the vector sum of the individual unit vectors, given by

$$R = \left[\sum l_i^2 + \sum m_i^2 + \sum n_i^2\right]^{\frac{1}{2}} \quad. \qquad (3.3.13)$$

The mean declination, D_R, and inclination, I_R, which together give the best estimate of the mean direction of the distribution, are given by

$$\tan D_R = \frac{\sum m_i}{\sum n_i} = \frac{Y}{X} \; ; \; \sin I_R = \frac{1}{R}\sum n_i = Z \quad . \tag{3.3.14}$$

Fisher (1953) showed that for $\kappa \geq 3$ the best estimate k of κ is given by

$$k = \frac{N-1}{N-R} \quad . \tag{3.3.15}$$

This estimate is best in the sense that k^{-1} is both a minimum variance estimator and an unbiased estimator for κ^{-1} (see McFadden, 1980). An angular standard deviation S can be defined as in the case of a normal distribution where

$$S^2 = \frac{1}{N-1}\sum_{i=1}^{N} \vartheta_i^2 \quad , \tag{3.3.16}$$

where ϑ_i is the angle individual vectors make with the sample mean. When the ϑ_i are small $\cos\vartheta \approx 1 - \frac{1}{2}\vartheta^2$, so, recognizing that $R = \sum\cos\vartheta_i$, (3.3.16) may then be written as

$$S^2 \approx \frac{1}{N-1}\sum_{i=1}^{N}(2-2\cos\vartheta_i) = 2\frac{(N-R)}{(N-1)} = \frac{2}{k}\,\mathrm{rad}^2 \quad . \tag{3.3.17}$$

In the limit of small scatter the *angular standard deviation* or *angular dispersion*, S, is thus simply related to the precision by

$$S = \frac{81}{\sqrt{k}}\,\mathrm{deg.} \tag{3.3.18}$$

This is of particular usefulness in the study of paleosecular variation discussed in §6.4.

Statistics in paleomagnetism are of a hierarchical type. Sample directions are averaged at the lowest level, then site directions are averaged to determine the paleomagnetic pole. Suppose N samples are collected at each of B sites, then following standard statistical analysis, the total angular dispersion, S_T, is related to the between-site dispersion, S_B, and the within-site dispersion, S_W, by the relation

$$S_T^2 = S_B^2 + \frac{1}{N}S_W^2 \quad . \tag{3.3.19}$$

It follows from (3.3.19) that the corresponding estimates of the precisions are related by

$$\frac{1}{k_T} = \frac{1}{k_B} + \frac{1}{k_W N} \quad .$$
(3.3.20)

In the case of lava flows the within-site scatter is often so small that one needs only $N = 4$ samples per site to make the last term of (3.3.19) or (3.3.20) much smaller than the first (see §6.4).

Given N unit vectors from a Fisher distribution with $\kappa \geq 3$, the true mean direction of the distribution will, with probability $(1-P)$, lie within a circular cone of semi-angle α_{1-P} about the resultant vector R where

$$\cos \alpha_{1-P} = 1 - \frac{N-R}{R} \left\{ \left(\frac{1}{P} \right)^{\frac{1}{N-1}} - 1 \right\} \quad .$$
(3.3.21)

In paleomagnetism it is common to use $P = 0.05$ and report values for α_{95}, commonly referred to as the circle of 95% confidence.

If a mean direction (D_m, I_m) is converted to a paleomagnetic pole the circle of confidence about this mean is transformed to an oval about the pole because of the nonconformal dipole transformation (3.3.4). The errors δD_m and δI_m associated with α_{95} are (Irving, 1956)

$$\alpha_{95} = \delta I_m = \delta D_m \cos I_m \quad .$$
(3.3.22)

From (3.3.3) the error δI_m corresponds to an error $\delta\theta$ in the colatitude given by

$$\delta\theta = \tfrac{1}{2}\alpha_{95}(1 + 3\cos^2\theta) \quad ,$$
(3.3.23)

and the declination error corresponds to an error δm in a direction perpendicular to the meridian given by

$$\delta m = \alpha_{95}\frac{\sin\theta}{\cos I_m} \quad .$$
(3.3.24)

Thus a circle of 95% confidence (α_{95}) for directions will transform to an oval of 95% confidence ($\delta\theta, \delta m$) for the paleomagnetic pole. The converse is also true. If a set of VGPs has a mean with a circle of confidence, the corresponding mean direction at any locality will be specified with an oval of confidence.

The form of the dipole transformation means that a set of directions having a Fisher distribution cannot in general transform into a set of VGPs having a Fisher distribution; the VGPs will have an oval distribution. Conversely, a set of VGPs with Fisher distribution will transform to a set of directions at any locality with an oval distribution. Random errors that occur at the site level due to experimental errors of one sort or another can be expected to produce a Fisher distribution of directions within-site. The between-site distribution will depend on the properties of the magnetic field variations and, depending on their nature, will tend to produce a Fisher distribution either of directions or of poles. In practice the result will be a combination of Fisher and oval distributions whose relative contributions might change with time and position on the Earth's surface. That is, a Fisherian or any other (e.g., a Bingham) distribution, will in general only be an approximation to the actual distribution of paleomagnetic data. In spite of this, Fisher statistics are extremely useful in describing the data, particularly if the statistics are calculated using VGPs, as is discussed in detail in Chapter 6.

3.4 Paleointensity Methods

3.4.1 The Problem

There are numerous paleointensity techniques that, it is argued, give reliable estimates for the ancient field strengths of the Earth and Moon. It is found experimentally for most rocks that in fields on the order of the Earth's magnetic field ($B \approx 10^{-4}$ T) the TRM, M_T, is proportional to the inducing field H, so that

$$M_T = C_1 H \ , \tag{3.4.1}$$

where C_1 is a constant of proportionality that depends on the magnetic properties of the rock. Equation (3.4.1) follows from single-domain theory by expanding the hyperbolic tangent of (3.1.17) and dropping the higher order terms. Thus the linear relation between M_T and H is expected only for small inducing fields, and this is what is usually observed experimentally.

Let M_L denote a TRM produced in the laboratory in a field H_L, so

$$M_L = C_2 H_L \tag{3.4.2}$$

and therefore

$$\frac{M_T}{M_L} = \frac{C_1}{C_2} \frac{H}{H_L} \ . \tag{3.4.3}$$

If the magnetic properties of the rock have not been altered since formation then $C_1 = C_2$, and it is clearly a simple matter to determine H, the ancient field strength. Unfortunately, this is seldom the case in practice. For example, chemical changes often occur in the laboratory when the rock is heated. Coe (1967a,b) has shown that even for historically erupted lava flows (cases in which the correct field strength is known in advance) only a very small percentage are suitable for paleointensity studies. It might be expected that for older rocks an even lower percentage would be suitable, because they are much more likely to have picked up unwanted remanences.

The basic idea behind all *reliable* paleointensity techniques is to develop a method by which several *independent* estimates of the ancient field strength can be obtained from the *same* sample. Consistency between such estimates provides some confidence regarding reliability. Unfortunately, many techniques do not do this. For example, several paleointensity estimates have been obtained by equating the ratio C_1/C_2 to the ratio of saturation magnetization of the sample before and after heating. The rock has undergone chemical change on heating and such a "correction factor" is used to account for it. This technique is becoming increasingly fashionable but is often very dangerous in practice because it typically involves elimination of some or all of the consistency checks.

Before accepting a published paleointensity value one should determine what *consistency checks* have been used to see whether they are acceptable. Two basic paleointensity techniques are now described that illustrate what is meant by consistency checks. There are many other reliable techniques, but all of them use similar principles to those described below.

3.4.2 Absolute Paleointensities I: The Modified Thellier Method

The classical method for paleointensity determination was developed by the Thelliers for studies of baked archeological material, but it applies equally well to any material whose magnetization was acquired as a TRM (see review by Thellier and Thellier, 1959a,b). The methods used by most workers are in fact modified versions of the original technique and these, with their associated problems, have been discussed by Coe (1967a,b), Smith (1967a,b,c), Levi (1975, 1989), Domen (1977), and Kono (1978). All these techniques rely on Thellier's law of additivity, which effectively states that a *partial TRM* ($M_{T_i}^{T_{i+1}}$) acquired in any temperature interval (T_i, T_{i+1}), is independent of the remanence acquired in any other temperature interval. Thus it follows (see Nagata, 1961) that

$$M_{T_0}^{T} = \sum_{T_0}^{T} M_{T_i}^{T_{i+1}} \qquad (3.4.4)$$

That is, the total TRM is equal to the sum of the partial TRMs. Equation (3.4.1) will usually hold for each temperature interval, although the constant C_1 will have to be replaced by C_i, because the constant of proportionality generally varies from temperature interval to temperature interval. Therefore (3.4.3) will also be applicable for each temperature interval, allowing one to obtain several separate estimates of the ancient field strength. It is these separate estimates that provide the consistency check in the Thellier method.

There are several ways of doing this in practice. One common way is to heat the sample from room temperature to T_i and cool it in a nonmagnetic space, after which the remaining remanence is measured. Subsequently, the sample is heated a second time to T_i, cooled in a known field, and the remanence again measured. The remanence lost in the first heating and cooling cycle is then plotted against the remanence gained in the second (Fig. 3.11). (A variation of this method by Kono (1978) involves heating and cooling a sample in a laboratory field that is perpendicular to the remanence. This eliminates the need for a double-heating procedure but one must be assured of the perpendicularity of the inducing field to the remanence.) These steps are then repeated to successively higher temperatures, ideally producing a straight line as shown in Fig. 3.11. From the slope of this line the ancient field strength can be estimated. Ideal behavior suggests the following:

(i)   the NRM is a pure unaltered TRM;

(ii) the TRM is linearly proportional to the inducing field (this can be checked in the laboratory);

(iii) Thellier's additivity law is satisfied (deviations are typically small); and

(iv) the physical–chemical properties of the magnetic minerals have remained unaltered since the acquisition of the initial TRM (that is, the constants C_i have remained invariant).

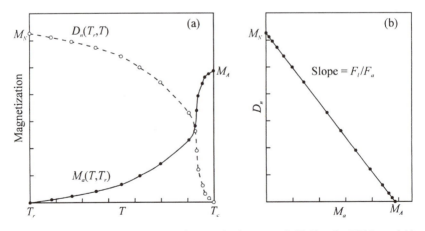

Fig. 3.11. (a) Ideal example of an NRM demagnetization curve $D_n(T_r,T)$ and a TRM acquisition curve $M_a(T,T_r)$ derived during the modified Thellier method for paleointensity determination. (b) The corresponding ideal NRM–TRM line obtained after T has been eliminated. The slope is the ratio of the paleofield intensity (F_a) to laboratory inducing field (F_l). After Coe (1967a).

Deviations from a straight line are referred to as nonideal Thellier behavior (see Coe (1967a) for possible origins of this behavior) and indicate that one of the above conditions has not been met. Nonideal behavior is, unfortunately, typical, so only a small percentage of rocks can be used for paleointensity estimates.

In the case of archeological material, a common problem is the effect of weathering causing chemical changes. In certain circumstances this can be modeled to provide field strength estimates under nonideal behavior (Barbetti and McElhinny, 1976; Barbetti *et al.*, 1977). Coe *et al.* (1978) have described statistical procedures for determining the quality of paleointensity data.

3.4.3 Absolute Paleointensities II: Shaw's Method

Van Zijl *et al.* (1962a,b) attempted to overcome the problem that most rocks contain secondary magnetization. If the initial paleomagnetic study showed that the secondary magnetization was very easily removed by AF demagnetization in peak fields of about 20 mT, it was supposed that (3.4.3) would apply if both NRM and total TRM were demagnetized in that field. However, there was no consistency check in this method so Smith (1967a) extended it by comparing the coercive force spectrum for NRM and TRM. The NRM is first demagnetized using increasing steps of peak alternating field. After completed demagnetization the sample is given a TRM that is subsequently demagnetized in the same way. A paleointensity estimate is obtained from the linear part of the NRM–TRM plot as shown in Fig. 3.12a. Thus the coercive force spectrum obtained on AF demagnetization is used to obtain the paleointensity estimate, rather than the blocking temperature spectrum as used in the Thellier method. The plot might deviate from a straight line at low fields because of the removal of secondary components from the NRM that are not present in the laboratory TRM. The existence of a comparable portion of the coercive force spectra suggests this portion has remained unchanged as a result of laboratory heating.

Unfortunately, there are ways in which rocks might be altered on heating yet maintain a comparable part of their NRM and TRM spectra. So although there is a consistency check of sorts, it is not necessarily rigorous. McElhinny and Evans (1968) improved on this by suggesting that, in addition, the coercive force spectrum of saturation IRM be compared before and after heating through AF demagnetization (Fig. 3.12b), thereby introducing an independent consistency check. Those parts of the spectrum that were not altered by the heating may then be used in the paleointensity estimate. Saturation IRM is not, however, a particularly good analogue of TRM, and this consistency check is too stringent and unrealistic in practice.

It has been shown that ARM (unlike IRM) often has an identical AF demagnetization spectrum to TRM. Shaw (1974) used this to introduce a consistency check by comparing the coercive force spectrum of ARM,

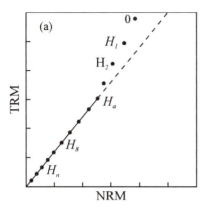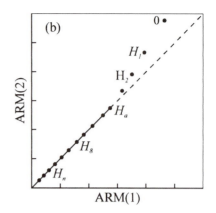

Fig. 3.12. Shaw's (1974) method for paleointensity determination. (a) Hypothetical NRM–TRM plot at various AF demagnetization fields (H_n). The slope of the straight line section at fields $H_n >$ H_a is the ratio of the laboratory inducing field for the TRM (F_l) to the paleofield intensity (F_a). For $H_n < H_a$ the points do not fall on the straight line due to removal of secondary components during demagnetization of NRM (but not present in the laboratory TRM) or to alteration of the coercive force spectrum in this region. (b) Hypothetical plot of ARM(2), determined after heating, against ARM(1), determined before heating, at various AF demagnetization fields (H_n). For $H_n \geq H_a$ the plot is a line of slope 1.0. For $H_n < H_a$ the points depart from the line because this part of the coercive force spectrum has been altered by the heating.

determined through AF demagnetization, before and after heating (Fig. 3.12b). This determines if any chemical changes during heating have affected parts of the coercive force spectrum, and thus which parts of the AF demagnetization curve may be used in the paleointensity estimate. A problem that has not yet been fully investigated is whether the AF demagnetization spectrum is really sensitive enough to distinguish CRM from TRM, and this remains the main query with the method. Some of the causes of nonideal behavior in Shaw's method have been examined by Kono (1978).

The main advantage of Shaw's method is that it is very much less tedious than the Thellier method. However, high failure rates are often involved with either method, so a basic problem is finding suitable samples that will give good results. Senanayake and McElhinny (1981) attempted to overcome this by using their observation that basic lava flows exhibiting high-temperature oxidation of their magnetic minerals (occurring during initial cooling above the Curie temperature) are also those most resistant to AF demagnetization. Oxidation is the most common form of chemical change that takes place during the laboratory heating, and this is naturally inhibited in the initially more oxidized samples. Such samples often have a high success rate using the Thellier method. Unfortunately, although proving helpful, the procedures suggested by Senanayake and McElhinny have not proven foolproof. A useful method is to select samples as above, test whether a good Thellier method can be obtained, and then use Shaw's quicker method to provide backup data (Senanayake *et al.*, 1982).

Stephenson and Collinson (1974) developed a method to avoid laboratory heating altogether and so eliminate the problem of chemical changes associated with the heating. They use the fact that, apart from often having an identical AF demagnetization spectrum to TRM, ARM is often a linear function of the inducing field and satisfies an additivity law analogous to Thellier's for TRM (Rimbert, 1959; Patton and Fitch, 1962; Levi and Merrill, 1976). Thus the ratio of ARM lost after AF demagnetization in some peak field H_p to the TRM lost after demagnetization in H_p should be constant irrespective of the value of H_p. The constancy of this ratio provides the necessary consistency check. The problem is how to relate the ARM intensity to the TRM intensity, assuming this has remained unchanged since formation. Levi and Merrill (1976) and Bailey and Dunlop (1977) have shown that, unfortunately, the ARM to TRM ratio varies by an order of magnitude depending on grain size (and also probably on mineralogy). This ARM method may yet prove usable in spite of this problem, but to be satisfactory it will have to be augmented by rock magnetic measurements such as hysteresis loop parameters (§3.1.2) from which estimates of the magnetic grain size distribution in a sample can be obtained.

Paleointensity information is clearly valuable, but caution is required because of the variable quality of paleointensity data. The cause of the problem is that chemical changes, which are difficult to identify, can cause erroneous paleointensity estimates. Both of the most popular and widely used techniques, the modified versions of the Thellier and Shaw techniques, were developed to detect when such chemical changes are present. These techniques are based on the premise that if NRM is not an unaltered TRM or if physical–chemical changes have occurred subsequently to initial cooling, then the blocking temperature distribution (Thellier method) or the AF demagnetization spectrum (Shaw's method) will be different for NRM- and laboratory-produced TRM. Walton (1988a,b), in a formalized debate with Aitken *et al.* (1988a,b), claims that this is not the case, and in particular that laboratory produced chemical changes in the Thellier technique can cause systematic errors in estimates. However, by reheating a sample to a lower temperature, as originally suggested by Thellier and Thellier (1959a,b), such affects should be detectable, since one should obtain the same paleointensity estimate on repeating the measurement (if chemical changes have not occurred at higher temperatures). Such procedures are being used in some studies (e.g., Pick and Tauxe, 1993). The problem is that this increases the number of reheatings (and the time required to carry out the experiments) and therefore the likelihood of chemical change.

The basic problem is one of determining whether a secondary component will, on thermal or AF demagnetization, exhibit a curve similar to that of the primary component (see also §3.3.1). Laboratory experiments are urgently needed to resolve this issue. When the spectra overlap, consistency checks can fail. Clearly the probability would be less if the entire thermal, or AF, demagnetization curve were to be used. There is an increasing tendency to use only part of these curves and to make corrections for laboratory-produced chemical changes. In such cases the probability for error clearly increases. It is worth pointing out that thermal demagnetization and AF demagnetization do not necessarily affect the same magnetic regions in a rock. Although rare, it is possible for a rock to exhibit high stability with respect to thermal demagnetization and low stability with respect to AF demagnetization or vice versa (see further discussion by Levi and Merrill, 1978). Thus, rocks most suitable for analysis by Shaw's method might not be suitable for the Thellier method and vice versa. Conversely, when both methods can be used on adjacent samples and the paleointensity estimates agree, there must be considerable confidence in the value obtained.

3.4.4 Relative Paleointensity Measurements

Comparison of paleointensities determined from historical flows and artifacts with the intensities known from historical records at the same location has shown (Coe, 1967a,b; Khodair and Coe, 1975) that absolute paleointensities can be obtained satisfactorily from lava flows and various archeological objects. To avoid oxidation effects it may be necessary to carry out Thellier measurements in vacuum rather than air for those lavas in which the Curie temperature of the titanomagnetites is low (Khodair and Coe, 1975). Even so, Prévot *et al.* (1990) note that all paleointensities measured on historical lava flows by the Thellier method are within 10% of the value known from observatory data. An exception involves samples from a single historical lava from Pagan, Micronesia (U.S.– Japan Cooperation Program in Micronesia, 1975; Merrill, 1975). However, these measurements are spot readings of the ancient field intensity and are discontinuous in time (§3.2.2); it is always desirable to try to obtain more continuous records.

Levi and Banerjee (1976) (also see King *et al.*, 1983; Constable and Tauxe, 1987) suggested a solution to this problem by obtaining relative intensities from carefully selected lake or marine sediment cores. The basic idea is to use NRM normalized to some sediment property as a surrogate for field intensity. One cannot simply use NRM alone because, for example, the source of the sediment and its mineralogy may change. Therefore one uses NRM/ARM (or NRM/SIRM where SIRM is a saturation IRM; see Table 3.1) to remove changes in intensity associated with changes in the volume of the magnetic mineral. The problem is that variations in ARM (or SIRM) do not always accurately reflect changes in the volume of the magnetic material; they can also be reflecting changes in domain state (as briefly discussed in §3.4.3). Therefore the procedure is to find sediments in which single-domain grains of a monomineralogy (usually magnetite) dominate. King *et al.* (1982) have suggested a method for the determination of particle size variations in magnetite based on a plot of anhysteretic magnetization (ARM) versus susceptibility (χ). Changes in the ARM/χ ratio imply changes in grain size, with higher ratios indicating smaller grain sizes, while changes in ARM and χ with constant ratio correspond to changes in concentration of magnetic material only. An alternative method proposed by Petersen *et al.* (1986) is based on a comparison of the median destructive field (MDF) values associated with ARM and SIRM. Many measurements are thus made to determine if the mineralogy is invariant down the core. If the mineralogy is invariant then some surrogate, usually NRM/ARM, is used to provide relative paleointensity estimates. In addition, there are consistency checks in that the top section of the core is expected to record accurately the known historical magnetic field. Good relative paleointensities appear to have been obtained in several cases (see review by Tauxe (1993) and §4.2.1) in spite of the stringent requirements.

3.4.5 Dipole Moments

For a geocentric dipole of moment p the magnetic field induction F observed at the surface of the Earth, radius R, at magnetic colatitude θ is given from (3.3.2) by

$$F = \frac{\mu_0 p}{4\pi R^3}(1+3\cos^2\theta)^{\frac{1}{2}} \quad , \tag{3.4.5}$$

where F is in tesla. Substituting from (3.3.3), the above can be expressed in terms of the magnetic inclination I as

$$F = \frac{2\mu_0 p}{4\pi R^3}(1+3\cos^2 I)^{-\frac{1}{2}} \quad . \tag{3.4.6}$$

Thus, under the assumption of a geocentric axial dipole field, paleointensity measurements are essentially a function of latitude or inclination. So that results from different sampling localities can be compared, Thellier arbitrarily referred all values to a paleoisocline of 65° using (3.4.6). This is acceptable for archeological results from a limited area, such as Europe and the Mediterranean region (Thellier and Thellier, 1959a,b), but is not suitable for global comparisons or extension to the geological time scale. For comparison of results from sampling localities at different latitudes, it is convenient to calculate an equivalent dipole moment p from (3.4.5) or (3.4.6). Such a dipole moment is called a *virtual dipole moment* (VDM) (Smith, 1967a) by analogy with the calculation of a VGP from paleodirectional data.

Paleointensities are determined in many archeomagnetic studies from broken pieces of earthenware (potsherds) and other baked material, whose exact position when originally fired is not known. The magnetic inclination at the site at the time of firing thus may not be known. In such cases (3.4.5) or (3.4.6) have often been used with the magnetic colatitude θ or the present value of the magnetic inclination, I. The dipole moment calculated in this way is referred to as a *reduced dipole moment* (RDM) (Smith, 1967b). However, even on the archaeological time scale the magnetic field at any place has varied widely, so it is more appropriate to use the geocentric axial dipole assumption and insert the geographic colatitude in (3.4.5) to calculate a *virtual axial dipole moment* (VADM) (Barbetti, 1977).

The advantage of the VDM calculation is that no scatter is introduced by wobble of the main dipole because the magnetic colatitude determined is independent of the orientation of the dipole relative to the Earth's axis of rotation. Calculation of the VDM is equally appropriate to the archeological or geological time scales as long as the ancient magnetic inclination has been determined. McFadden and McElhinny (1982) suggested that the observed VDMs for any geological epoch are derived from a set of *true dipole moments* (TDMs) plus the effect of nondipole components and errors in the paleointensity determinations. Assuming the nondipole field strength is proportional to the TDM (see §4.1 and §6.3 for further discussion), the distribution of TDMs can be calculated from the observed VDMs. The peak in the calculated TDM distribution is not the same as that found by simply calculating the average VDM, but has more physical significance. The peak value of the TDM is the preferred value about which the dipole moment fluctuates. This value is thus the intensity analogue of the paleomagnetic pole of directional data about which VGPs fluctuate, and is referred to as the *paleomagnetic dipole moment* (PDM). The PDM is thus the time-averaged value of the Earth's dipole moment (McFadden and McElhinny, 1982). Table 3.5 summarizes the various dipole moments referred to in paleointensity studies. Further discussion follows in §6.3.

3.5 Age Determinations

The direction and intensity of the ancient magnetic field at a location are not particularly useful unless one can date the rock from which the paleomagnetic field estimate was obtained. There are numerous absolute (radiometric) and relative dating techniques in practice today. The basis of potassium–argon dating, one of the common techniques used in conjunction with paleomagnetism,

TABLE 3.5
Summary of Various Dipole Moments Used in Paleointensity Studies

Reduced dipole moment (RDM)	Dipole axis assumed to be that determined from the *present* magnetic inclination at the site. Useful only for comparing archaeological data for the past few hundred years.
Virtual axial dipole moment (VADM)	Dipole axis assumed to be the axis of rotation (geographic axis). Used where no knowledge of magnetic inclination is available.
Virtual Dipole Moment (VDM)	Dipole axis corresponds to that determined by the measured magnetic inclination (*I*) at the site. analogous to the VGP for directional data. Dipole wobble does not introduce scatter in VDMs.
True dipole moment (TDM)	Observed VDMs for any epoch are derived from a set of TDMs plus the effect of nondipole components and errors in paleointensity determinations. Only the *distribution* of TDMs can be calculated under given statistical assumptions.
Paleomagnetic dipole moment (PDM)	The time-averaged value of the Earth's dipole moment, analogous to the paleomagnetic pole of directional data. Determined as the peak value in the distribution of TDMs.

is described briefly here along with the so-called astronomical relative age technique to illustrate some of the assumptions and uncertainties involved in the dating process.

3.5.1 Potassium–Argon and Argon Isotope Dating

All radiometric dating techniques are based on the linear decay equation

$$\frac{dN}{dt} = -\lambda N \tag{3.5.1}$$

or

$$N = N_0 e^{-\lambda t} \quad , \tag{3.5.2}$$

where N is the amount of the original element at time t and λ is the decay constant. λ is found to be independent of temperature and pressure, except for extreme conditions that do not occur in the Earth. N_0 is the initial amount of the parent isotope and is not generally known. The half-life, $t_{1/2}$, is that value of t when $N/N_0 = \frac{1}{2}$, so from (3.5.2), $t_{1/2} = (\ln 2)/\lambda$.

In the case of potassium there are three isotopes; ^{39}K, ^{40}K, and ^{41}K. Of these, ^{40}K is radiogenic and known to decay to two daughter products, ^{40}Ca (about 89% decays to this by β^- emission) and ^{40}Ar (decay by electron capture and γ emission). The measured value of the combined half-life is 1.250×10^9 yr.

The K-Ar dating technique of igneous rocks is based on two crucial assumptions: (i) all the argon is lost from the magma prior to crystallization and cooling; and

(ii) all ^{40}Ar generated subsequent to cooling has remained trapped in the rock. The basic equation for time is easily found to be:

$$t = \frac{1}{\lambda} \ln\left(1 + \frac{\lambda}{\lambda_{Ar}} \frac{^{40}Ar}{^{40}K}\right) \quad , \qquad (3.5.3)$$

where λ is the decay constant of ^{40}K and λ_{Ar}/λ is the proportion of ^{40}K decay that yields ^{40}Ar. Therefore, provided that the assumptions are valid, one can estimate the age of the sample by measuring the amount of ^{40}K and ^{40}Ar. Unfortunately both assumptions fail occasionally. For example, historical submarine basalts have yielded ages exceeding 10 million years because the overlying pressure of the ocean was sufficient that not all the argon escaped from the magma. Another example is that some older rocks have yielded ages far too young because chemical alteration has resulted in significant diffusion of argon out of the samples. Such problems can usually be avoided by dating fresh samples and dating different minerals in the same sample. If identical ages were obtained from, say, the minerals hornblende and biotite in a granite, one would have more confidence in the age than a whole-rock age determination (i.e., one for which minerals were not first separated and then individually dated.)

A variant of the K–Ar age is the ^{40}Ar/^{39}Ar method. The method is based on the observation that ^{39}K/^{40}K is essentially constant in nature. Instead of measuring ^{40}K, one bombards ^{39}K with fast neutrons which produces ^{39}Ar. ^{40}Ar and ^{39}Ar are then measured simultaneously in a mass spectrometer. Equation (3.5.3) is replaced with:

$$t = \frac{1}{\lambda} \ln\left(1 + J_a \frac{^{40}Ar}{^{39}Ar}\right) \quad , \qquad (3.5.4)$$

where J_a is a constant determined by measurement of a sample of accurately known K–Ar age irradiated by neutrons at the same time as the unknown.

There are two advantages to ^{40}Ar/^{39}Ar dating over the conventional K–Ar technique. The first is that ^{40}Ar and ^{39}Ar can be measured simultaneously from the same sample. The second is that a step-heating method can be used, which provides a consistency check similar to that used in the Thellier paleointensity method. The method involves heating the sample up to, say, 100°C, during which ^{40}Ar and ^{39}Ar in the less retentive sites will be released. An age is obtained from the released argon isotopes. This process is repeated at successively higher temperatures until the fusion point is obtained. Each time the process is repeated an age is obtained. The low-temperature age estimates will commonly be too young, indicating diffusion of argon occurred after crystallization in the least retentive sites. A plateau of constant ages over several temperature steps is taken to mean one has accurately dated the sample. These techniques are comprehensively discussed in the books of McDougall and Harrison (1988) and Dalrymple (1991).

3.5.2 Relative Age Determinations

One example of the partnership relationship between dating and paleomagnetism is the demonstration that the Earth's magnetic field has reversed polarity numerous times in the past and the development of a chronology of those reversals (Chapter 5). During the early stages of this development it was thought that the reversal chronology could not be extended back beyond about 5 Ma. The argument was that K-Ar dating was rarely more accurate than about 3%, so for rocks with ages around 5 million years the error was comparable to the mean life time of a given polarity state. However, this did not recognize the potential for a critical contribution from relative age determinations. A long sedimentary section of rock can exhibit many reversals with depth and so the relative position of older (deeper) reversals can be determined reliably, even if their absolute ages can not be determined directly. Similar to the study of tree rings, the order and pattern of these reversals can be determined, and excellent estimates of their absolute ages can be obtained by interpolation between a small number of accurate, directly determined absolute ages. Moreover fossils or other indicators can be used to align different rock sections (*stratigraphic correlation*) and thereby extend the reversal chronology back in time. An interesting consequence is that it is often possible (e.g., back at 150 Ma) to estimate the reversal rate with far greater accuracy than the estimates of absolute age from radiometric age determinations. In some instances the magnetic record is sufficiently well determined in time that it is used for relative age dating, which is referred to as *magnetostratigraphic dating* (see §5.1.5).

One potentially important relative age dating technique that has been used increasingly during the past decade is the so-called astronomical, or Milankovitch, age dating technique. Milankovitch (1941) proposed that variations in solar radiation associated with changes in the orientation of the Earth's rotation axis and in the shape of the Earth's orbit significantly affect climate. Variations in obliquity (tilt of the Earth's axis away from the plane of its orbit, presently about 23.5°) with a periodicity near 41,000 yr, precession of the Earth's axis of rotation with a periodicity near 23,000 yr, and eccentricity of the Earth's orbit with a periodicity near 100,000 yr are claimed to affect the climate significantly. A large amount of data, particularly oxygen isotope data, support this claim for the past million years (e.g., Imbrie and Imbrie, 1980; Jacobs, 1984). Interpretations of Milankovitch cycles are far more equivocal prior to a million years ago. The climatic effect appears most pronounced for the 100,000-yr period, which is surprising since solar radiation effects appear larger for the other two periodicities. The implication is that there must be nonlinear effects if the Milankovitch model is correct. Conversely, data have recently emerged that seriously challenge the Milankovitch model (e.g., Winograd *et al.*, 1992). However, regardless of the cause of the variations, there do appear to be significant changes in the oxygen isotope data (e.g., Imbrie and Imbrie, 1980) that can be used for relative dating for sediments less than a million years in age (i.e., the Milankovitch climate model need not be valid to use this relative age dating technique).

It might be supposed that relative age dating techniques are always inferior to absolute age dating techniques, but this is certainly not necessarily so. Johnson (1982) used the above "astronomical" relative age dating technique to date the Brunhes–Matuyama boundary, where the magnetic field switched to the present polarity (see Chapter 5), at 790 ka. However, this age was not generally accepted because K–Ar dating had "clearly indicated" that this polarity change occurred at 730 ka. Recently it has been found that the K–Ar ages were incorrect, and redating using ^{40}Ar/^{39}Ar now puts this boundary near 780 ka (Baksi *et al.*, 1992; Spell and McDougall, 1992). This example reflects the continual refinement in age estimates and the fact that valuable input comes from both absolute and relative age determinations. Naturally, interpretations of the magnetic field behavior in the past should allow not only for inaccuracies in the paleomagnetic data but also for inaccuracies in dating.

Chapter Four

The Recent Geomagnetic Field: Paleomagnetic Observations

4.1 Archeomagnetic Results

The Earth's magnetic field has exhibited a wide range of space–time variability during historical times. Westward, standing, eastward and meridional motions of the nondipole field have all been claimed to be dominant in the historical record (Chapter 2). In principle, archeomagnetic and paleomagnetic data, which cover much longer time spans than the historical data, should be sufficient to resolve this ambiguity. However, it will be shown in this chapter that none of the models can be eliminated on the basis of existing data. One reason is that it now seems likely more than one mechanism is contributing significantly to secular variation (see §4.2.8).

4.1.1 Evidence for Westward Drift

Pottery and, more usefully, bricks from pottery kilns and ancient fireplaces, whose last dates of firing can be estimated from carbon-14 contents of ashes, have a thermoremanent magnetization dating from their last cooling. The hearths of ancient fireplaces especially can provide measurements of declination and inclination at the time of their last firing, providing they have not since been disturbed. Bricks will often provide inclination data (assuming they were fired horizontally) and all such material may be used for determining the ancient field strength, typically through the use of the Thellier technique (§3.4.2). The most extensive archeomagnetic studies to obtain directional data have been carried out by workers in the former Soviet Union (see summary by Burlatskaya, 1983). However, a high proportion of these results give inclination data only. Detailed work has been carried out in Bulgaria and Yugoslavia by Kovacheva and Veljovich (1977) and Kovacheva (1980) covering the past 8000 yr. Extensive studies have also been made in Japan mainly for the past 2000 yr (see summary by Hirooka, 1983), in North America for the past 3000 yr (Du Bois, 1974, 1989; Sternberg, 1983; Sternberg and McGuire, 1990), and in England (see, for example, Aitken, 1970). Data are also available from France (Thellier, 1981), Australia (Barbetti, 1977, 1983) and China (Wei *et al.*, 1983). Dated lava flows have been used for studying secular variation at Mount Etna (Chevallier, 1925; Tanguy, 1970), at Mount Vesuvius (Hoye, 1981), in the Western United States (Champion, 1980), on Hawaii (Doell and Cox, 1963, Holcomb *et al.*, 1986) and in Japan (Yukutake, 1961).

Yukutake (1967) plotted the occurrences of the maxima and minima in declination and inclination for different regions of the northern hemisphere as a function of time and longitude of the observation. An updated version of his inclination analysis has been given by McFadden *et al.* (1985) and is shown in Fig. 4.1. The timing of the maxima and minima in each region depends critically

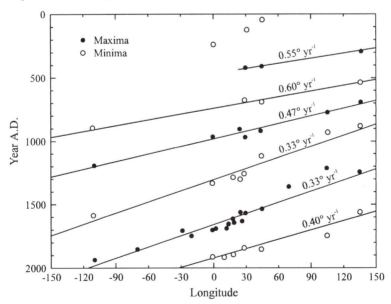

Fig. 4.1. Maxima (solid circles) and minima (open circles) in inclination as a function of longitude and age. Data come from midlatitude regions in the northern hemisphere. Straight-line fits provide westward drift rates estimates for the maxima and minima. After McFadden *et al.* (1985).

on the amount of data from which a complete inclination curve is drawn. The least precise variable in Fig. 4.1 is, therefore, the date of each maximum and minimum, and these become less well defined the greater the age. However, the westward drift of the first three maxima and minima in the inclination curves is clearly demonstrated for the past 2000 yr. Using the trends for the first two maxima and minima from Fig. 4.1 an average value of 0.38±0.07° longitude yr^{-1} is obtained (95% error), a drift rate similar to that deduced originally by Yukutake (1967). This is also similar to the value of ~0.3° yr^{-1} found by Bullard *et al.* (1950) for the *westward drift of the secular variation* from analyses of historic data for the nineteenth and twentieth centuries, but nearly twice the value of ~0.2° yr^{-1} found by Bullard *et al.* (1950) for the *westward drift of the nondipole field*. This difference led Yukutake and Tachinaka (1969) to propose that the nondipole field consisted of standing and drifting parts. The drifting field rotates at the faster rate and when the two are combined the lower value is then observed. It should be noted that, because of its nature, the analysis of Fig. 4.1 refers only to features of the drifting part of the field. The slightly lower values obtained from historic data are consistent with the view that the drift rate has been decelerating (Harwood and Malin, 1976; Hoye, 1981).

Historical records also suggest the westward drift velocity is a function of latitude. Yukutake (1967) plotted the drift rates for features of declination and inclination determined from Tokyo, the United States, Paris, London, Canada, and Oslo. For the equator he used the variation in longitude of the intersection of the west agonic line with the equator determined by Bauer (1895). The latitude variation is shown in Fig. 4.2. There is a clear latitude dependence, with values of around 0.4° yr^{-1} associated with the midlatitude regions from which the data of Fig. 4.1 were derived. Yukutake (1967) showed that a first-order fit to the data is obtained, as indicated by the dashed curve in Fig. 4.2, if a constant linear velocity of 0.058 cm s^{-1} is assumed at the core–mantle boundary for the drift.

Evans (1987) has analyzed archeomagnetic intensity results from Japan (Sakai and Hirooka, 1986) and Bulgaria (Kovacheva, 1980; Kovacheva and Kanarchev, 1986) for the past 2000 yr. He noted that three peaks in the intensity data for Japan could be aligned with three peaks in the data for Bulgaria merely by shifting the time scales by 350 yr. Thus if the intensity peaks reflect westward drifting sources then the 110° of longitude difference between the sites represents an average drift rate of 0.31° yr^{-1}, similar to the estimates from Fig. 4.1. However, this interpretation is nonunique. Note that eastward drift at 0.7° per year is an alternative interpretation of the data. Another possibility is that there are random changes in the intensity of the standing nondipole field for which the peak intensity happens to occur 350 yr later in Japan than it does in Bulgaria. Similarly, arguments can be constructed to show that the directional evidence presented earlier is also nonunique. In spite of this, the *simplest*

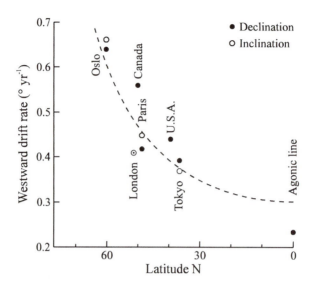

Fig. 4.2. Average westward drift rate estimates as a function of latitude for the northern hemisphere. Maxima and minima in inclination and declination data (cf. Fig. 4.1) have been used to construct this figure. The dashed line gives the latitude variation of angular velocity when the field rotates at a constant linear velocity of 0.058 cm s^{-1} at the surface of the core. After Yukutake (1967).

interpretation is that the dominant direction of drift of the nondipole field is westward.

4.1.2 Motion of the Dipole Axis

Spherical harmonic analyses of historical data back to A.D. 1600 have indicated a very slow movement of the dipole axis over the past 400 yr (see §2.4.5). To attempt to extend this motion back in time using archeomagnetic data, it is necessary to use some averaging technique that smoothes out the variations in the nondipole field so as to be able to see the main trends in the dipole field. Cox and Doell (1960) observed that the average of VGPs calculated from observatory data around the world plots very close to the present geomagnetic pole. Unfortunately, archeomagnetic data are not evenly spaced around the world, but are concentrated in the European region. Barbetti (1977) suggested that the effects of nondipole field variations could best be averaged out if VGPs were averaged over 100-yr intervals for a limited number of regions of the Earth's surface. The gross movement of the dipole axis could be obtained by further averaging these 100-yr interval means over a 500-yr interval and then finding the regional mean for the whole globe.

The suggestion made by Barbetti (1977) has been used in detail by Champion (1980), Merrill and McElhinny (1983), and more recently by Ohno and Hamano

(1992), who have calculated the position of the North Geomagnetic Pole for successive times at 100-yr intervals for the past 10,000 yr. The results of the analysis by Ohno and Hamano (1992) are illustrated in Fig. 4.3 for each 2000-yr interval as well as the entire 10,000 yr. Several factors lead one to believe that these 100-yr mean VGPs are close approximations to the successive positions of the north geomagnetic pole. The mean VGP calculated from the 1980 field values at the center of 12 regions covering the surface of the Earth lies close to the 1980 geomagnetic pole (Merrill and McElhinny, 1983). Also, the successive values for 1600 to 1900 A.D. lie close to, and have the same trend as, the positions of the geomagnetic pole calculated from historical observations (Barraclough, 1974; Fraser-Smith, 1987; see §2.4.5).

However, significant nondipole variations may occur on a time scale of 10^3 yr (e.g., westward drift of the nondipole field), so the averaging process above may not remove all nondipole effects. Figure 4.3 shows that the mean VGP for each 2000-yr interval does not always average to the geographic pole, whereas the mean over 10,000 yr appears to do so. Thus it appears that at least 10,000 yr are required for the motion of the dipole axis to average to the axis of rotation. Some caution is needed, however, because it is not at all clear that the motion of the dipole axis over the past 10,000 yr, as depicted in Fig. 4.3, can be regarded as a recurring feature, or that the average over the preceding 10,000 yr would also coincide with the geographic pole. There is further general discussion in §6.1 and §6.2 of the problems involved in averaging paleomagnetic results.

4.1.3 Variations in the Dipole Moment

Even though there have been many more archeomagnetic intensity determinations than directional determinations, the distribution of sites is still poor. They are grouped mostly in the northern hemisphere, and concentrated largely in Europe and the Middle East (Burlatskaya, 1983; Kovacheva and Kanarchev, 1986; Aitken *et al.*, 1983, 1989; Thomas, 1983; Walton, 1979), the western United States (Du Bois, 1989; Sternberg, 1989), Japan (Sakai and Hirooka, 1986) and China (Wei *et al.*, 1983). Determinations have typically been made using the Thellier method on ancient hearths and pottery, but there has been some controversy over the reliability of the Thellier technique in this context (Aitken *et al.*, 1988a,b; Walton, 1988a,b). In spite of this, the general view prevails that the Thellier method is by far the most reliable of the various methods that have been proposed. Indeed, many workers (e.g., Prévot *et al.*, 1990) strongly take the view that this is the only rigorous method for paleointensity determination.

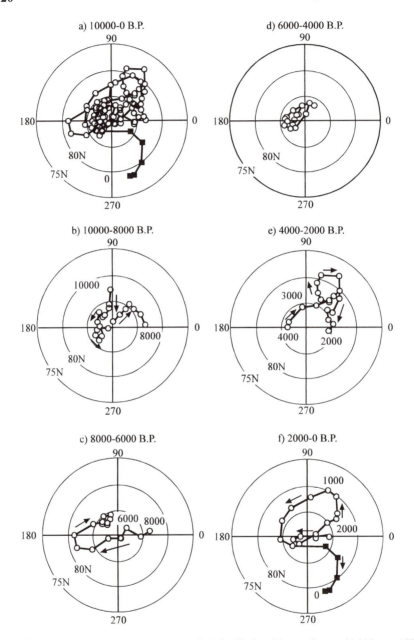

Fig. 4.3. The locations of the North Geomagnetic Pole (dipole axis) over the past 10,000 yr at 100-yr intervals as estimated by Ohno and Hamano (1992). Locations are given for each 2000-yr interval as well as for the entire 10,000 yr. After Ohno and Hamano (1992).

Figure 4.4 shows the variations in intensity over the past 4000 yr in Greece determined from a large number of samples of ancient pottery and other archeologically dated forms of baked clay (Aitken *et al.*, 1989). The samples used are well dated, usually to better than ±25 yr. Aitken *et al.* (1989) have drawn a "reference band" through the majority of points in Fig. 4.4 to show the overall trend of the data. With the exception of the points at A.D. 1300, data points lying outside the band are regarded as deviants, possibly a result of archeological mis-attribution. An alternative explanation of course is that rapid changes in intensity can occur locally over times of a hundred yr or less as has been observed in southeast Australia (see Fig. 4.5).

The variations in intensity determined in southeast Australia from aboriginal fireplaces and other baked clays and sediments (Barbetti, 1983) over the past 7000 yr are shown in Fig. 4.5. Intensities vary by a factor of two over this period, and substantial changes (up to 20 μT) have occurred over only a few hundred years. These variations are argued to be too large to be due to experimental error. Instead they appear to be reasonable estimates of field strengths, and if so, accuracy of the dating indicates that the changes sometimes occur on the order of a hundred years or less. That such rapid changes in intensity at any locality might be a real feature of the past field is supported by

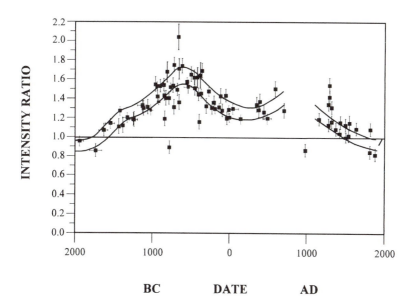

Fig. 4.4. Variation in intensity of the geomagnetic field in Greece from Thellier intensity measurements on pottery and baked clays dated archeologically, usually to better than ±25 yr. The absence of horizontal bars in some cases indicates very precise dating. A "reference band" has been drawn by the authors to indicate the general trend of the data. After Aitken *et al.* (1989).

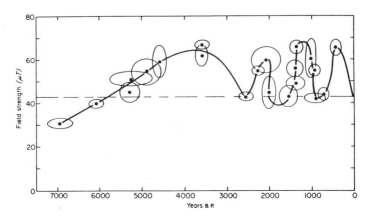

Fig. 4.5. Estimates of the intensity of the Earth's magnetic field as a function of time in southeast Australia using the Thellier technique. Data are from aboriginal fireplaces and baked clays and sediments dated using the carbon-14 method. After Barbetti (1983).

results from other parts of the world. It is noteworthy for example that the intensity of the present-day field varies by more than a factor of two at some latitudes (e.g., 30°S where the Australian data are located).

An unusual method for determining paleointensity has been proposed by Games (1977) using adobe (sun-dried) mud bricks in Egypt. He suggested that these bricks retain a record of the ancient field due to the construction method (throwing into a container) and subsequent drying in the Earth's magnetic field. By repeating the construction and drying process in the laboratory and applying Shaw's AF demagnetization method (§3.4.5) he claimed to have made estimates of the intensity variations in Egypt over the past 3000 yr (Games, 1980). Some support for the method was provided by independent Thellier intensity determinations on pieces of pottery found in the bricks. However, there has been no subsequent use reported for this method and its viability remains in question.

Barton *et al.* (1979) and Champion (1980) have analyzed global archeomagnetic data to determine possible time variations in the Earth's dipole moment. Because most of the data were concentrated in the European region it was difficult to judge the extent to which the data, when averaged, represent changes in the dipole moment. There appears to be an increase in intensity to a maximum about 2500 yr ago, preceded by a minimum about 6500 yr ago. Previously Bucha (1967, 1969) and Cox (1968) had suggested that this was indicative of variations in the dipole moment with a simple periodicity of between 8000 and 9000 yr with maximum and minimum respectively about 1.5 and 0.5 times the present dipole moment. Barton *et al.* (1979) argue strongly that inadequacies in the archeomagnetic data are such that this periodicity should be regarded as highly tentative. Furthermore, both Kono (1972) and McFadden and

McElhinny (1982) show that paleointensities in general are incompatible with a sinusoidal variation in intensity (see §6.3.2).

Barton *et al.* (1979) and Champion (1980) analyzed about 600 intensity results from around the world to determine possible time variations in the Earth's dipole moment. Because most of the data were concentrated in the European region it was difficult to judge the extent to which the data, when averaged, represent changes in the dipole moment. A much wider data set of 1160 worldwide results was used by McElhinny and Senanayake (1982) covering the past 10,000 yr. In an attempt to average out nondipole field variations it has been the practice to average results over 500- or 1000-yr intervals (Cox, 1968; Barton *et al.*, 1979; Champion, 1980). Table 4.1 summarizes the analysis of McElhinny and Senanayake (1982) using 500-yr averages back to 4000 yr B.P. and 1000-yr averages prior to that. The results are shown in Fig. 4.6 together with the 95% error limits. The mean dipole moment for the past ten 1000-yr intervals is 8.75×10^{22} Am^2 with an estimated standard deviation of 18.0%, which may be attributed to dipole intensity fluctuations. For the past 8000 yr the data from the European region are in general agreement with those from the rest of the world. There is a maximum about 2500 yr ago and a minimum about 6500 yr ago. The data prior to 10,000 yr B.P. clearly do not support the concept of a sinusoidal variation in archeomagnetic intensity (see §4.1.5 and §4.2.2 and Figs. 4.8 and 4.9).

TABLE 4.1
Global average dipole moments for the past 10,000 yr[a].

Years B.P.	N	Dipole moment (10^{22} Am^2)	σ	95% error
0-500	268	8.72	1.44	0.17
500-1000	187	10.30	1.89	0.27
1000-1500	205	10.90	2.01	0.27
1500-2000	131	10.94	2.18	0.37
2000-2500	89	11.10	2.63	0.54
2500-3000	60	11.28	2.49	0.63
3000-3500	43	9.64	2.85	0.85
3500-4000	17	9.21	1.90	0.90
4000-5000	34	8.87	2.20	0.74
5000-6000	44	7.20	1.94	0.57
6000-7000	36	6.73	2.00	0.65
7000-8000	18	7.08	1.44	0.66
8000-9000	15	8.61	2.32	1.17
9000-10000	14	8.26	2.39	1.25
Mean of 1000-yr intervals	10	8.75	1.58	0.97

Note. N, no of determinations averaged; σ, standard deviation.
[a] From McElhinny and Senanayake (1982).

Because there are contributions from rock magnetic and experimental errors and from the nondipole field variation, the nonuniqueness problems in the determination of the variations of the dipole moment appear formidable. The key to sorting out some of these problems lies in the observation from historical and paleomagnetic data that the dipole field changes more slowly than the nondipole field. In particular McElhinny and Senanayake (1982) utilized the observation that there is a fairly broad maximum in dipole moment between 1000 and 3000 B.P. (Fig. 4.6). By assuming a constant dipole moment in this interval, they obtained a first-order estimate of the scatter from other contributions. The estimated standard deviation of 472 values in this interval is 19.7%. If a similar analysis is done over the past 10,000 yr (assuming constant a dipole field in 1000-yr intervals), the corresponding estimated standard deviation is 21.2%. Note that these last two numbers refer to the within-1000-yr-interval standard deviation, whereas the value given earlier (18.0%) refers to the between-interval standard deviation. Using the present field as a first-order indicator of the magnitude of the nondipole contributions, McElhinny and Senanayake (1982) obtained a value of 17.5% for the standard deviation of the nondipole field contributions. This leaves an estimate of 10–12% for the scatter attributed to rock magnetic and experimental errors.

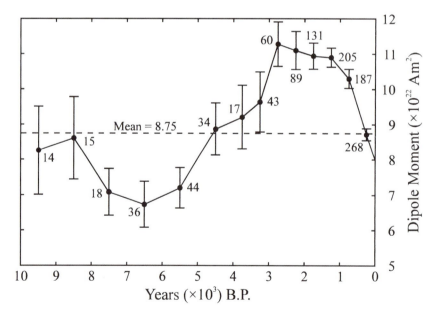

Fig. 4.6. Global dipole moment versus time estimates obtained from 500-yr and 1000-yr period averages as listed in Table 4.1. The error bars shown are for the 95% confidence level. After McElhinny and Senanayake (1982).

4.1.4 Deductions from Carbon-14 Variations

The isotope ^{14}C is one of several species of radionuclide produced by the interaction of cosmic rays with the Earth's atmosphere. ^{14}C is produced mainly in the stratosphere by the (n,p) reaction with ^{14}N and it forms $^{14}CO_2$ that is rapidly and thoroughly mixed with ordinary atmospheric carbon dioxide ($^{12}CO_2$ and $^{13}CO_2$). CO_2 is continually exchanged between the atmosphere and the oceans, which contain about 95% of the exchangeable carbon and where the mean residence time is on the order of 1000 yr. This exchange between reservoirs regulates the CO_2 content of the atmosphere. Carbon-14 dating is based on the fact that atmospheric CO_2, with its trace level of radiogenic ^{14}C, is absorbed and incorporated by living materials. After incorporation, the level of ^{14}C gradually diminishes through radioactive decay with a half-life of 5730 yr. Measurement of the residual ^{14}C allows a naive estimate of age to be made using the assumption that the atmospheric concentration in the past was constant, or a more sophisticated estimate by calibrating the past atmospheric ^{14}C concentration.

Changes in atmospheric ^{14}C concentration, apart from recent artificial perturbations (e.g., fossil fuel and atomic bomb effects) can arise mainly from reservoir changes or from changes in the production rate of ^{14}C. Damon (1970) predicts that the effects of reservoir changes are small and that changes in the production rate are mainly due to changes in the high-energy particle flux from the sun and to changes in the geomagnetic dipole moment. Observed variations in the atmospheric ^{14}C concentration on time scales of less than 500 yr are generally attributed to solar activity. The longer-term variations over several thousand yr or more are generally attributed to variation in the geomagnetic dipole moment.

A charged particle moving in the plane normal to the lines of force of a uniform magnetic field B will describe a circle of radius R. It can be shown that the product BR is equal to the ratio of the relativistic momentum m to charge ze, where z is the charge number and e is the charge on the electron. The *magnetic rigidity* of the particle (Wolfendale, 1963) is defined by the quantity

$$\frac{mc}{ze} = cBR \ \left(\text{volt}\right) \ , \qquad (4.1.1)$$

where c is the velocity of light.

The effect of the geomagnetic field on the cosmic ray particle is such that certain classes of trajectories of particles approaching the Earth remain trapped along field lines (a slight leakage does occur but will be neglected in this discussion). These bounded trajectories cannot, therefore, be followed by incoming cosmic ray particles (see, for example, Rossi, 1964). At a given geomagnetic latitude particles with rigidity less than a certain value cannot reach the Earth. Particles with a slightly greater rigidity may arrive at the Earth's surface from the west (for positively charged particles), but not from the east. The critical value of rigidity with which positively charged particles may arrive at the Earth's surface vertically or from the west, but not from the east, is known as the *vertical cutoff rigidity*, P_λ. Elsasser *et al.* (1956) give an expression for P_λ as a function of geomagnetic dipole moment p and latitude λ,

$$P_\lambda = \frac{\cos^4 \lambda}{4} \frac{\mu_0 pc}{4\pi r^2} \quad \text{(V)} \quad , \tag{4.1.2}$$

where r is the radius of the Earth. Taking the present value of the dipole moment p_0 as 8.0×10^{22} Am^2 and substituting values of μ_0, r and c, (4.1.2) reduces to

$$P_\lambda = 14.8 \frac{p}{p_0} \cos^4 \lambda \quad \text{(GV)} \quad . \tag{4.1.3}$$

This model of the modulation of the cosmic ray spectrum by the geomagnetic field has been used by several authors to calculate the ^{14}C production rate as a function of geomagnetic dipole moment (Elsasser *et al.*, 1956; Wada and Inoue, 1966; Bucha and Neustupny, 1967; Lingenfelter and Ramaty, 1970). Lingenfelter and Ramaty (1970) deduced a relation for the ^{14}C production rate, Q, as

$$Q = kp^{-\frac{1}{2}} \quad , \tag{4.1.4}$$

where k is a rather complex function derived from the model. Equation (4.1.4) is valid only for values of p close to its present value and the simple relationship breaks down for much smaller or much larger values. Applying this to the roughly sinusoidal variations in dipole moment suggested by Cox (1968) for the past 10,000 yr, they concluded that the gross features of ^{14}C variations could indeed be understood in terms of the variations in geomagnetic moment.

The comparison of ages measured using the radiocarbon method with those deduced from dendrochronology has enabled departures from the assumption of constant atmospheric ^{14}C concentration to be determined. Clark (1975) has listed estimates of the true age T_t, corresponding to radiocarbon ages T_c, back to 6500 yr B.P. The relative atmospheric ^{14}C concentration, C_A, can then be calculated from the expression

$$C_A = \exp\left\{\left(\frac{T_l}{5730} - \frac{T_c}{5568}\right)\ln 2\right\} \ , \qquad (4.1.5)$$

where the half-life of 5568 yr is the one conventionally used to determine radiocarbon age. Assuming these ^{14}C variations are solely of geomagnetic origin, and that the production and decay of ^{14}C are instantaneously balanced so that the atmospheric concentration varies linearly with production, the geomagnetic dipole moment can be estimated (Barton *et al.*, 1979) from (4.1.4). The dipole moment calculated in this way is referred to as the *radiocarbon dipole moment* (RCDM) and its deduced variation is shown in Fig. 4.7. The short-term variations in ^{14}C atmospheric concentration seen in Fig. 4.7 are probably due to changes in the solar high-energy particle flux, and only the broad trends should be attributed to changes in dipole moment.

Figures 4.6 and 4.7 show different time intervals, and it must be recognized that the estimates in Fig. 4.7 have been seriously affected (in the main, only for the past three or four hundred years) by anthropogenic factors (such as the large amount of fossil fuel burned). However, with these provisos, comparison of the RCDM (Fig. 4.7) with the dipole moment obtained from archeomagnetic data (Fig. 4.6) shows the same general features. The broad trends depict the variations in dipole moment, and a theoretical delay of about 1000 yr (Houtermans, 1966; Houtermans *et al.*, 1973) can be expected between changes in ^{14}C concentration and changes in the dipole moment. The ^{14}C concentration minimum between

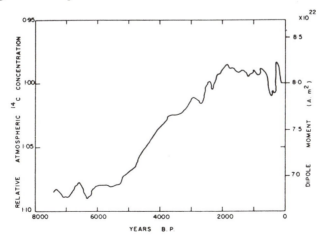

Fig. 4.7. Radiocarbon dipole moment (right vertical axis) versus age. The corresponding relative atmospheric ^{14}C concentration is shown on the left vertical axis. Data for the past three or four hundred years have been seriously affected by anthropogenic factors (e.g., burning of fossil fuel). After Barton *et al.* (1979).

1000 and 2000 yr B.P. implies a broad maximum in dipole moment between about 2000 and 3000 yr B.P., as is observed in Fig. 4.6. The high ^{14}C concentrations between 5500 and 7500 yr B.P. imply a broad minimum in dipole moment between 6500 and 8500 yr B.P., as is also observed in Fig. 4.6. This provides confidence that the longer period changes in Figs 4.6 and 4.7 are due to genuine variations in the dipole moment.

4.1.5 Dipole Moments before 10,000 yr B.P.

Relatively few estimates of the geomagnetic dipole moment in the time range 100,000 to 10,000 yr B.P. have been made from paleointensity measurements on volcanic rocks or archeological material. Thouveny *et al.* (1993) and Tanaka *et al.* (1994) have summarized the most recent estimates in this time range. Combining these data with those from McElhinny and Senanayake (1982) and excluding the high values associated with the Lake Mungo excursion and the low values from the Laschamp and Skalamaelifell excursions (see §4.3.2), the results are summarized in Table 4.2 and illustrated in Fig. 4.8. The data show that the dipole moment was much lower in the range 50,000 to 15,000 yr B.P. gradually increasing by 13,000 yr B.P. to higher values that dominate the past 10,000 yr. This confirms the observation made by McElhinny and Senanayake (1982) based on many fewer results.

If the trend to lower values in Fig. 4.8 is representative of the time between 50,000 and 15,000 yr B.P., then similar trends should be seen in a comparison between ^{14}C dates and other dating methods. Barbetti (1980) summarized ^{14}C and comparative thermoluminescence or uranium series (^{230}Th/^{230}U) ages between 40,000 and 10,000 yr B.P. Atmospheric ^{14}C concentrations deduced from these comparisons using (4.1.5) are almost all much higher than their

TABLE 4.2
Global Average Dipole Moments for the Period 50,000 to 10,000 yr B.P.

Years B.P.	N	Dipole moment (10^{22} Am^2)	σ	95% error
10,000-11,000	15	7.32	2.00	1.04
11,000-12,000	8	6.87	1.41	1.04
13,000-15,000	7	7.56	1.94	1.54
15,000-20,000	8	5.12	1.27	0.94
20,000-25,000	6	4.79	1.25	1.09
25,000-30,000	11	4.67	1.21	0.75
30,000-40,000	7	3.94	1.94	1.54
40,000-50,000	2	2.75	(0.55)	(1.08)
15,000-50,000 yrs combined	34	4.53	1.52	0.52

Note. Sources are as explained in the text. *N*, number of determinations averaged; σ, standard deviation.

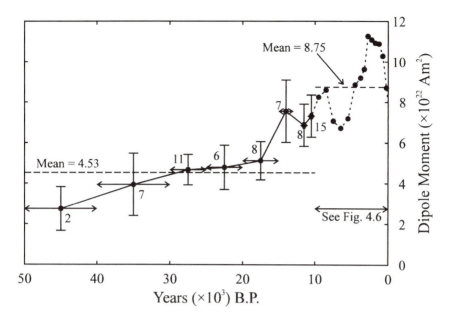

Fig. 4.8. Global dipole moment versus time estimates in the range 50,000 to 10,000 yr B.P. obtained from the period averages as listed in Table 4.2. Error bars are the 95% confidence limits. Data for the past 10,000 yr are as shown in more detail in Fig. 4.6.

present value, implying a prolonged period of reduced dipole moment as suggested in Fig. 4.8. A more recent analysis by Mazaud *et al.* (1991) supports this conclusion. These results indicate that the quasi-cyclic geomagnetic moment variation suggested by several workers for the past 10,000 yr (Cox, 1968; Bucha, 1969; Barton *et al.*, 1979) did not extend to previous epochs.

Between 50,000 yr and 15,000 yr B.P., the 34 values of dipole moment in Table 4.2 have an average value of 4.53×10^{22} Am^2, whereas the average over the past 10,000 yr has been 8.75×10^{22} Am^2 (Table 4.1). Allowing for a 10% standard deviation due to errors in paleointensity, the standard deviation of the dipole moments 50,000 to 15,000 yr B.P. is 32% due to a combination of dipole and nondipole field variations. For the past 10,000 yr this value is very similar at 25.9%. This observation led McElhinny and Senanayake (1982) to propose that the nondipole field intensity remains in the same proportion to the dipole field intensity. Global analysis of the dipole moment over the past 10 million yr (Kono and Tanaka, 1995b) gives a mean value of 7.84×10^{22} Am^2 with standard deviation due to dipole fluctuations of 43.7% (see §6.3.2). This suggests that the extended period of low dipole moment 50,000 to 15,000 yr B.P. is part of the expected variation over long time scales. The variation in relative paleointensities determined from ocean sediments tends to confirm this view (see

§4.2.2). Extended time spans of low dipole moment may contribute to other phenomena such as geomagnetic excursions (see §4.3.3).

4.2 Analysis of Recent Lake Sediments

4.2.1 The Recording Mechanism in Lake Sediments

Sediments deposited under suitably quiet conditions have been found to possess a stable remanent magnetization. In the case of slowly deposited deep-sea sediments this has provided a record of reversal history of the geomagnetic field (Chapter 5). There is also some promise that analyses of deep-sea sediments will eventually contribute to our understanding of paleosecular variation. On the basis of analyses of sediments from the Bermuda Rise (western North Atlantic Ocean), Lund and Keigwin (1994) have suggested that paleosecular variation records are reproducible on a regional scale. However, it appears that the magnetic recording process smoothes out the high-frequency component of the magnetic field variations because the remanence is locked in over a 10- to 20-cm thick zone. The deposition rate for these sediments varied from 10 to 30 cm per 1000 yr. Sediments deposited in lakes have provided a high-resolution record of the geomagnetic field over long periods of time. The typical high sedimentation rate in lakes (on the order of 1 m per 1000 yr) and their long life-span (on the order of 10,000 yr), make them ideal not only for extending the archeomagnetic record back in time but also because a continuous record is obtained.

The magnetization of lake sediments (and other types of sediments) is due to postdepositional DRM. Small magnetic particles (already magnetized from their previous history), slowly deposited in these lakes, are free to move in the water-filled interstices of a newly deposited sediment and their magnetic axes align (statistically) with the ambient magnetic field, probably over a depth of about 10 cm. During compaction and/or growth of the gels in the sediment, and sometimes during bioturbation, these magnetic particles become locked in position, giving the sediment an overall magnetization parallel to the geomagnetic field at that time. This recording process is particularly efficient in fine-grained homogeneous sediments, such as muds, that are characteristic of so many recent lakes. Particularly advantageous are the sediments deposited in lakes in volcanic craters (maars), because often these craters are unconnected with the local water table and, without inlet or outlet for water (other than rain or evaporation), the sediments have little chance of being disturbed. When these lakes finally dry up and cease to exist, the dry lake sediments still preserve their detailed record of the changes in the ancient geomagnetic field. Verosub (1977) has reviewed aspects of depositional and postdepositional processes in the magnetization of sediments. Barton *et al.* (1980) have carried out some detailed laboratory studies of the acquisition of magnetic remanence in slurries of fine-grained organic muds. Henshaw and Merrill (1980) have reviewed the effects chemical changes can have on the remanence in pelagic sediments and Karlin and Levi (1985) have reviewed the chemical effects for the continental shelf environment.

4.2.2 Relative Paleointensities

Over the past decade attempts have been made to determine *relative* changes in paleointensity from measurements of the magnetization of lake or ocean sediments (see §3.4.4). The processes accompanying the magnetization of sediments are neither clearly understood nor very well constrained because they depend to a large extent on environmental conditions, grain size, mineralogy, and postdepositional factors. However, the magnetization has been shown to be linearly related to the field intensity during deposition (e.g., Barton *et al.*, 1980) and it has been shown that where the sequences have invariant, homogeneous magnetic properties it is possible to determine relative paleointensity (Levi and Banerjee, 1976; King *et al.*, 1983). Results have been published showing good agreement with the archeological data set covering the past 15,000 yr (Constable, 1985; Constable and Tauxe, 1987; Tauxe and Valet, 1989).

Tric *et al.* (1992), Meynadier *et al.* (1992) and Schneider and Mello (1994) have made detailed studies of ocean sediment cores from the Mediterranean Sea, the Somali Basin (Indian Ocean) and the Sulu Sea (Philippines), respectively to determine variations in paleointensity over the past 140,000 yr. Figure 4.9 shows the variations in the dipole moment obtained by Schneider and Mello (1994) from four cores taken from two sites in the Sulu Sea. These sediments provide strong, stable magnetizations at a high sedimentation rate (~10 cm/kyr), a rare combination in the deep-sea environment. The absolute value of the dipole moment is normalized to the variation obtained by McElhinny and Senanayake (1982) shown in Fig. 4.6 for the past 10,000 yr. In the interval 45,000 to 30,000 yr B.P. the Earth's dipole moment was especially low and reached a minimum value of about 2×10^{22} Am^{-1}. The analyses of intensity variations in deep-sea sediments by Kent and Opdyke (1977), Tauxe and Wu (1990), Tric *et al.* (1992), and Meynadier *et al.* (1992) all strongly support this conclusion.

Tric *et al.* (1992) have suggested the presence of sinusoidal variations in the Earth's dipole moment with periodicities of 20 kyr, similar to those observed in the precession of the Earth's orbit. Although links between variations in the

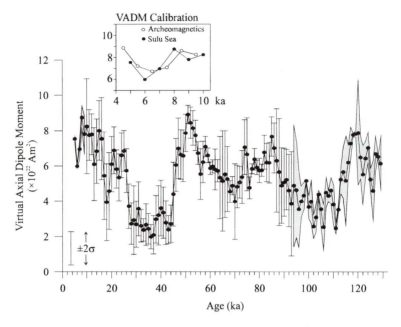

Fig. 4.9 Variation in the Earth's dipole moment over the past 130,000 yr obtained from deep-sea sediment cores in the Sulu Sea. Diagram provided by D. Schneider.

geomagnetic field and the Earth's orbital parameters have been suggested by several workers (Kent and Opdyke, 1977; Wollin *et al.*, 1978; Negi and Tiwari, 1984), the record is far too short to establish the validity of a periodic cycle. Meynadier *et al.* (1992) combined their data with those of Tric *et al.* (1992) to produce a "global" synthetic curve for the past 140 ka. Power spectrum analysis indicated the presence of two dominant peaks at 100 kyr and 22–25 kyr, with smaller peaks at 19 and 43 kyr. These periods are not significantly different from the Milankovitch periods and one might be tempted to argue that this represents evidence that external forcing (e.g., precessional torques) is having an affect on core dynamo processes. A simpler and more likely interpretation is that changes in climate have occurred with these periods and are reflected in undetected changes in the rock magnetic properties of these cores. Longer records have been analyzed by Kent and Opdyke (1977) and Constable and Tauxe (1987). Analysis by Tauxe and Wu (1990) of these data sets suggested that a dominant frequency of around 33 kyr prevailed for the past 350 kyr. However, these data sets have very much less resolution than those of Tric *et al.* (1992), Meynadier *et al.* (1992), or Schneider and Mello (1994) as shown in Fig. 4.9. Thus until data are available from many regions on a more global scale, any suggestion of periodic variations in the Earth's dipole moment remains highly speculative.

4.2.3 Analysis of Declination and Inclination

Mackereth (1971) first discovered long-period declination oscillations in cores taken from the postglacial organic sediments deposited at the bottom of Lake Windermere in England. Sampling was carried out using a pneumatically controlled Mackereth corer (Mackereth, 1958). These corers take 6-m-long cores and have been extensively used for studies of lake sediments throughout Europe and in Australia and Argentina. An extended version for 12-m cores has been developed by Barton and Burden (1979) and used in Australia. In North America wet lakes have been cored using conventional gravity corers or Livingstone piston corers, but extensive work has also been carried out on dry lakes.

Creer *et al.* (1972) subsequently confirmed the early work of Mackereth (1971) and found that the inclination variations in Lake Windermere were apparently of irregular period and amplitude. The results from Lake Windermere were later confirmed by further work in northwest England, Scotland, Finland,

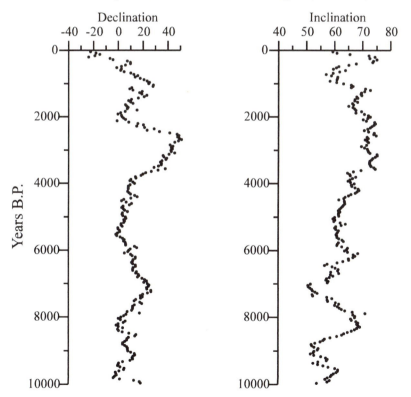

Fig. 4.10. Master curves for declination and inclination for Great Britain. After Turner and Thompson (1981, 1982).

Poland, Switzerland, and Greece by Creer and his colleagues (see, for example, summaries by Creer (1977) and Creer and Tucholka (1982c)). Creer (1981) clearly demonstrated the validity of lake sediments as reliable recorders of past geomagnetic secular variations by comparing the results obtained from Greece with the archeomagnetic data of Kovacheva (1980) from Bulgaria and Yugoslavia. More recently, records of the secular variation back to 120,000 yr have been obtained from extensive studies of cores from Lac du Bouchet in the Massif Central, France (Creer *et al.*, 1986; Smith and Creer, 1986; Thouveny *et al.*, 1990). The most detailed record for the past 7000 yr has been obtained from Loch Lomond, Scotland (Turner and Thompson, 1979), and this shows that there are identifiable features of variations in inclination as well as in declination. By combining all radiocarbon and palynological age determinations from Britain with the observed magnetic data, Thompson and Turner (1979) showed that the inclination data were repeatable between lakes. Turner and Thompson (1981, 1982) produced a master curve for Britain for the past 10,000 yr, which is illustrated in Fig. 4.10. It was suggested that this master curve would be applicable over the whole of Western Europe for the past 10,000 yr.

In North America cores from Lake Michigan and Lake Erie were examined by Creer *et al.* (1976a,b). Both records are superficially similar to those obtained at Lake Windermere with distinct oscillations in declination being observed and less pronounced changes in inclination. Further work in the Great Lakes and adjacent regions was summarized by Lund and Banerjee (1979, 1985) and Creer and Tucholka (1982a,c). More recent work on North American lake sediments has been summarized by Hanna and Verosub (1989) and Lund (1996). The reproducibility of magnetic features only a few hundred years in length in parallel cores from many of these lakes provides clear evidence of the fidelity of the recording process. Lund (1996) argues that distinctive features of maxima and minima of declination and inclination can be traced for more than 4000 km without serious change in pattern. He interprets these results as reflecting some long-term memory in the core flow, perhaps related to core–mantle coupling. Master curves describing secular variation in North America have been produced by Creer and Tucholka (1982a) for east central North America and Hanna and Verosub (1989) for northeast North America.

Very detailed studies have been made of volcanic crater lakes in southeast Australia (Barton and McElhinny, 1981) and northeast Australia (Constable and McElhinny, 1985), providing records of inclination and declination over the past 10,000 yr. Unlike the records in Europe and North America, the most pronounced swings are observed in inclination rather than declination. The master curves for southeast Australia derived from three volcanic crater lakes are shown in Fig. 4.11. They represent 100-yr means obtained by stacking profiles from the three lakes using the most pronounced features and the many radiocarbon dates determined for these lakes. Creer *et al.* (1983) have reported

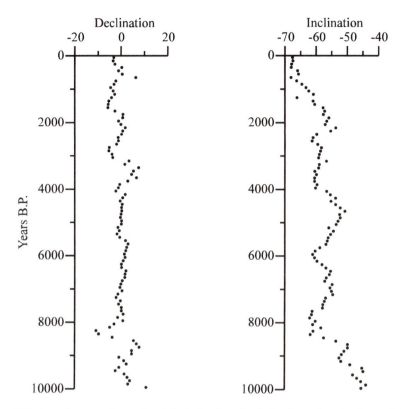

Fig. 4.11. A master curve for southeast Australia gives the declination and inclination variation recorded in lake sediments for the past 10,000 yr. Individual points represent 100-yr means obtained from stacked profiles from three lakes. After Barton and McElhinny (1981).

details of some studies in Argentina, Peng and King (1992) have made studies of Lake Waiau on Hawaii, and Turner and Lillis (1994) report results from New Zealand.

Dry lake sediments have been most extensively studied in North America, particularly nonvarved dry lake sediments from the western United States. Detailed secular variation records are available from Mono Lake, California, from several studies by Denham and Cox (1971), Denham (1974), and Liddicoat and Coe (1979). These relate especially to the Mono Lake excursion (see also §4.3.3). In Mexico, dry lake deposits have also been studied at Tlapacoya by Liddicoat *et al.* (1979). In New England varved sediments have been studied by Verosub (1979), and in British Columbia a section at Bessette Creek has been reported in detail by Turner *et al.* (1982).

4.2.4 Westward Drift and Runcorn's Rule

Runcorn (1959) has shown that a moving magnetic source (such as a dipole) in the Earth's core can cause the magnetic field vector at a fixed observatory to rotate clockwise (when viewed along the vector direction) for a westward motion of the source and anticlockwise for an eastward motion. If the total field at the surface is a function of time t and position $\mathbf{r}$ of the observatory, then

$$\mathbf{H}(\mathbf{r}, \mathbf{r}'(t)) = \mathbf{H}_0(\mathbf{r}) + \mathbf{H}'(\mathbf{r}'(t)) \quad , \qquad (4.2.1)$$

where $\mathbf{H}_0(\mathbf{r})$ is the steady component primarily due to the main axial dipole, and $\mathbf{H}'(\mathbf{r}'(t))$ is the perturbation field due to a small, moving, nonaxial dipole located at $\mathbf{r}'$. As the small dipole moves beneath the point of observation the motion of the total field vector $\mathbf{H}$ will generate an elliptical cone, and the tip of $\mathbf{H}$ will describe an ellipse. The situation is illustrated in Fig. 4.12. The sense of motion along this ellipse is opposite for a westward moving small dipole to an eastward moving one. It is important in visualizing this to recognize that one is only referring to the motion on the plane perpendicular to $\mathbf{H}$. This proposal has become known as *Runcorn's rule*.

As a result of the perturbation of the surface field by a moving dipole source, a plot of inclination versus declination (so-called Bauer plots) describes an elliptical curve corresponding to the motion above. The axes of the plot are of course oriented corresponding to geographical directions. The presence of several such curves (all clockwise) in geomagnetic records was first noticed by

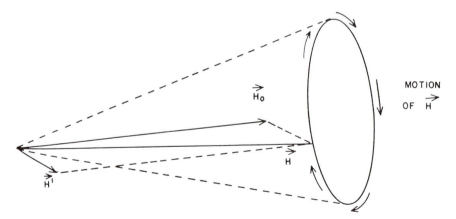

Fig. 4.12. Runcorn's rule can be described in terms of a change in the total instantaneous magnetic field $\mathbf{H}$ about its steady value $\mathbf{H}_0$. $\mathbf{H}$ is the vector sum of $\mathbf{H}_0$ and a small perturbation field $\mathbf{H}'$. The tip of $\mathbf{H}$ describes an ellipse about $\mathbf{H}_0$. A clockwise sense of motion around this ellipse, when viewed along the vector direction, indicates westward drift.

Bauer (e.g., Bauer, 1895). Skiles (1970) and Dodson (1979) have examined Runcorn's rule in some detail and shown that under certain limited conditions anticlockwise rotation of the vector could still be produced by westward drifting sources.

If the perturbation of the field is caused by a single westward drifting harmonic $P_l^m(\cos\theta)$, the perturbing vector traces out an ellipse whose plane is inclined at an angle ψ to the horizontal with

$$\tan\psi = \frac{-(l+1)P_l^m(\cos\theta)}{dP_l^m(\cos\theta)/d\theta} \ . \tag{4.2.2}$$

The magnetic vector corresponding to this harmonic then traces this ellipse with period $2\pi/m\omega$, where ω is the angular velocity relative to the mantle of the westward moving source in the core. Runcorn (1959) has shown that his clockwise rule will apply provided

$$I_0 - 180° < \psi < I_0 \ , \tag{4.2.3}$$

where I_0 is the inclination of the axial dipole field vector. This condition is satisfied by all but a few low-degree harmonics. If condition (4.2.3) is not satisfied then the magnetic vector will rotate anticlockwise for a westward drifting dipole source. However, any realistic source field might consist of several harmonics of different degree. Dodson (1979) has examined several such situations and shown that in certain circumstances Runcorn's rule will be violated. Even so, Runcorn's rule can still be regarded as applicable in a statistical sense.

Runcorn's clockwise rule has been used as a method of inferring the direction of drift of the geomagnetic field from data at a single observatory. It is particularly applicable to the paleomagnetic data from lake sediments. Over the past 10,000 yr there appears to be a predominant clockwise motion observed in the data so far available, although specific sections of some records show times when anticlockwise motion is seen. The Bauer diagrams corresponding to the younger sections of the master curves from Britain (Fig. 4.10; Turner and Thompson (1981, 1982)) are shown in Fig. 4.13. Although the dominant motion is clockwise there is an extensive anticlockwise portion between 1100 and 600 yr B.P. For older records Skiles (1970) observed clockwise motion as dominating results from Japan in the time range around 50,000 yr ago. The record from Mono Lake, California at around 25,000 yr is considered to be the best existing evidence for anticlockwise motion and eastward drift (Denham, 1974). However, Dodson (1979) points out that this motion could also be

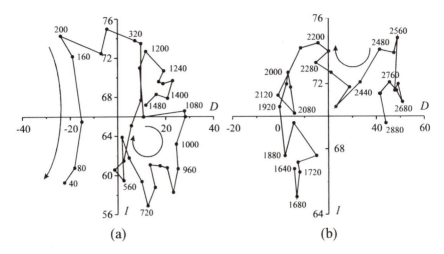

Fig. 4.13. Bauer plots of declination *D* and inclination *I* from the master curve for Great Britain. Individual points represent 100-yr intervals for (a) 1480 yr to present and (b) 2880 to 1640 B.P. Clockwise motions suggest westward drift according to Runcorn's rule. After from Turner and Thompson (1981, 1982).

produced by two westward drifting dipoles, rather than a single eastward drifting one.

Dodson (1979) has considered the two extreme cases of zero probability of eastward drift and equal probability of westward and eastward drift. In the case of no eastward drift the following should be observed:

(i)   anticlockwise motion would seldom be observed at high latitudes;
(ii) clockwise loops would be more numerous unless there is a bias toward high-latitude anomalies and/or toward positive anomalies;
(iii) if weak anomalies are more common than strong anomalies, clockwise loops would be larger on the average than anticlockwise loops;
(iv) clockwise VGP loops can be symmetric about either the observer's meridian or antimeridian, but anticlockwise loops can only be symmetric about the meridian; and
(v)   anticlockwise motion would always be accompanied by a decrease in the total field intensity, but clockwise motion could be accompanied by either a decrease or an increase in intensity.

If eastward and westward drift occur with equal probabilities, none of the above asymmetries would be expected.

The first possibility cannot yet be tested because there are still no continuous records of sufficient resolution from high latitudes. The data from Mono Lake seem consistent with the fifth possibility, but the interpretation of intensity measurements in sediments is a hazardous procedure. The Mono Lake data also appear consistent with the fourth possibility, but neither the British nor the older Japanese data show any conclusive dependence of the orientation of the VGP loops and their direction of precession. There are still too few continuous records of sufficient quality to test the second and third possibilities.

Creer and Tucholka (1982b,c) have demonstrated that clockwise motions could be produced by special configurations of fixed sources of oscillating moments. The record from Lac du Bouchet (Thouveny *et al.*, 1990) shows alternating clockwise and anticlockwise motions of the magnetic field vector between 71,000 and 29,000 yr B.P., followed by a clear dominance of clockwise motion between 29,000 and 12,000 yr B.P. For the most recent 8000 yr, clockwise looping (westward drift by Runcorn's rule) predominates in records from North America, Europe, Australia, and Argentina (Lund, 1994). On balance the evidence probably favors a dominant westward drift, but the existing data do suggest that for earlier times there were intervals during which eastward drift predominated.

4.2.5 Interpretations in Terms of Dipole Sources

Yukutake and Tachinaka (1968) noted that the westward drift of the historical field is most pronounced in the large-scale features of the nondipole field, corresponding to harmonics of degree $l = 2$ to 4. These large-scale features have been modeled by Lowes (1955) and Alldredge and Hurwitz (1964) by radial dipoles situated in the outer part of the core. The dipoles in the model of Alldredge and Hurwitz (1964) are situated at radii $r_0 = 0.28R$ from the Earth's center, where R is the Earth's radius. Although the dipoles may not be physically realistic, they provide a convenient way to describe the individual major nondipole features with nongeocentric sources for which only four parameters (dipole moment p, r_0/R, colatitude θ_0 and longitude ϕ_0) are required for each major nondipole feature. Such large anomalies dominate the nondipole field.

Each of the large nondipole anomalies may be approximated, nonuniquely, by an equivalent dipole in the outer core with parameters as described above. The coefficients of the potential of any such dipole (Hurwitz, 1960) given in SI units of tesla are

$$\left.\begin{aligned} g_l^m &= \frac{\mu_0 p}{4\pi R^3}\left(\frac{r_0}{R}\right)^{l-1} l\, P_l^m(\cos\theta_0)\cos m\phi_0 \\ h_l^m &= \frac{\mu_0 p}{4\pi R^3}\left(\frac{r_0}{R}\right)^{l-1} l\, P_l^m(\cos\theta_0)\sin m\phi_0 \end{aligned}\right\} . \qquad (4.2.4)$$

The potential of the dipole is thus

$$V = \frac{p}{4\pi R^2} \sum_{l=1}^{\infty} \sum_{m=0}^{l} \left(\frac{r_0}{R}\right)^{l-1} l\, P_l^m(\cos\theta_0)\left\{\cos m(\phi - \phi_0) P_l^m(\cos\theta)\right\} \ . \qquad (4.2.5)$$

A typical large feature of the present nondipole field as modeled by a dipole at the equator directed radially outward has $\mu_0 p/4\pi R^3 \approx 10$ μT ($p = 2.5 \times 10^{22}$ Am^2) (cf. Fig. 2.11 and Eq. (3.4.5)).

Creer (1977) has used the radial dipole model of Alldredge and Hurwitz (1964) to account for the geomagnetic secular variations observed in lake sediment cores from Europe and from Lakes Michigan and Erie in North America. For the present geomagnetic secular variation, Alldredge and Hurwitz (1964) have a nine-dipole model (central dipole plus eight radial dipoles) in which the positive radial dipoles drift predominantly eastward and the negative ones westward. The mean drift velocity of all the dipoles is westward. Creer (1977) argues that it is difficult to account for the observed variations in the lake sediment data by geomagnetic sources drifting either eastward or westward around the geographic axis. The observed pattern should be the same at sites occupying the same latitude but with a time lag due to the longitude difference. The observed differences do not appear to be as simple as this but include different oscillation periods for changes in declination and in inclination (see also §4.4.3).

Creer (1977) found that he could model the main features observed in the lake sediment declination data from Europe and the Great Lakes of North America by having the radial dipoles of Alldredge and Hurwitz (1964) remain essentially fixed in position over the time interval under study, but allowed to oscillate with the same amplitude and each with its own particular frequency. However, the inclination data showed good correlation across the North Atlantic with a phase shift of 500 yr, which Creer (1981) interpreted as evidence of westward drift. Why should standing sources appear to dominate secular variation declination records while at the same time drifting sources appear to dominate the inclination records? Creer and Tucholka (1982b,c) suggested that this was attributable to the geometry of the observation point in relation to the geomagnetic source. The observed secular variation could, in principle, be modeled as being derived from various combinations of standing but oscillating dipoles and drifting dipoles of constant intensity within the core, analogues of the standing and drifting parts of the nondipole field of Yukutake and Tachinaka (1968, 1969).

A more complete analysis of the above models has been given by Creer (1983) using computer models designed to synthesize the observed secular variation of the geomagnetic field. In this synthesis Creer derived the variations in declination, inclination, and intensity that would be expected at 45°N from the following models.

(a) *Precession of the main dipole.* This was simulated by allowing a geocentric equatorial dipole of moment p to rotate with uniform angular velocity about its axial component of moment P.

(b) *Drifting radial dipoles of fixed moment.* Following Yukutake and Tachinaka (1968, 1969), the drifting part of the nondipole field has an approximate quadrupolar symmetry and may be represented by four radial dipoles located at midlatitudes and separated by 180° longitude. Two of these radial dipoles would be situated in the northern hemisphere, one pointing up and the other pointing down, with the other two located in the southern hemisphere. Following Alldredge and Hurwitz (1964), the radial dipoles of moment $p=0.2P$ are located deep in the outer core at $r = 1750$ km from the center of the Earth.

(c) *Drifting current-loop sources of fixed moment.* Following the usual argument that, because of the screening effect of the conducting core, sources of the nondipole field must be located within the top few hundred kilometres of the outer core, the radial dipoles are replaced by current loops close to the outer core surface at $r = 3400$ km as modeled by Peddie (1979). Such current loops may be modeled by arrays of radial dipoles.

(d) *Oscillating dipoles fixed in position.* In this case the standing part of the nondipole field may be modeled simply by regular oscillations, with period T, of a fixed radial dipole of moment p.

Figure 4.14 illustrates the variations in declination, inclination, and intensity that would be observed from each of the above models. Creer (1983) concludes that the actual shape of the various curves may help to distinguish the type of source. For example, model (b) in Fig. 4.14 produces flat-topped declination perturbations (with slight dips in the middle) whereas the inclination maxima and minima are pointed, with the minima being of larger amplitude than the maxima. Model (c) on the other hand produces the opposite effects on declination and inclination maxima and minima. The corresponding VGP paths for each model also produce their own distinctive patterns.

Creer (1983) also noted that for each of the models used in Fig. 4.14 the mean direction *calculated using unit vectors* has an inclination that is (incorrectly) always less than the axial dipole field value by a few degrees. The mean direction calculated *using the full vector* information has an inclination that is (correctly) always equal to the axial dipole field value. In contrast, for the drifting sources, the declination of the mean is always zero using either method of calculation. In the case of oscillating sources the unit vector calculation will not in general give an average declination of zero, but the total vector method

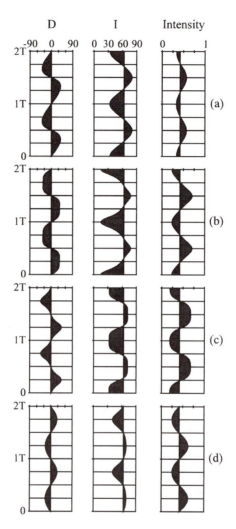

Fig. 4.14. Declination, inclination, and intensity variations produced by various source models for the nondipole field. (a) A precessing geocentric equatorial dipole with p/P=0.5. (b) A pair of drifting radial dipoles, one pointing up and one pointing down, separated by 180° longitude at 45°N, located near the base of the outer core at r=1750 km with p/P=0.20. (c) A pair of drifting current loops situated at 45°N at r=3400 km. The loops are modeled by an array of radial dipoles in two blocks each spanning 180° of longitude, one block up and the other down, around 45°N. The curves are calculated for p/P = 0.00086 per degree of longitude. (d) A single oscillating fixed radial dipole, located at r=1750 km 30° to the west of the observer, with both source and observer situated at 45°N. Sources (a), (b) and (c) drift to the west taking time T to lose one revolution relative to the mantle. Source (d) oscillates with period T. After Creer (1983).

always does. Thus, unit vector-based analyses of the various data sets synthesized by the models would lead to the conclusion that the steady background field was not purely that of an axial dipole of moment M, but that persistent nondipole components existed. The significance of this observation is discussed more fully in Chapter 6 (see §6.2.3).

Bloxham and Gubbins (1985, 1986) identified two pairs of magnetic foci that have remained stationary over the past few hundred years and so cannot contribute to drift of the nondipole field. Yukutake (1989) has modified his views on the nature of the nondipole field to include an eastward drifting component as well as westward and standing components. Although the details of Creer's (1983) models are physically unrealistic, they provide additional insight into the nonuniqueness of the problem and suggest ways to distinguish between competing models.

4.2.6 Interpretations Using Spherical Harmonic Analysis

If enough high-quality paleomagnetic and archeomagnetic data were available for successive times, one could, in principle, carry out a spherical harmonic analysis for each of these times and resolve the changes in the nondipole field. This has been attempted by Ohno and Hamano (1993) using data for the past 10,000 yr. The quality and quantity of the data vary considerably depending on the time interval used. For example, after applying various rock magnetic and dating selection criteria, Ohno and Hamano (1993) were left with data from Iceland, England, France, Bulgaria, Ukraine, Japan, Australia, southern United States, Central America, Hawaii, and China for the time interval 400 to 1600 A.D. From these data they obtain estimates of the dipole and quadrupole fields. The amount of usable data decreases further back in time and there are only six regions from which data are available for the past 10,000 yr. Note that it requires a minimum of eight distinctly different locations to obtain an estimate of all the coefficients in a spherical harmonic analysis out to only degree 2 (see §2.2.4 and Eq. (2.2.19)). Unfortunately at the present time the data are too sparse to discriminate accurately between most models for the nondipole field.

4.2.7 Interpretations in Terms of Dynamo Waves

In his pioneering work on the $\alpha\omega$-dynamo, Parker (1955a) introduced the concept of propagating wave-like instabilities that he called *dynamo waves* (see §8.2.1, §8.4, and §9.5). Parker (1979) elaborated and extended the theory of dynamo waves for a variety of cases. The propagation of dynamo waves is blocked and deflected (scattered) by boundaries, producing a wide variety of form and behavior. These waves generally migrate in a direction perpendicular to the fluid velocity shear direction, corresponding to north–south migration for rotationally symmetric shear. Parker (1979) shows that east–west drift can also occur if there is significant nonrotationally symmetric shear present or if the appropriate boundary conditions exist. In general, a complex wave number is required to satisfy the boundary conditions and so dynamo waves are transient features with exponential growth and decay phases. Olson and Hagee (1987) showed how this theory might be applied to explain certain features in the secular variation record from lake sediments.

Dynamo waves have only been extensively investigated for kinematic models: models in which the velocity is prescribed rather than derived. Hence, depending on the assumptions made, a wide variety of behavior can result, which can be exploited to provide interpretations of observational data. For example, in the $\alpha\omega$-dynamos (§8.3.2) the α-effect is usually assumed to change sign across the equator because of the Coriolis force. If the rotational shear is symmetric about the equator and positive, then dynamo waves will propagate toward the poles for α positive (i.e., $\alpha\omega$ positive) and toward the equator for α negative (Parker, 1955a). Parker (1955a, 1979) applied his theory to try to explain the solar cycle (Chapter 10). Olson and Hagee (1987) and Hagee and Olson (1989, 1991) showed how dynamo wave theory could be exploited in the case of the geodynamo. The sign of α often depends on whether the source of convection buoyancy originates near the core–mantle boundary or the inner-core boundary (Olson and Hagee, 1987). Thus dynamo waves may provide valuable information both on the processes powering the geodynamo and on the geomagnetic secular variation.

Olson and Hagee (1987) propose the existence of nonaxisymmetric waves in an $\alpha\omega$-dynamo and investigate their properties for appropriate boundary conditions. The apparent velocity of secular variation viewed at the Earth's surface will be the sum of the propagation velocity of the wave within the core plus the drift velocity of the outermost core fluid relative to the mantle. Once nonaxisymmetric wave motion is assumed, variations in declination and inclination will occur at any given site at the Earth's surface. Runcorn's rule can then be applied in the same fashion as before (and with the same exceptions) to determine westward (clockwise rotation) or eastward drift of the magnetic field. This allows one to determine the direction of propagation of the dynamo wave hypothesized to cause the variation.

Olson and Hagee (1987) deduced from their model time series of inclination and declination as would be observed in paleomagnetic secular variation data. Figure 4.15a illustrates synthetic inclination and declination records observed at 45°N produced by dynamo waves with a period of 2400 yr propagating at various azimuthal angles. Note that the inclination record is independent of the propagation direction. At this latitude the amplitudes of the declination maxima and minima are greater than those of the inclination record, and the phase, relative to the inclination record, depends on the propagation direction. If the wave propagation is due east or due west then the declination record is symmetrical about zero and almost a sinusoid. Otherwise, the declination record is asymmetrical about zero with cycloidal-like oscillations. These characteristics differ from those of Creer (1983) using moving and oscillating dipoles and current loops (Fig. 4.14). In Creer's models, large-amplitude cycloidal

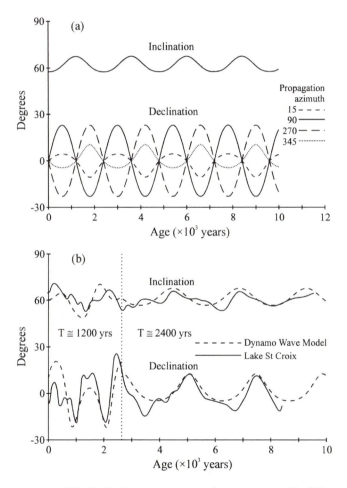

Fig. 4.15. Inclination and declination time series using the dynamo wave model of Olson and Hagee (1987). (a) Synthetic time series produced by dynamo waves propagating at azimuthal angles 15°, 90°, 270°, and 345° with a period of 2400 yr, phase angle zero and disturbance/dipole amplitude ratio of 0.3. Calculation is for a site latitude 45°N. (b) Best fitting synthetic dynamo wave model (dashed lines) to the smoothed paleomagnetic record (solid lines) of Lund and Banerjee (1985) from Lake St. Croix, Minnesota. See text for model parameters. After Olson and Hagee (1987).

oscillations in inclination and small-amplitude sinusoidal oscillations in declination are produced.

The phase relationship between the inclination and declination curves and the shape of the declination curves are diagnostic of the propagation direction of the wave. Olson and Hagee (1987) have exploited this in applying their model to the smoothed paleomagnetic record (wavelengths shorter than 120 yr and longer than 5000 yr have been removed) for the past 10,000 yr observed by Lund and Banerjee (1985) from Lake St. Croix, Minnesota (Fig. 4.15b). To match the inclination and declination records a two-wave model was used, the calculation being for a site at 45°N. A wave with a period of 2400 yr was used throughout the record and a second wave with a period of 1200 yr was used only over the past 2635 yr. The 2400-yr wave has a propagation angle of 345° (NW) with a phase angle of 130° and the 1200-yr wave has a propagation angle of 30° (NE) with a phase angle of 355°. Both waves have a disturbance/dipole amplitude ratio of 0.335. Thus the record is consistent with a dominantly poleward dynamo wave propagation, which requires that the product $\alpha\omega$ be positive. Since the westward drift implies $\omega<0$, the α-effect must be negative in the outer core. This implies that the α-effect is due primarily to the sinking of negatively buoyant fluid at the core–mantle boundary into the core.

Synthetic models of inclination–declination plots by Olson and Hagee (1987) show looping motions that are not symmetrical either as a function of the latitude of the observer or as a function of propagation direction of the dynamo waves. This concurs with the observation by Creer (1983) that the time-averaged field from his models of paleomagnetic secular variation does not in general average to the axial dipole inclination or to zero declination when unit vectors are used. From the equations of Olson and Hagee's (1987) model it is evident that a full vector time average of their field over a full cycle of the secular variation would give the axial dipole field direction.

A more extensive analysis of paleomagnetic secular variation records from nine sites located in North America, South America, Eurasia, and Australia by Hagee and Olson (1989) shows that sites on the same continent show a high degree of correlation, whereas correlation between sites on different continents is generally much lower. Models consisting of two waves, one with period ranging from 2400 to 3200 yr and the second with period ranging from 1200 to 1600 yr, were fitted to the data at eight of the nine sites. The propagation directions derived from the best fitting wave models have strong poleward components in both hemispheres. All propagation directions are within 40° of the pole. Hagee and Olson (1989) suggest this indicates that Holocene paleosecular variation may consist of a superposition of propagating dynamo waves with either eastward or westward components, but also with a strong poleward component of motion. They propose that these represent either nondipole fields advected poleward by a meridional circulation pattern in the outer core that is antisymmetric about the equator or, alternatively, propagating $\alpha\omega$-dynamo wave instabilities generated just beneath the core–mantle boundary.

4.2.8 Uncertainties in Interpretations of Secular Variation

The interpretations of secular variation given in this chapter have been shown to be essentially consistent with the conventional view of long-term westward drift of the nondipole field. However, several other interpretations are also broadly compatible with the data. For example, some of the lake sediment data from North America have been used to infer both westward drift of the nondipole field (§4.2.5) and significant poleward motion (§4.2.7). Lund (1994, 1996) points out that if you apply Runcorn's rule to lake sediment data spanning the past 30,000 yr, there apparently are significant time intervals (the data are few) of both "westward" and "eastward" drift. Moreover these uncertainties are not resolved by the direct measurements of the field during recent times. Bloxham and Gubbins (1986) suggest westward drift is very nonuniform and occurs primarily in the Atlantic hemisphere. In contrast, Yukutake (1993) presents evidence that the apparent low in the nondipole field in the Pacific and the low secular variation there is only a very recent phenomenon, and that on average there is no hemispherical preference for drift of the nondipole field. Hulot and Le Mouël (1994) argue that the secular variation is predominantly random.

Some of the differences in interpretations are almost certainly a consequence of data inadequacies. However, we suggest that the character of secular variation may also be affected by the length of the window used in examining it. For example, dynamo waves act on time scales of 10^3 yr or less. Quasi-random changes in intensity of nondipole fields described by higher degree spherical harmonics occur on a scale of a few tens to a few hundreds of years, probably reflecting turbulence in the outer core. Characteristic times of order 10^5 to 10^6 yr, associated with the dynamic processes that must balance decay of the dipole magnetic field (§9.1), are also likely to be present. Some of the longest characteristic times, of order 10^8 yr, are probably associated with changes in core–mantle boundary conditions (§11.4). That is, several mechanisms, operating on widely different time scales, could contribute to secular variation. Thus, for example, one should not expect the character of the secular variation to be the same when viewed on a 10^2-yr time scale as on a 10^6-yr time scale. This conjecture is supported by the spectral analysis of paleomagnetic data by Barton (1983) (§4.4.3 and Fig. 4.19). Finally, there is also the possibility that there is significant spatial variation in secular variation, as evidenced by the possibility that the fluid mechanical force balances acting within and without the so-called tangential cylinder may be different (§9.6). Thus there appears to be a strong need to model secular variation differently in the future to take into account its probable temporal, and possible spatial, variations.

4.3 Geomagnetic Excursions

4.3.1 Definition of Excursions

Apart from reversals of the geomagnetic field, which are discussed in detail in Chapter 5, wide departures from its usual value have been observed to occur in the geomagnetic field direction at a single locality. Such departures, when the field does not appear to change polarity but returns to its previous state, have been termed *geomagnetic excursions*. Geomagnetic excursions are generally defined to have occurred when the VGP calculated from the field direction at the locality departs more than 45° from its time-averaged position for that epoch and is not associated with a polarity transition. Sometimes it is difficult to distinguish whether the latter has occurred, because short polarity intervals (10^5 yr) referred to as *reversal events*, are known to be present in the geomagnetic record (§5.1.4). Examples of excursions have been observed in several studies where a succession of lava flows has recorded the time variations of the magnetic field in some detail (e.g., Doell and Cox, 1972). They are also recorded in sedimentary sequences but the interpretation is often equivocal because their time scale is short and only a narrow band of sediment is often involved. Also, the departures can be argued to be due to some effect of the sedimentation itself rather than the geomagnetic field (Verosub and Banerjee, 1977). This problem is discussed in more detail in §4.3.3.

An interesting, and as yet unanswered, question is whether reversals and secular variation are distinct phenomena, or whether reversals should be considered just as an extreme aspect of secular variation. For example, it is possible that reversals are associated with thermal or chemical blobs originating at one of the boundaries of the outer core (§5.4.4), whereas typical secular variation may be associated with dynamo waves as discussed in §4.2.7. In such a case one must distinguish between excursions associated with reversals of the main field and those that may occur because of some favorable situation during secular variation (e.g., if the ratio of the nondipole to dipole field became high during secular variation (see §4.3.5)). Whitney *et al.* (1971) have pointed out that it is theoretically possible for the nondipole field to become large enough that local reversals of the field could occur and thus both polarities would appear to exist simultaneously. To distinguish between such local reversals and excursions of a more modest nature, Merrill and McFadden (1994) suggest using the phrase *reversal excursion* to describe an excursion recording a VGP deviation in excess of 90°.

4.3.2 The Laschamp Excursion

The youngest recorded excursions are of particular interest because there is at least the potential of identifying them at several places on the Earth's surface and investigating their global morphology. For times much older than say 500,000 yr it is very unlikely that correlation would ever be possible with any certainty. For times less than 50,000 yr ago, however, the possibility does exist, and in this time range excursions have been documented as occurring in lava flows in several parts of the world. These records arouse particular interest because arguments about the viability of the sedimentary recording process do not arise. The first recorded and best documented excursion occurs in the Laschamp and Olby flows in lavas of the Chaine des Puys in France (Bonhommet and Babkine, 1967). The directions observed in these flows deviate about 140° from the axial dipole field direction. Initially Cox (1969) regarded this as a genuine reversal of the Earth's magnetic field (reversal event as defined above), but now it is usually referred to as the *Laschamp excursion* (a reversal excursion).

Heller (1980) and Heller and Petersen (1982) concluded that the Olby flow, and to a lesser extent the Laschamp flow, undergoes complete or partial self-reversal under laboratory conditions. This contrasted with earlier rock magnetic work by Whitney *et al.* (1971) who concluded that the remanence in the samples they analyzed resided in single-domain grains. Roperch *et al.* (1988) have pointed out that the experiments of Heller and Petersen show that above 200°C the remanence is always close to the characteristic "reverse" direction. For samples having more than one magnetic phase the experiments show that the phase with the lowest Curie point exhibits partial self-reversal by magnetostatic interaction. The "reverse" remanence has the higher blocking temperature. These observations do not imply the further proposition that the high-temperature "reverse" remanence was acquired by self-reversal. Furthermore Roperch *et al.* (1988) provide evidence that the baked clay underlying the Olby flow has a "reverse" direction close to that of the lava, thus providing a strong case for a geomagnetic origin of the Laschamp excursion.

Interest in the age of the Laschamp excursion has resulted in many attempts to date the Laschamp and Olby flows using a variety of techniques (K-Ar, ^{39}Ar/^{40}Ar, ^{230}Th/^{238}U disequilibrium, thermoluminescence, and ^{14}C). All these analyses indicate that the ages of the Laschamp and Olby flows lie between 35 and 50 ka (see summary by Roperch *et al.*, 1988). A compilation of 30 K–Ar ages of the Laschamp and Olby flows by three laboratories yielded a new age of

46.6±2.4 ka that is probably the best estimate of the age of the Laschamp excursion (Levi *et al.*, 1990). The actual duration of the excursion is estimated from sedimentary records (§4.3.3).

Paleointensity measurements have also been reported on the Laschamp and Olby flows and from "reverse" scoria inside the Laschamp crater (Roperch *et al.*, 1988) using the Thellier method (Thellier and Thellier, 1959a,b). Results from the two flows and the scoria are very similar and can be represented by a single mean paleointensity of 7.7 µT. This value is less than one-sixth of the present geomagnetic field, so it seems unlikely that the directions of the Laschamp and Olby flows were acquired during a stable reverse polarity interval. This conclusion is supported by a paleointensity determination of 12.9 µT on the Louchadiere flow that records an intermediate direction lying between the present field and the "reverse" Laschamp and Olby flows. The similarity in age suggests this flow also records a part of the Laschamp excursion (Chauvin *et al.*, 1989).

In southwest Iceland an excursion has also been recorded in late glacial basalts (Kristjansson and Gudmundsson, 1980) with shallow negative inclinations and westerly declinations, which they named the *Skalamaelifell excursion*. Further work by Levi *et al.* (1990) has identified the same excursion in a more extensive area of the Reykjanes peninsula. K–Ar dating of the excursion lavas give a mean age for 19 determinations of 42.9±7.8 ka and a Thellier paleointensity of 4.2±0.2 µT (Levi *et al.*, 1990). Independent determinations by Marshall *et al.* (1988) yielded a paleointensity of 4.3±0.6 µT during the same excursion, whereas the paleointensity from normal lavas outside the excursion was 30±9 µT. All these observations strongly suggest that the Laschamp and Skalamaelifell excursions record essentially the same event. The only other possible record of this event in lavas in other parts of the world is found in the Auckland volcanic field in New Zealand where lavas with northerly, downward dipping directions are observed (Shibuya *et al.*, 1992). Thermoluminesence and ^{14}C ages for these lavas suggest an age of between 25 and 50 ka.

The *Lake Mungo excursion* has been observed in prehistoric aboriginal fireplaces in southeast Australia (Barbetti and McElhinny, 1972) and is another example of an excursion apparently recorded through TRM, as in lava flows. However, Barbetti and McElhinny (1976) identified strong paleointensities associated with the intermediate directions. This unexpected result suggests that the record of the excursion may have been the result of a lightning strike. Although Barbetti and McElhinny (1976) dismissed the lightning hypothesis, they recognized that the shape of the AF demagnetization curve of the natural remanent magnetization was not incompatible with an isothermal origin for the Lake Mungo intermediate directions.

4.3.3 Excursions Observed in Lake and Deep-Sea Sediments

Verosub and Banerjee (1977) and Verosub (1982) have summarized the various excursions that have been observed in sedimentary sequences covering the past few hundred thousand years. Because there are many ways in which isolated distortions of the paleomagnetic recording process can arise, proof of the existence of an excursion must depend on consistent results from within a given lake or marine environment as well as from adjacent sedimentary environments. A critical appraisal of many of the proposed excursions reveals that there is not yet sufficient evidence to confirm their existence. In particular, they dismiss the evidence for the so-called Gothenburg event observed in Sweden at around 12,400 yr B.P. as being singularly unimpressive. This event could well be the manifestation of sedimentological (climatic?) effects. Various excursions observed in studies of the Great Lakes in North America are also similarly dismissed as even less impressive by Verosub and Banerjee (1977) and Verosub (1982).

At Lake Biwa, Japan, excursions have been observed in a deep core at 18,000 and 49,000 yr (Yaskawa *et al.*, 1973; Nakajima *et al.*, 1973). The older excursion is only very approximately dated by interpolation between widely separated and uncertain dates. However, the 18,000-yr excursion may also have been recorded in sediments for Imuruk Lake, Alaska (Noltimeier and Colinvaux, 1976). The large inclination changes coupled with only small changes in declination are consistent with the observations from Lake Biwa (Verosub and Banerjee, 1977). Verosub (1982) later dismissed the Imuruk Lake record as being due to sedimentological disturbances as evidenced by magnetic fabric studies. In the Gulf of Mexico, Clark and Kennett (1973) and Freed and Healy (1974) observed two excursions in at least a dozen deep-sea cores whose ages, based on foraminiferal zonation, lie at 17,000±1500 and 32,500±1500 yr respectively. The main problem is that the cores exhibit different magnetic signatures, so that the question of sedimentological disturbance arises. Abrahamsen and Knudsen (1979) have observed an excursion in a single core at Rubjerg, Denmark, with an age between 23,000 and 40,000 yr B.P.

The best documented excursion from lake sediments is the *Mono Lake excursion* recorded in dry lake sediments by Denham and Cox (1971) and Denham (1974), and subsequently investigated in more detail by Liddicoat and Coe (1979). The excursion has a duration of about 2000 yr beginning at 28,000 yr B.P. (Liddicoat, 1992). The major feature is an inclination swing from a normal value around +50° to -23° (Fig. 4.16). NRM/ARM ratios for one locality suggest the field intensity may have fallen well below the dipole field intensity during this excursion.

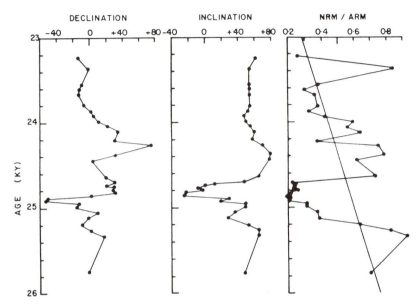

Fig. 4.16. Declination and inclination data plotted against age B.P. in Mono Lake, California. Sedimentary cores suggest that a magnetic excursion occurred near 25,000 yr ago (left and middle). NRM/ARM ratios are sometimes useful for obtaining a first-order estimate of the relative paleointensity down a core. The relative paleointensity during the excursion (right side) appears to be much lower than at other times. After Liddicoat and Coe (1979).

Liddicoat (1992) has summarized the various attempts to identify the Mono Lake excursion in other lakes in the region so that its existence could be firmly established. A restudy of the excursion in the Mono Basin at new localities 20 km apart shows that the record of the excursion in the basin is the most complete. The excursion is also recorded in northwest Nevada about 250 km north of the Mono Basin.

A unique inclination record for the past 60,000 yr has been obtained by Levi and Karlin (1989) from continuously deposited sediments in the Gulf of California at Site 480 on Leg 64 of the Deep-Sea Drilling Project. This core appears to record the Laschamp excursion between 51 and 49 ka and the Mono Lake excursion between about 29 and 26 ka. The close correspondence of the inclination fluctuations in the Gulf of California with those observed at Mono Lake more than 1000 km to the north provides impressive confirmation that the same excursion is being observed. Levi and Karlin (1989) note, however, that an excursion at around 17 ka, observed in Lake Biwa and the Gulf of Mexico, is not seen in the Gulf of California record. However, they do note that there is another excursion in their record at about 23 ka, slightly younger than the Mono Lake excursion by three to five thousand years and of considerably shorter duration. It has a paleomagnetic signature similar to an excursion observed at Summer Lake by Oregon by Negrini *et al.* (1984), who at that time considered this to be another record of the Mono Lake excursion. This *Summer Lake excursion* may thus be a distinct event.

The most surprising aspect of the Laschamp excursion is that it has not been detected in cores from Lac du Bouchet, at virtually the same locality in the Massif Central, France (Thouveny and Creer, 1992). The sedimentary recording process smoothes high-frequency secular variation, so the Lac du Bouchet record appears to indicate that the Laschamp excursion may have lasted for only a few centuries. Alternatively, excursions can exhibit reverse directions at one location but not at another. Moreover, more than one excursion could occur during an interval when the dipole intensity is low (Merrill and McFadden, 1994). Thus extreme caution must be exercised if attempting to use excursions as stratigraphic markers.

4.3.4 Reversals or Excursions during the Brunhes Chron?

It has been claimed (Champion *et al.*, 1988) that there are eight reversal events (subchrons; see §5.3.4) during the Brunhes Chron (the past 780 ka). Harrison (1974) analyzed 216 deep-sea sedimentary cores and showed that approximately 20% of these exhibited reverse directions of magnetization at some time during the Brunhes. To determine if these represented true reversal events, Champion *et al.* (1988) searched for reverse directions in lava flows. They reasoned that if there is evidence of reverse directions at the same time in both lava flows and sediments sampled at very different localities on the Earth's surface, then this constitutes good evidence for a reversal event. They found such evidence in several instances and so suggested that there were at least eight reversal events during the Brunhes. Their suggestion has gained some acceptance from geochronologists (e.g., Spell and McDougall, 1992). However, Merrill and McFadden (1994) point out that this would require the reverse polarity state to be substantially less stable than the normal polarity state during the Brunhes. This contradicts conclusions reached from robust statistical analyses of the reversal chronology record (§5.3.4) and is contrary to one of the few robust conclusions from theory (i.e., that the normal and reverse polarity states should be statistically equivalent apart from the inverted field direction, §8.6.2). Merrill and McFadden (1994) further point out that less than 0.1% of all Brunhes-age lava flows exhibit reverse directions. Assuming that the thousands of Brunhes lava flows distributed across the Earth are randomly distributed in time, then the minimum time for an individual reversal would be 390 yr if there were eight reversal events during the Brunhes. This estimate assumes the field reversed polarity and immediately reversed back again. If one makes the more likely assumption that the field spent some time in the reverse state, then the time for an individual reversal transition is reduced to approximately 100 yr or so. Both estimates are more than an order of magnitude below the time estimated for a reversal transition (§5.4.1). Merrill and McFadden (1994) conclude that of the events proposed by Champion *et al.* (1988), the Laschamp and Blake events are probably reversal excursions but that the other claimed events are reversal excursions or data artifacts. This illustrates the difficulty in distinguishing between reversal events and reversal excursions.

4.3.5 Models of Geomagnetic Excursions

Two possible explanations for geomagnetic excursions appear likely; they reflect either large-amplitude secular variation or aborted reversals of the geomagnetic field (see further discussion of this proposal in §5.4.4). Distinguishing between these alternatives may not be possible unless aborted reversals can be shown to have a different signature from large-amplitude secular variation.

Explanations in terms of large-amplitude secular variation generally involve models in which one of the nondipole sources of the Alldredge and Hurwitz (1964) type changes in strength and/or location to produce the desired effect (Creer, 1977, 1983; Liddicoat and Coe, 1979; Harrison and Ramirez, 1975). One might visualize three basic situations:

(i)   the dipole field underwent a dramatic change in direction;
(ii) the main dipole field decreased in strength and the nondipole field dominated over a large portion of the globe; or
(iii) one of the nondipole sources increased dramatically in strength.

The first two cases imply a global phenomenon. The extent of this in the second case is obviously related to how much the dipole field decreases. At the present time there is no evidence to suggest that case (i) applies. Cases (ii) and (iii) are really variations of the situation when the ratio of the nondipole field mean intensity to dipole field intensity is considerably larger than at present, allowing for nondipole field domination, at least on a local scale. Case (iii) is the most difficult to test because it implies that the excursion may be restricted to a very small region. Indeed, in a spherical harmonic description of the field, the field can be restricted mathematically to as small a region as one desires, simply by choosing high-enough degree harmonics to describe the restricted feature. This is also physically plausible.

Harrison and Ramirez (1975) have shown that the nondipole field can be represented by a dipole source in the outer core that produces a localized "reversal" of field direction at a site, while less than 15° (1665 km) away the field still exhibits its original polarity. In his computer syntheses of geomagnetic secular variation, Creer (1983) shows that significant shallowing of the inclination of a normal polarity field could be produced by deep-seated radial dipole sources that point upward at midlatitudes in the northern hemisphere or downward in the southern hemisphere. He found that changes in the sign of inclination could be produced by introducing quite moderate values of the moment p of the nondipole source. Particularly sharp inclination perturbations could be synthesized if p is made to fluctuate as it drifts. Such excursions would only be observed over a restricted region.

Liddicoat and Coe (1979) model the Mono Lake excursion with a single eccentric dipole of varying moment displaying a complex pattern of westward, eastward, and northward drift. The average moment of this dipole is comparable with the largest calculated for the present nondipole field and the maximum moment is almost twice as great. This model would predict a continental size effect observed only over a restricted portion of the globe.

Perhaps the most significant observation is that the geomagnetic dipole moment was abnormally low over the time span 50,000 to 20,000 yr ago (§4.1.5), the time interval in which both the Laschamp and the Mono Lake excursions occurred. With an overall low geomagnetic dipole moment, changes in the nondipole field sources will have much more significant effects on the magnetic field observed at the Earth's surface. It may be that a major requirement for the occurrence of reversal excursions is simply that the dipole field remain abnormally low so that nondipole effects are more readily seen (Valet and Meynadier, 1993; Merrill and McFadden, 1994).

4.4 The Geomagnetic Power Spectrum

4.4.1 Time Series Analysis

A geomagnetic record is typically a composite of a time-varying magnetic signal superimposed on a background of random noise and unwanted signals. Suppose for the moment the record is continuous and denoted by $f(t)$, where t is time. It is possible to extract information on dominant frequencies present in the record (e.g., associated with westward drift of the nondipole field). To do this it is convenient to transform from the time domain to the frequency domain,

$$A(\omega) = \frac{1}{2\pi} \int_{-\infty}^{\infty} f(t) e^{-i\omega t} \, dt \quad , \qquad (4.4.1)$$

where ω is the angular frequency and $A(\omega)$ is the Fourier transform of $f(t)$. The inverse transform is given by

$$f(t) = \frac{1}{2\pi} \int_{-\infty}^{\infty} A(\omega) e^{i\omega t} d\omega \quad . \qquad (4.4.2)$$

Usually the signals are analyzed in digital form and then the analysis is referred to as a *time series analysis*. A digitized signal can be handled analogously by simply using the discrete Fourier transform (e.g., Aki and Richards, 1980).

In practice the geomagnetic record is finite and requires a truncated series representation of $f(t)$,

$$f(t) = \sum_{j=0}^{N-1} A_j e^{i\omega_j t} \quad , \qquad (4.4.3)$$

in which $\omega_j = 2\pi j / N$ and the A_j are given by

$$A_j = \frac{1}{N} \sum_{j=0}^{N-1} f(t) e^{-i\omega_j t} \quad . \qquad (4.4.4)$$

Referring to §2.3.2, the A_j are the least-squares estimates of the corresponding parameters in the infinite Fourier series. The relative power in a given frequency can be defined by

$$W_j = 2|A_j|^2 \quad . \qquad (4.4.5)$$

Generally the power is spread over a continuum of frequencies and $W(\nu)$ is referred to as the *power spectrum* of the process. Here ν, the linear frequency, has been used instead of the angular frequency ($\nu = \omega/2\pi$). The variance of this process is given by:

$$\int_{-\infty}^{\infty} W(\nu) d\nu \quad . \qquad (4.4.6)$$

The method just described to obtain the power spectrum is essentially the periodogram method developed in the late nineteenth century (Schuster, 1898). Unfortunately (4.4.5) provides a poor estimate of the power, in that the variance of the frequency estimate is typically very large with respect to the mean (Childers, 1978). The consequence of this is poor spectral resolution.

The periodogram methods, modified by filtering and tapering, became very popular with the greatly improved computational procedures associated with the fast Fourier transform algorithm (Cooley and Tukey, 1965). However, several other methods have subsequently been introduced with the goal of achieving increased spectral resolution (e.g., Kanasewich, 1973; Childers, 1978). Perhaps none of the new methods has stirred more debate than the maximum entropy method (MEM) first introduced by Burg (1967, 1972). Only this method will be discussed further, because it is probably one of the most widely used and poorly understood methods in geomagnetism (e.g., Ulrych, 1972; Currie, 1974; Denham, 1975; Phillips and Cox, 1976; Barton and McElhinny, 1982).

Subsequent to its formulation, MEM has been shown to be directly related to least-squares error linear prediction and autoregression (Van der Bos, 1971). The discussion here of discrete autoregression follows that given by Barton (1978). This process assumes that $f(t)$, in digital form, is linearly dependent on the previous m values plus a Gaussian (white) noise,

$$f(t) = a_1 f(t-1) + a_2 f(t-2) + \cdots + a_m f(t-m) + B(t) \quad , \tag{4.4.7}$$

where m is referred to as the *order of the autoregressive process*, the a_j are constants, and $B(t)$ is called the *innovation* of the process. The mean values of $f(t)$ and $B(t)$ are taken to be zero and $B(t)$ is derived from a Gaussian process. The spectrum of the mth order autoregressive process can be viewed as providing a prediction of $f(t)$ based on its previous m values and $B(t)$ as the prediction error. The spectrum of the mth order autoregressive process is known to be (e.g., Ulrych and Bishop, 1975)

$$W(\nu) = 2\beta_m \left| 1 - \sum_{j=1}^{m} a_j e^{i2\pi\nu_j} \right|^{-2} \quad , \tag{4.4.8}$$

in which β_m is the variance of $B(t)$. $W(\nu)$ is to be evaluated between $\pm\nu_N$, the Nyquist frequency. Basically, the MEM involves fitting an autoregressive model to the data in such a way as to maximize the information in the actual geomagnetic record and minimize information about the series outside the observed record. Although this is clearly desirable, the method of doing it is still controversial. In particular, one of the fundamental problems in applying the analysis involves selection of the optimum value for the order m of the process. It is known that if m is chosen too low the resulting spectrum is too smoothed, so that information on the higher frequencies is lost. On the other hand, if m is chosen too large, spurious detail is introduced into the spectrum. Swingler (1979, 1980) has applied Burg's MEM algorithm to synthetic data and finds that the spectral peaks are often mislocated and their widths too broad. Thus both the frequencies and their associated power are inaccurately obtained. This indicates that the choice of a particular MEM algorithm must be done with care. There remains no consensus on an algorithm for optimum selection of the order of the process. Somewhat different results can thus emerge from the same record if analyzed by different workers (e.g., Burg, 1967, 1972; Akaike, 1969a,b, 1970; Smylie *et al.*, 1973; Ulrych and Bishop, 1975; Swingler, 1979, 1980; Barton, 1982, 1983).

4.4.2 Spectrum from Historical Records

There is considerable controversy over the interpretation of the geomagnetic variations with periods from 1 to 30 yr (e.g., analyses of Yukutake, 1965; Bhargava and Yacob, 1969; Currie, 1973, 1976; Rivin, 1974; Alldredge, 1976, 1977; Courtillot and Le Mouël, 1976a,b; Srivastava and Abbas, 1977; Chen, 1981; see also review by Courtillot and Le Mouël, 1988, and §2.4.2). Part of the reason for this controversy involves the methods used for analysing the power in particular periods, and part involves the interpretations. For example, the use of MEM on data from Hong Kong (Chen, 1981) indicates significant peaks in the spectrum of the horizontal component of the field near 20, 10.6, 7.1, and 5.8 yr; roughly consistent with worldwide peaks found by Currie (1973). The corresponding periods for the vertical component of the magnetic field in Hong Kong are at 20, 6.7, and 5.8 yr (Chen, 1981). The vertical component spectrum peaks at 6.7 and 5.8 yr can barely be resolved. It could be that this is due to peak splitting resulting from too large a choice for the order of the process (4.4.1). The strong correlation between the vertical and horizontal components argued by Chen (1981) would then be based on only two periods, one near 20 yr and one near 6 yr. The analysis of Chen (1981) is probably correct, but it shows how the interpretations and the methods of analyses are interwoven into a complex web that is difficult to untangle without more agreement on appropriate procedures for time series analyses.

Currie (1973) showed the tendency for spectral lines from several observations to cluster around a period of 21.4 yr. This highlights the controversial nature of the interpretation of the spectrum, for Currie suggested that this variation is of external origin and due to the 22-yr sunspot cycle (see §10.2.2), while Alldredge (1976, 1977) argues that it is of internal origin on the basis that there is no good

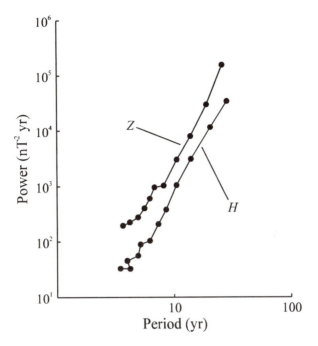

Fig. 4.17. Power in the vertical (*Z*) and horizontal (*H*) components of the geomagnetic field are shown as a function of period. After Currie (1967, 1968).

evidence for synchronous worldwide variation with a common period. It should be noted that these uncertainties in distinguishing between internal and external sources will be reflected in corresponding uncertainties in estimates of lower mantle conductivity (§2.4.2).

In spite of these problems with analysis and interpretation (there appears to be a consensus that most periods in the geomagnetic spectrum that are longer than a few years are predominantly of internal origin), it is possible to combine data at different sites to obtain an estimate of the power spectrum of internal origin (Currie, 1967, 1968) (see Fig. 4.17). Barton (1982, 1983) used a different algorithm for spectral analysis discussed in §4.4.3 (see Fig. 4.18). His spectral power estimates are significantly higher than those of Currie (1967, 1968) (even if *Z* and *H* in Fig. 4.17 are combined). Nevertheless a large increase in power with an increase in period is evident in both analyses. This indicates that most of the power in the geomagnetic field of internal origin probably resides in much longer periods than can be deduced from the analysis of historical records.

4.4.3 Spectrum from Lake Sediment Data

Denham (1975) first suggested that the quasi-periodic fluctuations in the geomagnetic field recorded by the magnetization of lake sediments (§4.2) could profitably be analyzed by power spectrum techniques. He showed that time series analysis could provide valuable information on the dominant periods present and could also be used to obtain the direction of drift of the nondipole field (Fig. 4.18). However, there are several problems associated with the time series analysis of paleomagnetic data that are not encountered with the analysis of historical data:

(i)   there might be changes in sedimentation rates that are not properly delineated by radiometric dating (see discussion by Clark and Thompson, 1979);

(ii) there might be dating errors;

(iii) the noise might increase with sediment core depths (time), as suggested to be the case in the Australian lakes studied by Barton and McElhinny (1981); and

(iv) significant changes in apparent periodicities might be present in the geomagnetic spectrum over the longer time periods sampled (nonstationary processes; see Chapters 6 and 10).

There is a basic problem in deciding the most appropriate variables to use for the spectral analysis of paleomagnetic data. Declination and inclination are the two most common variables measured in sediment cores. However, declination

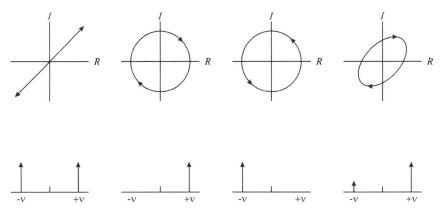

Fig. 4.18. Hypothetical power spectra (bottom) for four classes of simple two-dimensional complex (*I*, imaginary; *R*, real) fluctuations (top) at an arbitrary frequency v. Planar motion (far left), regardless of angle, yields a symmetric spectrum. Circular motion (center) produces a one-sided spectrum that can be used to distinguish the sense of rotation, while (far right) elliptical motion yields an asymmetric spectrum weighted to one side or the other depending on the sense of the looping. After Denham (1975).

and inclination time series can be radically different at a given site. Consider the simplified example of a magnetic field consisting of the sum of an axial geocentric dipole and an equatorial geocentric dipole. A change in the intensity of the axial dipole field affects only the inclination and not the declination, whereas a change in the equatorial dipole direction, constrained to lie in the equatorial plane, affects only the declination. Declination is a poor choice because it becomes highly unstable in the region of the dip poles. Although inclination is better, any spectrum obtained using this as a variable would be difficult to compare with that obtained from the historical record or to analyze for drift direction by Runcorn's rule. Barton (1982) has probably suggested the best compromise to the above problems by summing the contributions to the spectral power at a given frequency from each component,

$$W(v) = W_x(v) + W_y(v) + W_z(v) \quad , \tag{4.4.9}$$

where z is the vertical component and x and y are the orthogonal horizontal components. He bypasses the problem that absolute paleointensity data cannot be obtained from sedimentary cores by assigning a constant magnitude to the magnetic vector equal to that due to the present-day dipole field (8×10^{22} Am^2).

In an attempt to obtain a first-order estimate of the global power spectrum, Barton (1982, 1983) analyzed lake sediments data, deep-sea core data and archeomagnetic data from 10 locations within the latitudinal zones of 30°-55°. Three of these are in the southern hemisphere (all Australia) and the remainder are in the northern hemisphere. Barton also analyzed the historical records from 12 magnetic observatories reasonably well distributed over the Earth. The time series analysis used was a form of the periodogram method described by Bloomfield (1976), involving detrending, tapering and filtering methods that are significantly different from those used by Currie in his analysis of the historical record (§4.4.2).

Figure 4.19 presents a summary of the power spectrum as a function of period. The part of the spectrum obtained from historical records is similar in shape to that deduced by Currie (Fig. 4.18) using different data and a different method of spectral analysis. At longer periods the power generally increases, showing an apparently broad maximum near 10^3-10^5 yr. Structural detail in this spectrum can probably be attributed to the use of different types of data, to particular methods of analysis, and to the likelihood that there are several different mechanisms that contribute to geomagnetic secular variation (§4.2.8). For example, the dip near 10^2 yr coincides with the change from using short-period historical data to long-period paleomagnetic data.

Courtillot and Le Mouël (1988) have also produced a schematic power spectrum of the geomagnetic field, combining the analysis of Filloux (1980a) at

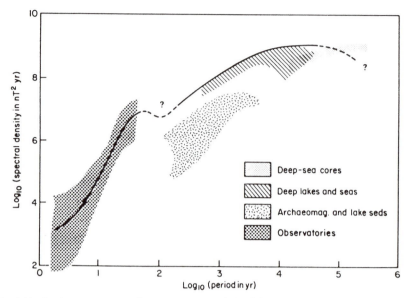

Fig. 4.19. Continuous geomagnetic power spectrum derived from observatory and paleomagnetic data. After Barton (1983).

short periods with those of Barton (1982, 1983). They note that the spectrum given by Barton has a shape similar to that of Filloux but is displaced by 6 orders of magnitude in period. Courtillot and Le Mouël (1988) suggest that the geomagnetic spectrum might have a roughly $1/v^2$ (where v is frequency) relation over the remarkably large range of periods from 1 minute to a million years (12 orders of magnitude), despite changes from internal to external sources somewhere in the spectrum. However, as they point out, there seems no particular reason why this should be so.

Chapter Five

Reversals of the Earth's Magnetic Field

5.1 Evidence for Field Reversal

5.1.1 Definition of a Reversal

The geomagnetic field property best documented by paleomagnetism is that the Earth's magnetic field has reversed its polarity many hundreds of times. Such field reversal is not an isolated phenomenon. The Sun, for example, also has a general magnetic field that appears to reverse its polarity frequently (§10.2). The magnetic field of the Milky Way (the galaxy in which our solar system resides) tends to be perpendicular to the plane of the galaxy but the sign of this field appears to vary on a large regional scale, suggesting reversals of that field.

Even though reversals of the geomagnetic field are so well documented that their existence can no longer be doubted, it is not a simple matter to provide a precise, generic definition of what is meant by a "reversal." See Jacobs (1994) for a comprehensive review. With a complex, time-varying field it is not, for example, feasible to say that it is just a change in the sign of the field. Thus we use a definition that is helpful in the context of the geomagnetic field, a field that at the surface of the planet is dominated by a geocentric axial dipole field (the g_1^0 term in a spherical harmonic analysis, §2.2). Therefore we are led to the basis of a definition in stating that a reversal occurs when the sign of the g_1^0

term changes. To see that this is not a generically useful definition, consider a field (such as that of Uranus; see Table 10.2 and §10.5) that is dominated by an equatorial dipole (the g_1^1 and h_1^1 terms in a spherical harmonic analysis) at the surface of the planet. A simple change in the sign of g_1^0 would then have a relatively small effect on the overall field, and could hardly classify as a reversal of the field.

However, this basic definition is not yet sufficient for the Earth. A paleomagnetic observation at one locality from rocks of a given age with field direction approximately opposite to the known mean field of that age is, surprisingly, not sufficient evidence for a reversal of g_1^0. This is because the nondipole field can produce very large local deviations of the field, even to the extent of appearing locally to be a reversal. Therefore it is necessary to observe inverted field directions globally before it can be concluded that g_1^0 has reversed. We also require that the change in sign exhibit some stability before it can be considered to be a reversal. Hence, we define a reversal as a globally observed 180° change in the dipole field averaged over a few thousand years. Because the Earth's mean dipole field is essentially a geocentric axial dipole field (§6.1), this definition essentially reduces to a stable change in sign of the axial dipole field.

5.1.2 Self-Reversal in Rocks

David (1904) and Brunhes (1906) were the first to observe magnetizations in lava flows that were roughly opposed to that of the present Earth's magnetic field. Later Matuyama (1929) produced the first (crude) reversal chronology by examining the magnetic directions in more than 100 lava flows from Japan and Manchuria. However, at that time these studies did not provide convincing evidence for reversals of the geomagnetic field. The problem was to decide whether the Earth's magnetic field itself had changed polarity or whether some self-reversal mechanism in the rock had produced a thermoremanent magnetization antiparallel to the applied field. The first theoretical models for self-reversal were produced by Néel (1955) and since then other possible self-reversing mechanisms have been proposed by several authors (e.g., Verhoogen, 1956; Uyeda, 1958; Ishikawa and Syono, 1963; O'Reilly and Banerjee, 1966; Hoffman, 1975, 1992a; Stephenson, 1975; Petherbridge, 1977; Tucker and O'Reilly, 1980; Nord and Lawson, 1992; McClelland and Goss, 1993).

All models for self-reversal require that there are (or were) at least two magnetic phases in the rock. One phase becomes magnetized first, parallel to the external magnetic field, and subsequently the second phase becomes magnetized antiparallel to the first phase. This occurs either because there is a negative exchange interaction acting between the two phases or because the magnetic field of the first phase swamps that of the external magnetic field (*magnetostatic*

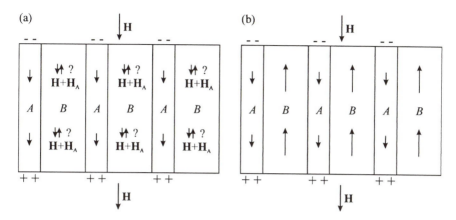

Fig. 5.1. Example of a self-reversal mechanism in rocks. There are two magnetic phases *A* and *B* for which *A* has the higher Curie temperature but a lower room-temperature value of saturation magnetization than *B*. (a) On cooling, *A* becomes magnetized first in a direction parallel to the external field **H**, and on further cooling phase *B* becomes magnetized in the total field **H**+**H**$_A$, where **H**$_A$, which will be antiparallel to **H**, is the field in phase *B* caused by the magnetization of *A*. The direction of the total field in which *B* becomes magnetized depends on the relative magnitudes of **H** and **H**$_A$. (b) If **H**$_A$ is larger than **H** then phase *B* will acquire a magnetization antiparallel to **H** on further cooling. Because *B* has a higher room-temperature saturation than *A*, the sample will then have a self-reversed magnetization.

interaction). An example of the latter is illustrated in the model of Fig. 5.1. Phase *A* has a higher Curie temperature than phase *B* and is magnetized parallel to the external field **H**. As shown in Fig. 5.1b, phase *B* becomes magnetized in the total field **H**+**H**$_A$, where **H**$_A$ is the magnetic field due to phase *A*, and which in this model would be opposite to the external field **H**. If $|\mathbf{H}_A| > |\mathbf{H}|$ then phase *B* will be magnetized opposite to **H**, and if at room temperature the total magnetization of *B* exceeds that of *A* then the sample will have *self-reversed*. Unfortunately, after the rock formed, chemical change (such as oxidation) can occur that alters one of the phases in such a way that the self-reversal is not detectable in a laboratory experiment.

Experimentally the best understood and most studied self-reversal mineral is a titanohematite containing roughly 50 mole percent ilmenite and hematite (Uyeda, 1958; Ishikawa and Syono, 1963; Westcott-Lewis and Parry, 1971; Hoffman, 1975, 1992a; Nord and Lawson, 1992). The self-reversal mechanism is complicated, involving Fe–Ti ordering and possibly exsolution. However, the essence of the mechanism is that a negative exchange interaction occurs between a weakly magnetic phase that has a high Curie temperature and a much stronger magnetic phase with a lower Curie temperature.

It is possible that self-reversal occurs in sediments as well as igneous rocks (McClelland and Goss, 1993; Merrill and McFadden, 1994). This could occur if two anticoupled phases coexist during chemical alteration. For example, McClelland and Goss (1993) show that in some circumstances the conversion of maghemite to hematite would result in self-reversal. Unfortunately, a self-reversal that occurred in a sediment at ambient temperatures would typically require so much time that it could not readily be reproduced in a laboratory experiment, so documentation of this effect is extremely difficult. Fortunately, self-reversal in sediments is probably very rare.

Although self-reversal mechanisms involve some challenging problems in geophysics, well documented examples of self-reversal in nature are rather rare. In addition to the titanohematite–hematite example given above, only a few pyrrhotites, titanohematite (near 90 mole percent hematite), and possibly a few restricted compositions of exsolved titanomagnetites have exhibited self-reversing tendencies (Schult, 1968; Carmichael, 1961; Ozima and Ozima, 1967; Readman and O'Reilly, 1970; Peterson and Bleil, 1973; Ryall and Ade-Hall, 1975; Petherbridge, 1977; Tucker and O'Reilly, 1980; Ozima et al., 1992). The bulk of evidence, as set out below, favors the view that the vast majority of reverse magnetizations observed in rocks arise from changes in the polarity of the Earth's magnetic field.

Because of the strong evidence in favor of field reversal, it was somewhat surprising that an apparent correlation between oxidation state and polarity was purported to have been found in the late 1960s (e.g., Wilson and Watkins, 1967; Smith 1970b). Although sometimes still referred to, this apparent correlation has been shown to be due to a data artifact (Merrill, 1985). It is very doubtful that the commonly used reversal chronologies discussed late in this chapter contain many (if any) erroneous magnetic field reversals arising from self-reversal. However, there is the possibility that suggested additions of short reversal events, or subchrons, to these chronologies (e.g., Champion et al., 1988) may on occasion be artifacts associated with self-reversal (Merrill and McFadden, 1994).

5.1.3 Baked Contacts

There are three field tests that provide compelling evidence regarding the hypothesis that the Earth's magnetic field has indeed undergone reversals. These tests are:
(i)   studies of baked contacts adjacent to intrusive igneous rocks or underlying lava flows should show agreement in the polarity of the magnetization between the igneous rock and the baked rock;
(ii) there should be worldwide simultaneous zones of one polarity; and
(iii) there should be records of the magnetic field observed in rock sequences showing the polarity changing continuously from one state to the other.
Each of these tests will be considered separately in subsequent parts of this chapter. The first two tests rely on simultaneous observations of different rock

types. In general a baked contact will have a very different magnetic mineralogy from the igneous rock that provides the thermal energy. Simultaneous zones of one polarity observed worldwide should also involve rocks of widely different magnetic mineralogy.

The evidence from the studies of baked contacts was first compiled by Wilson (1962b). This was later updated by Irving (1964) and McElhinny (1973). The successive increase in the overwhelming evidence showing the correspondence in polarity between the baked and baking rocks is given in Table 5.1. Conventionally, rocks whose magnetization is in the same sense as the present magnetic field are termed *normal* (N) and those in the opposite sense are termed *reverse* (R). Directions intermediate between these (defined as rocks giving virtual geomagnetic poles inclined at an angle greater than 45° to the axis of rotation) are termed *intermediate* (I). Comparisons of polarities for the case of reverse magnetizations were particularly important to confirm the reality of the field reversals and this is seen by the much greater emphasis given to this situation in Table 5.1. The three cases of disagreement reported in the table refer to measurements made before the technique of magnetic cleaning was used in paleomagnetic studies. The subject has advanced so much in the past few decades that baked contact studies are now used not so much as a check on the reality of field reversals, but as a check on the age of the magnetization of igneous rocks. This is especially true in studies of old Precambrian rocks where the importance of discovering evidence relating to the age of the magnetization being measured is paramount.

A particularly convincing study was made by Wilson (1962a) into the magnetic record in a doubly baked rock. The situation is illustrated in Fig. 5.2. A lava flow, reversely magnetized, had heated an underlying laterite. The baked laterite had a direction of magnetization the same as that of the lava – reverse. Subsequently, both lava and baked laterite were intruded by a dike whose magnetization was also reverse, but different by about 25° from that of the lava and baked laterite. A study of the zone of second heating of the laterite caused by the dike showed a gradual change from the dike direction, adjacent to the

TABLE 5.1
Comparison of Polarities Observed in Igneous Rocks and Their Baked Contacts.

Igneous	Baked contact	Number of observations		
		Wilson (1926b)	Irving (1964)	McElhinny (1973)
N	N	14	34	47
R	R	34	49	104
I	I	1	2	3
N	R	3	3	3
R	N	0	0	0

Note. Original analysis by Wilson (1962b); updated by Irving (1964) and McElhinny (1973). N, normal; R, reverse; I, intermediate.

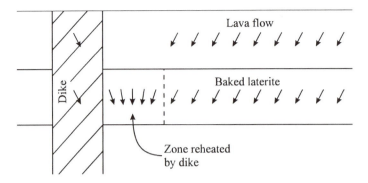

Fig. 5.2. Evidence for field reversal from the magnetic record in a doubly baked rock (Wilson, 1962a). A lava flow originally heated the laterite horizon. Subsequently a dike intruded both, reheating both lava flow and laterite in its adjacent region. In this region the laterite was thus first reheated by the lava and then reheated a second time by the dike.

dike, to the lava direction at a distance where the second heating had produced no effect. Thus in the same region both superimposed magnetizations were of reverse polarity. It is difficult to explain this observation by any reasonable mechanism other than that the Earth's magnetic field had changed polarity and was reverse during both heating episodes.

5.1.4 Development of the Polarity Time Scale for the Past 5 Myr

Mercanton (1926) first realized that if rocks containing reverse magnetizations were due to reversals of the Earth's magnetic field, then this should be registered in rocks worldwide. Rocks of the same age should have the same polarity of magnetization. Matuyama (1929) demonstrated that early Quaternary lavas had reverse polarity whereas younger lavas had normal polarity. Observations by Roche (1951, 1956) of volcanic rocks from the Chaîne des Puys in France led him to conclude that the most recent reversal of the Earth's magnetic field took place in the middle of the Early Pleistocene. A similar conclusion was reached by Opdyke and Runcorn (1956) working on rocks from the United States. These observations suggested that an ordered sequence of polarity inversions might exist in the geological record. This conclusion was further emphasized by the evidence that almost all rocks of Permian age are reversely magnetized (Irving and Parry, 1963).

The above observations were based upon the relatively imprecise methods used for dating rocks on the basis of fossil occurrences. It was only in the early 1960s that developments (by Evernden, McDougall, and Dalrymple) in the K–Ar isotopic dating method enabled quite young volcanic rocks to be dated with some precision (see §3.5). The first example of the combined use of magnetic polarity and K–Ar dating was that of Rutten (1959). He concluded that the present normal polarity extended from at least 0.47 million years ago and that an earlier period of normal polarity existed about 2.4 million years ago.

Systematic studies attempting to define a polarity time scale using joint magnetic polarity and K–Ar age determinations on young lava flows were undertaken both in the Uunited States and Australia. The first time scale put forward by Cox *et al.* (1963a) suggested a periodicity of magnetic reversals at about one-million-year intervals. However, as new data quickly appeared in the literature (Cox *et al.*, 1963b, 1964a; McDougall and Tarling, 1963, 1964) it soon became apparent that there was no simple periodicity; the lengths of successive polarity intervals were sometimes long (~1 Ma) or quite short (~0.1 Ma). Cox *et al.* (1964b) proposed that within intervals of predominantly one polarity lasting of the order of 1 Ma, there were short intervals of opposite polarity of the order of 0.1 Ma. The longer intervals were termed magnetic polarity epochs and the shorter intervals were called events. The epochs were named after pioneering scientists in geomagnetism (Brunhes, Matuyama, Gauss, and Gilbert) whereas the events were labeled from the location of their discovery. The terms *chron* and *subchron* have subsequently been officially adopted to replace the terms epochs and events (see Table in §5.1.5). However, the terms epochs and events are still used but with somewhat different meaning. For example, the term reversal event is often used to describe the phenomenon of two reversals that occur very close to each other in time while the term subchron refers to the

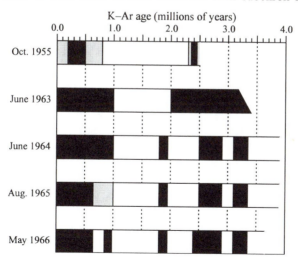

Fig. 5.3. The evolution of the geomagnetic polarity time scale 1959–1966. Black represents normal polarity, white represents reverse polarity, and gray indicates uncertain polarity. Abstracted with permission from Cox (1969). Copyright 1969 American Association for the Advancement of Science

stratigraphic record of an event.

A few of the earliest compilations of the geomagnetic polarity time scale (GPTS) for the past 5 million years covering the years 1959–1966 are shown in Fig. 5.3. It should be noted that these early studies combined conventional K–Ar dating with measurements of magnetic polarity from widely spaced localities worldwide, and were not carried out on continuous sequences. In general, the evolution of this time scale has been one in which more reversals have been included with time. An excellent history of this evolution has been given by Glen (1982) and more recently by Opdyke and Channell (1996). The development of high-precision $^{40}Ar/^{39}Ar$ dating techniques has shown that many of the ages determined by the conventional K–Ar method are too young. Furthermore, problems have arisen with the identification of some of the short subchrons as more and more measurements have been made. The problem that arises is to decide whether some of these represent true reversals or some other geomagnetic field behavior such as changes in intensity. Figure 5.4 shows the present state of the GPTS for the past 6 Myr following Cande and Kent (1995).

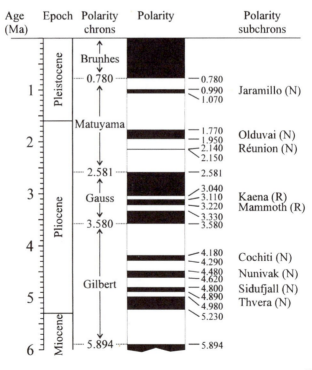

Fig. 5.4. Geomagnetic polarity time scale for the past 6 Myr based mainly on $^{39}Ar/^{40}Ar$ and paleomagnetic data on igneous rocks (after Cande and Kent, 1995). Black represents normal polarity; white represents reverse polarity.

This represents the time interval during which all the polarity chrons and subchrons have been named. Cande and Kent (1995) have also identified various cryptochrons (see §5.1.5, Table 5.2), which may be due to geomagnetic behavior such as geomagnetic intensity changes, that occur at 0.49, 1.20 (Cobb Mountain), and 2.42 Ma.

A continuous record of polarity changes can be obtained by investigating the paleomagnetic record preserved in deep-sea sediments by taking cores from the ocean bottom. Although the cores are usually not oriented, they are taken nearly vertically into the ocean bottom so that changes in sign of the magnetic inclination measured in the cores or changes of 180° in declination can easily be identified as records of polarity change. Pelagic sedimentation rates (e.g., in the northcentral Pacific Ocean) are relatively low and typically are on the order of 1 to 10 mm per 1000 years so that the most recent polarity change, corresponding to the Brunhes–Matuyama boundary at 0.78 Ma, will occur at depths of about 0.78–7.8 m. This slow sedimentation rate smoothes the magnetic signal but allows one to go further back in time for a given length of core. Oceanic sedimentation rates elsewhere (e.g., continental margins) can be much higher and in principle provide a more detailed record of magnetic changes for a shorter interval of time, if chemical changes are not severe (e.g., Karlin and Levi, 1985).

The earliest investigations of marine sediment cores soon established the reality of the polarity time scale observed on land (e.g., Ninkovich *et al.*, 1966; Opdyke and Glass, 1969). Using cores with suitable sedimentation rates up to 30 m long, the polarity sequences can be extended back to 10 Ma or more (Foster and Opdyke, 1970; Opdyke, 1972; Opdyke *et al.*, 1974). The largest errors in the marine sediment record are associated with dating (fossils) and variations in sedimentation rate. These errors generally increase with the age of the sediment.

5.1.5 Terminology in Magnetostratigraphy

The application of the well known principles of stratigraphy to sedimentary records of reversal sequences is referred to as *magnetostratigraphy* (see review by Opdyke and Channell, 1996). The terminology of magnetostratigraphy has been formalized by the IUGS Subcommission on the Magnetic Polarity Time Scale (Anonymous, 1979). The basic unit in magnetostratigraphy is the magnetostratigraphic polarity zone, which may be referred to simply as a magnetozone. Polarity zones may consist of strata with a single polarity throughout, they may be composed of an intricate alternation of normal and reverse units, or they may be dominantly either normal or reverse magnetozones containing minor subdivisions of the opposite polarity. The recommended hierarchy in magnetostratigraphic units and polarity chron (time) units is given in Table 5.2 following Opdyke and Channell (1996).

TABLE 5.2
Hierarchy in Magnetostratigraphic Units and Polarity Chron (Time) Units as Recommended by the
IUGS Subcommission on the Magnetic Polarity Time Scale.[a]

Magnetostratigraphic Polarity Units	Geochronologic (time) Equivalent	Chronostratigraphic Equivalent	Approximate duration (yr)
Polarity megazone	Megachron	Megachronozone	10^8-10^9
Polarity superzone	Superchron	Superchronozone	10^7-10^8
Polarity zone	Chron	Chronozone	10^6-10^7
Polarity subzone	Subchron	Subchronozone	10^5-10^6
Polarity microzone	Microchron	Microchronozone	$<10^5$
Polarity cryptozone	Cryptochron	Cryptochronozone	Existence uncertain

[a]After Opdyke and Channell (1996).

5.2 Marine Magnetic Anomalies

5.2.1 Measurement and Calculation

Magnetic anomaly is used here to mean the difference between the observed magnetic field intensity at a location and that expected from some reference field, such as that derived for the most recent magnetic epoch (e.g., the 1995 IGRF). Observed anomalies in total intensity are believed to be due to crustal sources and vary significantly over short distances ranging from a few metres to a few hundred kilometres in the case of some major marine magnetic anomalies. Such short wavelength features will not significantly affect the highest degree spherical harmonics used to describe the field for a given epoch.

The analysis of marine magnetic anomalies has made a major contribution to our understanding of the Earth's magnetic field in the past. The intensity of the magnetic field at sea is usually measured by a magnetometer towed behind a ship at the sea surface, although deep-tow magnetometers are sometimes used to measure the intensity close to the ocean bottom (e.g., Spiess and Mudie, 1970; Atwater and Mudie, 1973; Macdonald and Holcombe, 1978; Macdonald *et al.*, 1988). Vine and Matthews (1963) and Morley and Larochelle (1964) independently recognized that the linear marine magnetic anomalies observed at sea (Mason and Raff, 1961) were caused by alternating blocks of normally and reversely magnetized volcanic rocks that form the upper part of the oceanic crust. In this case the strong remanent magnetization swamps the induced component that occurs in the presence of the present Earth's magnetic field.

Because only the intensity of the magnetic field is measured at sea, some additional assumptions are required to determine the polarity of the magnetization from magnetic anomalies. Calculation of the external magnetic field can be made by first calculating the potential ψ at an external point distance $\mathbf{r}$ from a magnetized body from the relation

$$\psi(\mathbf{r}) = -\int_V \mathbf{M} \cdot \nabla \frac{1}{(r - r_0)} \, dr^3 \quad , \tag{5.2.1}$$

where the integration is carried out over the entire body of volume V, the dimensions of which are assumed to be known. The distribution of the magnetization $\mathbf{M}$ throughout the body needs to be assumed. The total magnetic field consists of the geomagnetic field $\mathbf{H_0}$ and the "anomalous" magnetic field, $\Delta\mathbf{H}$, where

$$\mathbf{H} = \mathbf{H_0} + \Delta\mathbf{H} = \mathbf{H_0} - \nabla\psi(\mathbf{r}) \quad . \tag{5.2.2}$$

Usually the field is measured with a total-field magnetometer, so that only the field magnitude is known, and not its direction.

Some of the more popular methods for determining magnetic anomalies were introduced by Bott (1967) and these have been augmented by Schouten and McCamy (1972), Parker (1973), Parker and Huestis (1974), and Parker (1994). When both the shape of the body and the direction of magnetization are known, it is possible to solve the linear problem to determine the magnetic anomaly. More general conditions (e.g., unknown shape) often lead to the necessity of solving nonlinear equations (e.g., Parker, 1994).

A development by Bott (1967) illustrates one method for handling the linear problem. Assume the problem to be two-dimensional, with the anomaly measured at the surface $z = 0$. Let ξ be the x coordinate of a point within the magnetized body and assume that the intensity of magnetization is $M(\xi)$, a function only of ξ within the body and zero outside the body. Assume also that the magnetic body is confined between the two planes $z = z_1$ and $z = z_2$. Under these conditions the anomaly $A(x)$ measured in the x direction is

$$A(x) = \int_{-\infty}^{\infty} M(\xi) \, K\big(z_1, z_2, \beta, (x - \xi)\big) \, d\xi \quad , \tag{5.2.3}$$

where β is the angle between $\mathbf{M}$ and the external field, here for simplicity assumed to be constant. K is a kernel that is independent of M, and since z_1, z_2, and β are assumed constant, it can be rewritten simply as $k(x-\xi)$, so that

$$A(x) = \frac{1}{\sqrt{2\pi}} \int_{-\infty}^{\infty} M(\xi) k(x - \xi) d\xi = M * k \quad , \tag{5.2.4}$$

where the asterisk denotes Fourier convolution. Let $A(s)$ be the Fourier transform of $A(x)$, with similar notation for the other functions; then

$$\overline{A}(s) = \frac{1}{\sqrt{2\pi}} \int_{-\infty}^{\infty} A(x) e^{-ixs} dx \quad . \tag{5.2.5}$$

Replace $A(x)$ by a smoothed anomaly $A'(x')$ given by

$$A'(x') = \frac{1}{\sqrt{2\pi}} \int_{-\infty}^{\infty} A(x) \omega(x' - x) dx = A * \omega \quad , \tag{5.2.6}$$

where ω is an appropriate weighting (smoothing) factor. Applying the Fourier transform theorem to (5.2.4) and (5.2.6) gives

$$\overline{A}'(s) = \overline{k}(s) \overline{M}(s)$$
$$\overline{A}'(s) = \overline{\omega}(s) \overline{A}(s) \quad , \tag{5.2.7}$$

so that

$$\overline{M}(s) = \frac{\overline{\omega}(s) \overline{A}(s)}{\overline{k}(s)} \quad , \tag{5.2.8}$$

where ω must be chosen so that $M(s)$ is a Fourier transform. Inverting (5.2.8) then gives

$$M(\xi) = \frac{1}{2\pi} \int_{-\infty}^{\infty} A(x) \left(\int_{-\infty}^{\infty} e^{-i(x-\xi)s} \frac{\overline{\omega}(s)}{\overline{k}(s)} ds \right) dx \quad , \tag{5.2.9}$$

and this allows M to be calculated formally from the anomaly $A(x)$.

Applications of the method and similar methods to data have been illustrated by Bott (1967), Schouten and McCamy (1972) and Parker (1973). Parker and Huestis (1974) pointed out that the linear formulation of the problem made it suitable for the inverse theory of Backus and Gilbert (1967, 1968, 1970) (see §2.3.5). In particular, they showed how a magnetic annihilator (which included topographical variations in the magnetized material) could be obtained if certain assumptions were met. These assumptions include the requirement that no polarity changes occur as a function of depth. Further discussion of the inverse problem can be found in Parker (1994).

5.2.2 Sea-Floor Spreading

The sea-floor spreading hypothesis was first formulated by Hess (1960, 1962). The midocean ridge system, which circulates the globe, is the site of the rising limbs of mantle convection cells where hot magma comes right through to the surface, and new oceanic crust is formed as the magma cools. It was thought that the intrusion of new material forced the cooling crust to move away from the ridge symmetrically on either side. However, it is now generally believed that the crust and part of the upper mantle, referred to as the *lithosphere*, are under tension at a spreading center. Thus the lithosphere is pulled apart, allowing magma to rise to the surface, and the whole oceanic crust is part of a conveyor belt system, rising up at the midocean ridges and eventually sinking down at the oceanic trenches.

Dramatic support for the sea-floor spreading hypothesis was provided independently by Vine and Matthews (1963) and Morley and Larochelle (1964). Interpreting the linear magnetic anomalies observed at sea as representing alternating blocks of normal and reversely magnetized oceanic crust, they realized that as new crust was formed at the midocean ridge according to the sea-floor spreading hypothesis, it would cool through the Curie temperature and become magnetized parallel to the Earth's magnetic field. If the Earth's magnetic field reversed itself periodically then strips of crust of alternating polarity would be produced parallel to and, in the simplest case, distributed symmetrically about the axis of the ridge. The model is illustrated in Fig. 5.5.

The hypothesis is, therefore, that the crust near the oceanic ridges acts as a crude tape recorder of the Earth's magnetic field in the past, with the polarity and smaller changes in the field being recorded horizontally away from the spreading center. This hypothesis was one of the most important leading to the plate tectonics model (McKenzie and Parker, 1967; Morgan, 1968; Le Pichon, 1968). The model is now accepted as a first-order explanation of global tectonics and is often described as having caused a revolution in the Earth sciences (see, for example, the review by McElhinny, 1973). Magnetic anomalies have also contributed significantly to our understanding of the Earth's magnetic field, as first became apparent in the work of Heirtzler *et al.* (1968) and which will be discussed in more detail later.

The first deep (0.5 km of basement or more) hole drilled into the igneous basement (layer 2 of the oceanic crust) in the Atlantic Ocean yielded polarity changes (Johnson and Merrill, 1978). Such down-hole reversals of magnetization were not expected on a model based on injection of material along an infinitely thin spreading center. Several years earlier Matthews and Bath (1967) and

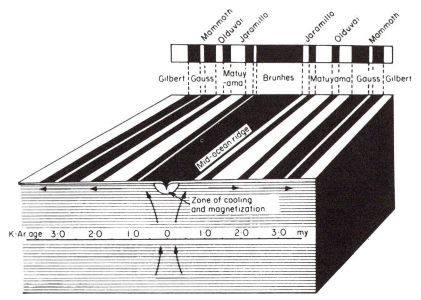

Fig. 5.5. Schematic representation of sea-floor spreading and the formation of linear magnetic anomalies due to reversals of the Earth's magnetic field. Normal polarity zones are shaded.

Harrison (1968) had suggested that the Vine–Matthews hypothesis required minor modification to account for certain features of magnetic anomalies. In particular, they suggested that the *average* magnetization changed smoothly between normal and reverse polarity anomalies, but that the detailed magnetization in any crustal section could be very complex, including the recording of reversals in the vertical direction. Other models, which predict complex magnetic vertical structures in oceanic crust (e.g., Blakely, 1976; Kidd, 1977; Johnson and Merrill, 1978; Schouten and Denham, 1979; also see reviews by Blakely and Cande, 1979; Johnson, 1979: Harrison, 1987), have subsequently been suggested. The various factors that must be considered in the construction of such models include:

(i)   the width of the magma injection zone;
(ii) the rate of spreading;
(iii) the cooling rate; and
(iv) the rate and extent of chemical alteration, particularly that associated with hydrothermal circulation.

Although no consensus yet exists as to the detailed magnetic structure of the oceanic crust, there is agreement that this structure is generally more complex than originally envisaged by Vine and Matthews. The thickness of the magnetized layer is generally greater than the 0.5-km-thick layer typically used in modeling magnetic anomalies (see reviews by Blakely and Cande, 1979;

Johnson, 1979; Harrison, 1987). Schouten and Denham (1979) have shown that complex magnetic structures can give rise to the observed magnetic anomalies. However, the core of the Vine–Matthews hypothesis remains intact, even though the magnetic structure is more complicated than originally envisaged.

5.2.3 Aspects of Magnetic Anomaly Interpretation

As suggested above, the inversion of magnetic anomaly data (intensity only) is highly nonunique. Because there is no field outside a uniformly magnetized infinite sheet (an example of a magnetic annihilator; see §2.3.5), it is immediately apparent that an infinite set of models exists for the distribution of magnetization for any anomaly pattern. It becomes necessary to use additional geological and geophysical considerations to constrain the problem (for a detailed discussion of how this is done, see Parker and Huestis, 1974; Parker, 1994).

Marine data are usually obtained at the sea surface, although they are sometimes obtained by a deep tow and even by satellites. Above the magnetic material in the sea floor the magnetic field can be derived from a scalar potential ψ that satisfies Laplace's equation (§2.2.1). For the scales involved here the sea floor may be considered flat rather than spherical, so for relatively small distances z above the sea floor it is more appropriate to use Cartesian coordinates instead of the spherical coordinates used in §2.2.1. Thus ψ must satisfy

$$\nabla^2 \psi = \frac{\partial^2 \psi}{\partial x^2} + \frac{\partial^2 \psi}{\partial y^2} + \frac{\partial^2 \psi}{\partial z^2} = 0 \quad . \tag{5.2.10}$$

If the field is being modeled over a rectangular area of length $2L$ in the horizontal x direction and $2M$ in the horizontal y direction, then a general solution to (5.2.10) is given by

$$\psi = \sum_{l=0}^{\infty} \sum_{m=0}^{\infty} \left(\cos\frac{l\pi x}{L} + \sin\frac{l\pi x}{L} \right) \left(\cos\frac{m\pi y}{L} + \sin\frac{m\pi y}{M} \right) e^{-k_{lm}z} \quad , \tag{5.2.11}$$

where

$$k_{lm} = \pi \sqrt{\left(\frac{l}{L}\right)^2 + \left(\frac{m}{M}\right)^2} = 2\pi \sqrt{\frac{1}{(\lambda_l)^2} + \frac{1}{(\lambda_m)^2}} \quad , \tag{5.2.12}$$

with λ_l and λ_m being the harmonic wavelengths in the x and y directions, respectively. This is of course just a bivariate Fourier series for any horizontal plane above the surface. The important point, though, is that the potential ψ

(and, therefore, in this instance, the field $H=-\nabla\psi$ itself) falls off exponentially with the order of the harmonic term, or the inverse of the harmonic wavelength. The situation is, therefore, similar to that discussed in §2.2.5: upward continuation is stable because shorter wavelengths are damped more rapidly, thereby smoothing the field; the greater the altitude of observation the greater the smoothing; and downward continuation (as is required with, for example, satellite observations) is unstable because short-wavelength noise is greatly amplified in the process. Thus a complete coverage as close to the source as possible would provide the most reliable data in principle (assuming equal errors in the measurements), but at substantially greater financial cost. Also one would not expect to gain much information about crustal magnetic anomalies that have wavelengths shorter than the distance from the anomaly source to the instrument. Therefore, the details of the magnetic field history are not as likely to be resolved from satellite data as they would be from anomaly data obtained at sea.

One method for reducing the noise problem discussed above was introduced by Blakely and Cox (1972), who adjusted the anomalies for different spreading rates and then stacked the resulting data obtained from several traverses across spreading centers. However, before stacking, a geocentric axial dipole field assumption (§6.1) was utilized and the anomalies were transformed to what they would be if the observation site were at the pole (referred to as reduction to the pole). This is necessary because of the way the anomalies vary with latitude. This is easily seen by considering the simple case of symmetric anomalies about a spreading ridge at the North Pole and about an E–W ridge at the equator, as shown in Fig. 5.6. Following the sea-floor spreading hypothesis the crustal blocks will be magnetized in opposite directions as the polarity changes, but only

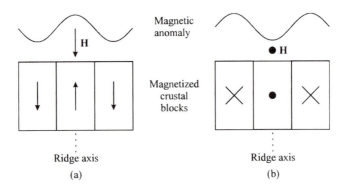

Fig. 5.6. Latitude variation of the shape of the magnetic anomaly due to the magnetized crustal blocks adjacent to the axis of a spreading ridge. The external magnetic field **H** due to an axial geocentric dipole is shown for the two cases. (a) At the North Pole the intensity of the field is increased over the ridge axis, but (b) at the equator it is decreased (dots and crosses indicate opposing horizontal fields perpendicular to the page).

the first magnetic reversal is shown. During normal polarity time the magnetic anomaly is positive at the pole (so reinforcing the field's intensity) but it is negative at the equator. Without reduction to the pole, stacking of the data from these two observations of the same event would destroy the signal, but stacking after reduction to the pole reduces the noise while retaining the signal.

It might be added that the assumption that anomalies are symmetric about a spreading axis if the magnetization is symmetric is not necessarily true and not always observed. The observed symmetry about several spreading centers played an important role in convincing geophysicists that plates were spreading about this axis, but the symmetry is partly an accident of the strike of ridges under investigation. Following the dike model of Vine and Matthews (1963), consider the magnetization of two dikes symmetric in all respects, including magnetization, that are displaced different distances from the spreading axis. The external field due to sources in the Earth's core will usually be different (in intensity and direction) above the two dikes and this will cause the anomalies to be different. Again, this represents no particular problem after reduction to the pole, when symmetric magnetic anomalies do imply symmetric magnetization and vice versa. This symmetry exists, however, only if the axial geocentric dipole field assumption is correct. Insight into this symmetry problem is provided by Fig. 5.7.

Paleointensity data may also provide valuable insight into the reversal process (see §5.3 and §5.4). For example, Valet and Meynadier (1993) have suggested that, on average, the paleointensity of the dipole field is greater immediately

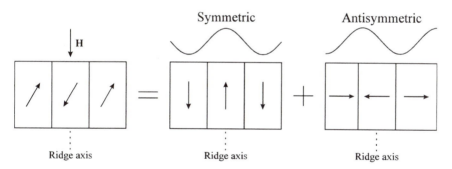

Fig. 5.7. Symmetry of magnetic anomalies about a ridge axis. A symmetric magnetization of the crustal blocks (left) is not parallel to the external field **H**, taken to be vertical. The magnetization can be divided into a component parallel (center) and perpendicular (right) to the external field. These components produce symmetric and asymmetric anomalies, respectively. Normally the magnetization associated with any given anomaly will not be parallel to the present magnetic field in the region because spreading has displaced the lithosphere from its place of origin. Thus there will normally be both a symmetric and an asymmetric component. The observed symmetry in many marine magnetic anomalies over oceanic ridges is because of the predominant north–south orientation.

following a reversal. If so, this should be manifested as a skewness in magnetic anomalies. Unfortunately, skewness of anomalies can occur from other sources (e.g., Cande and Kent, 1976). The extraction of paleointensity information from stacked magnetic anomalies seems straightforward, but in practice is very difficult and nonunique. Unlike an infinite uniformly magnetized sheet (which has no external field and so is an annihilator; see §2.3.5), a *finite* uniformly magnetized sheet has an external magnetic field with maximum intensity near the edges of the sheet. As the dimensions of this sheet increase, one must approach the infinite dimensional case of no external field. This means that even if the intensity were constant between reversals (which is certainly not the case), magnetic anomalies would exhibit a sagging effect in the middle of a polarity interval, the magnitude of which would depend on the rate of spreading and on the length of the polarity interval.

There are numerous other factors that can affect the magnitude of an anomaly. These include:

(i)   the present intensity of the magnetic field and the intensity of the paleofield in which the anomaly producing rocks formed;

(ii) alteration of the magnetization with time;

(iii) structure of the ridge when it formed and any subsequent changes in structure due to tectonic processes; and

(iv) spreading rates and the symmetry of the spreading.

In addition, spreading centers often exhibit discontinuities other than transform faults and these can have significant effects on the magnetic structure. The most common discontinuities are propagating rifts, overlapping spreading centers and saddle points (for further discussion see Macdonald *et al.*, 1988). Careful work can, however, still provide valuable information on ancient paleointensities, as seen in the study of Cande and Kent (1992a). Cande and Kent (1992b) introduced the term *cryptochron* (see Table 5.2) to describe those tiny wiggles in magnetic anomaly records that are clearly related to paleomagnetic field behavior, but may not be short polarity subchrons or microchrons since they have not been confirmed in magnetostratigraphic sections. They can be modeled either as very short subchrons (or microchrons) or more likely as being due to longer period (50-200 kyr) global changes in the intensity of the Earth's magnetic field.

5.2.4 Extension of the Polarity Time Scale to 160 Ma

There was some early skepticism that the GPTS could be extended beyond a few million years using absolute dating methods alone. If the reversal time scale were extended back beyond a few million years using the methods of §5.1.4, errors in dating would become too large to delineate reversals. A 3% error in radiometric dating of a lava flow with age near 10 Ma is larger than the mean polarity interval for the past few million years. Therefore, it is not possible to date a lava flow accurately enough to conclude that it erupted in the same polarity chron as another lava flow in a different locality. This problem was surmounted by using long sequences of lava flows on land (McDougall *et al.*, 1976a,b, 1977) and long sediment sections exhibiting several reversals on land and beneath the oceans (Kennett, 1980; Lowrie and Alvarez, 1981). One or more absolute ages are usually also required to compare a magnetostratigraphic section with a magnetic anomaly sequence.

Using these methods the GPTS for the past 10 or 20 Myr has stabilized significantly in recent times, but minor improvements continue. This is evident in Fig. 5.8, which compares the Baksi (1993) time scale for the past 18 Myr with the time scale of Cande and Kent (1995) only two years later. Only those subchrons that are recorded both in marine magnetic anomaly records and in magnetostratigraphic section have been included. These time scales involve the integration of marine magnetic anomaly data with biostratigraphy and age dating information. The difference between the two scales is minor for the past 10 Myr, but they become significant in the time interval 10-18 Ma. Thus it needs to be appreciated that published time scales for times older than 10 Ma are still in the process of adjustment as new data come to hand.

A series of papers was published in 1968 showing that the same sequence of magnetic anomalies parallel to the ridge crest is present in much of the Pacific, Atlantic, and Indian Oceans (Pitman *et al.*, 1968; Dickson *et al.*, 1968; Le Pichon and Heirtzler, 1968; Heirtzler *et al.*, 1968). The most prominent positive magnetic anomaly peaks were numbered from 1 at the midocean ridges to 32, the oldest anomalies examined. Following the lead of Vine and Matthews (1963), Heirtzler *et al.* (1968) extrapolated from the well dated polarity time scale for the past few million years (§5.1.4) by assuming uniform spreading rates in the different oceans. However, this extrapolation led to different estimates of the polarity time scale for each of the four regions involved, the South Atlantic, the North and South Pacific, and the South Indian Oceans. Various geological and geophysical criteria were used to decide that the South Atlantic magnetic profile should be selected as a standard. A magnetic reversal chronology extending back to 80 Ma was then obtained by assuming a uniform axial spreading rate at midlatitudes in the South Atlantic. This spreading rate was estimated to be 1.9 cm yr^{-1} from the known polarity sequence for the past few million years (§5.1.4). This reversal chronology was subsequently found to be consistent with ages obtained by paleontological dating of basement sediments obtained by the Deep-Sea Drilling Project (Maxwell *et al.*, 1970; LaBrecque *et al.*, 1977). Even today this South Atlantic time scale provides a reasonable first-order approximation to the reversal chronology during the Cenozoic. One comparison of paleontological ages of basal sediments with magnetic anomaly ages is shown in Fig. 5.9.

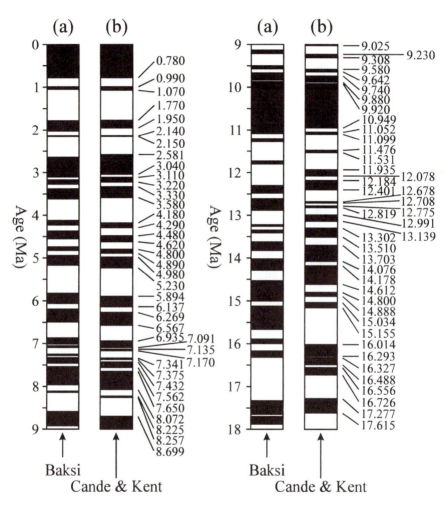

Fig. 5.8. Comparison of (a) the Baksi (1993) geomagnetic polarity time scale for the past 18 Myr with (b) the most recent time scale of Cande and Kent (1995), including only those subchrons that are recorded both in marine magnetic anomaly records and in magnetostratigraphic sections.

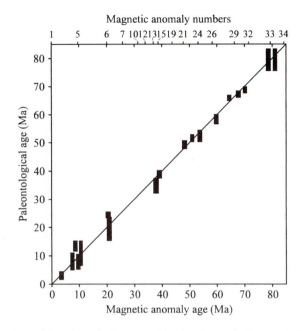

Fig. 5.9 Comparison of the paleontological ages of basal sediments in deep-sea drill holes with the basement ages predicted from the magnetic anomalies. The 45° line is that expected for perfect agreement. After Lowrie and Alvarez, (1981).

Reversal chronologies have been updated and improved over the past decade since the widely cited chronology of Cox (1982) for the past 170 Myr. This chronology did not incorporate the recent proposed changes based on $^{40}Ar/^{39}Ar$ dating (§5.1.4). Kent and Gradstein (1986) presented an integrated geomagnetic and geologic time scale for the past 160 Myr based on the integration of marine magnetic anomaly data with biostratigraphic, magnetostratigraphic, and radiometric data. Later significant changes in the Late Cretaceous and Cenozoic reversal chronology were proposed by Cande and Kent (1992a) from a major analysis of magnetic anomaly profiles from the world's ocean basins, the first since that of Heirtzler *et al.* (1968). A revised calibration of this time scale has been made by Cande and Kent (1995) that is consistent with the reversal chronology for the entire Mesozoic proposed by Gradstein *et al.* (1994). This Mesozoic time scale is an integrated geomagnetic polarity and stratigraphic time scale whose framework involves observed ties between radiometric dates, biozones, and stage boundaries, and between biozones and magnetic reversals observed from marine magnetic anomalies and in sediments. Unfortunately the Gradstein *et al.* (1994) time scale does not provide listings of chron and subchron boundaries. Therefore, in Fig. 5.10 we have combined the chronologies

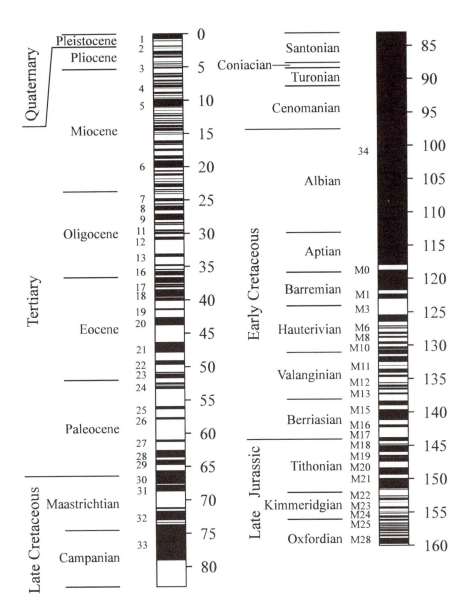

Fig. 5.10. Geomagnetic polarity time scale with magnetic anomaly numbers for the past 160 Myr using the combined scales of Kent and Gradstein (1986) and Cande and Kent (1995).

of Kent and Gradstein (1986) and Cande and Kent (1995) to give a complete reversal time scale for the past 160 Myr. However, we have omitted any suggestion of the presence of two zones of mixed polarities between 100 and 107 Ma (Gradstein *et al.*, 1994), since these are not seen in marine magnetic anomalies, and in any case their existence in magnetostratigraphic data can be disputed (Opdyke and Channell, 1996).

For times older than 5 Ma, the normal polarity chrons are designated by numbers that are correlated with marine magnetic anomalies. For example, polarity chron C26r represents the time of reverse polarity between the normal polarity chrons correlated with magnetic anomalies 26 and 27. Polarity chrons for the pre-Aptian sequences are generally described by the designation M0 through M29 as assigned to magnetic anomalies in order of increasing age, where the prefix M stands for Mesozoic. However, these M sequence anomalies were mainly (but not exclusively) assigned to reverse polarity anomalies, which results in a somewhat confusing situation. The identification and correlation of low-amplitude magnetic anomalies older than M29 from aeromagnetic profiles in the western Pacific has enabled the sequence to be extended to anomaly M38 (Handschumaker *et al.*, 1988). For consistency, Opdyke and Channell (1996) have used the prefix C (e.g., CM29) to distinguish polarity chrons as observed in magnetostratigraphy from the magnetic anomaly numbers themselves.

The nomenclature has quite naturally evolved to accommodate additional polarity chrons (LaBrecque *et al.*, 1977; Harland *et al.*, 1982, 1990; Cande and Kent, 1992b). Following Cande and Kent (1992b), the longest intervals of predominantly one polarity are referred to by the corresponding anomaly number followed by the suffix n for normal polarity, or r for the preceding reverse polarity interval. When these chrons are subdivided into shorter polarity intervals, they are referred to as subchrons and are identified by appending, from youngest to oldest, a 0.1, 0.2 etc, to the polarity chron identifier, and adding an n or r as appropriate. For example, the three normal polarity intervals that make up anomaly 17 (chron C17n) are called subchrons C17n.1n, C17n.2n and C17n.3n. Similarly, the reverse interval preceding (older than) C17n.1n is referred to as subchron C17n.1r. This nomenclature enables every chron and subchron to be identified uniquely. For more precise correlations, the fractional position within a chron or subchron can be identified by the equivalent decimal number appended within parentheses. For example, the younger end of chron C29n is C29n(0.0), and a level 3/10ths from this younger end is designated C29n(0.3).

Although certain statistically determined properties of the reversal chronology may have changed in the past decade, there are certain robust features that have always been present. They all show a decreasing rate of reversals entering into the Cretaceous and an increasing rate of reversals following the Cretaceous.

5.3 Analysis of Reversal Sequences

5.3.1 Independence of Polarity Intervals

Our current lack of understanding of the physical processes producing reversals, and the apparent randomness of the lengths of polarity intervals, leads us to adopt statistical characterizations and to undertake statistical analyses in an attempt to provide robust constraints on our understanding of the underlying process. Paleomagnetic directional data indicate that the time required for a polarity transition is typically about 4 to 5 kyr (e.g., Merrill and McElhinny, 1983; Hoffman, 1989; Laj, 1989). In geological terms this is a very short time, and is short relative to the typical length of a polarity interval. Thus within the GPTS it is reasonable, to a first approximation, to think of a reversal as a rapid (almost instantaneous) event. This then naturally leads to the concept of the reversal process being a general renewal process.

Suppose t is the lapsed time since the last reversal and $p(t)dt$ is the probability that a reversal will occur between t and $(t+dt)$. In a *general* renewal process $p(t)$ can increase or decrease with an increase in t, but the probability does not depend on the length of prior polarity intervals. If reversals are generated by a general renewal process, the lengths (duration) of polarity intervals will be independent. Such independence was first claimed without convincing proof (Cox, 1968, 1969; Nagata, 1969) and subsequently challenged (Naidu, 1974). One reason to suspect polarity intervals are not entirely independent is that, although relatively rapid, reversals do not occur instantaneously and so the magnetic field must have some memory. If the fluid motions powering the Earth's magnetic field were to cease, the magnetic field would decay away with a characteristic decay time of about 10^4 yr (§9.1). The fluid motions operating the geodynamo appear moderately complex and therefore this time is probably a reasonable estimate of the memory time of the core processes. This is more than an order of magnitude shorter than the mean polarity time during the Cenozoic, suggesting that any departures from independence might be small.

A second reason for suspecting the independence assumption is that reversal sequences are not stationary, as illustrated by Fig. 5.11. The mean polarity interval increases as one goes back in time during the Cenozoic. This means there has to be *some* statistical correlation between the lengths of successive polarity intervals. However, the change is slow relative to the mean polarity interval in the Cenozoic. Again this suggests that departures from independence might be small. The significance of the above effects can be tested statistically as described by Phillips *et al.* (1975) and Phillips and Cox (1976). Let t_i be the length of time between two successive reversals of the Earth's magnetic field.

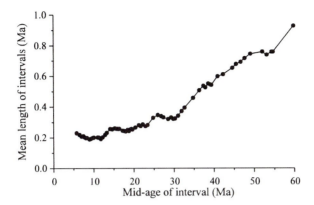

Fig. 5.11. Nonstationarity of the geomagnetic reversal rate. The mean length of polarity intervals appears to have changed continuously since 85 Ma ago. Each point is for 25 intervals of each polarity. After McFadden and Merrill (1984); updated using the time scale of Cande and Kent (1995).

Unbiased estimates $\hat{p}_j$ of the correlation coefficients can be obtained from a sequence of polarity intervals with lengths t_i, $i = 1,....,n$,

$$\hat{p}_j = \frac{\hat{C}_j}{\sqrt{C'_{0,j}C''_{0,j}}} \,, \qquad (5.3.1)$$

where

$$\hat{C}_j = \frac{1}{n-j}\sum_{i=1}^{n-j}(t_i - t'_j)(t_{i+j} - t''_j) \quad t'_j = \frac{1}{n-j}\sum_{i=1}^{n-j}t_i \quad t''_j = \frac{1}{n-j}\sum_{i=1}^{n-j}t_{i+j}$$

$$\hat{C}'_{0,j} = \frac{1}{n-j}\sum_{i=1}^{n-j}(t_i - t'_j)^2 \qquad \hat{C}''_{0,j} = \frac{1}{n-j}\sum_{i=1}^{n-j}(t_{i+j} - t''_j)^2 \,. \qquad (5.3.2)$$

If the observed polarity intervals are independent, then the expected values for the correlation coefficients are all zero for j not equal to zero and one for j equal to zero. This was found to be the case within 95% confidence limits. Spectral analysis techniques were later used by Phillips and Cox (1976) to show that no significant periodic components in the reversal frequency appeared to be present in the reversal record. Thus, to a very good first-order approximation, reversals of the Earth's magnetic field can be regarded as a purely stochastic process in which the lengths of polarity intervals can be treated as being independent. This is in marked contrast to the Sun, in which reversals appear to occur approximately every 11 yr (§10.2).

5.3.2 Statistical Analysis of Reversal Sequences

Statistical tools for analyzing the reversal sequence in terms of a renewal process are given by McFadden (1984a), McFadden and Merrill (1986), and McFadden and Merrill (1993). Let x be the time since the last reversal and recognize that reversals are, from a process perspective, rare events, which simply means that the probability of two events occurring in any very short interval $x \rightarrow (x+\delta x)$ is negligible. Define P$\{x\}$ as

$$P(x) = \Pr\{\text{no event in } 0 \rightarrow x\} \quad , \qquad (5.3.3)$$

where Pr is short for probability. Further, define f(x) through

$$f(x)\delta x = \Pr\{\text{event in } x \rightarrow (x + \delta x) \,|\, \text{no event in } 0 \rightarrow x\} \quad , \qquad (5.3.4)$$

where the vertical bar means "conditional upon the event to its right." Hence the function f(x) describes how the instantaneous probability for the occurrence of another reversal varies with the time since the last reversal. From these definitions it follows that

$$(1 - f(x)\delta x) = \Pr\{\text{no event in } x \rightarrow (x+\delta x) \,|\, \text{no event in } 0 \rightarrow x\} \quad , \qquad (5.3.5)$$

so that

$$P(x + \delta x) = P(x)(1 - f(x)\delta x) \quad , \qquad (5.3.6)$$

from which, on taking the limit as $\delta x \rightarrow 0$, it follows that

$$\frac{dP(x)}{dx} = -P(x)f(x) \quad . \qquad (5.3.7)$$

If p(x) is the probability density of interval lengths x, then

$$p(x)\delta x = \Pr\{\text{interval between events having length between } x \text{ and } (x+\delta x)\}$$
$$= \Pr\{\text{no event in } 0 \rightarrow x\} \cdot \Pr\{\text{event in } x \rightarrow (x+\delta x) \,|\, \text{no event in } 0 \rightarrow x\}$$
$$= P(x)f(x)\delta x \quad ,$$

so that

$$p(x) = P(x)f(x) \quad . \qquad (5.3.8)$$

Equations (5.3.7) and (5.3.8) give

$$\frac{dP(x)}{dx} = -p(x) \qquad (5.3.9)$$

or

$$P(x) = 1 - \int_0^x p(t)dt \quad , \tag{5.3.10}$$

so that

$$f(x) = \frac{p(x)}{P(x)} = \frac{p(x)}{1 - \int_0^x p(t)dt} \quad . \tag{5.3.11}$$

Thus, given f(x), the density p(x) can be obtained from (5.3.7) and (5.3.8), or given p(x), f(x) can be obtained from (5.3.11).

Poisson Process

A Poisson process is one in which the probability of a future event does not depend on the time since the last event, so $f(x) = \lambda$, a constant. Equations (5.3.7) and (5.3.8) and the boundary condition (from the definition) that P(0)=0, give

$$\frac{dP(x)}{dx} = -\lambda\, P(x) ; \quad P(x) = e^{-\lambda x} ; \quad p(x) = \lambda\, e^{-\lambda x} \quad . \tag{5.3.12}$$

Using $<x>$ to denote the expectation, or mean, of x and Var(x) to denote the variance of x, we have

$$< x >= \mu = \frac{1}{\lambda} \tag{5.3.13}$$

$$\mathrm{Var}(x) = \mu^2 \quad ,$$

showing that λ is just the rate of the process.

Gamma Process

For a gamma process the density p(x) of interval lengths is commonly written in one of the forms

$$p(x)dx = \frac{1}{\Gamma(k)}(k\Lambda)^\kappa x^{(k-1)} e^{-k\Lambda x}$$

$$= \frac{1}{\Gamma(k)}\lambda^\kappa x^{(k-1)} e^{-\lambda x} \quad ; \lambda = k\Lambda \quad , \tag{5.3.14}$$

where $\Gamma(k)$ is the gamma function of k. This gives

$$< x >= \mu = \frac{k}{\lambda} = \frac{1}{\Lambda}$$

$$\mathrm{Var}(x) = \frac{\mu^2}{k} = \frac{1}{k\Lambda^2} = \frac{k}{\lambda^2} \quad , \qquad (5.3.15)$$

Thus the Poisson distribution is quite clearly just a gamma distribution with $k=1$.

Figure 5.12a shows $p(x)/\Lambda$ as a function of $(x\Lambda) = (x/\mu)$ for $k = 1$ (Poisson), 2, 3, and 4, showing clearly how the variance decreases as k increases. For the function $f(x)$ it is more convenient to scale against $(x\lambda) = (kx/\mu)$, so Fig. 5.12b shows $f(x)/\lambda$ (effectively the normalized reversal rate) as a function of $(x\lambda)$. It is a simple matter to show (McFadden and Merrill, 1986) that as $x\rightarrow\infty$, $f(x)\rightarrow\lambda$.

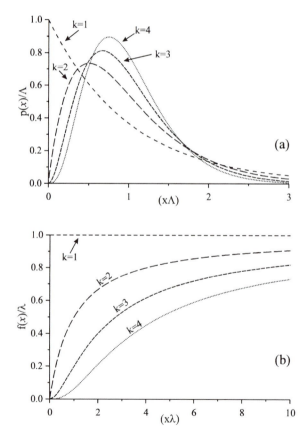

Fig. 5.12. Scaled probability density, $p(x)/\Lambda$, and normalized reversal rate, $f(x)/\lambda$, as a function of the relevant scaled time since the last reversal for gamma distributions with $k = 1$ (Poisson), 2, 3, and 4.

The plots of f(x) are particularly illuminating. For a Poisson process the occurrence of an event (reversal) has no impact at all on the probability of a future event; the system has no memory at all. For $k>1$, the probability for a future reversal drops to zero as soon as a reversal occurs, and then gradually rebuilds to its undisturbed value. Thus the process has a memory of the previous event, and for $k>1$ this memory causes inhibition of future events by depressing their probability of occurrence. Conversely, $k<1$ produces unstable encouragement ($p(0)=\infty$).

The simple fact that a reversal requires a finite amount of time ($\approx$5 kyr for the directional change) means the process cannot be truly Poisson. However, by using an extreme model for the function f(x), in which $f(x)=\lambda$ everywhere except for 5000 yr after a reversal when $f(x)=0$, McFadden and Merrill (1993) were able to show that within the GPTS sequence the impact is not resolvable.

McFadden (1984a) shows that estimation of μ is a simple matter. Given n observations from a gamma distribution the estimate

$$\hat{\mu} = \frac{1}{n}\sum_{i=1}^{n} x_i \qquad (5.3.16)$$

is a minimum variance unbiased estimator for μ with variance given by

$$\mathrm{Var}(\hat{\mu}) = \frac{\mu^2}{nk} \ . \qquad (5.3.17)$$

Conversely, effective estimation of k is not so simple. McFadden (1984a) shows that the most appropriate estimate for k is k_m, the solution of

$$\ln(k_m) - \Psi(k_m) = \frac{n}{n-1}\ln(\hat{\mu}) - \frac{1}{n-1}\sum_{i=1}^{n}\ln(x_i)$$

$$\Psi(k_m) = \frac{d}{dk}[\ln(\Gamma(k))] \quad , \qquad (5.3.18)$$

where Ψ is known as the digamma function. Furthermore, the statistic

$$\frac{\ln(k_m / k)}{\sqrt{\mathrm{Var}\{\ln(k_m)\}}}$$

$$\mathrm{Var}\{\ln(k_m)\} = \frac{1}{nk_m^2\left\{\dfrac{d\Psi(k_m)}{dk} - \dfrac{1}{k_m}\right\}} \qquad (5.3.19)$$

is, to a good approximation, distributed as a standardized normal variate. From this it is apparent that it is often more appropriate to consider $\ln(k)$ rather than k itself.

In practice, analysis of the reversal sequence is plagued by three serious problems:

(i)   the process is nonstationary (Fig. 5.11 shows clearly that there is a long-term variation in μ);

(ii) it is only for the past 120 million years or so that we have a reasonably reliable GPTS (see McFadden *et al.*, 1997b), meaning that we actually have a very short record in terms of numbers of intervals; and

(iii) even with the most recent time scales we are seeing a filtered version of the true time scale, in that it is likely some short intervals have still been missed (and perhaps some included erroneously) and the precise ages of several reversals will still need adjustment.

Nonstationarity of the Process

Nonstationarity in the process is of central geophysical interest (see §5.3.5), but causes substantial difficulties in analysis of the statistical properties of the GPTS. It certainly precludes taking all the intervals of the past 83 Ma and considering them as a random sample from a common distribution. Analysis then requires one of four basic approaches.

(i)   Incorporate the structure of the nonstationarity into the parameters of the distribution, as done by McFadden (1984a) and McFadden and Merrill (1984).

(ii) Estimate the structure of the nonstationarity by fitting a smooth curve to point estimates of the parameter (e.g., fitting a smooth curve to Fig. 5.11), remove this trend from the data, and analyze the resulting interval lengths as if they were observations from a common distribution.

(iii) Use a fixed-length sliding window to obtain separate point estimates of the parameters over intervals in which the parameters do not vary a lot.

(iv) Use a sliding window with a fixed number of intervals (as done to generate Fig. 5.11).

Removal of a trend from the observations has the problem that it changes the relative structure of the interval lengths, so errors in the trend will be reflected in incorrect structure in the resulting sequence. For example, an overfitted trend will follow the changes in interval length too closely and so reduce the variance in interval lengths too much. Equation (5.3.15) shows that this will lead to an estimate of k that is too large. Thus, by overfitting a trend, it is easy to make a nonstationary Poisson process appear more like a gamma process ($k>1$).

Use of a fixed-length sliding window initially appears to be an obvious choice. However, because such a window does not cover a fixed number of intervals, the estimates will be based on a smaller number of intervals when the reversal rate is low. Equation (5.3.17) shows that both the fractional and the absolute error in estimation will increase as μ increases. Thus, for example, estimates for the separate reverse and normal sequences can seem very different when the reversal rate is low, even if the two sequences are drawn from common distributions. Without extremely careful statistical analysis this can lead to erroneous conclusions, such as a perceived polarity bias (McFadden *et al.*, 1997a). Another unfortunate aspect is that a fixed-length window tends to create a false pseudoperiodicity that depends on the rate of change of a nonstationary μ, because one is forced to count the number of events within the window rather than considering a whole number of intervals. Consider a set of events that uniformly get slightly closer together. If a set of complete intervals fit exactly into the window then the estimated mean will be correct. As the window slides along the events there will gradually be more and more of an extra interval included in the window, but the number of events does not change so the estimate gradually falls behind the trend, until eventually an extra event is included, when the estimate ratchets back up to the correct value. This ratcheting effect can be greatly amplified when the interval lengths are random but their mean length μ is nonstationary, as is the case with the reversal process.

A sliding window with a fixed number of intervals produces estimates with a constant fractional error as μ changes, and has less tendency to produce false periodicities than a constant-length window. In common with a fixed-length window, it does have the problem that a nonstationary parameter will vary within the window. Sliding window techniques also have the problem that they create a visual image of providing far more information than is actually available. For example, despite the apparent detailed information in Fig. 5.11, there are in fact only three (nearly four) independent estimates. Thus, when testing null hypotheses (such as "the normal and reverse sequences have a common value of μ" or "the data do not require a periodicity"), great care is needed to ensure that these hypotheses are tested appropriately (e.g., McFadden, 1984a; McFadden and Merrill, 1984).

Shortness of Record

In the most recent time scale (Cande and Kent, 1995) there are only 184 polarity intervals documented for the past 83 Myr. Combined with the problems of nonstationarity, this represents a small sample, particularly if one is attempting to compare the normal and reverse polarity sequences (§5.3.4), which have only 92 intervals. This combination of nonstationarity and a short record also causes severe problems in interpretation of structure (e.g., periodicities) in the sequence (§5.3.5).

Filtering of the Record

As discussed in §5.2.3, it is not a trivial matter to invert the observations to a reliable GPTS. As shown in Fig. 5.8, even the sequence for the past 18 Myr has evolved in just the past 2 yr. The major problems have been accurate dating of the individual events and reliable recognition of the shorter intervals. If a short polarity interval is missed, then this represents a gross error in the sequence. Consider a short sequence NRN in which the reverse polarity interval is very short. If the short R interval is not identified then the NRN sequence is concatenated into a single long N interval. This means that the short R interval is not included in the reverse sequence, and one very long N interval incorrectly appears in the normal sequence instead of two shorter intervals. McFadden (1984a) shows that if n independent observations x_i (i.e., lengths of intervals) from a Poisson distribution with rate λ are concatenated into a single interval, of length z, then the probability $p(z)dz$ that the length of this interval (i.e., the sum of the n lengths x_i) will lie between z and $(z+dz)$ is given by:

$$p(z)dz = \frac{1}{(n-1)!}\lambda^n z^{(n-1)} e^{-\lambda z} dz$$

$$= \frac{1}{\Gamma(n)}\lambda^n z^{(n-1)} e^{-\lambda z} dz \qquad (5.3.20)$$

$$z = \sum_{i=1}^{n} x_i \quad ,$$

which is just a gamma distribution with $k=n$. Thus if short intervals in one polarity are missed, this tends to make the intervals in the sequence of the *opposite* polarity appear as if they are drawn from a gamma distribution with relatively large values of k (when short intervals are missed, at least three real intervals are concatenated into a single apparent interval).

McFadden and Merrill (1984) analyzed the NLC time scale (Ness *et al.*, 1980) and, on the basis of the sensitivity of the k_m to the inclusion of short intervals, concluded that some short intervals had been missed. Similarly, McFadden *et al.* (1987) showed that the LKC time scale (LaBrecque *et al.*, 1977) was extremely sensitive (see Figs 5.13a and 5.13b) to the inclusion of a single short interval into the sequence of 198 intervals.

Contrary to previous conclusions (e.g., Phillips, 1977), McFadden and Merrill (1984) and McFadden *et al.* (1987) concluded that the perceived differences in k for the normal and reverse sequences were a consequence of errors in the available time scales, and that they did not represent genuine differences between the two polarity states. This conclusion was based on an improved understanding of the consequences of errors in the time scale, a recognition that

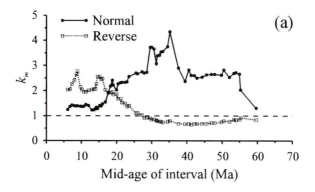

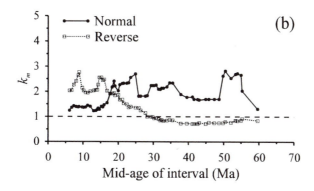

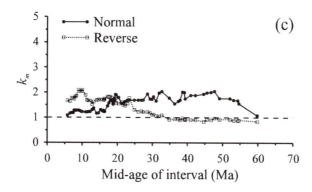

Fig. 5.13. Sensitivity of the estimate k_m to errors in the time scale. (a) The LKC time scale (LaBrecque *et al.*, 1977). (b) The LKC time scale with a single interval added at 34.00 to 34.01 Ma (McFadden *et al.*, 1987). (c) The most recent time scale (Cande and Kent, 1995), showing the improvement in the statistical properties with improved reliability of the time scale. After McFadden and Merrill (1984) and McFadden *et al.* (1997b).

different values of k was inconsistent with other statistical properties of the sequences, and careful numerical analysis of the confidence limits on $\ln(k_m)$. Furthermore, asymmetry in k would be inconsistent with one of the few truly robust results from dynamo theory that there should be no differences in the statistical properties of the normal and reverse polarity states (§8.6.2). The validity of this view has been borne out by recent time scales, as shown in Fig. 5.13c. This improvement in the statistical properties of the GPTS implies a large improvement in the reliability of the time scale, and, as noted by McFadden *et al.* (1997b), the most recent time scale will probably require only minor modifications in the future.

5.3.3 Superchrons

As indicated in Table 5.2, superchrons refer to intervals of near constant polarity in the range 10^7–10^8 yr. Two superchrons were suggested before the reversal chronology was well developed: the Permo-Carboniferous (*Kiaman*) Reverse Superchron (Irving and Parry, 1963) and the *Cretaceous* Normal Superchron (Helsley and Steiner, 1969). The Cretaceous Superchron, which extends from approximately 118 to 83 Ma (Cande and Kent, 1995), is the best documented of these two, since it is evident both in marine magnetic anomalies and in magnetostratigraphic data. The Kiaman Superchron appears to be longer, extending approximately from 312 to 262 Ma (Opdyke and Channell, 1996). A substantial body of magnetostratigraphic data supports its existence, but it is too old to be seen in any marine magnetic anomaly record. Figure 5.14 shows a histogram of the lengths of all polarity intervals from 330 Ma to the present as compiled by Opdyke and Channell (1996). The majority of polarity intervals lie in the time range 0.1 to 1.0 Myr and there are very few with duration longer than 2 Myr. It has been proposed that each of the two superchrons contains a short interval of the opposite polarity (see Opdyke and Channell, 1996), which would mean that each of the superchrons would then be made up of two exceptionally long intervals of the same polarity separated by a very short interval. We do not yet feel that the evidence for these short intervals is convincing in either case. The lengths of the Cretaceous and Kiaman Superchrons shown in Fig. 5.14 are their total lengths; they are clearly outliers of the general distribution with or without inclusion of the short interval of opposite polarity. Analysis of the sequence shows that the Cretaceous Superchron cannot be accommodated in the sequence, but implies a fundamental transition in the dynamo process (§5.3.5).

The concept of "polarity bias" was introduced by McElhinny (1971) and Irving and Pullaiah (1976) when knowledge of the polarity time scale was poor, analysis techniques for the sequences were poorly developed, and a basic constraint from dynamo theory (that the statistical properties of the normal and reverse polarities should be the same; see §8.6.2) was not appreciated. The

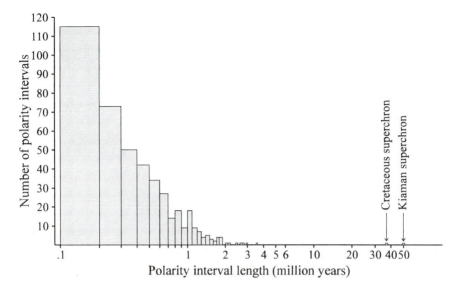

Fig. 5.14. Histogram of polarity intervals for the past 330 Myr. After Opdyke and Channell (1996).

hypothesis was that one or other of the polarities dominated for times of order 10^7 to 10^8 yr, and arose because of the then perceived long-term differences in the time spent in one polarity or the other. Indeed, Cox (1982) formalized this into polarity bias superchrons. As shown by McFadden *et al.* (1997a), and as outlined in §5.3.4, there is no substantiating evidence for polarity bias over the time for which we now have a reasonable polarity time scale. There is no reason to suspect that polarity bias might have existed during times when we do not have a reasonable polarity time scale, and indeed there are good theoretical reasons for expecting that there would not have been any bias.

Johnson *et al.* (1995) used data from the global paleomagnetic database (McElhinny and Lock, 1991, 1993a,b) and interpreted indirect statistical arguments as showing that there was another superchron during the Ordovician. This conclusion is contrary to the best published magnetostatigraphic data, which do not show evidence for a superchron at that time (e.g., Ogg, 1995; Opdyke and Channell, 1996). However, the two known superchrons give reason for expecting the existence of others, and the work of Johnson *et al.* (1995) highlights the need for more high-quality magnetostratigraphy.

5.3.4 Stability of Polarity States

McFadden (1984a) showed that the parameter k provides little information regarding the relative stabilities of the normal and reverse polarity states – the apparent rate of reversals is a far more indicative parameter. This conclusion matches intuition, for if the field spends (on average) equal times in each of the two polarity states, how can it be claimed that the stability of one state is less than that of the other? McFadden and Merrill (1984), using the distribution provided by McFadden (1984a) for the fraction of time spent in one polarity, showed that the null hypothesis of a common reversal rate, λ, could not be rejected for the past 170 million years. The existence of the 35-million-year Cretaceous Superchron might seem to contradict this but, as discussed in §5.3.7, it would appear that the reversal process simply ceased during the superchron, so λ was zero for both the normal and the reverse polarity sequences.

This conclusion of a common stability contradicted earlier conclusions of Phillips (1977) and Lowrie and Kent (1983), based on earlier time scales and inappropriate hypothesis testing techniques for the unstable estimates k_m, that the normal and reverse polarity states had significantly different stabilities. The conclusion by McFadden and Merrill (1984) of common reversal rates and stabilities for the two polarities was reaffirmed by McFadden *et al.* (1987), and confirmed by Merrill and McFadden (1994) for the Cande and Kent (1992a) time scale and by McFadden *et al.* (1997b) using the Cande and Kent (1995) time scale. Figure 5.13c shows that the estimates k_m have stabilized significantly in the most recent time scales, and certainly give no cause for suggestions of different values of k for the two polarities.

The fact that the normal and reverse polarity sequences appear to have been drawn from the same distribution is in accord with theory (§8.6.2) and provides a powerful constraint on magnetostratigraphy. Reverse directions in sediments during the Brunhes Chron are not uncommon. However, Verosub and Banerjee (1977) note that this may not reflect true magnetic field behavior, and so many paleomagnetists do not accept the existence of a subchron unless it also appears in igneous rocks. This has led to extensive searches for reverse magnetization in igneous rocks of Brunhes age, as illustrated by the work of Champion *et al.* (1988) who carried out work on Brunhes-age lavas. They have suggested the existence of several subchrons during the Brunhes Chron, which has fundamental implications (Merrill and McFadden, 1994). The reason is that the reverse polarity intervals would have been remarkably short compared with the preceding and succeeding normal polarity intervals. This would seem to imply that, at least for the past 780 kyr, the reverse polarity state has been substantially less stable than the normal polarity state, yet the evidence is strong that the two polarity states have shared a common stability right up to the beginning of the Brunhes. It would require a change in the boundary conditions for this to occur, but the boundary conditions are controlled by characteristic times for the mantle (typically about 10^8 yr). Consequently, Merrill and McFadden (1994) conclude that the proposed subchrons are likely to be excursions or data artifacts and should not at this stage be included in reversal chronologies.

5.3.5 Nonstationarity and Inhibition in the Reversal Record

The variation in the reversal rate, as determined by analysis of the reversal chronology of Cande and Kent (1995) for the interval 0–118 Ma and of Kent and Gradstein (1986) for the interval 118–160 Ma, is shown in Fig. 5.15. As already noted, there is clear nonstationarity in this record. Although not statistically significant, McFadden and Merrill (1984) noted that the reversal rate appears to have peaked around 12 Ma, a feature that has become more pronounced in more recent time scales, and is obvious in Fig. 5.15. The Cretaceous Superchron is very obvious in this figure. The Kiaman Reverse Superchron occurs approximately 200 m.y. prior to the Cretaceous Superchron (a dominant feature in Fig. 5.15), which might suggest a characteristic time of approximately 200 m.y. Such a characteristic time is essentially the same as that associated with mantle processes (Chapter 7), leading to suggestions that nonstationarities in the reversal chronology record are associated with changes in the core–mantle boundary conditions (e.g., Jones, 1977; McFadden and Merrill, 1984, 1993; and Courtillot and Besse, 1987). These and other interpretations are discussed in Chapter 11.

The nonstationarity in the reversal chronology record, but not its details, is so well established that it should no longer be doubted. The chronology is rich with information but, because of the problems outlined in §5.3.2 (and as illustrated by the discussion on the sensitivity of the estimates k_m), great care is needed to avoid conclusions that are not in fact supported by the data. Thus one should

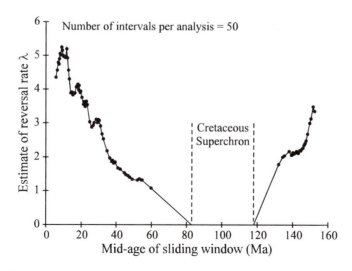

Fig. 5.15. Estimated reversal rate for the geodynamo. Constructed from the time scales of Kent and Gradstein (1986) and Cande and Kent (1995), following the methods of McFadden (1984a).

seek conclusions that are robust to changes in the GPTS as our knowledge of the chronology improves.

Having established that the normal and reverse polarity sequences were drawn from the same distribution, it is sensible to consider the normal and reverse sequences as a single sequence of intervals and examine the statistics of that sequence. Figure 5.16 shows the estimates k_m for the Cenozoic using the Kent and Gradstein (1986) time scale. k_m is presently greater than 1 and generally tends to decrease back in time toward the Poisson value of 1. This broad pattern, but not the magnitude or details, is observed for all Cenozoic chronologies published during the past two decades. As discussed in §5.3.2, and as is made clear in Fig. 5.12, $k>1$ implies inhibition of a future reversal by the most recent reversal, and so it appears somewhat contradictory that k appears to increase slightly as the reversal rate *increases*. McFadden and Merrill (1986) noted that this meant the Cretaceous Superchron could not be a consequence of inhibition and that the triggering of reversals is to some extent decoupled from the geodynamo process itself. McFadden and Merrill (1984) interpreted the increase in k_m with increasing λ as an indication of very short polarity intervals that had not been identified in the GPTS (Eq. (5.3.20)). Despite the fact that this remains a viable interpretation in its own right (McFadden *et al.*, 1997b), subsequent

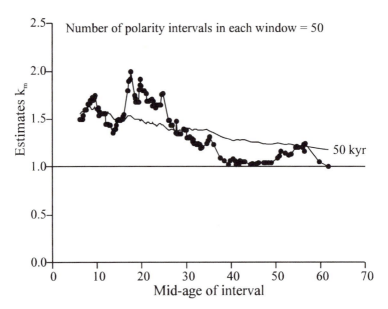

Fig. 5.16. Estimates k_m for the combined normal and reverse polarity sequences from the Kent and Gradstein (1986) time scale together with the predicted expectation value for k from a model for inhibition in which at the start of a reversal the probability for a future reversal drops to zero for 5 kyr and then recovers exponentially for a further 45 kyr. After McFadden and Merrill (1993).

polarity time scales have not changed the situation, and so McFadden and Merrill (1993) sought an alternative interpretation.

If there is a constant amount of time $\mathfrak{I}$ (i.e., independent of the reversal rate) after a reversal during which the probability for a future reversal is depressed, then as λ increases $\mathfrak{I}$ becomes a greater fraction of the mean interval length. From Fig. 5.12b it is apparent that k will increase. Thus the change in inhibition is driven by the change in λ, rather than a change in λ being driven by a change in inhibition. As discussed in §5.3.2, a time of $\mathfrak{I}=5000$ yr (the time that seems to be needed for the directional change in a reversal) is insufficient to account for the effect. However, McFadden and Merrill (1993) show that a reasonable fit is obtained if the probability for a future reversal is set to zero for a time $\mathfrak{I}=28$ kyr from the start of a reversal.

The simple on–off model of a constant probability for a future reversal except for a time $\mathfrak{I}$ after the start of reversal during which the probability is set to zero is extreme. Thus the time of 28 kyr is a minimum. McFadden and Merrill (1993) developed a general method for modeling the expected value of k given a function for the decrease and recovery of the probability for a future reversal after the start of a reversal. Their preferred model is that following the start of a reversal the probability drops to zero for the 5 kyr needed for the directional change and then recovers exponentially over a further 45 kyr, so that the probability is depressed for a total of about 50 kyr. The expected k (about which the estimates k_m will vary) for this model is also shown in Fig. 5.16. McFadden and Merrill suggest that this recovery time might be related to subsequent diffusion of the magnetic field through the core.

Analysis of the Cretaceous and Cenozoic reversal chronology reveals an interesting diagnostic about the Cretaceous Superchron (Merrill and McFadden, 1994) when an attempt is made to consider the superchron as part of the reversal sequence. This result is shown in Fig. 5.17. The value for k_m drops well below 1, which should be associated with encouragement but is patently not the case for the superchron. The value of $\ln(k_m)<-0.5$ is the consequence (Eq. (5.3.17)) of the very large variance of the intervals included in the window, caused by having the analysis window over a discontinuity in the process. That is, the analysis was inappropriate because it included intervals from different types of processes. Similar results are obtained if the Cretaceous Superchron is included within the reversal chronology data that precedes it (Merrill and McFadden, 1994). The simplest and most consistent interpretation is that the Cretaceous Superchron was caused by a cessation of the reversal process. As noted by McFadden and Merrill (1995a), the geodynamo appears to have two basic states: a reversing state and a nonreversing state. McFadden and Merrill (1995a) interpret this as requiring a fundamental transition in the geodynamo at the beginning and end of the superchron. This is discussed further in §11.3.2.

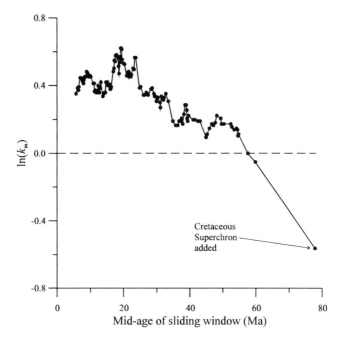

Fig. 5.17 Estimates $\ln(k_m)$ for the combined normal and reverse polarity sequence for the Late Cretaceous and Cenozoic using the Cande and Kent (1992a) time scale. After McFadden and Merrill (1995a).

The question arises naturally as to whether the fine structure in such curves as shown in Fig. 5.11 is meaningful. For example, several studies (Negi and Tiwari, 1983; Mazaud *et al.*, 1983; Mazaud and Laj, 1991; Marzocchi and Mulargia, 1990; Rampino and Caldeira, 1993; Raup, 1985; Stothers, 1986) have suggested either a 15- or a 30-million-year periodicity in the post-Cretaceous Superchron reversal chronology record. McFadden and Merrill (1984) had not observed these periodicities and McFadden (1984b) showed that similar apparent periodicities were produced by the use of fixed-length sliding windows (see §5.3.2) to analyze a Poisson process with a linear (i.e., no periodicities) nonstationarity in λ. Lutz (1985) showed that the 30-Myr signal found by Raup (1985) is sensitive to the length of the sequence. Stothers' (1986) analysis was questioned by Stigler (1987) and McFadden (1987). Lutz and Watson (1988) showed that the long-term trends can produce most of the apparent spectral evidence for the claimed 30-Myr periodicity. They also note that the record is short (only 5.5 cycles of a 30-Myr periodicity, split by the Cretaceous Superchron so that the longest continuous section is shorter than 3 cycles), making estimation unreliable. Mazaud and Laj (1991), following a suggestion by McFadden (1987) for determination of appropriate confidence limits, concluded that there is reasonable evidence for a 13- to16-Myr periodicity in the reversal rate. The evidence of Mazaud and Laj (1991) is, however, marginal, and the debate continues. At this stage, we conclude that claims of periodicities in the reversal chronology record are not compelling.

Finally, it should be noted that the use of a gamma (including Poisson) distribution, or even a power-law-type distribution (Seki and Ito, 1993), does not preclude the possibility that reversals are a manifestation of a deterministic chaotic process. Indeed, Gaffin (1989) showed the contrary: the data are consistent with the process being either chaotic or a nonstationary Poisson process. Although reversals may be a manifestation of chaos in the core, as often claimed (e.g., see Cortini and Barton, 1994), Gaffin has also shown the great difficulty in demonstrating this claim.

5.3.6 Paleointensity and Reversals

It seems intuitive that changes in intensity should somehow be related to the probability for a reversal. The first model explicitly discussing this was presented by Cox (1968), who suggested that the intensity of the dipole part of the geomagnetic field exhibited a 10^4-yr sinusoidal variation and the probability for reversal was greatest when the dipole intensity was least. A sinusoidal variation in the dipole intensity predicts a distribution of global paleointensities significantly different from that observed (§6.3). Pal and Roberts (1988) and Larson and Olson (1991) argued from kinematic dynamo theory that the intensity during the Cretaceous Superchron would have been generally higher than that observed at other times and this explained the absence of reversals. This prediction is not supported by paleointensity data (§6.3.4, Fig. 6.12).

Valet and Meynadier (1993) have argued, on the basis of relative paleointensity estimates from sedimentary cores, that the dipole intensity increases after a reversal and then decreases in an approximately linear manner. They also argue that the data show a positive correlation between the magnitude of the increase in intensity following a reversal and the length of the *subsequent* polarity interval. The prescience required for such a correlation implies that a large increase in paleointensity after a reversal imposes an inhibition on the probability for a future reversal. Concomitant with this is the implication that changes in the reversal rate are driven by changes in inhibition, which is precisely the opposite (see discussion in §5.3.5) of the conclusion drawn by McFadden and Merrill (1993). If the change in reversal rate were driven by inhibition it would have had a pronounced effect on the gamma distribution parameter k, which would have had to be large at the end of the Cretaceous Superchron (around 12; see McFadden and Merrill (1986)) and small now, the opposite of that observed (see Fig. 5.16 and McFadden and Merrill (1997)). However, it might be that the paleointensity does increase after a transition and that this is associated with the 50-kyr reversal inhibition interval discussed in §5.3.5.

5.3.7 Summary

In our view, the simplest broad-scale interpretation consistent with theory and with the polarity data for the past 160 Myr is as follows. The normal and reverse polarities are equivalent quasi-equilibrium states, with the same statistical characteristics apart from the different signs of g_1^0. From about 160 Ma onward the reversal rate gradually slowed, probably as a consequence of changing boundary conditions imposed by the mantle (§11.4.2), until at some time soon after 118 Ma the reversal process ceased, thereby producing a superchron. The cessation of the reversal process marks a fundamental transition in the dynamo process (discussed further in §11.3.2), but there is nothing special about the normal polarity; the field simply remained in the polarity state it occupied at the time the reversal process went into abeyance. Some time soon before 83 Ma the dynamo process underwent another fundamental transition and the reversal process restarted, the first reversal occurring at 83 Ma. From then until the present, the reversal rate gradually increased again, perhaps peaking at about 12 Ma. When operating, the reversal process is well described by a general renewal process, and is closely approximated by a nonstationary Poisson process, or at least by a gamma process with k only slightly greater than 1, which would indicate either that we have missed several short intervals or that the probability for a future reversal is depressed for about 50 kyr following the start of a reversal.

5.4 Polarity Transitions

5.4.1 Recording Polarity Transitions

Polarity transitions are recorded in sedimentary rocks, intrusive igneous rocks, and extrusive igneous rocks (lava flows). The major problem in gaining reliable, detailed information on transitional fields is that even though they take several thousand years, this is extremely short on geological time scales. The best records in sedimentary rocks occur when the remanence is a DRM or postdepositional remanent magnetization (§3.1). In cases of rapid sedimentation, a good continuous record of a transition can often be obtained. Unfortunately, extracting this record is often difficult. There is no absolute method available for obtaining paleointensities from sediments, although normalizing procedures utilizing anhysteretic remanent magnetization, isothermal remanent magnetization, or susceptibility can provide relative intensity values in some cases. Chemical changes, physical disturbances such as bioturbation, changes in sedimentation rates, and sampling problems are common in sediments and result in distortion of the geomagnetic record in many cases. Because the magnetic record is locked-in over a 10- to 30-cm range in marine sediments, the magnetic signal recorded by them is a filtered one and the nature of the filter is often not well understood (e.g., Hyodo, 1984; Okada and Niitsuma, 1989; Langereis *et al.*, 1992). In spite of these problems, sediment records have proved to be one of the most valuable sources for information on the transitional behavior of the field (e.g., Niitsuma, 1971; Opdyke *et al.*, 1973; Hillhouse and Cox, 1976; Koci and Sibrava, 1976; Freed, 1977; Burakov *et al.*, 1976; Valet and Laj, 1981; Herrero-Bervera and Theyer, 1986; Valet *et al.*, 1986, 1988a,b, 1989; Clement and Kent, 1991; Clement, 1991; Tric *et al.*, 1991a,b; Laj *et al.*, 1991, 1992a,b; Valet *et al.*, 1992; van Hoof and Langereis, 1992a,b; Clement and Martinson, 1992; Herrero-Bervera and Khan, 1992; McFadden *et al.*, 1993).

If intrusive igneous rocks are formed during the time of a polarity transition they may cool slowly enough to provide continuous records of reversal transitions. The magnetic field is recorded as the cooling front sweeps through the intrusive. However, minerals in igneous rocks are seldom, if ever, in chemical equilibrium during cooling and cooling rates are slow enough in intrusives for chemical changes to occur even below the Curie temperature. Also, reliable estimates of absolute times within a polarity transition are difficult to obtain because they must be based on cooling models that are generally poorly determined. Because of these problems only a few successful studies of reversal transitions in intrusive igneous rocks have been completed (e.g., Dunn *et al.*, 1971; Dodson *et al.*, 1978).

Terrestrial lava flows have the fewest rock magnetic problems, primarily because they cool very rapidly. Although they likely contain minerals that are in chemical disequilibrium with their environment, alteration often occurs very slowly for kinetic reasons. Although lava flows have desirable rock magnetic properties, unfortunately they only provide instantaneous records of the Earth's magnetic field. Hence good records of reversal transitions in lava flows usually require the eruption of a large number of flows in rapid succession. In addition, the spacing in time between successive flows erupted during a transition cannot be determined accurately because the errors associated with radiometric ages are typically much greater than the duration of a polarity transition. In spite of this, lava flow sequences have provided valuable records of the magnetic field changes during polarity transitions (e.g., Watkins, 1969; Larson *et al.*, 1971; Dagley and Lawley, 1974; Bogue and Coe, 1981, 1982, 1984; Mankinen *et al.*, 1985; Prévot *et al.*, 1985; Hoffman, 1986, 1989, 1991, 1992b; Coe and Prévot,

1989; Chauvin *et al.*, 1990; Bogue and Paul, 1993; Prévot and Camps, 1993; Bogue and Hoffman, 1987; Laj, 1989; Clement and Constable, 1991; Bogue and Merrill, 1992; Jacobs, 1994).

Estimates for the length of a polarity transition range from 1000 to more than 10,000 yr (e.g., Bogue and Merrill, 1992). Because of radiometric dating problems, the best estimates come from oceanic cores, for example, 4000 yr by Harrison and Somayajulu (1966), 4700 yr by Niitsuma (1971) and 4600 yr by Opdyke (1972). These estimates are based on changes in direction and most other estimates also cluster around 4000 to 5000 yr, a figure that has not changed over the past decade (Bogue and Merrill, 1992). Statistical analyses of lava flow data from Iceland (based on the ratio of the number of transitional directions to nontransitional directions) led to an estimate of 5500 yr, but the procedures used include transition times for excursions and hence this estimate is probably too high (Kristjansson, 1985). The data are inadequate to determine if the reported variations in transition times are due to differences in the geographic location of the recording site, differences in duration for different transitions, recording artifacts (e.g., filtering processes in marine sediments), or experimental errors.

5.4.2 Intensity Changes

Absolute paleointensities determined from lavas and relative paleointensity estimates from both sedimentary and igneous rocks recording transition directions indicate that the field intensity decreases substantially during a polarity change (van Zijl *et al.*, 1962a,b; Momose, 1963; Coe, 1967b; Lawley, 1970; Larson *et al.*, 1971; Dagley and Wilson, 1971; Wilson *et al.*, 1972; Opdyke *et al.*, 1973; Hillhouse and Cox, 1976; Dodson *et al.*, 1978; Bogue and Coe, 1981; Prévot *et al.*, 1985). These data indicate reductions sometimes to only 10% of the usual field intensity outside the transition. Some notable exceptions have been suggested, including examples of very strong fields observed right in the middle of transitions (Shaw, 1975, 1977; Prévot, 1977), but these may well be artifacts (Shaw, personal communication, 1992).

Figure 5.18 gives a summary of absolute paleointensities relating to polarity transitions for lavas less than about 10 Myr old. Absolute paleointensities for this time interval are given in the paleointensity database of Tanaka and Kono (1994) and have been analyzed by Tanaka *et al.* (1995), using only those determinations made by the Thellier or Shaw methods. The 323 values include many from the central part of polarity transitions that are characterized by low-latitude VGP. In order to allow for the different site locations all intensity values have been converted to virtual dipole moments (VDM; see §3.4). Even if the geomagnetic field is nondipolar during transitions, the use of VDM (as with VGP) attempts to take account of the effects of site location on the surface of the Earth. Average

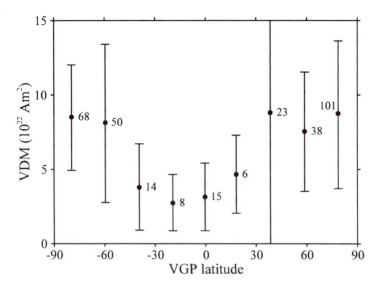

Fig. 5.18. Mean VDM versus VGP latitude for data covering the past 10 Myr averaged over 20° VGP latitude bands. After Tanaka *et al.* (1995).

VDM for VGP latitude bands of width 20° between -90° and +90° are shown in Fig. 5.18. Note that when the VGP latitude lies between -45° and +45°, there is a dramatic drop in VDM values. These absolute intensity values indicate that *on average* the field at the central part of transitions (VGP latitude 0°) is about 25% its usual value.

Kristjansson and McDougall (1982) and Kristjansson (1995) have examined the relative intensities of lavas from Iceland in a similar way to show the variation in intensity that occurs during polarity transitions. Assuming that the intensity of magnetization of each lava is proportional to the field in which it cooled, they calculated the geometric mean remanence intensity M_{10} (after alternating field demagnetization in 10 mT) for lavas within successive 10° VGP latitude bands. They argued that with sufficient numbers the variation in their magnetic properties would average out. The values are transformed to the VGP for each lava by multiplying the intensity by $2/(1+3\cos^2 p)^{1/2}$, where p is the geomagnetic paleomagnetic colatitude for the particular flow. These results are shown in Fig. 5.19 where the geometric mean intensity M_{10} is plotted against VGP latitude (irrespective of sign) for the two studies made by Kristjansson and McDougall (1982) for lavas in the range 1–14 Ma (three samples per flow) and Kristjansson (1995) for lavas in the age range 7–14 Ma (four samples per flow). Both of these studies of relative paleointensities again show that *on average* the intensity of the field at the central part of a polarity transition is about 25% of its usual value.

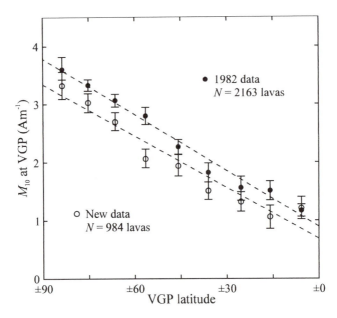

Fig. 5.19. Geometric mean value of remanence intensity (M_{10}) in Icelandic lavas after transformation to the virtual geomagnetic pole for each lava. The means are calculated for 10° VGP latitude bands with normal and reverse units combined. Solid circles are data from Kristjansson and McDougall (1982) for lavas in the range 1–14 Ma (three samples per flow), and open circles are from Kristjansson (1995) for lavas in the range 7–14 Ma (four samples per flow). Vertical bars indicate standard errors on the assumption of log-normal distributions. Simple regression lines for both data sets are shown. After Kristjansson (1995).

Beginning with suggestions by Hillhouse and Cox (1976) and especially Dodson *et al.* (1978) the possibility that a decrease in intensity occurs over a longer time interval than the direction change during a transition has been examined, but the data are too few and the rock magnetic problems too pervasive to document this suggestion convincingly. Bogue and Paul (1993) have presented a case for an average increase in intensity immediately following a polarity transition (see also Valet and Meynadier, 1993). In any case, estimates for the duration of a transition are shorter than the free-decay time estimates for the Earth's magnetic field (Chapters 8 and 9) and indicate that reversals are probably not associated with times when convection ceases in the outer core (McFadden *et al.*, 1985). This conclusion is strengthened by observations that much of the change in direction (at least at a site) associated with a reversal sometimes may occur on a time scale of hundreds of years or less (e.g., Laj *et al.*, 1988; Okada and Niitsuma, 1989).

Dodson *et al.* (1978) were the first to suggest that the intensity decrease associated with a transition occurs before the start of the directional transition, and Valet and Meynadier (1993) and Bogue and Paul (1993) suggest an increase in intensity after the transition. These suggestions are supported by the most detailed study yet of a polarity transition recorded in lava flows (a Miocene transition recorded from Steen's Mountain, Oregon; see Mankinen *et al.* (1985) and Prévot *et al.* (1985)). These interpretations are also compatible with inhibition data discussed in §5.3.5.

5.4.3 Directional Changes and Interpretations

During the past 30 yr there have been four major model types suggested for polarity transition (McFadden and Merrill, 1995b). The first model type was proposed by Creer and Ispir (1970) who suggested that a reversal occurred either by the decay of the dipole field and subsequent build-up in the opposite direction or by the rotation of the dipole field, without variation in intensity (Fig. 5.20). This dipole could be approximated by a virtual geomagnetic pole, or VGP (an

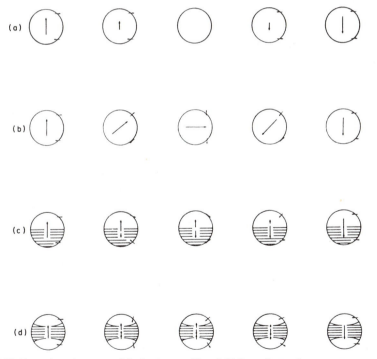

Fig. 5.20. Examples of some models for the transition field for a change from reverse to normal polarity. (a) Decrease in dipole moment which remains axisymmetric during the process. (b) Rotation of the main dipole without changes in moment. (c) A quadrupole transition model in which the polarity change is initiated in the southern hemisphere of the core. (d) An octupole transition model in which the polarity change is initiated in the low-latitude zone of the core.

estimate of the north geomagnetic field from a local field direction obtained by assuming a geocentric axial dipole).

If the field is dominantly that of a geocentric axial dipole throughout a transition, then observations from anywhere on the Earth's surface should give coincident VGP transition paths for that reversal. In contrast, if a transition is characterized by a significant nondipolar field, then one should expect that observations of a single transition from different positions (specifically, different longitudes) on the globe would give noncoincident VGP paths (but see Gubbins and Coe, 1993). Thus, when it later appeared that there were different VGP paths for the same reversal, a nondipole transition model emerged (Hillhouse and Cox 1976).

Comparison of transition records from different localities is difficult. For example, for a simple dipole rotation it is possible for one observer to see only changes in inclination, while an observer 90° away would see mainly changes in declination during the transition. Hoffman (1977) circumvented this problem by replotting transition VGP paths with respect to a common longitude and noticed an apparent systematic behavior in the records. Hoffman found for reverse to normal transitions the VGP paths tended to lie in the hemisphere centered about the site meridian (this has not held up with additional data). Hoffman (1977) referred to such VGP as near poles, and their corresponding VGP paths as near-sided paths. Far poles and far-sided paths then refer to VGP paths that lie in hemispheres opposite to the site meridian. This was a precursor to the next major conceptual advance, pioneered by Hoffman (1979) and Fuller et al. (1979) who argued that the transitional field was dominated by nondipolar zonal harmonics (Fig. 5.20). Although the axially symmetric component part of the Hoffman–Fuller hypothesis is no longer widely accepted, these papers ushered in the era of "modern" polarity transition studies. The ultimate aim of transition studies must be to develop a spherical harmonic description of the field throughout a transition. Given such a description we would learn a tremendous amount about the reversal mechanism. However, in order to achieve such a description, we would need many data well distributed around the globe for each time slice. At present we have enough trouble tracking down transition records at all, let alone achieving good global distribution of observations for a transition. Beyond that, there is the further problem that one cannot currently date rocks accurately enough to be able to separate the data from different locations into common time slices throughout the transition. Thus we are restricted to a narrow, and somewhat confusing, perspective when viewing transitions. What the papers of Hoffman and Fuller showed is that if the magnetic field is predominantly axially symmetric during reversal transition, then certain significant systematics should be observable through the perspective that is available to us. Today the majority of paleomagnetists seem to agree that those particular systematics are not present. However, their approach formalized the use of as many high-quality transitional data sets as possible to determine if transitional systematics exist and then to determine the simplest model that can describe any such systematics.

The most recent, and currently controversial, model is that transitional VGP paths are strongly biased to lie near the Americas or near an antipodal path (Laj *et al.*, 1991, 1992a,b; Tric *et al.*, 1991b). It is argued that this preference exists not only for observations from different positions for the one transition, but also for different transitions. This is an exciting hypothesis, for if true it implies mantle control of geodynamo processes during polarity transitions. Because of its potential importance, it is critical that it be subjected to rigorous testing.

There is considerable controversy over this hypothesis (e.g., Valet *et al.*, 1992; Prévot and Camps, 1993), and even quite opposing views of the data (Fig. 5.21).

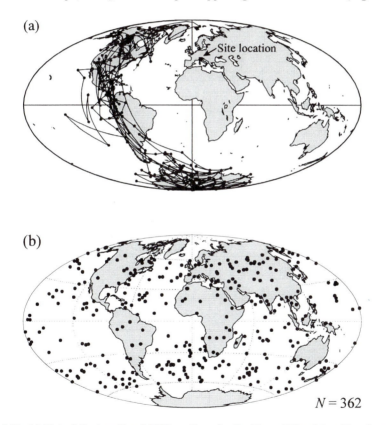

Fig. 5.21. (a) Plot of the transitional VGP positions for the Upper Olduvai transition from the Crostolo sediment. (b) 362 transitional VGPs from 121 volcanic records of reversals less than 16 Myr old. Constructed from Tric *et al.* (1991a) and Prévot and Camps (1993).

The data in Fig. 5.21a are for the Upper Olduvai transition and come from the Crostolo sediment (Tric *et al.*, 1991a). They show quite strong longitudinal confinement of the transitional VGP positions, and a path that passes through the Americas, the region claimed to be a preferred region for transitional VGP paths. This may be contrasted with transitional VGP data that come from lava flows with ages that extend back into the Miocene (Fig. 5.21b), and which do not exhibit any obvious clustering or preferred longitude sector (Prévot and Camps, 1993).

In yet a different type of approach, Hoffman (1991, 1992b) argues that transitional VGPs from lava flows anomalously clump in two patches: one on the American path, lying between central South America and Antarctica, with the other cluster falling in western Australia and the eastern Indian Ocean. Similar to other claims of systematics, alternate interpretations of these clusters are that they result from a statistical artifact, or are real, but of a rock magnetic origin (van Hoof, 1993).

Figure 5.22 shows a plot of equator crossings for the available VGP transition paths from sedimentary data for the past 12 Myr, and appears to show fairly strong clustering of the paths in two approximately antipodal positions. Certainly the clustering is significantly greater than would be expected from a uniform random distribution. However, this apparently obvious conclusion is implicitly based on the assumption that each observation in Fig. 5.22 is an independent random observation. This is not in fact so for two main reasons. First, there are very different numbers of observations from the different transitions. Second, the observation sites are very poorly distributed, being strongly grouped in two antipodal latitude bands. This leads to some concerns about the apparently obvious conclusion drawn from the data set.

There is the question of the structure of the field during a single transition; that is, do observations from well distributed sites on the globe give coincident VGP transition paths? This can only be resolved with observations from many

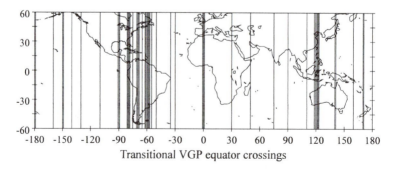

Transitional VGP equator crossings

Fig. 5.22. Plot of equator crossings for the available sedimentary VGP transition paths. After McFadden and Merrill (1995b).

different sites for each transition. A major problem is that the total number of records is barely sufficient to decide this question for a single transition, let alone for the multiple transitions involved. Within each transition, each transition path from an individual site may be considered as an independent observation. If the conclusion is that the VGP paths are not coincident for observation sites at different longitudes, then the model is void. If the conclusion is that the VGP paths are coincident, then it will be possible to estimate the location of the preferred path for each reversal, and then determine whether different reversals share the same "preferred path." Only then can the hypothesis of mantle control on the geodynamo polarity transitions be considered as well established.

The antipodal clustering of VGP paths and the antipodal clustering of site longitudes mean that the angles between the VGP paths and the longitude of their observation site naturally fall into two strong clusters. The interesting, and worrying, point is that this angle clusters strongly around ±90° (Valet *et al.*, 1992; McFadden *et al.*, 1993). As shown by McFadden *et al.* (1993), the data at that time actually favored the hypothesis that the VGP paths are clustered because the site longitudes are clustered and some artifact causes a bias of VGP equator crossings toward ±90° away from the site longitude over the hypothesis of a genuine geophysical preference for VGP longitudinal paths. Because the data are from sediments, an obvious candidate is inclination shallowing, as suggested by McFadden *et al.* (1993). Quidelleur and Valet (1994) considered some aspects of this concept, and it was restated by Courtillot and Valet (1995). A major attraction of an explanation such as inclination shallowing is that VGP paths lying outside the preferred paths are quite acceptable, but under the hypothesis of a genuine geophysically preferred path, there is an uncomfortable number of such exceptions. Barton and McFadden (1996) used the relation $\tan I_R = f \tan I_A$, where f is the flattening factor, I_R is the inclination of the magnetic remanence, and I_A is the inclination of the applied field, to investigate how much inclination shallowing is required for this to be a dominant feature in the clustering of VGP transitional paths. They concluded that inclination shallowing becomes important when f is below about 0.3 and is pronounced when f approaches 0.1. This is a fairly severe amount of shallowing, but in recent redeposition experiments Quidelleur *et al.* (1995) obtained values of f as low as 0.23 with an ambient field of 4.6 μT, similar to the field in the middle of a transition. Further experimental work is needed to resolve the impact of inclination shallowing, but at this stage it cannot be ruled out as a major factor in the longitudinal clustering.

Gubbins and Coe (1993) point out from symmetry considerations that any preferred VGP transition path on one side of the Earth (e.g., through the Americas) should be accompanied by an equally probable antipodal path on the opposite side of the Earth. The reason for this is that there appears to be no difference between the mean reverse and normal polarity states; i.e., the core fluid cannot sense the direction of the magnetic field (§8.5). Although there is an antipodal peak in the distribution that occurs opposite to the Americas, the magnitude of this peak is clearly smaller than of that through the Americas (see Fig. 5.22). Gubbins and Coe (1993) also point out that the hypothesis of preferred transitional VGP paths does not necessarily imply dominantly dipolar fields during transitions. Merrill and McFadden (1994) point out that even if there are systematics in the VGP paths for polarity transitions, there may not be for excursions that do not give rise to a complete reversal, and vice versa.

The question of systematics in transitional directions remains unresolved and requires substantially more data. McFadden and Merrill (1995b) point out that it should be possible to use reversals that occurred within the past 25 million years (there are more than 100) to determine if systematics exist. The use of data over a longer time span is unwise, since significant changes of the core–mantle boundary may then have occurred.

5.4.4 Reversals and Secular Variation

During the 1960s Allan Cox (personal communication) often raised the question as to whether reversals should be considered as an end member of the secular variation spectrum. This question remains unanswered in any convincing manner. The nonstationarity in reversal rate (Fig. 5.11) is remarkably similar to nonstationarities in secular variation with time (McFadden and Merrill, 1995a), suggesting a positive answer to Cox's question. On the other hand, as shown by Watkins (1969), the VGP appears occasionally to deviate several tens of degrees from the rotation axis, only to return quickly ("rebound"). This gives rise to the possibility that there are unsuccessful attempts at reversals, or *aborted reversals* (Doell and Cox, 1972; Barbetti and McElhinny, 1976; Hoffman, 1981; McFadden, 1984a). There is some ambiguity over the definition of "aborted reversal", with some paleomagnetists using the phrase to describe any large geomagnetic excursion. However, it is possible that reversals occur by a process that is not part of the normal geomagnetic secular variation spectrum and that such a hypothetical process could give rise to failed reversal attempts. Hereafter we will use the phrase aborted reversal to refer to this possibility. For example, reversals, and aborted reversals, could be triggered by boundary layer instabilities (e.g., hot or cold blobs (McFadden and Merrill, 1986)). Such mechanisms are different from those that usually produce the secular variation (§8.5). In contrast, reversals may occur from processes acting entirely within the dynamo, as they do in the Sun (McFadden and Merrill, 1993), and there is some evidence to suggest that this may be happening (Merrill and McFadden, 1994). The only conclusion that can be safely drawn at this time is that excursions, as defined by VGPs that deviate by more than 45° from the rotation axis, should neither be considered as part of the normal secular variation record nor be included as aborted reversals in statistical analyses of transition paths. One or the other of the above (but not both) may become acceptable once Cox's question is definitely answered.

Another unanswered question is whether the power spectrum of secular variation is similar to the power spectrum during a polarity transition. This is a very difficult question to answer. Transitions are geologically very short and there is no consensus as to what are the dominant frequencies in the typical secular variation spectrum (Chapter 4). However, Mankinen *et al.* (1985), Prévot *et al.* (1985), and Coe and Prévot (1989) argue that there are some very rapid changes in intensity and (independently) in the direction of the main field during polarity transitions that have not been observed in the historical (past 350 yr or so) data. Indeed, these authors argue there have been astonishingly rapid changes in the field, up to a 21° change in direction per week in a field with an intensity 30% or more of the usual value. (Note that in the limiting case of zero intensity, a directional change of 180° could occur instantaneously.) These rapid changes are claimed to have occurred during the Steen's Mountain reversal at 16 Ma. The evidence for such rapid change comes from two lava flows that, although having cooled very rapidly, exhibit significant changes in the paleomagnetic directions and intensities in their interiors. Coe and Prévot (1989) have examined these flows closely and argue against the possibility that rapid changes are due to changes in the external magnetic fields. Merrill and McFadden (1990) and Bogue and Merrill (1992) point out that changes on this time scale should be screened out by the conducting mantle (§2.4.2), even if the conductivity shows no increase below the 670-km transition. They suggest that remagnetization probably has occurred, a possibility considered by Coe *et al.* (1995; personal communication, 1994) who argue that this possibility is not supported by rock magnetic analyses that show no significant change in the nature of the magnetic carrier throughout the flows in question. Another possibility is that three-dimensional conductivity variation in upper mantle structures distort magnetic fields of core origin in such a way as to produce rapid variations at the Earth's surface. This is possible because the rate of diffusion of magnetic fields through electrical conductors is inversely proportional to their conductivity (§8.2). A quantitative assessment of this possibility has yet to be done. We conclude that the origin of the rapid changes observed in the two flows at Steen's Mountain remains unresolved, but that its resolution has important ramifications.

Chapter Six

The Time-Averaged Paleomagnetic Field

6.1 Geocentric Axial Dipole Hypothesis

6.1.1 The Past Few Million Years

The interpretation of paleomagnetic results has always depended upon the fundamental hypothesis that the time-averaged geomagnetic field is that of a geocentric axial dipole. The instantaneous field at the present time deviates substantially from that of a geocentric axial dipole, but because the field undergoes significant long-term secular variation (e.g., the westward drift), the time-averaged field can be expected to be significantly different from the instantaneous field. Hospers (1954b), using the newly developed statistical methods of Fisher (1953) (see §3.3.3), was the first to demonstrate that, averaged over several thousands of years, the virtual geomagnetic poles centered on the geographic pole in recent times. Creer *et al.* (1954) first introduced the concept of the apparent polar wander path for the interpretation of paleomagnetic results from Great Britain. In so doing they explicitly invoked the geocentric axial dipole hypothesis. It is now standard procedure in paleomagnetism to calculate paleomagnetic poles (§3.3.2) which, on the basis of the geocentric axial dipole hypothesis, are then assumed to coincide with the paleogeographic poles.

It has been customary to illustrate the validity of the geocentric axial dipole hypothesis by plotting all paleomagnetic poles for the past few million years on the present latitude–longitude grid and showing that they center about the geographic pole. Successive reviews in earlier texts by Cox and Doell (1960), Irving (1964), and McElhinny (1973) have used this method and indeed this is clearly shown to be the case. Exploration of the oceans has enabled a large number of deep-sea cores to be collected worldwide. Assuming the cores are taken vertically, the magnetic inclination can be determined. Opdyke and Henry (1969) first compared these inclinations from different parts of the world with those expected from a geocentric axial dipole and found a very good fit with the model.

Unfortunately, the observation that worldwide paleomagnetic poles determined for the past few million years center about the geographic pole does not by itself demonstrate that the time-averaged field is purely a geocentric axial dipole (g_1^0 only). Worldwide data from a time-averaged field that consists of a series of zonal harmonics (g_1^0, g_2^0, g_3^0, etc.), when plotted in this way, will always produce a set of paleomagnetic poles that center about the geographic pole. Wilson and Ade-Hall (1970) first noted that the paleomagnetic poles from Europe and Asia for the past few million years all tended to plot too far away from the observation site along the great circle joining the site to the geographic pole. Successive analyses by Wilson (1970, 1971), McElhinny (1973), Wilson and McElhinny (1974), Merrill and McElhinny (1977, 1983), Quidelleur *et al.* (1994), and McElhinny *et al.* (1996a) have confirmed that this occurs on a global scale. This effect is shown in Fig. 6.1 in which global data for the past five million years have been averaged by 45° longitude sectors. The position of the pole is related to the sector in each case, but the average of all eight sector poles still gives the geographic pole within a degree. Note that the reverse data tend to plot further over the pole from the sampling region than do the normal data.

Wilson (1971) introduced the concept of the common-site-longitude pole position as a convenient way to analyze the overall far-sided effect. The method is to place all observers at zero longitude by replacing the pole longitude with the common-site longitude given by the difference between the pole and the site longitudes. McElhinny *et al.* (1996a) have considered the data for the time interval 0–5 Ma in an analysis that involves 4455 spot readings of the normal polarity field (267 separate studies) and 2488 spot readings of the reverse polarity field (163 separate studies). These are summarized in Table 6.1 where the global mean paleomagnetic pole positions are given both as an overall mean and in common-site longitude. The mean common-site-longitude poles for the normal and reverse global data are shown in Fig. 6.2. As is apparent in Fig. 6.1, the reverse poles plot more far-sided than the do the normal poles. When all the data are used (sedimetary and igneous rocks), the means are significantly

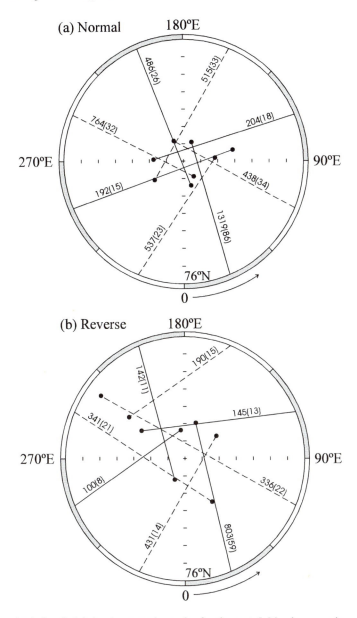

Fig. 6.1. Analysis of global paleomagnetic results for the past 5 Myr by averaging over 45° longitude sectors: (a) normal; (b) reverse. The paleomagnetic poles tend to plot too far away and over the geographic pole. The number of spot readings of the field is indicated for each average. Polar stereographic projection at 76°N. After McElhinny *et al.* (1996a).

TABLE 6.1
Mean Paleomagnetic Pole Positions for the Time Interval 0–5 Ma Both as the Overall Mean and as
the Common-Site-Longitude Mean.

Data set	Polarity	N (groups)	Overall pole position			Common-site-longitude pole position		
			Lat N	Long E	A_{95}	Lat N	Long E	A_{95}
All data	N+R	6943(430)	88.9	172.8	0.8	86.4	143.7	0.7
	N	4455(267)	89.4	105.4	0.8	87.2	143.9	0.8
	R	2488(163)	87.3	195.7	1.5	85.0	143.6	1.4
Igneous rocks	N+R	4408(273)	89.0	143.4	0.8	87.4	148.3	1.4
	N	2986(169)	89.4	107.1	0.9	87.6	152.4	0.9
	R	1422(104)	87.7	162.3	1.5	86.8	142.0	1.5
PSV lavas[a]	N+R	2947	88.7	105.5	0.6	87.4	146.0	0.6
	N	1978	89.1	85.0	0.6	87.9	146.3	0.6
	R	971	87.6	120.7	1.1	86.3	145.6	1.0

Note. N is the number of sites. A_{95} is the radius of the circle of 95% confidence about the mean calculated using the group analysis technique of McFadden and McElhinny (1995). After McElhinny *et al.* (1996a).
[a]Selected paleosecular variation data set of Quidelleur *et al.* (1994) (see §6.4).

different at the 95% confidence level (Fig. 6.2a), whereas they are not when only
igneous rocks are considered (Fig. 6.2b). This is discussed further in §6.2.2.
Note that the mean poles always fall to the right of the line joining the zero
common-site longitude to the geographic pole (particularly for the reverse data),
a feature Wilson (1971, 1972) referred to as the right-handed effect (see §6.2.3
for further discussion).

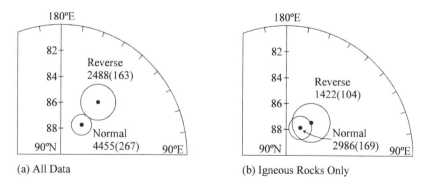

Fig. 6.2. Common-site-longitude representation of the normal and reverse data shown in Fig. 6.1
and listed in Table 6.1 with 95% circles of confidence about each mean. Polar stereographic
projection at 80°N. After McElhinny *et al.* (1996a).

Constable (1992) found that the histogram of VGP longitudes from a database of 2244 lava flows for the past 5 Myr had two roughly antipodal peaks. These peaks lay near the longitudes claimed to be preferred paths for reversal transition VGPs (§5.4.3). This was taken to imply that there were persistent nonzonal components of the geomagnetic field. Egbert (1992) then showed that, for very simple statistically homogeneous models of secular variation, the distribution of VGP longitudes peaks 90° away from the sampling locality. The effect is not large, but is a possible explanation of the right-handed effect of Wilson (1971, 1972). Quidelleur *et al.* (1994) and Johnson and Constable (1995a), using independently derived and much improved lava databases covering the past 5 Myr (see Table 6.1), both noted that the VGP longitude distribution is now rather uniform and does not show the preferential longitude bands suggested by Constable (1992).

Wilson (1970, 1971) modeled the far-sided effect as originating from an axial dipole source that is displaced northward along the axis of rotation rather than being geocentric, but was unable to explain the right-handed effect satisfactorily. However, as was demonstrated in §2.3.3, modeling sources of the geomagnetic field in this way is nonunique and it is always more appropriate to use spherical harmonics. The time-averaging of the paleomagnetic field is performed over time intervals much longer than the longest periods of secular variation ($\sim 2 \times 10^3$ yr). Thus it might be supposed that the tesseral and sectoral components of the magnetic scalar potential ψ (Eq. (2.2.16)) will be eliminated, leaving only the zonal potential. At the surface of the earth, Eq. (2.2.16) with $a = r$ then reduces to

$$\psi = \frac{a}{\mu_0} \sum_{l=1}^{\infty} g_l^0 \, P_l^0 (\cos\theta) \quad , \tag{6.1.1}$$

where $P_l^0 = P_l$, the Legendre polynomial of degree l (§2.2.2). Following James and Winch (1967), the potential (V) at the Earth's surface of a dipole of strength m displaced a distance x along the axis of rotation is

$$V = \frac{m}{4\pi a^2} \cos\theta + \frac{mx}{4\pi a^3} (3\cos^2\theta - 1) + \cdots \quad . \tag{6.1.2}$$

If this is related to the expansion of (6.1.1), namely

$$\psi = \frac{a}{\mu_0} g_1^0 \cos\theta + \frac{a}{\mu_0} g_2^0 \left(\frac{3\cos^2\theta - 1}{2} \right) + \cdots \quad , \tag{6.1.3}$$

then, on equating V and ψ,

$$g_1^0 = \frac{\mu_o m}{4\pi a^3} \quad \text{and} \quad g_2^0 = \frac{2\mu_0 mx}{4\pi a^4} \quad . \tag{6.1.4}$$

For a small offset, therefore, Wilson's offset-dipole model is (nonuniquely) equivalent to a geocentric axial dipole (g_1^0) plus a geocentric axial quadrupole (g_2^0), and the displacement x of the offset dipole can then be expressed in terms of the zonal coefficients (Wilson, 1970) as

$$\frac{x}{a} = \frac{g_2^0}{2g_1^0} \quad . \tag{6.1.5}$$

Paleointensities for the past 10 Myr also conform to the geocentric axial dipole field model. Tanaka *et al.* (1995) have shown that the data provide a good fit to the expected latitude variation in intensity (F) for a dipole field in which the polar value of F is twice its equatorial value (see §6.3.2).

Although the time-averaged paleomagnetic field has, to a first order approximation, been that of a geocentric axial dipole over the past few million years, the above shows there are second-order departures from the model. A more detailed examination of these second-order terms is given in §6.2.

6.1.2 The Past 600 Million Years

Over the past few million years it is reasonable to assume that continental drift has been small and that the relation of present continents to the axis of rotation has remained unchanged. However, for older times the relationship of the continents to the axis of rotation is usually determined from paleomagnetic data by invoking the geocentric axial dipole hypothesis. So it is possible only to test the *dipolar* nature of the paleomagnetic field in the past from paleomagnetic data. Testing the *axial* nature of the field requires the use of paleoclimatic information, as will be described in §6.1.3.

If paleomagnetic poles for a given geological epoch are consistent for rocks sampled over a region of continental extent (e.g., the Permian of Europe or North America), this provides compelling evidence that the dipole assumption used in calculating the poles is essentially correct. In determining the apparent polar wander paths for the different continents, it is usually observed that the mean poles for successive geological epochs are tightly grouped within each continent. This in itself is strong evidence in favor of the dipole assumption. In favorable circumstances the paleomagnetic meridian crosses the length of a continent as, for example, in Africa during the Mesozoic (McElhinny and Brock, 1975). Extensive paleomagnetic studies in north–west Africa may be compared with comparable studies in southern Africa. The two regions are separated by an angular great circle distance of 55.4°, almost exactly along the paleomagnetic meridian. The paleomagnetic inclinations are +32.5° and -55.5°, which correspond under the dipole assumption to paleomagnetic latitudes of +17.7° and -36.0° for the two regions. The paleolatitude difference of 53.7° is not discernibly different from the great circle distance, thus confirming the validity of the dipole assumption.

Evans (1976) suggested a most effective test of the dipolar nature of the geomagnetic field throughout the past 600 Myr. For any given magnetic field a definite probability distribution of magnetic inclination (I) exists for measurements made at randomly distributed geographical sites. This is easily obtained by simply estimating the surface area of the globe corresponding to any set of $|I|$ classes. For example, the dipole field is horizontal at the equator and has $|I|=10°$ at latitude 5.0°. At the poles the field is vertical and has $|I|=80°$ at latitude 70.6°. The surface areas of these two zones imply that if sampling is sufficient and geographically random the $0° \le |I| \le 10°$ band would make up 8.8% of results and the $80° \le |I| \le 90°$ band would make up only 5.7% of results.

The present uneven distribution of land on the earth's surface and the existence of areas of relatively intense study means that in present-day terms a random geographical sampling of paleomagnetic data has not been undertaken. However, over the past 600 Myr considerable polar wander and continental drift have taken place and it might be assumed that this has been sufficient to render the paleomagnetic sampling random in a paleogeographical sense. Evans (1976) therefore compared the observed frequency distribution of $|I|$ from paleomagnetic data over the past 600 Myr with that expected for the first four axial multipole fields. This analysis was later updated by Piper and Grant (1989) to cover data for the past 3000 Myr. The dependence of $|I|$ on colatitude θ for axial multipoles can be obtained from the relationship

$$\tan I_l = \frac{-(1+l)P_l}{(\partial P_l / \partial \theta)} \quad .$$ (6.1.6)

The frequency distribution of $|I|$ for the first four multipoles is compared in Fig. 6.3 with the observed values in 10° bands (Piper and Grant, 1989). A chi-square test suggests that the fit to the dipole curve is acceptable and is favored over higher order multipole models for the past 3000 Myr. A similar level of acceptance is displayed by data covering only the past 600 Myr. Although the analysis shown in Fig. 6.3 is based upon 4787 results worldwide (more than 10 times that used by Evans, 1976), no effective data selection was carried out (see Van der Voo (1993) for a discussion on data selection criteria). As a result the data set is contaminated to an extent that almost certainly degrades the effectiveness of the test.

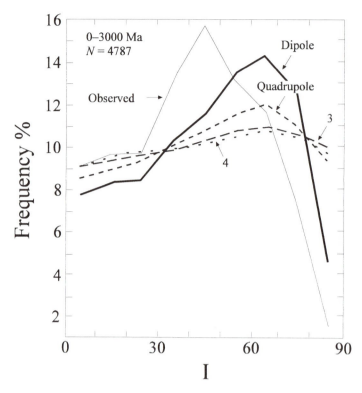

Fig. 6.3. Probability of observing magnetic inclination |*I*| within 10° bands for the first four axial multipole fields. The observed values are based on 4787 results covering the past 3000 Myr. After Piper and Grant (1989).

6.1.3 Paleoclimatic Evidence

Paleomagnetic and paleoclimatic data provide independent evidence for past latitudes. The factors controlling climate are quite independent of the Earth's magnetic field, but depend essentially on the fact that the net solar flux reaching the Earth's surface has a maximum at the equator and a minimum at the poles. At present the mean annual equatorial temperature is +25°C and the polar value is about -25°C. The determination of paleolatitudes from paleomagnetic data depends on the axial geocentric dipole assumption. Whereas the dipole assumption can be well demonstrated (as in the previous section), proof of the axial assumption requires a comparison with some independent measure of past latitudes. Comparison of paleoclimatic and paleomagnetic data provides the essential test, as suggested by Irving (1956).

There are many paleoclimatic indicators in the geological record. For example, at the present time coral reefs, evaporites, and carbonate deposits have a distribution with a maximum at the equator and either a polar minimum or a high-latitude zone from which they are totally absent. The comparison between paleomagnetic latitudes and the occurrences of paleoclimatic indicators has been made in several ways (Irving, 1956; Opdyke and Runcorn, 1960; Blackett, 1961; Runcorn, 1961; Opdyke, 1962; Irving and Gaskell, 1962; Irving and Briden, 1962; Briden and Irving, 1964; Irving and Brown, 1964; Briden, 1968; Stehli, 1968; Drewry *et al.*, 1974; Cook and McElhinny, 1979). The most useful method is to compile paleolatitude values for a particular occurrence in the form of equal angle or equal area histograms to give the paleolatitude spectrum of the particular indicator.

Figure 6.4 gives an analysis of the paleolatitudes of fossil reefs. The distribution of modern coral reefs is symmetrical about the equator (Fig. 6.4a), the maximum frequency being between 10° and 20° latitude and most occurrences lying within 30° of the equator. The present latitudes of fossil reefs (Fig. 6.4b) do not show this distribution, but when referred to their paleolatitudes (Fig. 6.4c) the spectrum is very similar to that of modern reefs, over 95% of the occurrences falling within 30° of the equator (Briden and Irving, 1964; Briden, 1968). Other paleoclimatic indicators that have been studied include evaporites (Blackett, 1961; Runcorn, 1961; Irving and Briden, 1962; Opdyke, 1962; Drewry *et al.*, 1974), glacial deposits (Blackett, 1961; Runcorn, 1961; Opdyke, 1962; Drewry *et al.*, 1974), coral deposits (Briden and Irving, 1964; Drewry *et al.*, 1974), oil fields (Irving and Gaskell, 1962), paleowinds (Opdyke and Runcorn, 1960; Opdyke, 1961), phosphate deposits (Cook and McElhinny, 1979), and the distribution of various fossils (Runcorn, 1961; Irving and Brown, 1964; Stehli, 1968). There is broad agreement from all these studies that the paleolatitude distribution of these paleoclimatic indicators as determined from paleomagnetism is essentially symmetrical about the equator. This provides the strongest indication that the time-averaged paleomagnetic field is both axial and dipolar.

6.1.4 Longevity of the field

There are numerous paleomagnetic results that confirm the existence of the geomagnetic field in Precambrian times. The oldest results are from rocks greater than 2.5 Ga (Archean age). In North America (the Laurentian Shield), a particularly good piece of evidence is found where Matachewan dykes (2.6 Ga) intrude Archean rocks. A baked contact test confirms the Archean rocks were remagnetized at the time of intrusion. Away from the dykes a consistent magnetization, that must predate the dykes, is observed. Irving and Naldrett (1977) estimate the age of this Archean magnetization at about 2.8 Ga.

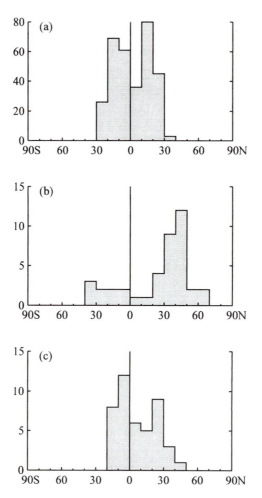

Fig. 6.4. Equal angle latitude histogram for organic reefs. (a) Present latitude of modern reefs; (b) present latitude of fossil reefs; (c) paleolatitude of fossil reefs. After Briden and Irving (1964).

   In southern Africa the 2.7-Ga Modipe Gabbro is intruded by a 2.3-Ga granite. At the contact the gabbro has been reheated by the granite, providing a baked contact test for the magnetization of the granite (Evans and McElhinny, 1966). The consistent magnetization in the gabbro is thus older than 2.3 Ga and presumably relates to its time of intrusion 2.7 Ga ago. Paleointensity measurements on this gabbro (McElhinny and Evans, 1968) give a mean VDM about 1.5 times the present dipole moment of the Earth.

   The oldest magnetizations so far observed are from the Duffer Formation Volcanics of northwest Australia (McElhinny and Senanayake, 1980) and the

Komati Formation Lavas of southern Africa (Hale and Dunlop, 1984). Both of these formations are 3.5 Ga in age. The Duffer Formation rocks were folded about 3.0 Ga ago and a positive fold test (§3.3.1) confirms that the magnetization predates the folding. Hale and Dunlop (1984) have also made paleointensity measurements on the Komati Formation rocks that provide the oldest estimate of the strength of the Earth's magnetic field in the past with a mean VDM about 0.6 of the present dipole moment (see also §6.3.4 and §11.3.1 for further discussion). The existence of the geomagnetic field in Archean times is thus well established since at least 3.5 Ga.

6.2 Second-Order Terms

6.2.1 The Problems in Time Averaging

It is now well established that, at least for the past 600 Myr, the first order description of the time-averaged paleomagnetic field is that of a geocentric axial dipole. However, there is good evidence (§6.1.1) that there are important second-order terms. Before attempting to define and then calculate these second-order terms it is necessary to ask the fundamental question as to just what is meant by a time-averaged field and how does one define it. Most debate on the subject of the second-order terms centers around this question.

As a very simple example of the problems involved consider the situation of Fig. 6.5. A traveling wave of the form $A\cos(wt - kx)$ in Fig. 6.5a is modified by the standing amplitude modulation $B(x)$ in Fig. 6.5b resulting in the function of Fig. 6.5c. An observer at P, taking observations at different times, would observe the wave of Fig. 6.5a unmodified. An observer at Q, however, also taking observations at different times, would observe a wave of the same form as the observer at P but with a greatly reduced amplitude. The observer at Q would then claim a far smaller scatter in observations than the observer at P, and an analogous situation in the observation of secular variation will produce effects such as the so-called Pacific dipole window of Doell and Cox (1972). Finally, an observer making all observations at the same time but distributed in space would observe the function of Fig. 6.5c. Disagreements between the three observers regarding the form of the function observed are entirely a consequence of the differing restricted perspectives. The only manner in which the true form of the function can be determined is by taking observations distributed in both space and time.

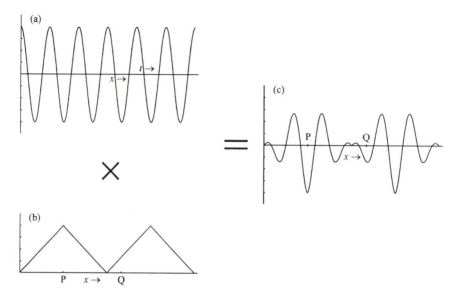

Fig. 6.5. To illustrate the concept of space–time averaging. (a) Traveling wave of form $A\cos(wt-kx)$. (b) Standing amplitude modulation $B(x)$ on (a). (c) Resulting motion. Observers at P and Q see very different amplitudes, but when space–time averaging is performed a consistent result is obtained.

The preceding illustrates why it is important to define the time-averaged field in a *global* sense, and why the properties of such a field can only be specified through a combination of space–time averaging. Virtually all disagreements between results obtained by different authors stem from a lack of appreciation of this time-averaging problem.

6.2.2 Spherical Harmonic Analyses

The representation of the time-averaged paleomagnetic field in terms of an offset dipole (§6.1.1), although a reasonable first- and second-order model, is obviously not strictly correct and in general it becomes more effective to analyze the data in terms of spherical harmonics. This was first performed by Wells (1969, 1973) on late Tertiary and Quaternary data. Benkova *et al.* (1971) first analyzed paleomagnetic declination data, and later Benkova and Cherevko (1972) and Benkova *et al.* (1973) analyzed inclination data. Creer *et al.* (1973) and Kono (1973) analyzed both declination and inclination data, and subsequently Georgi (1974) and Adam *et al.* (1975) analyzed inclination data only so as to be able to use the results from deep-sea cores (Opdyke and Henry, 1969). Different methods of analysis were then used by Cox (1975), Merrill and McElhinny (1977, 1983), Coupland and Van der Voo (1980), Livermore *et al.* (1983), and Schneider and Kent (1990). The different analytical techniques have produced markedly varying results and much argument has been generated as to the most appropriate technique. The discussion that follows attempts to explain the reason for these differences and to produce a technique that is internally self-consistent.

Any spherical harmonic expansion of paleomagnetic data will be severely truncated because the data are finite. As shown in §2.3.2, a better representation of a given data set cannot in general be obtained by using higher degree harmonics at the expense of lower degree ones. Wells (1973) shows that the accuracy with which the truncated series represents the actual situation cannot be represented statistically unless particular models for the magnetic field are invoked. In particular, unevenly distributed data or the use of poor quality data can result in very inaccurate spherical harmonic descriptions. After careful consideration of the errors involved, Wells (1973) concluded that only the zonal harmonics (g_1^0, g_2^0, g_3^0) were significant. However, Creer *et al.* (1973) and Georgi (1974) concluded that the nonzonal harmonics (such as g_1^1 and g_2^1) were often comparable in size to the zonal ones. Georgi (1974) concluded it was not possible to say that the zonal terms dominate the time-averaged paleomagnetic field. Livermore *et al.* (1983) also believed that in their analysis they had reliably detected a small (3.1%) h_2^1 term.

Neither Creer *et al.* (1973) nor Georgi (1974) placed any statistical limits on their estimates of the various spherical harmonic coefficients. Their procedure was to determine regional averages of the paleomagnetic data at (respectively) 20 and 32 control points on the surface of the Earth. However, the data distribution is so uneven in both quantity and quality that such a procedure can easily produce nonzonal harmonics as an artifact, so it is not surprising that apparently significant nonzonal harmonics were determined. The analysis of Wells (1973) strongly suggests that only the zonal coefficients are significant in such analyses and that the large and variable nonzonal coefficients determined are a data distribution problem. Merrill and McElhinny (1977, 1983), Coupland and Van der Voo (1980), Livermore *et al.* (1983), Schneider and Kent (1990), and McElhinny *et al.* (1996a) came to the same conclusion using independent methods.

Cox (1975) first realized that the difference in inclination, ΔI, between the observed field and that from a geocentric axial dipole may be used to estimate the value of zonal harmonics. Suppose for a given region there is a departure from pure axial symmetry (i.e., the declination D departs from zero). In this case ΔI will be slightly different from the value ΔI_z expected from purely zonal harmonics. Although it is not possible to give a general formula that is accurate in all instances, for the problem at hand projection of the direction onto $D=0$ will typically give a good approximation, so

$$\sin \Delta I_z \approx \sin \Delta I \cos D \quad . \qquad (6.1.7)$$

Even for moderate values of D this correction will be small. If $\Delta I = 5.0°$, then D can be as large as $10°$ before ΔI_z differs from ΔI by $0.1°$. Thus the observed values of ΔI can be used to obtain estimates of the zonal harmonics with negligible error.

Merrill and McElhinny (1977, 1983) and McElhinny *et al.* (1996a) have taken this type of analysis one step further. They argue that if the time-averaging of the field has been maximized by using a combination of both time and space averaging (see §6.2.1), then it is to be expected that the nonzonal terms will be eliminated, leaving only the zonal potential as in (6.1.1). Figure 6.6a shows how the ΔI inclination anomaly would look in two simple cases. In the first case the observed field consists solely of an axial geocentric dipole and quadrupole with $g_2^0 / g_1^0 = 0.05$. In the second case the observed field consists solely of an axial geocentric dipole and octupole with $g_3^0 / g_1^0 = 0.05$. The ΔI anomaly is always negative in the first case, but changes sign across the equator in the second case.

In order to check that the assumption of a purely zonal time-averaged field is acceptable, Merrill and McElhinny (1977, 1983) and McElhinny *et al.* (1996a) analyzed the declination anomaly ΔD (departure of D from zero) to determine whether nonzonal harmonics (g_1^1 and h_1^1) are required to explain the observations. For an axial geocentric dipole field the northward component of the Earth's magnetic field (X) varies with latitude (λ) according to the relation

$$X = X_0 \cos \lambda \quad , \qquad (6.2.1)$$

where X_0 is the northward component at the equator. Thus ΔD is also a function of latitude for a geocentric axial dipole field and varies according to the relation

$$\tan \Delta D_0 = \tan \Delta D \cos \lambda \quad , \qquad (6.2.2)$$

where ΔD_0 is the equatorial equivalent of ΔD. All values of ΔD can then be normalized to the equivalent equatorial value to determine if the lowest nonzonal harmonics are present, which as shown in Fig. 6.6b would have a sinusoidal variation with longitude.

If the nonzonal terms from a combination of space–time averaging can be shown to be small, then the initial assumption of a purely zonal time-averaged field is shown to be a reasonable one. Merrill and McElhinny (1977, 1983), Coupland and Van der Voo (1980), Livermore *et al.* (1983), Schneider and Kent (1990), and McElhinny *et al.* (1996a) all identified the zonal harmonics as the most important. In particular, McElhinny *et al.* (1996a) show that nonzonal harmonics are not required to explain the observations. Thus it is concluded that the departures from the best fitting $g_1^0 + g_2^0 + g_3^0$ model are essentially random and not due to nonzonal terms.

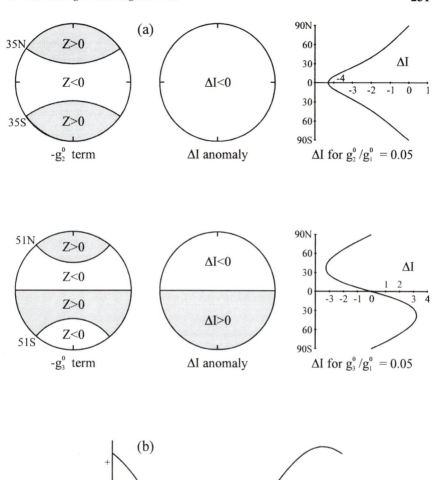

Fig. 6.6. (a) Latitude variation of the vertical component Z of the zonal harmonics $-g_2^0$ and $-g_3^0$. When these terms are added to the present geocentric axial dipole field $(-g_1^0)$ they produce an inclination anomaly ΔI. (b) When added to the geocentric axial dipole field, g_1^1 and h_1^1 produce a declination anomaly ΔD that is a function of latitude. This may be normalized to the equator to produce ΔD_0, that has a sinusoidal variation with longitude. After McElhinny *et al.* (1996a).

6.2.3 The Past Five Million Years

In the early analyses of paleomagnetic data each result as published in the literature was taken as a datum. In looking at the time-averaged field in some detail it is necessary to take into account the variability in the amount of data included in each result. For example, Merrill and McElhinny (1977) examined 366 data covering the past 5 Myr, equivalent to about 3500 spot readings of the field over the Earth's surface. Merrill and McElhinny (1983) used a database of all individual measurements of the field, and had 4968 measurements (sites) derived from 538 results (364 normal and 174 reverse). In the most recent analysis McElhinny *et al.* (1996a) used the *Global Paleomagnetic Database* of Lock and McElhinny (1991) and its updated versions (McElhinny and Lock, 1993a, 1994, 1995) as well as the ocean core data summarized by Schneider and Kent (1990). This entire data set comprised 9490 sites (5831 normal and 3659 reverse) derived from 730 results (442 normal and 288 reverse). Because the results contain widely differing numbers of sites, they were combined using the method of combining groups of paleomagnetic directions or poles of McFadden and McElhinny (1995).

For each measurement the equator normalized declination anomaly ΔD_0 and inclination anomaly ΔI were calculated. Note that only ΔI values are determined from ocean core data. These ΔI anomalies were then averaged in time and space by using $10°$ latitude strips at low and intermediate latitudes and wider latitude strips at high latitudes depending on the amount of data. For a geocentric axial dipole at colatitude θ,

$$\tan I = 2\cot\theta \quad . \tag{6.2.3}$$

If g_2^0, g_3^0, and g_4^0 terms are included in the field, and setting

$$G2 = g_2^0 / g_1^0; \quad G3 = g_3^0 / g_1^0; \quad G4 = g_4^0 / g_1^0 \quad , \tag{6.2.4}$$

then

$$\tan I' = A / B \quad , \tag{6.2.5}$$

where (see Appendix B)

$$A = 2\cos\theta + G2(\tfrac{9}{2}\cos^2\theta - \tfrac{3}{2}) + G3(10\cos^3\theta - 6\cos\theta)$$
$$+ G4(\tfrac{175}{8}\cos^4\theta - \tfrac{75}{4}\cos^2\theta + \tfrac{15}{8})$$

$$\tag{6.2.6}$$

$$B = \sin\theta + G2(3\cos\theta\sin\theta) + G3(\tfrac{15}{2}\cos^2\theta\sin\theta - \tfrac{3}{2}\sin\theta)$$
$$+ G4(\tfrac{35}{2}\cos^3\theta\sin\theta - \tfrac{15}{2}\cos\theta\sin\theta) \quad ;$$

then

$$\Delta I = I' - I \quad . \tag{6.2.7}$$

McElhinny *et al.* (1996a) subdivided their data set into results from igneous rocks, sediments, and ocean cores and then used (6.2.3) to (6.2.7) to determine weighted least-squares model fits to the data. The analyses showed that in all cases the G4 term is small and not significant, so that only the G2 and G3 terms need be considered further. McElhinny *et al.* (1996a) considered that the combined data for Brunhes- and Matuyama-age igneous rocks and ocean sediment cores probably provide the best estimate of G2 and G3, although it is doubtful whether any terms other than G2 have any geomagnetic significance. These results are illustrated in Fig. 6.7 and values of G2 and G3 determined from various analyses are summarized in Table 6.2.

The data indicate that there is no difference between the G2 values deduced for the normal or reverse fields within the error limits quoted regardless of how the data are combined. For the G3 term, however, there are several ways in which the value can be affected by some artifact of the data. These have been summarized by McElhinny *et al.* (1996a) as follows:

(i)   inclination errors in sediments arising from DRM processes (King and Rees, 1966) or compaction effects (Blow and Hamilton, 1978);

(ii)   inclination error in lavas arising from shape anisotropy (Coe, 1979);

(iii) The use of unit vectors in paleomagnetism (Creer, 1983).

(iv)   incomplete magnetic cleaning of Brunhes-age overprints in reversely magnetized rocks (Merrill and McElhinny, 1983; McElhinny *et al.*, 1996a);

(v)   nonorthogonality of spherical harmonics in a disconnected space and aliasing problems (McElhinny *et al.*, 1996a).

All of the above effects cause magnetic inclinations to be deflected toward the horizontal; that is, ΔI will be negative for positive I and positive for negative I. This is the equivalent of a positive G3 term as illustrated previously in Fig. 6.6. The wide variation in G3 values shown in Table 6.2 confirms the existence of some or all of these effects. The removal of sedimentary data from the continental data causes a large drop in G3 and the reverse data consistently show larger G3 values compared with the normal data. The only puzzling feature is the curious fact that the G3 values for the ocean sediments are always negative whereas for the continental data they are always positive. There appears to be no obvious explanation for this. However, the above suggests that only the G2 values have any significance in discussing permanent second-order terms in the time-averaged paleomagnetic field (Quidelleur *et al.*, 1994; Quidelleur and Courtillot, 1996; McElhinny *et al.*, 1996a).

Almost all papers on the time-averaged field published to date have proposed that there are apparent statistically significant differences between the time-

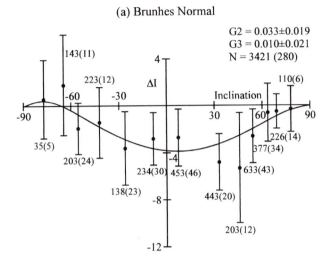

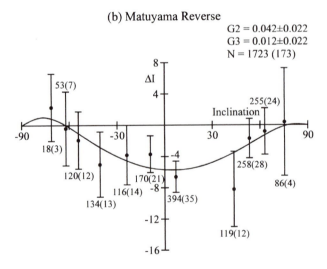

Fig. 6.7. ΔI anomalies, with 95% confidence limits, from Brunhes- and Matuyama-age igneous rocks and ocean sediment cores averaged over latitude strips using data for the past 5 Myr: (a) Brunhes Normal and (b) Matuyama Reverse data. The anomalies are plotted against the average axial dipole field inclination value for each latitude strip. This has the effect of spreading the diagram horizontally. For each latitude strip the number of sites is given with the number of groups (cores) in brackets, N being the total number. Model fits using three terms have been used with only the G2 and G3 terms being significant. After McElhinny *et al.* (1996a).

TABLE 6.2
Normalized zonal Gauss Coefficients (G2, G3) for the Time-Averaged Field over the Past 5 Myr
Using Various Combinations of Data.[a]

Data combination	Polarity	Sites (studies)	G2	G3
All data combined	N	5831(442)	0.043±0.030	0.017±0.031
	R	3659(288)	0.063±0.035	0.056±0.037
Continents all data	N	4455(267)	0.041±0.037	0.031±0.035
	R	2488(163)	0.054±0.026	0.094±0.056
Continents igneous rocks	N	2986(169)	0.039±0.028	0.029±0.027
	R	1422(104)	0.059±0.048	0.058±0.056
Brunhes igneous rocks	N	2045(105)	0.032±0.025	0.030±0.023
Matuyama igneous rocks	R	552(48)	Insufficient data available	
Oceans (drift corrected)	N	1376(175)	0.026±0.021	-0.027±0.019
Schneider and Kent (1990)	N	(175)	0.026±0.010	-0.029±0.015
Oceans (drift corrected)	R	1171(125)	0.035±0.012	-0.014±0.013
Schneider and Kent (1990)	R	(125)	0.046±0.014	-0.021±0.020
Oceans (drift uncorrected)	N	1376(175)	0.027±0.021	-0.026±0.019
	R	1171(125)	0.046±0.012	-0.011±0.013
Continents igneous + oceans	N	4362(344)	0.038±0.025	0.013±0.027
	R	2593(229)	0.049±0.022	0.028±0.024
Brunhes igneous + oceans	**N**	**3421(280)**	**0.033±0.019**	0.010±0.021
Matuyama igneous + oceans	**R**	**1723(173)**	**0.042±0.022**	0.012±0.022
Combined B-M ign.+ oceans	**N+R**	**5144(453)**	**0.038±0.012**	0.011±0.012

N(Groups) gives the number of sites (groups or cores) used in each calculation.
[a]After McElhinny *et al.* (1996a).

averaged normal and reverse polarity states (e.g., Merrill and McElhinny, 1977, 1983; Schneider and Kent, 1990; Merrill *et al.*, 1990; Johnson and Constable, 1995b). Quidelleur and Courtillot (1996) suggested that conclusions that there are differences in certain structures of the time-averaged reverse and normal states were premature. McElhinny *et al.* (1996a) point out that in none of the normal and reverse pairs (Table 6.2) is the estimate for the normal polarity G2 significantly different from that of the reverse polarity G2. Thus it can now be concluded that there is no discernible difference between the time-averaged normal and reverse polarity fields. The data can thus be combined to obtain an overall best estimate for G2 of 0.038±0.012 as shown in the last line of Table 6.2.

To determine if the time-averaged field really does consist predominantly of zonal harmonics, it is necessary to check that the declinations do not deviate significantly from zero. For this purpose Merrill and McElhinny (1977, 1983) tested for the existence of the first term in the nonzonal harmonics given by

g_1^1 and h_1 as indicated in Fig. 6.6. In practice there are larger errors in measuring ΔD than in measuring ΔI. For example, there are often small tectonic rotations that go undetected when sampling occurs. These have little effect on ΔI but can have a significant effect on ΔD. Also the errors in ΔD increase at points closer to the pole. Using (6.2.2) it is possible to normalize all values of ΔD to the equator to take this into account and produce equator-normalized values ΔD_0.

McElhinny *et al.* (1996a) noted that these small tectonic effects could readily be recognized from spuriously large values of ΔD_0. Therefore they restricted their analysis to values of $|\Delta D_0| < 10°$. Their analysis is shown in Fig. 6.8. In no case can the null hypothesis that ΔD_0 is zero for all longitudes be rejected. That is, a straight line with $\Delta D_0 = 0$ provides an acceptable fit to the data. This confirms previous analyses that it is not possible to determine even the first-degree nonzonal harmonics in a spherical harmonic analysis of paleomagnetic data for the time interval 0–5 Ma (Merrill and McElhinny, 1977, 1983; Quidelleur *et al.*, 1994). However, Gubbins and Kelly (1993) and Johnson and Constable (1995b) have concluded that there may be persistent nonzonal terms in the time-averaged field. McElhinny *et al.* (1996a) argue that this conclusion does not take into account the difficulties of time-averaging (see §6.2.1), second-order tectonic effects, and the many ways in which some harmonics can be contaminated by the nature of the paleomagnetic measurements themselves as discussed earlier in this section.

6.2.4 Extension to 200 Ma

It is possible to use the sea-floor spreading magnetic anomalies (§5.2.2) to reconstruct the relative positions of the continents through time back to 200 Ma. Most of the world's oceans have now been studied in some detail and the relative rotation parameters are well known. Using this information, paleomagnetic data can be compared with the reconstructed globe and, in principle, the departures from the geocentric axial dipole model can be determined (see also §6.4.6). Coupland and Van der Voo (1980) first attempted this procedure and later Livermore *et al.* (1984) and Lee and Lilley (1986) repeated it.

The continents can be relocated relative to a chosen fixed continent (Coupland and Van der Voo, 1980; Lee and Lilley, 1986) or relative to a fixed hot-spots reference frame (Livermore *et al.*, 1983, 1984). The advantage of the latter procedure is that the paleogeographic coordinates may be fixed without reference to the paleomagnetic data. The general conclusion of these studies is that it is only possible to attempt to define the $G2 = g_2^0 / g_1^0$ term in a spherical harmonic analysis and that there are insufficient data to attempt a subdivision into normal and reverse components. In any case, the problems associated with determining the G3 term over the past 5 Myr (§6.2.3) alone suggests that estimates for terms higher than G2 would be meaningless.

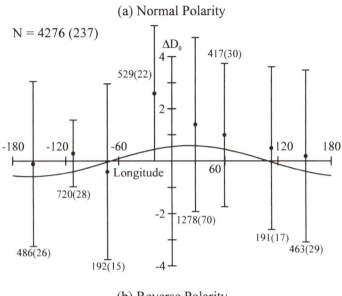

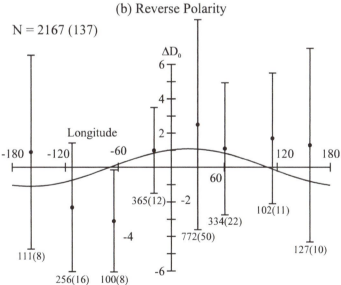

Fig. 6.8. ΔD_0 anomalies, with 95% confidence limits, averaged by longitude sectors using all data for the past 5 Myr. The number of sites ($|\Delta D_0| < 10°$), with the number of groups given in parentheses, is indicated at each point, N being the total number. The curves show the best fit amplitude using the direction of the *present field* equatorial dipole. After McElhinny *et al.* (1996a).

Coupland and Van der Voo (1980) and Lee and Lilley (1986) both included a G3 term in their analysis and obtained differing results. Coupland and Van der Voo (1980) found that G3 was not significant prior to 50 Ma, whereas Lee and Lilley (1986) found large G3 values, with G2 becoming insignificant prior to 100 Ma. Livermore *et al.* (1984), however, did not include a G3 term in their analysis, but their method has the advantage that the paleogeographic coordinates do not depend on the paleomagnetic data and that a larger database was used. Figure 6.9 summarizes the determinations made by Livermore *et al.* (1984) covering the past 200 Myr. The data were grouped into 20-Myr windows at intervals of 10 Myr for the past 100 Myr, and 40-Myr windows at intervals of 30 Myr for the 100- to 200-Myr time span. The G2 values are negative for pre-Cenozoic times and are positive during the Cenozoic. There is a clear negative G2 term during the Cretaceous Normal Superchron but Livermore *et al.* (1984) indicate that the values determined for older ages may not be significantly different from zero.

6.3 Variation in the Earth's Dipole Moment

6.3.1 Paleointensities and Dipole Moments

For the purpose of comparing data from sampling sites at different latitudes it is convenient to calculate the equivalent dipole moment that would have produced the measured intensity at the calculated paleolatitude of the sample. Smith (1967a) termed such a dipole moment the virtual dipole moment, VDM, as was

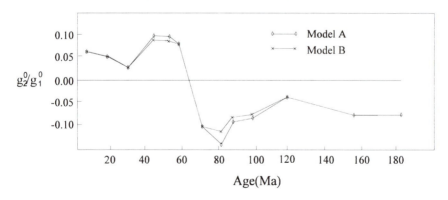

Fig. 6.9. Variation of $G2 = g_2^0 / g_1^0$ with time for the past 200 Myr. Models A and B assume two alternative models of Pacific plate motions. After Livermore *et al.* (1984).

discussed in §3.4.5. This has the advantage that no scatter is introduced in an analysis of VDMs by any wobble of the main dipole because the determined magnetic paleolatitude is independent of the orientation of the dipole relative to the Earth's rotation axis. However, it is not possible to identify that portion of the observed paleointensity (and therefore of the VDM) caused by nondipole components. Hence a true dipole moment, TDM, cannot be obtained directly from paleointensity studies. The rapidly varying part of the nondipole field changes over periods up to about 10^3 yr, whereas the dipole field changes over periods of about 10^4 yr (Cox and Doell, 1964). These changes can be considered to be superimposed on a slowly varying part. This difference in time constants makes it possible, at least in principle, to separate out the statistical properties of the TDM.

McFadden and McElhinny (1982) have discussed the problem of estimating the mean dipole moment for a given period of time from the observed VDM data. There are three types of means usefully associated with VDM data.

(i)   The arithmetic mean of the observed VDMs can be calculated and this is referred to as the *mean VDM*. The scatter of observed VDMs about the mean is caused by fluctuations in the TDM, the effect of nondipole components and errors in the paleointensity determinations.

(ii) By making assumptions about the effect of nondipole components and errors in paleointensity determinations it is possible to extract the distribution of TDMs from the observed distribution of VDMs. The peak of the determined TDM distribution is then the inherently preferred value of the TDM about which the Earth's dipole moment fluctuates. This preferred value of the TDM is then the intensity analogue of the Paleomagnetic Pole (§3.3.2) of directional data about which the VGPs fluctuate, and is referred to as the *paleomagnetic dipole moment*, PDM, (§3.4.5) after McFadden and McElhinny (1982).

(iii) The third value of interest is the *mean TDM*, which may be different from the PDM depending on the distribution assumed for the TDM. Generally the mean TDM will be equal to the mean VDM but the scatter of TDMs will be much less due to the removal of the effects of nondipole components and errors in the paleointensity measurements. Because only the distribution of TDMs may be determined from a given set of VDMs, it can be a lengthy procedure to determine the scatter of TDMs.

6.3.2 Absolute Paleointensities — The Past 10 Million Years

McFadden and McElhinny (1982) analyzed all paleointensities determined from rocks less than 5 Myr in age and Kono and Tanaka (1995b) have updated this analysis covering the past 10 Myr using a new paleointensity database compiled by Tanaka and Kono (1994). Data from polarity transitions were excluded by using only those data where the latitude of the associated VGP is greater than 45°. Kono and Tanaka (1995b) used only measurements determined by the Thellier or Shaw methods (see §3.4) and excluded results whose age was less than 30,000 yr B.P. Their analysis included 267 VDMs (145 normal polarity and 121 reverse polarity) that are heavily biased to middle and high latitudes in the northern hemisphere.

A histogram showing the distribution of these VDMs is shown in Fig. 6.10. In their analysis of a much smaller data set, McFadden and McElhinny (1982) originally suggested there was a fairly sharp cutoff in the observed VDMs at low values and modeled the distribution of VDMs to take such a truncation into

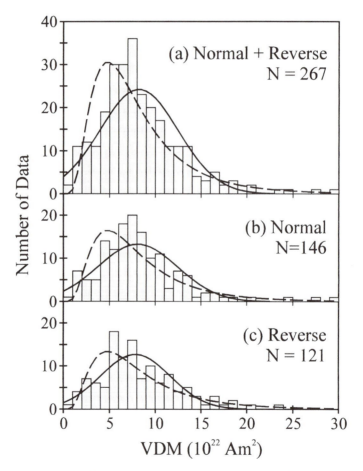

Fig. 6.10. Histogram of VDMs determined from rocks less than 10 Myr old: (a) all data, (b) normal polarity; (c) reverse polarity. Solid curve, best fitting normal distribution. Dashed curve, best fitting log-normal distribution. After Kono and Tanaka (1995b).

account. Kono and Tanaka (1995b) argue that this apparent truncation is caused by the exclusion of transitional data that contain many low VDM values. If these transitional data are included, with appropriate weighting to represent the percentage of time occupied by the transitional state, then the gap near zero VDM will disappear. Kono and Tanaka (1995b) show that the distribution of VDMs appears to be neither normal nor log-normal, but argue that the presence of a few outliers essentially determines the outcome of any significance test. They suggest that these high values could come from low-quality data. After removing six high-value outliers the distribution of VDMs has a normal distribution with a mean value of 7.84×10^{22} Am^2 and a standard deviation of 3.80×10^{22} Am^2 (48.5%). In this case the mean TDM will then be equal to the mean VDM. Separate analyses of the normal and reverse VDMs show no differences between the mean VDMs or their standard deviations.

Cox (1968) suggested that the most useful analogue to geomagnetic polarity reversals was provided by the double-disk dynamo of Rikitake (1958). From this he developed a probabilistic model for reversals in which the dipole moment oscillates sinusoidally about some mean value during stable polarity times. Such a model predicts a distribution of dipole moments that groups with minimum frequency about some central mean (Kono, 1972; McFadden and McElhinny, 1982). This is virtually opposite to that observed (Fig. 6.10). The data thus quite clearly do not support a model of an oscillating dipole moment during stable polarity times.

McElhinny and Senanayake (1982) analyzed archeomagnetic data discussed previously in §4.1. For the past 10,000 yr they deduced that the scatter of VDMs due to nondipole components and errors in the paleointensity determinations had a standard deviation of 21.1%. They also noted that in the period 15,000 to 50,000 yr B.P., even though the dipole moment was half its mean value for the past 10,000 yr, the percentage standard deviation due to the nondipole components was the same as for the past 10,000 yr. The more recent data set summarized in Table 4.2 still supports this general conclusion. Apparently the intensity of nondipole components can be considered in the first-order approximation to have been linearly proportional to the value of the TDM.

Senanayake *et al.* (1982) have shown that the error in paleointensity determinations has a standard error of about 10%. Using this value the archeomagnetic data therefore suggest the standard deviation due to nondipole components is 18.6%. McElhinny and Senanayake (1982) analyzed the present field by looking at the standard deviation of VDMs calculated from field values at points randomly distributed over the Earth's surface. The scatter in VDMs would be due entirely to nondipole components and equivalent to an analysis of worldwide VDMs as in Fig. 6.11 (see §6.3.3). The average standard deviation derived from 50 samplings each of 100 points on the Earth's surface was 17.5%. Kono and Tanaka (1995b) have carried out a similar analysis for the 1990 IGRF and obtained a value of 16.3%. This lends strong support to the deductions from archeomagnetic data.

The VDMs for the past 10 Myr have a standard deviation of 48.5%, which can be considered as arising from variations in the dipole field, the nondipole field, and errors in the paleointensity measurements. Using the values deduced by McElhinny and Senanayake (1982) from archeomagnetic measurements (10% for paleointensity errors and 18.6% for nondipole field variations), this means that variations in the dipole moment over the past 10 Myr must have been 43.7%, which in this case will be equivalent to the variations in TDMs. A comparison of the parameters of the Earth's dipole moment for the past 10 Myr with those for the past 50,000 yr is given in Table 6.3. The fluctuations in the dipole moment, as represented by the standard deviation of TDMs, are very much greater over the 10^7-yr time scale than those observed over 10^4–10^5 yr. This suggests that the fluctuations in the Earth's dipole moment occur over long time scales of 10^6 yr or more.

Using the same database for the past 10 Myr, Tanaka *et al.* (1995) have shown that the observed paleointensities are consistent with that expected for a geocentric axial dipole field. When the data are averaged in 20° latitude bands, the mean values provide a good fit to the expected variation in intensity (F) for a dipole field, where F is proportional to $(1+3\cos^2\theta)^{\frac{1}{2}}$, where θ is the paleomagnetic colatitude (see §3.4.5 and Eq. (3.4.5)). The best fitting curve shows the expected pole to equator variation of 2 to 1 with an equatorial paleointensity of 31.3 µT, corresponding to a dipole moment of 8.20×10^{22} Am^2.

6.3.3 Relative Paleointensities – The Past Four Million Years

Relative paleointensities can be determined in favorable circumstances from sediment cores using the methods described in §3.4.4 and §4.2.1. Such methods have been successful in determining relative geomagnetic field intensity variations over the past several hundred thousand years (§4.2.1). To extend these relative paleointensities to the past several million years requires long and continuous sedimentary records with precise time control and resolution.

TABLE 6.3
VDM Scatter – Dipole and Nondipole Contributions.

Time span	Average dipole moment (10^{22} Am^2)	Standard deviation (%)			
		Dipole	Nondipole	Exp.	Dip + ND
(a) Present Field	7.91	-	17.5	-	-
(b) 0-10,000 yr	8.75	18.0	18.6	10	25.9
(c) 15,000-50,000 yr	4.53	-	-	10	32.0
(d) 0-10 Ma	7.84	43.7	18.6	10	47.5

Sources: (a) McElhinny and Senanayake (1982); (b) Table 4.1; (c) Table 4.2; (d) Kono and Tanaka (1995b).

Sequences drilled during Leg 138 of the Ocean Drilling Program (ODP) offered the first opportunity to obtain such continuous records covering the past 4 million years (Valet and Meynadier, 1993).

The deep-sea sediment cores studied by Valet and Meynadier (1993) show excellent characteristics for relative paleointensity determinations. There are no large changes in lithology, mineralogy, or deposition rates and the magnetic homogeneity down the cores is such that normalization of the NRM for the amount of magnetic material using ARM, IRM, or low field susceptibility gives identical results. The composite relative paleointensity variation between 0.5 and 4 Ma determined by Valet and Meynadier (1993) represents 80 m of sediment and is shown in Fig. 6.11. Because the mean deposition rate does not exceed 2.5 cm ka^{-1} it is reasonable to assume that nondipole variations have been averaged out and that the resulting record is largely dominated by changes in the intensity of the axial dipole moment.

Valet and Meynadier (1993) converted the relative paleointensities to VADMs after matching the data with the synthetic record of Meynadier *et al.* (1992) for the past 140 kyr. This synthetic record had been calibrated to the absolute VADMs for that period determined by McElhinny and Senanayake (1982) and Tric *et al.* (1992). The mean VADM for the past 4 Myr is $3.9 \pm 1.9 \times 10^{22}$ Am^2 and is not significantly different from the value of $5 \pm 2 \times 10^{22}$ Am^2 calculated for the past 140 kyr by Meynadier *et al.* (1992), but is only about one-half the mean VDM of 7.84×10^{22} Am^2 calculated by Kono and Tanaka (1995b) from absolute measurements on volcanic rocks over the past 10 Myr (§6.3.2). Also the mean normal polarity VADM ($4.1 \pm 2.0 \times 10^{22}$ Am^2) and mean reverse polarity VADM ($3.3 \pm 1.5 \times 10^{22}$ Am^2) are not significantly different.

The data of Fig. 6.11 show that there are occasions of variable duration during which the intensity drops but which do not coincide with polarity changes. In many of these cases the drop in intensity coincides with the age of various geomagnetic excursions (see §4.3), lending support to the view that a major requirement for the occurrence of reversal excursions is that the dipole intensity was low, enabling nondipole effects to dominate (Merrill and McFadden, 1994).

6.3.4 Variation with Geological Time

The problem of determining paleointensities in the geological past becomes increasingly difficult the older the rocks studied. The presence of secondary components and the decay of the original magnetization all serve to complicate the problem. Also some of these determinations have been made using dubious techniques with no consistency checks (see §3.4). Because of these inherent difficulties some recent reviews of the subject have taken the view that only measurements carried out using the Thellier method have sufficient rigor to be

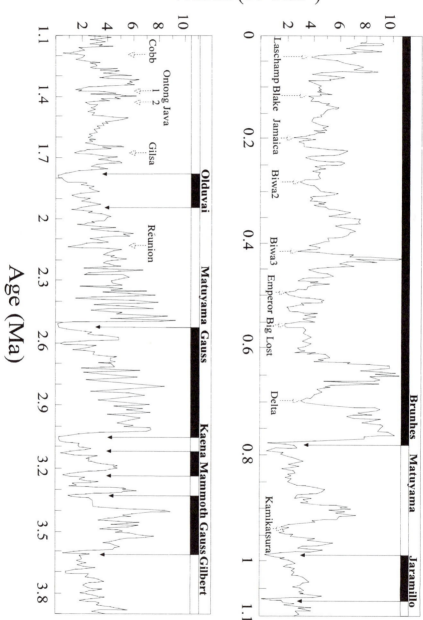

acceptable for discussions of magnetic field strength variations in the geological past (Prévot *et al.*, 1990; Prévot and Perrin, 1992). Tanaka and Kono (1994) compiled a database of global paleointensities for geological time and, in an analysis of the data, Kono and Tanaka (1995b) considered that results determined by either the Thellier or the Shaw method were acceptable.

If dipole fluctuations with standard deviations of 40% or more, as observed for the past 10 Myr, have persisted for geological time, then sufficient numbers of determinations would be required to average out this paleosecular variation to calculate an average dipole moment for any particular geological epoch. Figure 6.12 summarizes the available paleointensity data in terms of virtual dipole moments following Kono and Tanaka (1995b), in which only results obtained using the Thellier or Shaw method are used.

Prévot *et al.* (1990) first suggested that there was an extended period during the Mesozoic when the Earth's dipole moment was low, at about one-third of its Cenozoic value. The general coherence of the data set and its trend with time suggested that the effects of paleosecular variation have at least to some degree been averaged out. Further measurements for the Early Jurassic by Perrin *et al.* (1991) confirmed the trend to low values. This is clearly seen in the results for the past 400 Myr shown in Fig. 6.12a. Prévot *et al.* (1990) suggest that the Mesozoic dipole low corresponds roughly with the progressive decrease in the average frequency of reversals that occurred prior to the Cretaceous normal superchron (Chapter 5). The near present dipole moment was apparently established during or at the end of the Cretaceous normal superchron, with the rise in the moment occurring during the 120- to 80-Myr time interval. During the Cenozoic the dipole moment was similar to its present value.

For the period prior to 400 Ma, the number of paleointensity measurements is much fewer and, following Kono and Tanaka (1995b), Fig. 6.12b summarizes all the available data for the past 3500 Myr. Of particular interest is the oldest paleointensity measurement, which is for the 3500 Ma Komati Formation Lavas in South Africa (Hale and Dunlop, 1984). Two Thellier experiments give an average VDM of $4.7 \pm 0.1 \times 10^{22}$ Am^2, about 60% of the present dipole moment. This result clearly demonstrates the existence of the Earth's magnetic field 3.5 Ga ago. The range of variation of the Earth's dipole moment is 2 to 12×10^{22} Am^2 and is approximately the same for Phanerozoic and Precambrian

---

Fig. 6.11. Relative variations in the Earth's dipole moment for the past 4 Myr determined from deep-sea sediment cores. Polarity chrons and subchrons are indicated by the horizontal bars (black/white shows normal/reverse polarity) and the positions of the reversals are shown by solid arrows. Various excursions are named and indicated by the open arrows and correlated with minina in the dipole moment. Note that the upper figure for the past 1.1 Myr is plotted on a more expanded horizontal scale so as to see more detail of the Brunhes Chron. After Valet and Meynadier (1993).

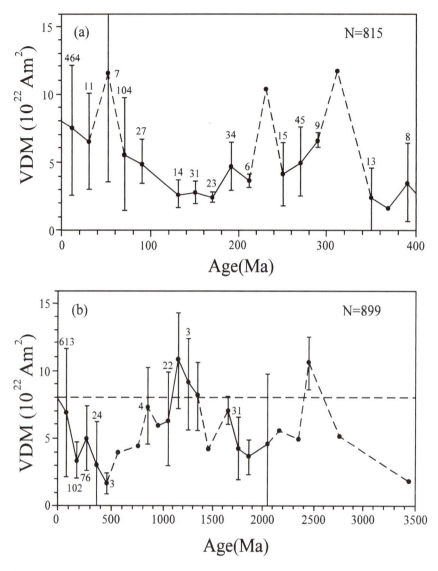

Fig. 6.12. Variation of VDM with geological time calculated from Thellier and Shaw paleointensity determinations. (a) The past 400 Myr averaged over 20-Myr intervals. (b) The past 3500 Myr averaged over 100-Myr intervals. Error bars give the standard error of the mean for each interval, and the number of measurements in each interval is indicated. The horizontal dashed line in (b) indicates the present dipole moment (8×10^{22} Am^2). After Kono and Tanaka (1995b).

times (Prévot and Perrin, 1992). With the present data set, no very-long-term change in dipole moment is apparent. Kono and Tanaka (1995b) point out that it is remarkable that the dipole intensity seems to have been within a factor of 3 of its present value for most of geological time.

6.4 Paleosecular Variation from Lavas (PSVL)

6.4.1 Angular Dispersion of the Geomagnetic Field

The angular dispersion of the geomagnetic field over long periods of time is the aspect of geomagnetic secular variation most accessible to paleomagnetic observation. This is because most paleomagnetic measurements comprise a set of random observations of the geomagnetic field over a given time. The angular dispersion of paleomagnetic measurements may be calculated in two ways, either by finding the dispersion of paleomagnetic field directions or by finding the dispersion of the corresponding virtual geomagnetic poles. The variations of angular dispersion as a function of latitude calculated in these two ways are quite different. The dispersion of field directions decreases with increasing latitude whereas the dispersion of VGPs increases with increasing latitude (see, for example, Cox, 1970; McElhinny and Merrill, 1975). Because of the statistical requirement that spot readings of the paleomagnetic field need to be analyzed, these studies have been restricted to measurements of lava flows. Such studies of the angular dispersion of the geomagnetic field are generally referred to as *paleosecular variation from lavas* (PSVL) in order to distinguish them from studies of recent lake sediments (§4.2), which are referred to more generally as *paleosecular variation* (PSV).

Suppose a set of paleomagnetic measurements consists of N paleodirections of the geomagnetic field. The set of field directions maps into a set of VGPs, and conversely. Let δ_i be the angle between the ith field direction and the mean direction, and let Δ_i be the corresponding angle between the ith VGP and the mean VGP. The best estimate of the angular variance of field directions is

$$s^2 = \frac{1}{N-1} \sum_{i=1}^{N} \delta_i^2 \qquad (6.4.1)$$

and that for the VGPs is

$$S^2 = \frac{1}{N-1} \sum_{i=1}^{N} \Delta_i^2 \quad . \tag{6.4.2}$$

Most analyses of paleomagnetic data use the statistical method of Fisher (1953) (see §3.3.3). If k is the best estimate of the precision of field directions, and K the best estimate of the precision of the VGPs, then the best estimate of the angular variance of field directions is

$$s^2 = 6561/k \tag{6.4.3}$$

and that of the VGPs is

$$S^2 = 6561/K \quad . \tag{6.4.4}$$

However, it should be noted that the Fisher (1953) distribution function has azimuthal symmetry about the mean. Field directions with azimuthal symmetry do not in general map into VGPs that possess azimuthal symmetry, and conversely. Cox (1970) has determined the relationship between angular variances and precision parameters of field directions and VGP in the separate cases that either one or the other has a Fisherian distribution.

When the *field directions are assumed to have a Fisherian distribution*, there will be a conversion factor C_F relating the angular variances and precisions of the field directions and VGPs. If k_F and s_F are the precision and angular dispersion of field directions and K_F and S_F are the corresponding parameters for the VGPs, then

$$C_F = k_F / K_F = S_F^2 / s_F^2 = \tfrac{1}{8}(5 + 18\sin^2 \lambda + 9\sin^4 \lambda) \quad , \tag{6.4.5}$$

where λ is the mean geomagnetic latitude calculated from the mean inclination I from the usual dipole field equation (3.3.5). Conversely, when the *VGPs are assumed to have a Fisherian distribution*, there is a conversion factor C_P relating the angular variances and precisions of the field directions and VGPs. If k_P and s_P are the precision and angular dispersion of field directions and K_P and S_P are the corresponding parameters for the VGPs, then

$$C_P = k_P / K_P = S_P^2 / s_P^2 = \frac{2(1 + 3\sin^2 \lambda)^2}{(5 + 3\sin^2 \lambda)} \quad . \tag{6.4.6}$$

Creer (1962) has determined a slightly different equation for this case as

$$k_P / K_P = S_P^2 / s_P^2 = \frac{2(1 + 3\sin^2 \lambda)}{(5 - 3\sin^2 \lambda)} \quad . \tag{6.4.7}$$

Cox (1970) noted that these transformations are approximate and only valid for small values (<10°) of s and S, but claimed that the differences between these and the full transformation functions were only small and could be neglected. Most models of PSVL used the above approximations until McFadden and McElhinny (1984) showed that the omission of the full transformation function was significant in defining models of PSVL (see *Model F* below).

6.4.2 Models of PSVL

Two types of models have been proposed for analyzing the variation of angular dispersion with latitude. Until recently all these models can best be described as *parametric*, in which variations in the intensity and direction of the dipole and nondipole field have been modeled in a variety of ways. The most recent model is a *phenomenological* model based upon spherical harmonics. In the parametric models it has been convenient to separate contributions to angular dispersion of the geomagnetic field into three factors:
(i)   variation in the intensity and direction of the nondipole field;
(ii) variation of the geomagnetic dipole moment with time;
(iii) changes in the orientation of the geomagnetic dipole such that on average the dipole axis coincides with the axis of rotation (dipole wobble).

The various parametric models have traditionally been given letters of the alphabet following the summary given by Irving (1964). These models may be summarized as follows:

Model A (Irving and Ward, 1964)
   The model considers only variations in the nondipole field. A geocentric axial dipole of fixed dipole moment is perturbed by a single randomly directed vector of constant magnitude. The model predicts an angular dispersion of *field directions* of the form

$$s_A = s_N (1 + 3\sin^2 \lambda)^{1/2} \quad , \qquad (6.4.8)$$

where $s_N = 46.8(f/F_o)$ and (f/F_o) is the ratio of the perturbing nondipole field to the equatorial strength of the dipole field.

Model B (Creer *et al.*, 1959; Creer, 1962)
   In this case it is postulated that the main dipole wobbles such that the resulting VGP follow a Fisher (1953) distribution. The angular dispersion of VGP S_D due to this dipole wobble will be latitude invariant, but expressed as the angular dispersion of field directions, and following (6.4.7), the model predicts

$$s_B = S_D \left(\frac{(5 - 3\sin^2 \lambda)}{2(1 + 3\sin^2 \lambda)} \right)^{1/2} . \tag{6.4.9}$$

Model C (Cox, 1962)

Models A and *B* involve the study of the angular dispersion of field directions and its variation with latitude. In the more general case where there are components from dipole wobble as well as from the nondipole field, it is generally much simpler to consider the angular dispersion of VGPs and their variation with latitude. This is because the angular dispersion of VGPs caused by dipole wobble is latitude invariant and the total angular dispersion S is simply calculated from

$$S^2 = S_D^2 + S_N^2 , \tag{6.4.10}$$

where S_N is the angular dispersion of VGPs caused by the nondipole field. Cox (1962) used a value of $S_D = 11.5°$ for dipole wobble and calculated values of S_N from an analysis of the present geomagnetic field (see §6.4.3).

Model D (Cox, 1970)

In this model Cox assumed that the nondipole field was generated by a set of randomly direction vectors f of varying length that are added to dipole field vectors F of varying length but constant direction. The angular variance of the population of vectors $f + F$ is given by

$$\sigma^2 = (f_R / F)^2 , \tag{6.4.11}$$

where f_R is the most probable length of f. The resulting population has a Fisherian angular distribution (Cox, 1964) and describes the angular variance of the nondipole field. Cox (1970) then assumed that variations in the Earth's dipole moment were sinusoidal with amplitude r about some mean value so that F in (6.4.11) is modified and the angular variance of the population $f + F$ becomes

$$\sigma^2 = \left(\frac{f_R(\lambda)}{F_0} \right)^2 \frac{1}{(1 - r^2)^{3/2}} \cdot \frac{1}{(1 + 3\sin^2 \lambda)} , \tag{6.4.12}$$

where F_0 is the mean equatorial field value and f_R may vary with latitude.

The total angular variance of the geomagnetic field is found by adding the variance from dipole wobble to that from the nondipole and dipole oscillations as in (6.4.10). The angular variation of field directions is found by converting S_D to s_D using (6.4.6) and combining with (6.4.12) to obtain the relation found by Cox (1970):

$$s^2 = C_P S_D^2 + \sigma^2 \quad . \tag{6.4.13}$$

However, it is the angular dispersion of VGPs that is of most interest in *Model D*. In this case there are two possibilities depending on whether field directions or their corresponding VGPs are considered to have a Fisherian distribution. Cox (1970) obtained two relations by using the relationships determined in (6.4.5) or (6.4.6). For the case of Fisherian field directions, the total angular dispersion becomes

$$S^2 = S_D^2 + C_F \sigma^2 \quad . \tag{6.4.14}$$

For the case of Fisherian VGPs, the total angular dispersion becomes

$$S^2 = S_D^2 + C_P \sigma^2 \quad . \tag{6.4.15}$$

Model E (Baag and Helsley, 1974)

It is supposed that there is a linear relationship between the dipole and nondipole fields, and the total angular variance of VGP may be expressed as

$$S^2 = S_D^2 + S_N^2 + 2c_{DN} S_D S_N \quad , \tag{6.4.16}$$

where S_N is the angular dispersion of the nondipole field and c_{DN} is the correlation coefficient between S_D and S_N ($0 \leq c_{DN} \leq 1$) and is a measure of the degree of the assumed linear relation between S_D and S_N. It is suggested that c_{DN} may assume different values for each latitude. The model therefore describes the variation of the angular dispersion of the geomagnetic field with latitude as a region bounded by the limits $(S_D^2 + S_N^2)^{1/2}$ and $(S_D + S_N)$.

Model M (McElhinny and Merrill, 1975)

This is an extension of *Model D* and assumes that the nondipole field might arise from two sources, the first due to some sort of interaction with the dipole field (analogous to the drifting part of the present nondipole field) and the second due to fixed eddies near the core–mantle boundary (analogous to the standing part of the present nondipole field). The first source of the nondipole field above will result in both a Fisherian distribution of nondipole VGP and an increase of the average intensity of the nondipole field with latitude, as is in fact observed for the present geomagnetic field. The intensity of the geocentric axial dipole field varies with latitude as $(1 + 3\sin^2 \lambda)^{1/2}$, so that this will also describe the variation in intensity of the nondipole field. The function $f_R(\lambda)$ in (6.4.12) is then replaced with

$$f_R(\lambda) = f_R^0 (1 + 3\sin^2 \lambda)^{1/2} \quad . \tag{6.4.17}$$

The second source of the nondipole field would, however, be expected to result in a Fisherian distribution of nondipole field directions with the same intensity at all latitudes with f_R constant in (6.4.12). Noting that the first source gives Fisherian VGPs whereas the second gives Fisherian directions, then

$$S^2 = S_D^2 + aC_P\left(\frac{f_R^0}{F_0}\right)^2 q + bC_F\left(\frac{f_R}{F_0}\right)^2 \frac{q}{(1+3\sin^2\lambda)} \quad , \tag{6.4.18}$$

where a and b are constants giving the proportion of the nondipole field attributed to each source and $q = 1/(1-r^2)^{3/2}$.

Harrison (1980) later pointed out that this model did not fit the latitude variation of the angular dispersion of the present geomagnetic field. To overcome this difficulty he modified (6.4.17) to include a $\cos^2\lambda$ term. However, both McElhinny and Merrill (1975) and Harrison (1980) failed to take into account the imprecise nature, as pointed out by Cox (1970), of the relationships between angular dispersion of VGPs and field directions resulting from the mapping functions.

Model F (McFadden and McElhinny, 1984)

All of the models discussed above suffer from several assumptions that are now known to be incorrect. In all cases it was assumed that time variations in the Earth's dipole moment are sinusoidal. This would produce a distribution of VDM totally at variance with that observed (Fig. 6.11a). In *Model F* it is assumed that the nondipole field originates from an interaction with the poloidal field. On such a model the VGPs would be expected to be distributed with spherical symmetry about the geographic pole. Cox (1970) noted that the angular dispersion of field directions of the present field is not symmetric but that of the corresponding VGPs is nearly so. The model also implies that the time-averaged energy density of the nondipole field at any given latitude will be linearly related to the energy density of the dipole field at that latitude so that

$$\langle f^2 \rangle = k^2 F^2 \quad , \tag{6.4.19}$$

where f is the intensity of the nondipole field, F is the intensity of the dipole field, k is a constant, and $\langle f^2 \rangle$ is the expectation value, or time average, of f^2. Analyses of VDMs over the past 50,000 yr by McElhinny and Senanayake (1982) (see §4.1) and over the past 5 Myr by McFadden and McElhinny (1982) (see §6.3.2) are entirely consistent with the above in that they show a correlation between the intensity of the nondipole and the dipole field.

Imagine the distribution of VGPs arising from the nondipole field to be a set of "nondipole" VGPs of constant length but uniformly distributed in direction, superimposed upon a "dipole" VGP of constant direction. Then, from (6.4.19), as the nondipole field changes in proportion to the dipole field, there will be a set of VGPs with spherical symmetry about the geographic pole. This set of VGPs can then be mapped back to the direction and intensity observed at latitude λ. If **J** is the vector denoting this direction and intensity, then the tip of **J** will be on a closed surface that is in general not a sphere but an ellipsoid. The relative lengths of the axes x, y, and z of this ellipsoid will change with latitude. Thus the angular dispersion of directions will change with latitude even though the energy density of the nondipole field remains a constant proportion of the energy density of the dipole field. This is an important factor that was ignored in all previous models.

From the preceding the nondipole field vector **f** is simply given by

$$\mathbf{f} = \mathbf{J} - \mathbf{F} \quad . \tag{6.4.20}$$

McFadden and McElhinny (1984) have shown that the time-averaged angular variance of *directions* is then given by

$$\sigma_d^2(\lambda) = \frac{2k^2[1 + a\exp(-b\lambda^2)]}{3\sqrt{1 + 3\sin^2\lambda}} \quad , \tag{6.4.21}$$

where k is given by (6.4.19) and a and b are functions of $\sigma_d(0)$, but over the range of interest they are essentially constants. Thus $\sigma_d(\lambda)$ is independent of the distribution of either f or F! The overall variance of VGPs is given by

$$S^2 = S_D^2 + S_N^2 W^2 \quad , \tag{6.4.22}$$

where S_N is the angular dispersion of VGPs at the equator caused by variations in the nondipole field and W is given by

$$W^2 = \frac{[1 + a\exp(-b\lambda^2)]}{(1 + a)\sqrt{1 + 3\sin^2\lambda}} \cdot \frac{H[\lambda, \sigma_d^2(\lambda)]}{H[0, \sigma_d^2(\lambda)]} \quad , \tag{6.4.23}$$

where $H[\lambda, \sigma_d^2(\lambda)]$ is the transformation from the angular variance of directions to the angular variance of VGPs at latitude λ. Tables of appropriate values of a, b, and $H[\lambda, \sigma_d^2(\lambda)]$ are given by McFadden and McElhinny (1984).

Model G (McFadden *et al.*, 1988)

This is the only *phenomenological* model of PSVL that has been proposed. It makes no assumptions about the relation of the nondipole to dipole field but is based upon the theoretical work of Roberts and Stix (1972) in relation to dynamo theory (§8.5). Roberts and Stix have shown that under certain symmetry conditions the magnetic field solutions for a spherical dynamo separate into two completely independent families referred to as the antisymmetric ("dipole") and symmetric ("quadrupole") families. Spherical harmonic terms of degree n and order m (that is, with Gauss coefficients g_n^m and h_n^m) belong to the antisymmetric family if $(n-m)$ is odd and to the symmetric family if $(n-m)$ is even. Table 8.3 (§8.5) shows how some of the low-degree Gauss coefficients are separated into the two families.

McFadden *et al.* (1988) have shown that a separation into these two families can be a powerful tool for modeling PSVL. This separation is, however, only valid for a mean velocity field that is symmetric with respect to the equator and if the α-effect of dynamo theory (§8.3.1) is antisymmetric with respect to the equator. A strongly dominant g_1^0 term is of course central to the concept of a VGP, so the parameter of interest in determining a model for PSVL is the angular dispersion of VGP positions about the g_1^0 position. The total angular variance of VGPs on this model is thus

$$S^2 = S_A^2 + S_S^2 \ , \tag{6.4.24}$$

where S_A is the angular dispersion due to the antisymmetric family and S_S is that due to the symmetric family. From an analysis of the present geomagnetic field (§6.4.3 below) it is seen that the dispersion at the equator is caused entirely by the symmetric family terms and is effectively independent of latitude. The latitudinal variation comes from the antisymmetric family terms and, except for the leveling off above 70°, the dispersion is proportional to the latitude. The model therefore predicts

$$S^2 = (a\lambda)^2 + b^2 \ , \tag{6.4.25}$$

where $S_A = a\lambda$ and $S_S = b$, and a and b are constants to be determined.

Model H (Constable and Parker, 1988)

Here the nondipole field has been modeled by noting that the spatial power spectrum of the present-day nondipole field is consistent with a white source near the core–mantle boundary having a Gaussian distribution. After suitable scaling, the spherical harmonic coefficients may be regarded as statistical samples from a single giant Gaussian process. Assuming that this characterization holds for the magnetic field in the past, this model can then be combined with an arbitrary statistical description of the dipole. The corresponding probability density function of and cumulative distribution functions for declination and inclination at any site on the Earth's surface can then be computed.

Constable and Parker (1988) use the paleomagnetic data spanning the past 5 Myr to constrain the free parameters of the model that describe the dipole field. The final model then has the following properties:

(i) The axial dipole (g_1^0) distribution is bimodal and symmetric, resembling a combination of two normal distributions whose peaks are close to the present-day value and its reverse counterpart.

(ii) The standard deviation of each of the nonaxial dipole terms (g_1^1 and h_1^1) and that of the magnitude of the axial dipole are each about 10% of the present-day g_1^0 term.

(iii) The axial quadrupole (g_2^0) reverses sign with the axial dipole and has a mean magnitude of 6% of that of the axial dipole.

(iv) Each Gauss coefficient for the nondipole field is independently normally distributed with zero mean and standard deviation, consistent with a white source at the core surface.

Kono and Tanaka (1995a) have analyzed *Model H* to assess the predictions of the model for VGP dispersion with latitude by mapping the Gauss coefficients to the pole. They show that the model predicts a nearly constant VGP dispersion irrespective of latitude and conclude that the assumption of homogeneous randomness in the Gaussian process is probably too simple. Hulot and Gallet (1996) and Quidelleur and Courtillot (1996) have also analyzed *Model H* and concluded that a substantially different variance structure is required for a model of this type to match the observed field. Harrison and Huang (1990) and Harrison (1994) noted that the lower order harmonics of the present field tend to be larger than the higher order harmonics of the same degree *n*. This alone shows that the assumptions in *Model H* are not consistent with the present field.

As noted by Quidelleur and Courtillot (1996), *Models G* and *H* introduced the language of spherical harmonics into the modeling of paleosecular variation. This has facilitated analyses such as those of Kono and Tanaka (1995a), Hulot and Le Mouël (1994), Hulot and Gallet (1996), and indeed that of Quidelleur and Courtillot (1996). This has greatly reduced the separation in analysis techniques of the paleomagnetic and geomagnetic communities, and will undoubtedly lead to improved understanding of the geomagnetic field.

6.4.3 Angular Dispersion of the Present Geomagnetic Field

Cox (1962) and Creer (1962) first attempted to analyze the dispersion of the present for comparison with paleomagnetic results. The method used is to suppose that at any latitude, over the time span represented by a typical paleomagnetic data set, the observed dispersion either in directions (Creer, 1962) or in VGPs (Cox, 1962) will correspond to the average of the present field values observed around that latitude. The procedure is to calculate either the direction or the VGP at 10° intervals of longitude from a current IGRF model and then calculate the dispersion of directions or VGPs for each latitude. Cox (1962) calculated the dispersion with respect to the geomagnetic dipole axis in an attempt to isolate from the total VGP dispersion that component caused by the nondipole field (the rest of the dispersion being attributed to dipole wobble). However, if the PSVL dispersion is separated into its dipole wobble and nondipole field components, the nondipole field component is still with respect to the spin axis; it is not translated to some other axis (McFadden *et al.*, 1988).

The VGP dispersions for IGRF90 about the Earth's spin axis have been calculated as described by McFadden *et al.* (1988) and above, and are shown in Fig. 6.13. There is a clear asymmetry in the dispersion between the southern and the northern hemispheres. Rotations of the spherical harmonics about the spin axis affect this symmetry. For example, rotation of the g_2^1 and h_2^1 harmonics through 90° effectively eliminates this asymmetry, and a rotation through 180° reverses the sense of the asymmetry. Hence, in the first-order approximation, it is reasonable to assume north–south symmetry of the time-averaged paleomagnetic field. McFadden *et al.* (1988) made a least-squares fit of their *Model G* to the average north–south values of dispersion and found an excellent fit with model parameters of $S_A = (0.24\pm0.02)\lambda$ and $S_S = 13.5°\pm0.6°$ (see §6.4.2).

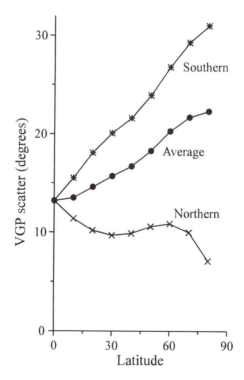

Fig. 6.13. Latitude variation of VGP scatter (in degrees) calculated from IGRF90. Asterisks, southern hemisphere; crosses, northern hemisphere; solid circles, average (considered a representative equivalent of the time-averaged field). Following the analysis of McFadden *et al.* (1988).

6.4.4 The Past Five Million Years

McElhinny and Merrill (1975) discussed the problems of analyzing paleomagnetic data to obtain estimates of the variation of angular dispersion with latitude. The problem relates to how one defines a time-averaged field, and the basic arguments have already been set out in §6.2.1 and illustrated in Fig. 6.5. The properties of the time-averaged field can only be specified through a combination of space–time averaging. The field needs to be defined in a global sense. Thus observations at any one locality can produce spurious effects such as the so-called Pacific dipole window (Doell and Cox, 1972).

Only data from lava flows or thin intrusions are presently adequate for studying the paleosecular variation. Data from individual flows or thin intrusions are spot readings of the geomagnetic field at one locality. The angular dispersion of VGPs from a sequence of lava flows can be measured with respect to the present geographic pole using (6.4.2) when data for the past 5 Myr are being considered. It is assumed that for the past 5 Myr the present axis of rotation has remained fixed with respect to all landmasses (i.e., continental drift has been neglected). However, several samples are usually collected from each lava flow and it is necessary to make a correction for the within-lava angular dispersion S_W to determine the angular dispersion of the geomagnetic field S_T (the between-lava angular dispersion). If S_T is the total angular dispersion, then

$$S_F^2 = S_T^2 - S_W^2 / \bar{n} \quad , \tag{6.4.26}$$

where $\bar{n}$ is the average number of samples measured in each lava flow. In order to make sure that the correction term for the within-site dispersion is small, it has been traditional to make detailed measurements for each lava flow (e.g., Doell and Cox, 1965; Doell 1970, 1972; Watkins, 1973). However, S_W is usually considerably smaller than S_F, and in such circumstances it is possible to use results where only three or four samples per flow have been studied.

Figure 6.14 shows the fit of *Model G* to PSVL data for the past 5 Myr as determined by McFadden *et al.* (1991), combining northern and southern hemisphere results as originally proposed by McElhinny and Merrill (1975). Normal and reversely magnetized data have been combined because the data are inadequate to attempt any subdivision. In any case Merrill and McElhinny (1983) have pointed out that this combination has the advantage of canceling out the effects of inadequate magnetic cleaning in the paleomagnetic data. These effects tend to increase the scatter observed in reversely magnetized rocks and reduce the scatter in normally magnetized rocks so their mean probably represents the best estimate of the angular dispersion of the paleomagnetic field. The estimated values of the *Model G* parameters for the past five million years are $S_A = (0.23\pm0.02)\lambda$ and $S_S = 12.8°\pm0.4°$, which are indistinguishable from the

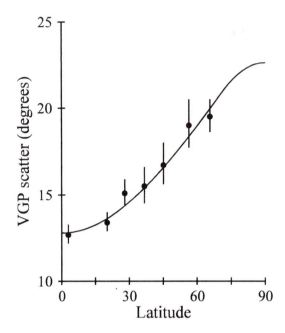

Fig. 6.14. Least-squares fit of *Model G* to the combined polarity VGP scatter for the past 5 Myr. After McFadden *et al.* (1991).

values estimated for the present field (§6.4.3 and Fig. 6.13). However, as noted in §6.4.3, rotation of individual harmonics can have a significant effect on the "equivalent" VGP scatter. McFadden and McElhinny (1984) discuss the effect of the varying shape of equivalent VGP scatter curves as the field changes deterministically. Hulot and Le Mouël (1994) find that the correlation times of individual harmonic coefficients of the field appear to be less than 300 yr (the correlation time decreasing with increasing degree), and Hulot and Gallet (1996) note that the equivalent VGP scatter for the 1800 and 1900 epoch fields is different from that for recent years. Thus, even though the present field is a typical sample of the field over the past 5 Myr, the close similarity of its equivalent VGP scatter curve to the time-averaged curve must be considered as coincidental.

Johnson and Constable (1995a) have made a careful selection of all available paleomagnetic data taking into account the relevant concerns. To exclude data from polarity transitions or geomagnetic excursions from the analysis, it is necessary to decide on some arbitrary cutoff angle from the pole beyond which VGPs are not included in the data set. McElhinny and Merrill (1975) used a cutoff angle of 40° (VGPs with latitudes <50° were excluded) based upon the observation by Wilson *et al.* (1972) that there is a rapid fall in intensity in lava flows with VGP latitudes <45°. Johnson and Constable (1995a) used a stricter criterion, excluding all VGPs with latitudes <55°. As an arbitrary overall cutoff this criterion is possibly rather severe. Vandamme (1994) has analyzed the effects this can have on the resulting statistical distribution. He concludes that the cutoff point should really be varied according to the observed dispersion. More dispersed data sets, such as are observed at higher latitudes, may need to have the cutoff point widened to avoid truncating the data set and thus reducing the observed dispersions to values lower than the true ones.

The data set compiled by Johnson and Constable (1995a) consists of 2187 records from 104 distinct locations (1528 normal and 659 reverse) covering the past 5 Myr and was more rigorously selected than previous data sets. They analyzed their data set primarily to test the Constable and Parker (1988) statistical model for paleosecular variation (*Model H*) but noted that there appeared to be a significant difference between data for the northern and southern hemispheres. Although the data set compiled by Johnson and Constable (1995a) is superior to those analyzed previously and some of the older data have (correctly) been discarded, the careful selection criteria used have resulted in an overall reduction in the amount of data from the previous analyses. Thus the problems with time-averaging discussed in §6.2.1 persist and our view is that it is not possible, even with the current data set, to substantiate the existence of hemispherical differences.

Johnson and Constable (1995a) showed that *Model H* (Constable and Parker, 1988) does not predict the latitudinal variation of VGP dispersion, in concert with Kono and Tanaka (1995a) who demonstrated this from a mathematical analysis of the model.

6.4.5 A Pacific Dipole Window?

Fisk (1931) noted many years ago that the historical rates of secular variation in the Pacific region were anomalously low. Doell and Cox (1972) summarized the then available information relating to historic spherical harmonic analyses. The nondipole field intensity in the Pacific is presently very low and appears to have been that way since at least 1829. The rate of change of the nondipole field is also somewhat lower in the Pacific than elsewhere. Bloxham and Gubbins (1985) substantially extended the available database of direct field observations by recovering and corrrecting mariners' observations, and produced field models at the core–mantle boundary from 1715 onward. Under the central Pacific they found low secular variation but a significant nondipole component (a flux bundle that has only a small manifestation at Earth's surface because of its short wavelength). The problem that arises is whether this low in nondipole components and their secular variation has been a persistent or a transient phenomenon. Most recently Yukutake (1993) reexamined historical declination data and found an intense focus of the nondipole field in the north Pacific in the 17th century. He concludes that the near absence of any nondipole field and its secular variation in the Pacific is only a very recent, and therefore transient, phenomenon.

Doell and Cox (1972) summarized their extensive investigations of Brunhes-age lava flows from different parts of the world including, especially, data from the Hawaiian islands. The notable feature of their data was the anomalously low angular dispersion for Hawaii, the most extensively sampled region at that time. They argued that their data provided strong evidence for subdued secular variation in the central Pacific for the past 0.7 Myr, consistent with pronounced attenuation in the geomagnetic spectrum in the period range 200–2000 yr. The implication is that there exists a lateral inhomogeneity in the lowermost mantle that interacts with the core in such a way as to suppress the generation, beneath the central Pacific, of nondipole fluctuations in this range. The absence of nondipole field variations thus allows only the variations in the dipole field to be seen. This observation has led to the central Pacific region being referred to as the *Pacific Dipole Window* (or Pacific nondipole low).

Several mechanisms, each involving lateral variations in the physical or chemical properties in D" (Lay, 1995), might be able to produce a dipole window. For example, Doell and Cox (1972) suggested that core convection was altered because of differences in the core–mantle thermal boundary conditions beneath the Pacific region. Gubbins and Richards (1986) suggest that topography on the core–mantle boundary associated with subduction zones may explain the low secular variation in the Pacific. Runcorn (1992) suggests that most of D" beneath the Pacific has very high (metallic) electrical conductivity, which explains both the dipole window and the claimed systematics in reversal transition paths (§5.4.3). There appears to be mounting evidence that the seismic properties of the lowermost mantle beneath the Pacific are different from elsewhere (Dziewonski *et al.*, 1977; Dziewonski and Woodhouse, 1987; Woodward and Masters, 1991; Su *et al.*, 1994; Grand, 1994; Lay, 1995) and that this may be a consequence of subduction processes (Richards and Engebretson, 1992; Grand, 1994). Changes in the core–mantle boundary conditions occur over long (tens of millions of years) time scales (Loper, 1992; McFadden and Merrill, 1995b), so if the present Pacific Dipole Window is caused by core–mantle boundary conditions, then it should have existed for millions of years.

However, several studies have cast considerable doubt on the hypothesis of a Pacific Dipole Window. The large proportion of data used by Doell and Cox (1972) from the Puna and Kau volcanic series extensively biases the observed angular dispersion to a low value. It now appears that this low dispersion arises because the flows were erupted in such a short time that they did not record anything like the full range of the secular variation. In their global analysis of paleosecular variation data from lavas, McElhinny and Merrill (1975) proposed that any low nondipole field variation in Hawaii can only be a recent phenomenon and restricted at most to Holocene time. Duncan (1975) studied lava flows for the past 5 Myr in French Polynesia and similarly concluded that there was no evidence for unusually low secular variation in that region of the Pacific prior to the Holocene. Coe *et al.* (1978) concluded from directional and paleointensity studies on Hawaii that the low angular dispersion on Hawaii arises not because the lavas are of Holocene age, but because they were erupted so rapidly as to record very little secular variation. They concluded that their data strongly suggest that sizable nondipole geomagnetic fields existed in the vicinity of Hawaii during the Holocene epoch and perhaps earlier.

In contrast, McWilliams *et al.* (1982) reopened the possibility of a Pacific Dipole Window based on their reexamination of the Hawaiian data using new ^{14}C dates for many of the lavas. They concluded that the angular dispersion of the VGPs had been unusually low during the Brunhes Epoch when compared with global data. However, McWilliams *et al.* (1982) made the error of comparing the dispersion of VGPs about their mean on Hawaii with the global dispersion that was calculated about the geographic pole. Given the implications of the hypothesis, McElhinny *et al.* (1996b) reanalyzed all the data both for the Pacific region and elsewhere for the Brunhes Epoch. Their analysis is quite unequivocal in showing that the angular disperion of VGP about the geographic pole in the Pacific region is almost identical to that observed outside the Pacific. Thus all the evidence now strongly suggests that the hypothesis of a Pacific Dipole Window can confidently be rejected for prehistoric times.

6.4.6 Variation with Geological Time

Brock (1971) first attempted to see if time variations in PSVL could be seen in paleomagnetic data. He analyzed 83 selected paleomagnetic results extending back to 2.7 Ga. Because of the nature of the data he had to analyze the angular dispersion of directions about the calculated mean for each result rather than about any known mean direction. Brock then used Cox's (1970) *Model D* in the form

$$s^2 = s_N^2 W_N^2 + S_D^2 W_D^2 \qquad (6.4.27)$$

to describe the latitude variation of the angular dispersion of directions. S_D and s_N are the angular dispersion arising from dipole wobble and the nondipole field, respectively (see §6.4.3), and W_N and W_D are transformation functions that are functions of latitude. The first term is the equivalent of *Model A* (Irving and Ward, 1964) given by Eq. (6.4.8). A least-squares fit to the paleomagnetic data showed that $S_D = 2.0°$, suggesting that dipole wobble must have been very low in the past. This led Irving and Pullaiah (1976) to propose that *Model A* could be used as a good approximation for PSVL over geological time. Unfortunately, it is known that the nondipole field cannot be represented by a randomly oriented vector of fixed length independent of latitude as is presumed in *Model A*.

McFadden *et al.* (1991) have used *Model* G to make estimates of the variation of PSVL with time over the past 190 Myr. To carry out such an analysis it is necessary to be able to reconstruct the positions of the tectonic plates using marine magnetic anomalies and sea-floor spreading. Africa was considered fixed for each reconstruction and the mean global paleomagnetic pole for Africa is effectively determined. It is then a simple matter to rotate this mean pole in present-day coordinates to the present geographic north pole (see also §6.2.4). The dispersion of the VGPs about the pole can then be determined. Figure 6.15 summarizes these results in terms of the *Model G* parameters S_S and S_A/λ (see Eqs (6.4.24) and (6.4.25)). McFadden *et al.* (1991) show that the results obtained for the PSVL in this way are remarkably robust and do not depend precisely on the plate reconstructions used.

The most obvious result in Fig. 6.15 is the fact that the contribution to PSVL from the symmetric dynamo family was much reduced during the Cretaceous Normal Superchron (80–110 Ma). Indeed there is a near mirror imaging of the variations in S_S and S_A/λ with time, possibly suggesting that an increase in one

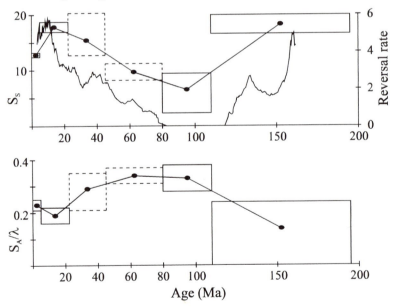

Fig. 6.15. Variation of S_S and S_A/λ from the present back to 190 Ma together with the reversal rate. The boxes around the plotted points indicate the range of ages contributing to the estimated parameter and the 95% confidence limits on the estimate. A solid box indicates a good fit to the data and a dotted box indicates a poor fit. After McFadden *et al.* (1991).

family is accompanied by a decrease in the other. At this point we merely note this result, the significance of which is discussed in §11.3.2.

Chapter Seven

Processes and Properties of the Earth's Deep Interior: Basic Principles

The properties of the Earth's interior, particularly the core, are important to the geomagnetist for evaluating theories dealing with the Earth's magnetic field. Considering the difficulties involved, geophysicists have done a remarkable job in obtaining information about the Earth's interior. However, the accuracies quoted for some of the important physical parameters are often far higher than warranted. Reviews of some of the properties of the Earth's core have been given by Brett (1976), Ringwood (1979), Ahrens (1979), Stevenson (1981), Jeanloz (1990), Poirier (1991), Stacey (1992), Jacobs (1992), and Anderson (1995).

7.1 Seismic Properties of the Earth's Interior

Inversion of seismic data has provided the best information on the elastic properties and the density of the Earth's interior. The magnitudes of compressional and shear wave velocities for an isotropic homogeneous material are respectively given by

$$v_P = \sqrt{\frac{\kappa_s + \frac{4}{3}\mu_e}{\rho}} \qquad (7.1.1)$$

$$v_{sh} = \sqrt{\frac{\mu_e}{\rho}} \quad , \tag{7.1.2}$$

where κ_s is the adiabatic incompressibility, μ_e the shear modulus or rigidity (not to be confused with permeability), and ρ the density. Additional elastic constants are required to describe an anisotropic medium, ranging from a total of 3 for a cubic material to 21 for a triclinic solid. There are three unknowns in (7.1.1) and (7.1.2), so the system is underdetermined. However, this problem can be overcome if an equation of state, relating density to pressure, for the Earth is available, and there are several such theoretical and empirical equations used in geophysics (e.g., Poirier, 1991). One useful empirical equation of state, which can even be used across phase transitions (providing significant changes in bonding type or coordination number do not occur), is Birch's law,

$$v_P = a_m + c\rho \quad , \tag{7.1.3}$$

where c is an empirical constant near 3 for the Earth and a_m is a constant that depends on mean atomic weight. For example, the mean atomic weight of Mg_2SiO_4 is 20.13 g, and Birch found the mean atomic weights for the Earth's mantle and core to be close to 21 and 49 g, respectively.

In addition to body wave seismic data, surface wave and free oscillation data also provide valuable information. In particular, free oscillations (eigenvibrations or normal modes) can be separated into spheroidal and torsional modes. The spheroidal modes have displacement in the radial direction, so the gravitational force must be included in the equations of motion. One consequence is that normal mode data can provide direct information on density without an equation of state from mineral physics. However, only the longest period normal modes (the longest of which is approximately 54 min) sample the core and so these modes can only provide estimates of elastic parameters and density averaged over many hundreds of kilometres. Normal mode data can be used to show that the inner core has rigidity (Dziewonski and Gilbert, 1971) and to provide constraints on the Earth's density distribution.

Although the inner core has finite rigidity, it is important to understand that this does not necessarily imply that the material is solid in the conventional sense: it only requires that long-period shear waves can pass through the material. Poisson's ratio,

$$\sigma_p = \frac{3\kappa_s - 2\mu_e}{6\kappa_s + 2\mu_e} \quad , \tag{7.1.4}$$

is sometimes used by seismologists to provide an estimate of fluid content in a material. Relatively high values of σ_p are reported for the inner core and are sometimes interpreted to imply that the inner core is partially molten. However, Stacey *et al.* (1981) have shown that finite strain theory can be used to predict large values for σ_p for "solid" (long-range atomically ordered) material at core pressures. Therefore, the seismic data bearing on the inner core cannot be interpreted unambiguously. Nevertheless, based on parsimony, normal mode data, and the increase in the compressional wave velocity, it appears safe to conclude that the inner core is mostly, if not completely, solid.

The core–mantle boundary, at a depth of 2891 km, represents the largest discontinuity in the Earth: the shear wave velocity vanishes below this boundary, there is a sharp drop in compressional wave velocity, and there is a density contrast across the boundary of 4.4×10^3 kg m^{-3}. Although the vanishing of shear waves in the outer core is (reasonably) assumed to show that the outer core is liquid, the possibility of solid particles suspended in the liquid cannot be dismissed.

A parameterized Earth model (PREM, Preliminary Reference Earth Model) developed by Dziewonski and Anderson (1981), based on astronomical, free oscillation and surface wave, and body wave data, is a good first-order estimate of the Earth's seismic structure. Table 7.1 gives some of the core's parameters from this model. These parameters are adequate for most, but not all, geomagnetic studies. One crucial parameter needed to estimate core convection energy sources is the density contrast across the inner core boundary. Estimates for the density contrast can be obtained from the reflection of compressional waves off the inner core boundary and from normal mode data. The best estimate for this contrast appears to be that given by Shearer and Masters (1990) and is 0.55×10^3 kg m^{-3}. However, estimates both substantially lower and substantially higher are published in different seismic models.

Many seismic studies over the past decade or so have shown that the older

TABLE 7.1
Some Parameters for the Earth's Core from the PREM Model[a]

Location	Radius (km)	Density (10^3 kg m^{-3})	Pressure (10^{11} Pa)	Gravity (ms^{-2})	Incompress-ibility (10^{11} Pa)	Rigidity (10^{-11} Pa)
Earth's center	0	13.0885 -8.8381 x^2	3.6385	0	14.253	1.761
Inner core–outer	1221.5	12.7636	3.288	4.40	13.434	1.567
core "boundary"	1221.5	12.5815 -1.2638 x -3.642 x^2 -5.5281 x^3	3.288	4.40	13.047	0
Outer core–mantle	3480.0	9.9040	1.3575	10.68	6.441	0
"boundary"	3480.0	5.5650	1.3575	10.68	6.556	2.938
Top of D"	3630.0	—	1.2697	10.48	6.412	2.899

Note. $x \equiv r/a$, where r = radius and a = 6371 km.
[a]Dziewonski and Anderson (1981).

isotropic, homogeneous, and radially symmetric models for the Earth's deep interior are inaccurate. Instead, three-dimensional seismic velocity, or tomographic, images for the Earth's lower mantle are needed (e.g., Dziewonski *et al.*, 1977; Dziewonski and Anderson, 1984; Dziewonski and Woodhouse, 1987; Woodward and Masters, 1991; Su *et al.*, 1994; Grand, 1994; Lay, 1995). One result is that the seismic velocities appear to be relatively slow in the lower mantle in a broad region beneath the Pacific Ocean. It is possible that this reflects relatively higher temperatures there and that those temperatures might affect core processes, perhaps in such a way as to reduce the magnitude of the nondipole field beneath the Pacific as observed in the historical record (§2.4.4). If so, however, then this so-called Pacific Dipole Window should have been present for the past several million years. McElhinny *et al.* (1996b) show that this has not been the case (see also §6.4.5).

All recent global seismological models indicate a layer approximately 200 to 300 km thick at the base of the mantle, the so-called D" layer, that exhibits reduced, or even negative, shear wave velocity gradients (e.g., Lay, 1989). The reasons for this decrease in velocity gradient are believed to include an increase in the temperature gradient and a change in chemistry (e.g., Lay 1989). Morelli and Dziewonski (1987) have used data from compressional waves reflected off the core–mantle boundary and from waves that pass through the core to infer that there is large-wavelength (greater than 1000 km) topography (±4 km) on the core–mantle boundary. In contrast, Creager and Jordan (1986) compare data from compressional waves that pass through the inner core with waves that pass through the outer core but not the inner core, to find large-scale lateral variations in seismic velocities within D". Unfortunately, the seismic residuals (variations relative to PREM) found by Morelli and Dziewonski and by Creager and Jordan are not always compatible, suggesting that aliasing and/or biasing effects are present. Nevertheless, there is little doubt that topography at the core–mantle boundary and/or lateral chemical heterogeneity in D" are indicated by the seismic data (e.g., Young and Lay, 1987, 1990; Lay, 1989; Gaherty and Lay, 1992; Rekdal and Doornbos, 1992). In summary, D" can probably best be thought of as a thermal boundary layer in which a chemical boundary layer is embedded (Lay, 1989). Some implications for geomagnetism are discussed in Chapter 11.

Seismic data suggest that the inner core may exhibit cylindrical anisotropy about the rotation axis (e.g., Creager, 1992). It is unclear whether this anisotropy is confined to the outermost part of the inner core (Morelli *et al.*, 1986) or is present throughout the inner core (Shearer *et al.*, 1988). Interpretations of these data include preferential crystal growth on freezing (Morelli *et al.*, 1986) and crystal alignment associated with convection (Jeanloz and Wenk, 1988). These interpretations may need to be reconsidered if the Saxena *et al.* (1993) analyses of the phase diagram of iron stand up: a *bcc* structure for iron is preferred over the *hcp* structure previously favored. Anisotropy of the magnitude favored in some interpretations of the seismic data (e.g., Creager, 1992) may be incompatible with the *bcc* structure, a conclusion supported by recent mineral physics theory (Stixrude and Cohen, 1995a).

There are suggestions that the outer core may also exhibit structure. It has been suggested from seismic body wave studies that the outermost 200 km (Lay and Young, 1990) or 800 km (Souriau and Poupinet, 1991) is chemically different from the rest of the core; i.e., the outermost core is stratified. Finally, large-scale aspherical structure in the outer core is called on to explain the anomalous splitting of core-sensitive multiplets in the normal mode data (Widmer *et al.*, 1992). Contrary to common supposition, data from geomagnetism, paleomagnetism and mineral physics do not exclude the preceding possibilities, as will be discussed further in this chapter, Chapter 9, and Chapter 11.

7.2 Chemical and Physical Properties

7.2.1 Composition

Candidate materials for the Earth's deep interior are obtained primarily from geochemical considerations. Subsequently, one attempts to subject these materials to mantle and core pressures and temperatures to determine their seismic velocities and densities. If the properties (often extrapolated to appropriate pressure and temperature conditions) match those obtained from seismology, the material remains an acceptable candidate; otherwise it is rejected. Pressures in the lowermost mantle and core have been achieved in the laboratory using diamond anvil (static) techniques and shock wave (dynamic) techniques, both of which have recently been described by Poirier (1991).

Diamonds become brittle above 700 to 900°C, so it is difficult to obtain uniform high temperatures in the laboratory. Transmitting a laser beam through a diamond anvil to heat a spot a few microns in diameter in small samples (a few hundred microns in width) bypasses the diamond brittleness problem, but at the expense of introducing strong nonhydrostatic stress. Because of this, and other experimental problems (e.g., fugacity control), there are conflicting results from, and interpretations of, diamond anvil experiments recently published. Nevertheless, all the data seem to suggest that perovskite (($MgFe$)SiO_3, with about one Fe cation for every nine Mg cations) is the most abundant mineral in the Earth, constituting approximately 80 vol% of the lower mantle (e.g., Liu,

1976; Jeanloz and Thompson, 1983; Poirier, 1991). Magnesiowustite ((MgFe)O) is widely believed to be the next most common phase in the lower mantle.

The precise chemical composition of D" is unknown. Chemically dense material may sink to the base of the mantle as "dregs" and produce (along with temperature instabilities) a D" layer that has substantial lateral variation in thickness and composition (Davies and Gurnis, 1986). As mentioned in §7.1, it appears D" includes a chemical boundary layer imbedded in a thermal boundary layer. It is also possible that there are metallic regions (FeO and FeSi) with enhanced electrical conductivity near the base of D" (Knittle and Jeanloz, 1989, 1991; Goarant *et al.*, 1992). Determination of the existence of the hypothesized metallic regions in D" is extremely important to geomagnetism, for such regions would partially screen variations in the magnetic field (§2.4), increase electromagnetic coupling between the core and the mantle (§11.2), affect the outer core velocity estimates (§11.1), and change the toroidal and poloidal magnetic fields near the core–mantle boundary (Chapters 8 and 9).

Shock waves (pressure pulses that travel faster than sound and are associated with a nearly stepwise increase in pressure) provide the bulk of the experimental high-pressure data for the core. Shock waves are initiated either by exploding a charge behind a flyer plate that travels a short distance at very high speed and then impacts a second plate to which the samples are attached, or by firing a flat-faced projectile down the long barrel of a gas gun into a target plate. Shock waves can be generated in solids because incompressibility, κ_s, typically increases with pressure and compressional waves have a velocity proportional to the square root of κ_s (7.1.1), so most waves will quickly catch up with the shock front and a steady-state shock wave will propagate through the solid. The shock wave is characterized by a pressure rise that occurs on the order of a nanosecond and lasts on the order of a microsecond. There is a lower limit on pressures for shock experiments determined by the dynamic yield strength of the material; above this pressure the rigidity vanishes and the material is in hydrostatic equilibrium. The particle velocity $((\kappa_s/\rho)^{-\frac{1}{2}}$; see (7.1.1)) and the shock wave velocity are always measured. Temperatures (using outgoing radiation and assuming a black body) have recently been measured (e.g., Ahrens 1987) and other parameters, such as electrical conductivity, have sometimes also been measured.

A shock experiment yields but one point in pressure–density space, representing the final shocked state. The locus of final states is referred to as the Hugoniot equation of state. It is important to recognize that the pressure–density curve is neither an adiabatic nor an isothermal curve. Temperatures are estimated either from pyrometry or from a conventional equation of state in which the resulting values depend exponentially on the thermodynamic Grüneisen parameter (defined in the next section). Shock wave experiments are difficult, and some measurements (and interpretations), such as the determination of melting temperatures of relevant core materials, are controversial.

There is general agreement that the core has Fe as its major constituent (e.g., Birch, 1964, 1972; Ringwood, 1977, 1978; Brett, 1973, 1976; Jeanloz, 1979; Ahrens, 1979; Jacobs, 1987, 1992; Anderson, 1995). Based on iron meteorite studies it is commonly believed that the core also contains about 4% Ni (e.g., Brett, 1976). Figure 7.1, modified after Birch (1968), gives the bulk sound speed $(\kappa_s/\rho)^{1/2}$ versus density as determined from shock experiments for various metals and for the outer core. The mean atomic weight (§7.1) affects the slopes of the curves to some degree, but it is clear that lighter elements are needed to reduce the density by about 10% below that of Fe (Fig. 7.1). Lighter elements that could be present in the core in sufficient volume and that could significantly reduce the density are few: H, He, O, S, Mg, and Si. Which of these elements is present in the core at a significant level is not clear; this subject has a history of elements being rejected and later reinstated as candidates. For example, H was once rejected because it is not readily soluble in iron (true at low pressures but false at high pressures) and then, after reinstatement, later rejected again because it was

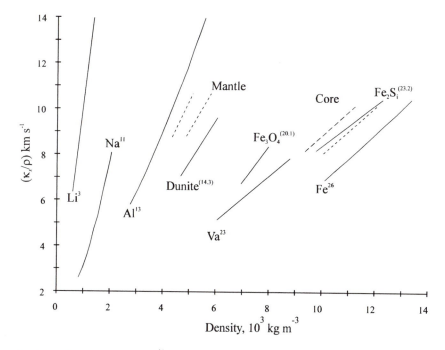

Fig. 7.1. Bulk sound speed $(\kappa_s/\rho)^{1/2}$ versus density along the Hugoniot compression curves. Mean atomic weights are attached to each curve. The areas within which the corresponding parameters for the Earth's mantle and core must lie are shown by the pairs of dotted lines. After Birch (1968).

thought that it would enter interstitial positions in Fe and not affect the density much (e.g., Verhoogen 1980). However, the density of FeH is about 0.75 that of Fe, casting serious doubt on the validity of this last argument. It is also possible that small amounts of elements heavier than Fe are present in the core, but this would require a concomitant increase in the percentage of lighter elements. Although the resolution of the many ongoing controversies concerning core composition cannot be resolved here, some principles can be established that help to evaluate properties useful to the geomagnetist.

Sulfur (Murthy and Hall, 1970) and oxygen (Ringwood, 1979) have received the most attention as light element candidates and both may be present in the core. It remains controversial whether O and S form solid solutions with Fe (i.e., are mutually soluble in Fe) or are present as a eutectic system (i.e., mutually insoluble in Fe) at core pressures (e.g., Knittle and Jeanloz, 1991; Anderson *et al.*, 1989). Figure 7.2 shows a eutectic interpretation. The composition of the inner core clearly depends on the location of the eutectic point E, the temperature and composition of which is sensitive to changes in pressures. Two different initial liquid compositions help to illustrate what happens during cooling in a eutectic system. If the initial composition were that consistent with point A, crystals of pure Fe would first form upon cooling to T_A (equilibrium

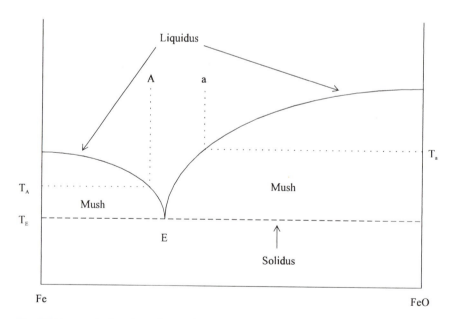

Fig. 7.2 Nonquantitative phase diagram for Fe and FeO showing a eutectic "E". The position of "E" is pressure dependent. Only liquid exists above the liquidus and only solid below the solidus. Both partial melt and liquid exist as a "mush" between the liquidus and the solidus.

assumed). As the crystals form, the liquid becomes more O rich (the composition is indicated by point A in Fig. 7.2). In contrast, if the initial composition were that indicated by point a in Fig. 7.2, the first crystals to appear would have FeO composition. Anderson *et al.* (1989) claim that if O were the only light element present, the composition would be on the right side of the eutectic point in Fig. 7.2, and, since solid FeO has density less than that of the inner core, O cannot be the sole light element in the core. Although this is quite possibly true, one should recognize the great uncertainty that exists: not only is the eutectic point not well determined experimentally, but it is not even clear (Knittle and Jeanloz, 1991; Anderson, 1995) that one is dealing with a eutectic system! The only solid conclusion on composition that can be drawn is that about 10% of elements lighter than Fe is needed in the outer core and that the relative partitioning of these elements between the inner and the outer core is not well known. Nevertheless, the density contrast across the inner core boundary (§7.1) suggests that the inner core is more Fe rich than the outer core.

Unless the core's composition is precisely at a eutectic point (a possibility that is unlikely given the large apparent density contrast across the boundary), there will be a transition region in which both liquid and solid coexist (referred to as "mush" in Fig. 7.2). This is also true for a solid solution series. Jacobs (1953) first suggested that the inner core boundary might be a freezing boundary, a suggestion that appears to have become dogma. However, theories on the core's temperature and thermal history (next sections) depend on whether one chooses this boundary as the liquidus or solidus, or somewhere in between. If one chooses the liquidus, then at least the outer part (and perhaps all) of the inner core must be partially molten as suggested by Loper and Roberts (1978, 1980). On the other hand, if one chooses the solidus, the bottom of the outer core must be "mush," i.e., it will be layered. It is likely the inner core boundary lies between the liquidus and solidus, but closer to the solidus since long-period normal mode data indicate the inner core exhibits rigidity (§7.1).

7.2.2 Physical Properties

Two parameters that can have a pronounced effect on geomagnetism, and on the thermal history of the core, are electrical conductivity and viscosity. The electrical conductivity of Fe, in both its solid and its liquid states, has been measured by Secco and Schloessin (1989) as a function of temperature and pressures up to 7 GPa. Making the approximately twentyfold extrapolation to conditions at the top of the core, they conclude that the most probable range of electrical conductivity for the core is 6.7 to 8.3×10^5 Sm^{-1}.

The effects of impurities are not as straightforward to estimate as might be supposed. Matthiessen's rule applies to impurities dilutely distributed in a lattice: this rule states that the total electrical impedance (inverse of the conductivity) is the sum of the metallic impedance and the impurity impedance. The theory behind this rule is based on phonon scattering in solids (e.g., Poirier, 1991) and it is not clear whether the rule should be applied to liquids. Nevertheless, it is likely that this rule is at least approximately correct and that the estimates for the conductivity for the core should be lowered slightly to account for the presence of impurities.

Shock wave experiments have been conducted on Fe and on Fe containing various amounts of Si in solid solution (Matassov, 1977). Values of conductivity are found to be near 5×10^5 Sm^{-1} at 180 GPa (core pressure). These values are consistent with those found earlier by Keeler and Mitchell (1969). However, values half or twice this are acceptable. Table 7.2 summarizes estimates of some properties of the Earth's core, and we suggest a conductivity of 6×10^5 Sm^{-1}. It should be pointed out that at a given temperature the electrical conductivity and the thermal conductivity are related, so the choice of electrical conductivity affects interpretations of the Earth's thermal history. This is discussed further in §7.3 and §7.4.

Estimates for the viscosity of the Earth's core vary by at least 11 orders of magnitude (e.g., Suzuki and Sato, 1970; Gans, 1972). It is important to know the magnitude of viscosity to estimate the state of core convection (neither laminar convection nor turbulence can currently be rejected) and to access a variety of other important parameters, such as boundary layer thickness. Poirier (1988) has reviewed the theory and the systematics in the data to arrive at an estimate of 0.06 poise for the outer core (the value presented in table 7.2). For comparison, this viscosity is about that of molten iron and only six times that of water at the Earth's surface. Surprisingly, considering the range of viscosities given in the literature, this value is likely within 1 or 2 orders of magnitude of

Table 7.2
Estimates of Some Properties of the Earth's Core

Mean radius of Earth	6371 km
Depth to outer core	2891 km
Depth to inner core	5049 km
Composition	Primarily Fe, probably a few percent Ni, and approximately 10% of lighter elements, H, S, O
Electrical conductivity	6×10^5 Sm^{-1}
Adiabatic temperature gradient in outer core	Between 0.7 and 1.0 C° km^{-1}
Temperature core–mantle boundary	Probably near 4000 ±1000°C
Viscosity of outer core	0.06 poise (not well known)
Mean density of outer core	11×10^3 kg m^{-3}
Density contact across inner core boundary	0.5 to 0.6 ±3 × 10^3 kg m^{-3}
Pressure, core–mantle boundary	136 GPa
Pressure, inner core boundary	328 GPa
Pressure, Earth's center	367 GPa

the actual viscosity of the outer core (Poirier, 1988; personal communication 1992).

The seismically determined inner core boundary likely differs from the rheological boundary. As is well illustrated by the Earth's mantle, a material can appear elastic if a stress is applied over a short time (e.g., the passage of a seismic wave) but viscous if the stress is applied over a long time (e.g., times associated with isostatic rebound or mantle convection). Based on the finding that extrusive igneous rocks rarely exhibit more than 50% crystals, it has been suggested that there is a sharp increase in viscosity in magma chambers when the volume ratio of melt to solid is equal (Marsh, 1981; Brandeis and Marsh, 1989). As the present consensus seems to be that even at the top of the inner core there is a substantially smaller melt fraction than this, core convection processes probably do not extend all the way to the seismically determined inner core boundary. That is, it again (§7.2.1) seems likely there is a mush zone at the base of the outer core that has a chemical composition intermediate between that of the rest of the outer core and that of the inner core.

7.2.3 Electrical Properties of the Lower Mantle

The picture of electrical conductivity in the lower mantle is confusing. Time series analyses indicate that magnetic signals of much less than about a year are rapidly filtered out by electrical screening of the mantle. This led to an estimate of electrical conductivity in the lower mantle near 3×10^2 Sm^{-1} (§2.4). However, several recent studies imply a lower value. The study implying the most extreme change is that of Coe *et al.* (1995) who claim to see evidence of changes, of internal origin, in the direction of the Earth's magnetic field that occur at an average rate of 6° per day over several consecutive days! A mantle conductivity of 3×10^2 Sm^{-1} would screen such rapid changes in the internal field very effectively and they would be unobservable at the Earth's surface (§2.4). The claims of Coe *et al.* (1995) are based on paleomagnetic studies of Miocene lava flows that record a magnetic field reversal in Oregon, USA; this is an extension of previous work on Steen's Mountain, as discussed in §5.4. Coe *et al.* (1995) document very rapid changes in the direction of the remanent magnetization within a few lava flows at Steen's mountain, flows that erupted during a reversal of the field about 16 Ma. Coe *et al.* (1995) argue that the rapid directional changes are caused neither by rock magnetic artifacts nor by changes in the external field. Instead, they argue that these changes record amazingly rapid changes in the internal field. For the magnetic field signal to be of sufficient magnitude at the Earth's surface, Coe *et al.* (1995) estimate that the lower mantle conductivity is about 1 Sm^{-1} and that the internal source is identified with a field described by a spherical harmonic of degree 6 or higher.

Recent numerical hydromagnetic dynamo calculations (Glatzmaier and Roberts (1995a,b; see §9.6) can model some aspects of reversals, but the changes in the velocity and magnetic fields found are several orders of magnitude smaller than those required by Coe *et al.* (1995). The careful experimental procedures followed by Coe *et al.* (1995) certainly reduce the likelihood of alternative explanations of the data, but they do not eliminate them. For example, Starchenko and Shcherbakov (1994) point out that the magnetosphere would likely have a very different structure during a reversal and that this could give rise to external fields with properties similar to the those inferred for the unusual internal fields at Steens Mountain. Thus, additional methods are required to determine if the results of Coe *et al.* (1995) require modification of the conventional view of mantle electrical conductivity.

One-dimensional models of electrical conductivity using induction techniques currently give the electrical conductivity down to about 1000 km, where the conductivity is near 1 to 2 Sm^{-1} (e.g., Schultz and Larsen, 1987; Egbert and Booker, 1992; Constable, 1993). Most electromagnetists suggest the resolution is too poor to resolve the conductivity below about 1000 km. Conversely, McLeod (1994), using only a forward model, has determined the transfer function from the external to the internal field for zonal harmonics and estimates the mean conductivity of the entire lower mantle to be 10 Sm^{-1}.

Additional information on conductivity from the lower mantle comes from mineral physics. An influential paper by Shankland *et al.* (1993) combines theory with diamond anvil cell (DAC) measurements. They find only a weak dependence of electrical conductivity on temperature, pressure, and composition for conditions appropriate to the lower mantle. They conclude that the conductivity gradually increases below the 660-km transition from about 1 Sm^{-1} to a maximum of 3 to 10 Sm^{-1} in the lowermost mantle (excluding the 200- to 300-km thick D" layer at the mantle's base). In contrast, Wood and Nell (1991) show that parallel conduction in the dominant lower mantle mineral phases perovskite and magnesiowustite can give conductivity estimates consistent with the higher values determined from time series analyses of the geomagnetic data (§2.4). This partially reflects a history of conflicting conclusions from DAC measurements of electrical conductivity (for additional examples see Peyronneau and Poirier (1989), and Li and Jeanloz (1990)). Duba and Wanamaker (1994) help to explain many of the conflicting results: most, and perhaps all, of the above-cited DAC results are influenced by grain boundary effects. When highly conducting phases are connected to each other, say along grain boundaries, they can produce a relatively high conductivity relative to a material in which the conducting phases are not connected. One advantage the Shankland *et al.* (1993) study has over previous studies is that the theory given in it can be used to extrapolate from the observed values for conductivity below the 660-km transition to obtain estimates of conductivity throughout most of the lower mantle. However, there still remain several sources of uncertainty: The "observed" values may change when there are more accurate two- and three-dimensional inversion models and the extrapolation can fail if there is any change in chemistry or phase with depth. In addition, conduction in semiconductors can be significantly affected by impurities (e.g., donor or acceptor ions) and the effects of minor constituents have not been adequately explored for the lower mantle. We conclude that the mantle conductivity in the lowermost mantle above D" is not well known, but probably falls in the range of 3 to 100 Sm^{-1}.

The electrical conductivity within D" is even more poorly known. The problem is that the molten iron in the core may react chemically with the silicate perovskite and magnesiowustite (Knittle and Jeanloz, 1986, 1987, 1991; Urakawa *et al.*, 1987; Guyot *et al.*, 1988). Such reactions can produce metallic regions of FeO or FeSi with very high conductivities of order 10^5 Sm^{-1} (e.g., Knittle and Jeanloz, 1989). The size and distribution of these regions remains speculative.

However, the bulk of D" cannot consist of high (metallic) conductivity, because observations of phenomena such as the 1969 jerk (§2.4) would not be possible. The determination of the size and distribution of the hypothesized metallic regions is essential since such regions can have significant effects on geodynamo models (through changes in the boundary conditions), on geomagnetic secular variation, and on core–mantle coupling estimates (§11.2).

7.3 Thermodynamic Properties of the Earth's Deep Interior

A parameter that repeatedly appears in calculations of thermodynamic properties of the Earth's interior is the thermodynamic Grüneisen parameter,

$$\gamma_T \equiv \frac{\alpha_p \kappa_s}{\rho C_p} \quad , \tag{7.3.1}$$

where α_p is the coefficient of thermal expansion and C_p is the specific heat at constant pressure. This Grüneisen parameter is dimensionless and is of order unity for all materials, be they solid, liquid, or gas. Its utility in geophysics is a consequence of the fact that it is related to the pressure derivatives of elastic moduli. Using the definition of the seismic parameter,

$$\phi_p \equiv (v_p)^2 - \frac{4}{3}(v_{sh})^2 = \frac{\kappa_s}{\rho} \quad , \tag{7.3.2}$$

one obtains an alternate relation for γ_T in the outer core (see (7.1.1) and (7.1.2)):

$$\gamma_T = \frac{\alpha_P \phi_P}{C_P} \quad . \tag{7.3.3}$$

A necessary, but not sufficient, condition for the onset of thermal convection is that the absolute magnitude of the temperature gradient exceed that of the adiabatic temperature gradient, given (e.g., Poirier, 1991) by

$$\left[\frac{\partial T}{\partial r}\right]_s = \frac{-g\gamma_T T}{\phi_P} \quad , \tag{7.3.4}$$

where T is the temperature and the subscript s indicates constant entropy (i.e., adiabatic). Moreover, in a convecting system the actual temperature gradient is closely approximated by the adiabatic one, except at the top and bottom where conduction through the boundary layers is important, so the adiabatic temperature gradient is a critical parameter in a variety of geophysical problems in geophysics. For the convecting systems considered here, there are no lateral boundaries. The rising and falling limbs in thermal convection will actually exhibit slightly different temperatures at equal distances from the boundaries, but such differences, at least in the core, are negligible.

Because shear waves are transmitted through the mantle it is believed to be solid, except possibly for certain localized regions containing partial melt. In contrast, the outer core is believed to be liquid because it does not transmit shear waves. Therefore the melting temperature provides an upper bound for the temperature of mantle material and a lower bound for the temperature of the outer core. Thus, knowledge of the melting temperature gradient, which again involves γ_T, is also crucial. Modern condensed-matter theories for melting often involve particular materials for which the bond type in the solid (long-range ordered) phase is reasonably well known. Because the composition and the nature of bonding in the Earth's deep interior are not well known, geophysicists often rely on more general, but less precise, models for melting. An example is Lindemann's law (Lindemann, 1910); melting occurs when lattice vibrations exceed a critical value, presumably because then a sufficient number of defects are created that break down the long-range order and make the lattice unstable to shear. Stacey *et al.* (1989) have shown how Lindemann's law can be placed on a solid thermodynamic foundation and provided a relation between melting temperature, T_m, and γ_T:

$$\frac{d \ln T_m}{d \ln \rho} = 2\gamma_T \quad . \tag{7.3.5}$$

The relationship between depth and ρ then gives the melting temperature as a function of depth. This is just one example of a melting temperature theory; there are several others, with T_m depending on γ_T (see, e.g., Poirier, 1991).

The major unknown in (7.3.5) for the melting temperature gradient in the Earth is γ_T, while in (7.3.4) for the adiabatic temperature gradient γ_T and T are the major unknown parameters. Use of γ_T has also been mentioned in §7.2 with regard to temperature corrections in shock wave experiments (where the result depends exponentially on γ_T, so it is crucial to have an accurate value). Each of these important applications demands knowledge of γ_T, so it is critical to obtain accurate estimates of γ_T in the Earth's deep interior and numerous theories have been constructed and measurements made on solids to achieve this.

A useful empirical law, first determined by Anderson (1979), is

$$\gamma_T \, \rho^q = \text{constant} \quad , \tag{7.3.6}$$

where q is a constant close to one. It is usually argued that γ_T varies between 1.1 and 1.4 in the Earth's core, leading to estimates of the adiabatic temperature gradient in the outer core between 0.7 and 1.0 C° km^{-1}. In contrast, the adiabatic temperature gradient in the lower mantle is commonly determined to be about half that of the outer core. However, values for γ_T in fluids can be very different from those measured for solids (Verhoogen, 1980), and M. Brown (personal communication, 1995) recently estimates γ_T at about 1.8 to 1.9 in the outer core, which increases the estimates of the magnitudes of the adiabatic and melting temperature gradients by approximately a third. Further implications will be discussed in §7.4.

To estimate core temperatures, one uses both the adiabatic and the melting temperature gradients. Typically one obtains an estimate of iron's melting temperature at a high pressure using experimental data, and then uses a melting theory, such as given by (7.3.5), to determine the melting temperature of iron at the inner core boundary. This melting temperature estimate needs to be lowered because of the presence of less dense elements, but the magnitude of this decrease is uncertain (see §7.2). Decreases in the melting temperature of Fe by a few hundred to more than 1000°C have been assumed in order to provide estimates of the temperature at the inner core boundary.

The temperatures estimated for the core–mantle boundary (CMB) vary from 3000°C to more than 6000°C. Several years ago this large range of estimates could be attributed to the fact that measurements were made at low pressures and extrapolated to core conditions using different theoretical techniques. Today such differences occur because of differences in the interpretations of experimental results. Most experiments attempt to determine the melting temperature of iron at 330 GPa, the pressure at the inner–outer core boundary. The temperature is usually assumed to be somewhat less due to the presence of a less-dense element and this lower temperature is extrapolated upward to the CMB along an adiabat. Brown and McQueen (1980, 1986) made the first shock wave measurements of iron and, combined with theory, estimated the melting of iron at 330 GPa to be around 5800 K. In contrast, Williams *et al.* (1987) used extrapolation of diamond anvil measurements and shock wave data to estimate the melting of iron at 7600±500 K. Boehler (1992, 1993) found substantially lower values for the melting point of iron in diamond anvil experiments from that of Williams *et al.* (1987). Boehler's results have been duplicated by Saxena *et al.* (1994), leading Duba (1992, using preprint information) to conclude that the CMB temperature is close to the 4000 K value often cited (e.g., Poirier, 1991; Jacobs, 1992). Recently Yoo *et al.* (1993) have conducted new shock wave experiments to estimate a value near 6830±500 K. Anderson (1995) and Duffy and Hemley (1995) have provided summaries of current views.

Surprisingly, for geomagnetism it is not important to resolve the above differences. With the exception that the differences affect the energy budget for the Earth's magnetic field, there is little effect on dynamo theory. However, to determine mantle processes, it is crucial to resolve such differences because the higher temperatures either require a very large thermal boundary layer in D" or additional thermal boundary layers elsewhere in the mantle. This is illustrated below by assuming a 4000 K temperature for the CMB.

It appears that CMB conditions, such as temperature, are controlled primarily by processes in the mantle. The deepest point in the mantle at which the temperature can be estimated with moderate safety from a phase change is the 670-km seismic discontinuity (Gamma spinel to Magnesiowustite and Perovskite structures) and the temperature there appears to be close to (or slightly below) 1700°C (e.g., Poirier 1991, Jacobs, 1992). (The largest uncertainties in this estimate are in the estimation of the chemical composition, such as the Fe–Mg partitioning between phases.) If thermal convection persisted throughout the lower mantle, and thermal boundary layers could be ignored, the temperature at the CMB could be estimated at about 2588°C by assuming an adiabatic gradient (0.4 C°/km) throughout the lower mantle. However, this value is too low to be consistent with melting temperature estimates for the top of the core, implying that there is one or more thermal boundary layer(s) in the lower mantle.

There appears to be consensus that there is a thermal boundary layer in the lowermost mantle, taken to be the roughly 200-km-thick seismic D" layer. The temperature change across this layer is typically taken to be 800 to 1000°C (e.g., Loper and Stacey, 1983). This provides a self-consistent model for temperatures down to the core, but one vulnerable to assumptions. For example, the thermal diffusivity in D" may be substantially higher than typically assumed (Brown, 1986) and this could lower the estimated temperature drop across D". Alternatively, the temperatures at the top of the core may be much higher (Jeanloz, 1990), requiring either a much greater temperature drop across D" or another thermal boundary layer elsewhere in the lower mantle. Our knowledge of the temperature drop across D", which could be as low as 300°C or as high as 1500°C (e.g., Lay, 1995), is very imprecise. The possibility of a second thermal boundary below the 670-km discontinuity is likely if upper-mantle convection is largely separate from lower-mantle convection. In contrast, such a layer may be absent if whole-mantle convection occurs. Resolution of conflicting convection models is beyond the scope of this book (see reviews by Tackley, 1995; Kincaid, 1995). However, it is noteworthy that there does not appear to be any geomagnetic or paleomagnetic information that can be used to definitively distinguish between whole-mantle and layered-mantle convection models.

7.4 Thermal History Models

The generalized heat flow equation,

$$\frac{\partial T}{\partial t} = \eta_T \nabla^2 T + \nabla \eta_T \cdot \nabla T + \frac{\varepsilon_T}{\rho c_\rho} - \mathbf{v} \cdot \nabla T + \text{(radiative transfer term)} \ , \quad (7.4.1)$$

describes how thermal energy can be transferred through the Earth. Here η_T is the diffusivity and ε_T represents the thermal energy source function, which is a constant only if the sources are uniformly distributed and do not vary with time (in the Earth ε_T varies in both space and time). Radiative heat transfer is believed to play a negligible role in the deep interior, but it probably played an important role during the Earth's formation. It is taken into account by assuming the Earth is a black body for which radiative heat transfer goes as T^3. The first two terms on the right side of (7.4.1) are associated with conduction, the third term is associated with the generation of thermal energy by internal sources (predominantly radioactive sources), and the fourth term describes the advected heat component associated with convection.

This equation can be used to obtain order-of-magnitude estimates for certain characteristic times of processes acting in the Earth. For example, a very crude estimate for mantle heat transfer by conduction processes can be made from dimensional analysis using only the first term on the right side of (7.4.1),

$$t_c \approx \frac{L_c^2}{\eta_T} \approx \frac{(2900 \times 10^3)^2}{1 \times 10^{-6}} \ \mathrm{s} \approx 3 \times 10^{11} \ \mathrm{yr} \ , \quad (7.4.2)$$

where L_c is the depth to the CMB. This value is larger than the age of the Earth (4.6×10^9 yr). In contrast, for a convection speed of 0.015 mr/yr, the characteristic time for the transfer of heat from the core–mantle boundary to the crust can be estimated using the fourth term on the right side of (7.4.1) to be about 2×10^8 yr. Because convection (either whole or layered) appears to be the dominant process for thermal energy transfer in the mantle, this value will be taken as the order-of-magnitude estimate for the characteristic time of mantle transport of thermal energy.

The thermal diffusivity is essentially constant around 1.1×10^{-6} m^{2}s^{-1} throughout the mantle except within D″ where it may be a factor of 2 or 3 higher (Brown, 1986). Therefore, the second term of the right side of (7.4.1) can be neglected except within D″. The diffusivity in the core can be obtained from the Wiedmann–Franz law linking electrical to thermal conductivities, in metals (e.g., Poirier, 1991),

$$K_T = \eta_T \rho^{-1} c_P^{-1} = W\sigma_e T \quad , \qquad\qquad (7.4.3)$$

where K_T is the thermal conductivity and W is the Wiedmann constant equal to 2.45×10^{-8} W Ω K^{-2}. Thus the value of K_T increases with depth in the core as the temperature increases. (As discussed in §7.2, the electrical conductivity can be well approximated as a constant throughout the core and its value is probably within half an order of magnitude or so of 6×10^5 Sm^{-1}).

There is an interesting energy/efficiency tradeoff between Earth cooling and magnetic field dissipation that does not appear to have been previously recognized, and which depends on the choice for σ_e. As evident from (7.4.3), a higher value of σ_e implies a higher value for η_T, and consequently a greater amount of heat will be conducted down the core's adiabat (out of the core). One might expect that higher values of σ_e, and hence higher rates of Earth cooling, might make it more difficult to sustain a magnetic field. However, as will be discussed in more detail in Chapter 8, the rate at which the magnetic field decays is inversely proportional to σ_e; that is , it requires far less energy to sustain the magnetic field for larger values of σ_e.

The solution of the generalized heat flow problem not only involves choosing values for the parameters in (7.4.1) and solving that equation, but it also depends on the choice of boundary and initial conditions. Like many problems in geophysics, several assumptions are required to do this. It is important to remember that if one makes three independent assumptions to solve a problem, each of which is 80% likely to be true, then the probability is only roughly a half that the solution is correct. In constructing a thermal history model for the Earth and its core, the uncertainties are greater and the assumptions required more numerous. This will be illustrated by first discussing the problem of the Earth's initial condition, following by briefly discussing the uncertainties in the time of core origin, and its boundary conditions.

The first problem concerns the Earth's initial conditions and whether there was a terrestrial magma ocean. To gain some appreciation of the uncertainties involved in this aspect of the problem, consider the gravitational energy converted to thermal energy on the formation of a hypothetical incompressible homogeneous planet. The increment of gravitational potential energy $d\psi_g$ for a spherical shell of mass m_r at a distance r from the planet's center is

$$d\psi_g = G\frac{m_r \, dm_r}{r} \quad , \tag{7.4.4}$$

where G is the gravitational constant and $dm_r = 4\pi \, r^2 \, \rho \, dr$ is the mass of the shell. The gravitational potential energy released on forming the planet is then

$$\psi_g = \frac{GM^2}{2R} + \frac{G}{2} \int_0^R \frac{m_r^2}{r^2} dr \quad , \tag{7.4.5}$$

which for a planet with constant density is

$$\psi_g = \frac{2GM^2}{5R} \quad , \tag{7.4.6}$$

where M and R are the total mass and radius of the planet, respectively. The gravitational potential energy released in forming the Earth is more than that given by (7.4.6) because the density increases with depth; indeed it is more than enough to melt the entire Earth (e.g., Birch 1965a,b). Clearly, a substantial amount of thermal energy must have been radiated away as the Earth formed, but the estimated amount is model dependent. The moon is believed to have had a magma ocean early in its history (Chapter 10) and the Earth may also have had one (e.g., Ohtani, 1985), depending on the size and timing of the accreting bodies: less energy is radiated away from larger accreting bodies, increasing the possibility of a terrestrial magma ocean. It is now widely believed that a large planetesimal collided with the Earth to form the moon, and as one consequence a terrestrial magma ocean was formed. However, the depth of this ocean, and even its existence, is controversial. In short, the initial conditions are not well known and our estimates are strongly model dependent.

As a corollary, there are also many uncertainties concerning the core's origin and evolution. It is widely believed that most of the core formed after the Earth accreted, probably mostly within the first 10^8 yr. A considerable amount of gravitational energy was converted to thermal energy at this time (e.g., Shaw, 1978); our perception of the early thermal history depends on assumptions made about the nature of this core formation "event". Another problem is that the distribution of radioactive heat sources in the Earth in general, and in the core in particular, is poorly known – including the distribution of short-lived radioisotopes (e.g., [26]Al) that could have contributed significantly to thermal sources in the past, but not today (e.g., Jacobs, 1992).

It should be clear that the construction of a thermal history of the Earth's core cannot be done independently from the construction of a thermal history for the entire Earth. In principle, all one needs to do is to prescribe either the temperature or the heat flux (both would amount to overprescription) at the CMB boundary throughout time to separate the thermal history of the core from that of the overlying mantle and crust. However, it is not clear how to do this without solving the thermal history of the mantle and crust first. One attempt to do this has been to assume that the core's heat flux is small and approximately constant with time, but reasonable questions can be raised concerning both of these assumptions. For example, almost all thermal models for the Earth have the present heat flux out of the core to be about 10% (e.g., Verhoogen (1980) has 10%; Poirier (1991) 10%; Stacey (1992) 8%) of the total heat flux (44×10^{12} W) across the Earth's surface. Because these models contain some significant differences, it might appear that the flux of core heat is a robust result. However, the presence of only 20 ppb of U or 72 ppb of Th, either of which is possible (e.g., Murrell and Burnett, 1986) and neither of which was considered present in any of the thermal models referenced, would approximately double the estimate of core heat flux. Indeed, Tackley *et al.* (1993) estimate that 40% of the heat flow at the Earth's surface comes from the core.

In spite of this, it is useful to give one thermal model (Buffet *et al.*, 1992) explicitly to illustrate how paleomagnetic results might eventually help to place constraints on the problem. Three curves are shown in Fig. 7.3, each of which assumes a different value for the heat flux (held constant) at the CMB. Buffet *et al.* (1992) find that the thermal buoyancy associated with cooling at the top of the core is presently equal to the compositional buoyancy associated with freezing at the inner-core boundary. As will be discussed further in §7.7, chemical buoyancy appears to be a far more efficient source for generating the magnetic field than thermal buoyancy. However, one expects the ratio of thermal buoyancy to compositional buoyancy in the past to have been larger. Considering this and core thermal models, such as shown in Fig. 7.3, one could expect there to have been a marked change in intensity of the Earth's magnetic field in the past. This has been speculated on before with the reasonable, but not rigorously justified, expectation that the intensity would have undergone a marked increase sometime (model dependent) prior to one billion years ago (Stevenson, 1983; Hale, 1987). Unfortunately, the paleointensity data are too few and their quality too poor to adequately test these possibilities at this time (Merrill, 1987).

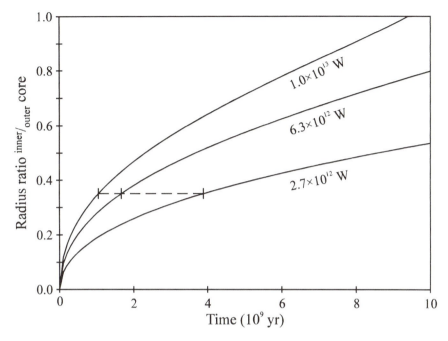

Fig. 7.3. Changes in the ratio of the inner-core radius to the outer core radius with time for different assumed values for the total heat flux emanating from the core (modified after Buffet *et al.*, 1992). The dotted line indicates the present radius ratio.

7.5 Nondynamo Models for the Earth's Magnetic Field

7.5.1 Permanent Magnetization

The concept that permanent magnetization in the Earth's deep interior contributes significantly to the origin of the Earth's magnetic field is usually dismissed on the basis of high temperatures in the core. Although this is a good argument against permanent magnetization, it is not conclusive. At a pressure of 1 atm the Curie temperatures of rock-forming minerals are all well below 1000°C (e.g., 580°C for magnetite, 696°C for hematite, and 770°C for pure iron). Measurements of the change of Curie temperature with pressure are not extensive but do show that it is highly variable: in some substances the Curie temperature increases with pressure and in others it decreases. Predictions from theory of the effects pressure has on the Curie temperature are not good, primarily because of difficulties in predicting how the overlap of the electron orbitals of the various magnetic minerals changes with pressure. In addition, reliable empirical predictions are not likely to be forthcoming, simply because the present measurements are at such low pressures that extrapolations to core pressures are not justified. It might, therefore, be argued that material in the deep interior, such as the inner core, could be above some critical pressure at which it is permanently magnetized. However, such a contrived argument runs into serious problems with respect to reversals of the Earth's magnetic field. It is difficult to conceive of a mechanism that uses permanent magnetization to explain reversals and the observed properties of secular variation. Therefore, the presence of permanent magnetization in the deep interior appears highly contrived and improbable.

7.5.2 Thermoelectric Effects

Thermoelectricity concerns the direct generation of an e.m.f. when an electrically conducting material is subjected to a thermal gradient. Physically, the phenomenon arises because electrons at the hot end of the conductor have a higher kinetic energy and consequently more of them diffuse to the cold end of the conductor than vice versa. This implies a charge separation across the Earth's core; electric currents will be produced by this charge as the Earth rotates. The magnetic field will be proportional to the rotation rate, consistent with the speculation by Blackett (1947) and others (e.g., Brecher and Brecher, 1978) that astronomical bodies have dipole moments proportional to their angular momentum. However, the magnetic fields generated by this mechanism are many orders of magnitude too small to produce any of the observed planetary magnetic fields in the solar system (e.g., Inglis, 1955; Merrill and McElhinny, 1983)

Although thermoelectric effects can be safely eliminated as the primary cause of the Earth's magnetic field, they might constitute a source of the initiating magnetic field required in dynamo models. The most likely mechanism for doing this involves currents generated when two different electrical conductors are involved.

Figure 7.4 illustrates how an e.m.f., ε_{AB}, can be generated by the Seebeck effect when a temperature gradient is maintained along two conductors, A_o and B_o. ε_{AB} is given by

$$\varepsilon_{AB} = Q_s(T_1 - T_2) \equiv Q_s \Delta T \quad , \tag{7.5.1}$$

where T_1 is taken to be greater than T_2 and Q_s is a constant called the thermopower. If the ends of material B_o are connected (where the detector is shown in Fig. 7.4), a current will flow producing a magnetic field. This magnetic field is powered by the thermal energy required to sustain the temperature difference, ΔT. (The inverse effect, called the Peltier effect, is used in thermoelectric refrigeration, and involves the use of electric currents to change the relative temperatures.)

It is straightforward to show (e.g., Merrill *et al.*, 1979, 1990) that the magnitude of the magnetic field is given by

$$H \approx C\sigma_e Q_s \Delta T \quad , \tag{7.5.2}$$

where C is a constant that depends on the geometry of the current. To apply this to the Earth one assumes the two materials, A_0 and B_0 in Fig. 7.4, correspond to materials close to the core–mantle boundary; at least one of these materials must lie within D". The major uncertainties involve the conductivities and thermopowers of these materials; ΔT; and the size and geometry of the lateral heterogeneities present in D". These uncertainties are formidable and only very crude estimates can be provided on the magnitudes of magnetic fields that might be produced from the Seebeck effect acting near the core–mantle boundary. Conversely, the Peltier effect may act to reduce lateral temperature differences within D".

7.5.3 Other Mechanisms

Although there is now consensus that the Earth's magnetic field is a consequence of a dynamo acting in the core (see subsequent chapters), there are several magnetic effects that may play secondary roles in the generation processes. One of the most intriguing of these is the battery effect, first

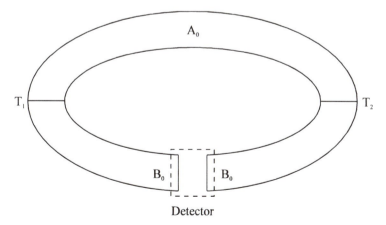

Detector

Fig. 7.4. A Seebeck e.m.f. will occur at the detector if the temperatures T_1 and T_2 are different and the electrical conductivities of A_0 and B_0 are different.

suggested by Ringwood (1958) and further discussed by Merrill *et al.* (1990). The basic idea is that D" is undergoing chemical interactions (e.g., Knittle and Jeanloz, 1986, 1991) and that the transport of positive ions in one direction is balanced by the transport of electrons in the opposite direction (e.g., $Fe \rightarrow Fe^{2+} + 2e^-$). Estimating the size of the fields generated by such an effect is very difficult because of the numerous uncertainties involved (e.g., the geometry of inhomogenities in D" and the nature of the chemical interactions). Nevertheless, Stevenson (personal communication, 1991) suggests that fields created in this manner could be a significant fraction of the observed field.

Other suggested mechanisms for generating magnetic fields in the core include:

(i) residual electric currents set up in the Earth's interior early in its history (Lamb, 1883);
(ii) gyromagnetic effect (Barnett, 1933);
(iii) the Hall effect (Vestine, 1954);
(iv) galvanomagnetic effect (Stevenson, 1974);
(v) differential rotation effect (Inglis, 1955);
(vi) electromagnetic induction by magnetic storms (Chatterjee, 1956); and
(vii) the Nernst–Ettinghauser effect (Hibberd, 1979).

7.6 Fluid Mechanics Fundamentals

7.6.1 The Navier–Stokes Equation and Boundary Conditions

The Earth's outer core is fluid and rotating. The appropriate equation of motion is the Navier–Stokes equation

$$\frac{\partial \mathbf{v}}{\partial t} + \mathbf{v} \cdot \nabla \mathbf{v} = -\frac{1}{\rho} \nabla P - \mathbf{\Omega} \times (\mathbf{\Omega} \times \mathbf{r}) - 2\mathbf{\Omega} \times \mathbf{v} + \nu \nabla^2 \mathbf{v}$$
$$+ \text{(body forces)} \quad , \tag{7.6.1}$$

where Ω is the rotation of the Earth (core), $\mathbf{r}$ the position vector, $\mathbf{v}$ the velocity, ν the viscosity, and P the pressure. Sometimes the left side is written as

$$\frac{D\mathbf{v}}{Dt} \equiv \frac{\partial \mathbf{v}}{\partial t} + \mathbf{v} \cdot \nabla \mathbf{v} \quad , \tag{7.6.2}$$

which is just the acceleration relative to the rotating frame. Pressure gradient effects on the fluid are reflected by the first term on the right of (7.6.1), the centrifugal and Coriolis forces by the second and third terms respectively (both of which would vanish in a nonrotating system), and viscous effects in the fourth term on the right side. The last "term," labeled body forces (but, more precisely, "body fields"), can be a composite of many terms. For the geodynamo we will need to include the gravitational field, **g**, and the Lorentz Force divided by density, $(\mathbf{J} \times \mathbf{B}) / \rho$.

Vorticity,

$$\mathbf{\omega} \equiv \nabla \times \mathbf{v} \quad , \tag{7.6.3}$$

defined as the curl of the velocity field, plays a central role in core processes. If the fluid were rotating as a rigid body then $\mathbf{v} = \mathbf{\Omega} \times \mathbf{r}$ and so $\omega = 2\Omega$; i.e., the vorticity is the same at every point and equal to twice the angular velocity. Insight into vorticity in a nonrigid body can be obtained by applying Stokes' theorem to vorticity to give

$$\int \mathbf{\omega} \cdot \mathbf{n} \, dS = \int \nabla \times \mathbf{v} \cdot \mathbf{n} \, dS = \oint_{\mathcal{L}} \mathbf{v} \cdot d\mathbf{r} \quad , \tag{7.6.4}$$

where dS is an increment of area and **n** is the unit normal to that increment. The last integral is the definition of the circulation around any arbitrary closed loop $\mathcal{L}$. That is, vorticity is closely related to circulation: there must be vorticity within a loop around which circulation occurs.

To gain further insight into vorticity, take the curl of the Navier–Stokes equation in a nonrotating system to get

$$\frac{D\mathbf{\omega}}{Dt} = -(\mathbf{\omega} \cdot \nabla)\mathbf{v} + \nu \nabla^2 \mathbf{\omega} \quad . \tag{7.6.5}$$

This is obtained by remembering that the curl of a conservative field, e.g., ∇P (or for that matter gravity, since gravity can be written as the gradient of a scalar potential), vanishes. The significance of the second term can be seen by analogy with the scalar diffusion equation, except that the vorticity, ω, is a vector. The action of viscosity, ν, produces diffusion of vorticity down a vorticity gradient. The first term on the right side of (7.6.5) describes how the vorticity is stretched and twisted by the velocity field. In a two-dimensional inviscid flow (ν=0) it is straightforward to show from (7.6.5) that

$$\frac{D\mathbf{\omega}}{Dt} = \mathbf{0} \quad , \tag{7.6.6}$$

which states that vorticity is a conserved quantity in a two-dimensional flow.

While providing insight into vorticity, this also illustrates the dangers of extrapolating to three dimensions from theorems proved in two dimensions. According to (7.6.6), one may not compress vorticity; physically this occurs because the compression of vorticity lying in a plane would move the fluid out of that plane (i.e., it would involve the third dimension).

A common practice in many calculations in fluid mechanics is to assume the fluid is inviscid in the interior of the body (e.g., the core) and that viscosity need only be considered near the boundaries. A major mathematical advantage of this assumption can be appreciated be inspecting (7.6.1): the only term containing second-order derivatives in (7.6.1) can be dropped if the fluid can be assumed inviscid. Unfortunately, to adopt this commonly made assumption, one must assume there is no diffusion of vorticity in the core on geological time scales.

The quantitative study of processes becomes much more difficult if the Lorentz force is included, as it must be for the core. So far all published solutions (and solutions that are likely to be forthcoming in the next several years) of fluid processes in the core involve several simplifying assumptions. One test of whether such simplifications are reasonable is of course to compare predictions from theory with observations, observations that come predominantly from geomagnetic or paleomagnetic data.

To understand core processes, one must solve (7.6.1) along with the appropriate boundary and initial conditions. Commonly, and reasonably, no-slip boundary conditions are assumed at the top and bottom of the core, i.e., $\mathbf{v}=\mathbf{0}$ at the boundaries. The velocity increases from zero across a boundary layer in which viscosity cannot be neglected. To obtain insight into the nature of this boundary layer, consider a layer of liquid in the northern hemisphere moving at some velocity, $\mathbf{v}'$. The fluid motion in this layer affects the fluid in the layer "immediately" below it. The velocity of this next layer will be slightly less than $\mathbf{v}'$ and, because of the Coriolis force, the velocity will be directed to the right of that in the top layer. This process continues downward in the fluid to produce a velocity spiral, an Ekman spiral (as depicted in Fig. 7.5). Throughout this spiral the velocity decreases and an Ekman layer is formed. Of course, in the core the velocity increases away from the boundaries but the principles remain the same, since all one needs to do is to move upward from the reference layer rather than downward. The size of the Ekman layer can be approximated by balancing the viscous term in the Navier–Stokes equation against the Coriolis term, i.e.,

$$\nu\nabla^2\mathbf{v} \approx \Omega \times \mathbf{v} \quad . \tag{7.6.7}$$

Writing $\nabla^2 \sim L^{-2}$ where L is the length scale that will be identified as the boundary layer thickness, we obtain

$$L \sim \sqrt{\nu/\Omega} \quad . \tag{7.6.8}$$

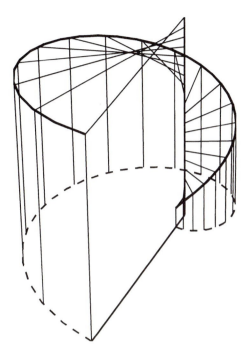

Fig. 7.5. An Ekman spiral in the northern hemisphere.

Substituting reasonable values for the Earth, one finds the boundary layer to be less than a metre thick, an estimate compatible with more precise calculations (including when the Lorentz force is included, producing an Ekman–Hartmann boundary layer; e.g., Jackson (1975)). The possibility of an eddy viscosity is considered in §7.6.4, and this can increase the core viscous boundary layers to 50 m or so. Nevertheless, the conclusion is that the core's viscous boundary layers are very thin.

  Boundary layers serve as interfaces between different regions in the Earth. Thus they are typically very complex, and often a crucial element in our understanding of processes in the Earth. This may be illustrated by considering the great sensitivity of the core to minute lateral variations in temperature at the core–mantle boundary. To obtain a rough estimate for the sustainable lateral temperature difference ΔT per length scale along the core–mantle boundary without creating *thermal winds*, equate the magnitude of the Coriolis force to thermal buoyancy

$$2|\Omega \times \mathbf{v}| = \rho \alpha_P \Delta T g \quad , \tag{7.6.9}$$

where α_p is the thermal expansion. Substitution of values appropriate to the CMB (e.g., $v \approx 3 \times 10^{-4}$ m s^{-1}) gives an estimate of order 10^{-6} K km^{-1}. Hence if there is a layer at the top of the core free from turbulence, very strong *thermal winds* can be generated by very small variations in temperature along the CMB. This great sensitivity to temperature variation shows that the CMB can be considered to be essentially isothermal. If strong magnetic fields are present near the CMB they can significantly affect the thermal winds, as will be discussed in the next two chapters. However, the temperature variation along the CMB is still probably less than a degree, unless there is strong radial convection very close to the boundary (e.g., turbulence). In summary, if there is any stratification in the outer core, the CMB is isothermal, and even without stratification the isothermal boundary condition is likely met.

7.6.2 Dimensionless Numbers

The value in using dimensionless numbers in dynamo problems can be illustrated by considering the fluid mechanical Reynolds number. For simplicity, consider the case of an infinitely long cylinder placed in a fluid initially flowing uniformly between two plates (see Fig. 7.6). The corresponding Navier–Stokes equation (8.1.14) is

$$\rho \left(\frac{\partial v_i}{\partial t} + v_j \frac{\partial v_i}{\partial x_j} \right) = \frac{\partial P}{\partial x_i} + \eta \frac{\partial^2 v_i}{\partial x_j \partial x_j} \quad , \tag{7.6.10}$$

where i and j represent either of the two coordinates in the problem, x and y, with x in the direction of the initial flow. Summation is carried out over the repeated roman indicies. Let L be some representative length scale for the flow behind the obstacle and u some representative velocity. The following nondimensional parameters are used: $v' \equiv v/u$, $t' \equiv tu/L$, and $x'_i \equiv x_i/L$. In addition, let $P' \equiv (P-P_o)/\rho \mu^2$, where P_o is some representative value of the modified pressure. Under these

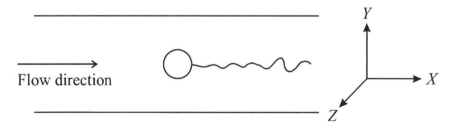

Fig. 7.6. Distortion of fluid flow behind an obstacle. An infinite cylinder is parallel to the z axis. The fluid flow is in the x direction and is laminar before it is affected by the cylinder. A stream line is shown behind the cylinder.

conditions, the Navier–Stokes equation becomes

$$\frac{\partial v'_i}{\partial t'} + v'_j \frac{\partial v'_i}{\partial x'_j} = \frac{\partial P'}{\partial x'_i} + \frac{1}{R_e} \frac{\partial^2 v'_i}{\partial x'_j \partial x'_j} \quad , \tag{7.6.11}$$

where $R_e \equiv \rho L u / \eta$ is the *Reynolds number*.

The Reynolds number provides a great deal of information on the class of nonunique solutions of (7.6.11). Any three of the four parameters, velocity, density, length scale, or viscosity, can be varied arbitrarily, while the fourth is varied in such a manner to keep R_e constant. The solutions to (7.6.11) are clearly independent of these variations, so long as R_e is constant. All fluid flows having the same value for R_e in (7.6.11) are said to be *dynamically similar*. Changes in R_e will be reflected in changes in the dynamical flow of the system. In the case of the infinite cylinder the flow is laminar for $R_e = 25$ and turbulent for $R_e = 250$.

At first it might seem an easy task to determine whether the Earth's outer core is turbulent by determining the appropriate Reynolds number. Note that even if the equations used in the Earth's core are more complicated than (7.6.11), the Reynolds number is still effectively the ratio of the inertial to viscous forces; the larger R_e becomes the more complicated the flow structure will be in any given situation. Unfortunately, on closer examination, the problem becomes far more complicated. The magnitude of the *critical Reynolds number,* that value above which the flow becomes turbulent, depends both on (7.6.11) and on the *boundary and initial conditions.* All too often a particular "critical value" for some characteristic dimensionless number determined for a particular boundary value problem will be erroneously used to characterize a different boundary value problem. For example, a very strong magnetic field in the x direction could result in laminar flow in the above example, even for very high values of R_e. However, once one has a solution for a particular problem, then the use of an appropriate dimensionless number can be very helpful.

A great deal of effort is expended to determine which dimensionless number is the most useful to characterize a particular problem, and so the literature is filled with a variety of such numbers. A few of the common ones used in fluid mechanics are given in Table 7.3, along with a brief description of their meanings.

Most of the parameters needed to estimate these numbers have been given in Table 7.3. One exception is the velocity, which is difficult to determine and is critical for most estimates (this is discussed further in §11.1). A velocity of $\approx 10^{-3}$ to 10^{-4} m s^{-1} appears to be a good first-order estimate. A value for the thermal expansion (α) of the order of 10^{-5} K^{-1} seems reasonable. A final word of caution is needed in reference to the physical intuition provided in Table 7.3: the intuition is not necessarily appropriate when the magnetic forces are significant, as may well be the case in the core.

TABLE 7.3
Some Common Dimensionless Numbers Used in Dynamo Theories

$R_e \equiv \dfrac{uL}{v}$	*Reynolds number*. This gives the ratio of inertial to viscous effects. Flow becomes turbulent for large R_e and is laminar for small R_e. See text of this section for further discussion
$R_a \equiv \dfrac{g\alpha\beta L^4}{kv}$	*Rayleigh number*. A fluid, heated from below, will become unstable at the critical Rayleigh number, above which convection occurs. The size of this critical number depends on the specific circumstances, but is often on the order of 10^3.
$R_b \equiv \dfrac{u}{\Omega L}$	*Rossby number*. The time it takes a fluid element moving with speed u to traverse a distance L is L/u. If that time is less than the period of rotation of the core then the fluid cannot sense the rotation. Thus, if $R_b \gg 1$, rotation can be neglected, while rotation plays a central role if $R_b \ll 1$.
$P_r \equiv \dfrac{v}{k}$	*Prandtl number*. This is a constant for a given material and does not depend on the properties of the flow. The viscous response will be instantaneous relative to the thermal response for a very large Prandtl number and vice versa for a very small one.
$E \equiv \dfrac{v}{\Omega L^2}$	*Ekman number*. This describes the relative importance of frictional effects. If E is small, frictional effects will be small and the Ekman boundary layer thickness at solid–liquid boundaries will be small.
$P_e \equiv P_r R_e = \dfrac{uL}{k}$	*Peclet number*. $P_e \gg 1$ implies that convective heat transport is much more important than conductive heat transport.

Notation:
u = estimate of the magnitude of some characteristic velocity
L = some representative length scale
v = kinematic viscosity = η/ρ
ρ = density
g = gravitational field
α = thermal expansion
β = temperature gradient in excess of the adiabatic temperature gradient
k = thermal diffusivity
Ω = angular rotation of the core

7.6.3 Instabilities

Understanding instabilities is central to understanding how waves are generated in fluids and how various boundary layers can break down. Entire books have been written on instabilities in fluids (e.g., Turner, 1979) and only very brief comments are made here. In dealing with instabilities in the mantle and core that lead to convection, the Boussinesq approximation is made. In this approximation, variations in density are ignored except for the gravitational body force; that is, the fluid is assumed to be incompressible. The conditions under which the Boussinesq approximation is applicable are given in detail by Tritton (1988), and Verhoogen (1980) suggests that the Boussinesq approximation may not be strictly applicable in the Earth's outer core. This approximation is further discussed by Gubbins and Roberts (1987). However, given the numerous assumptions and approximations made in dynamo theory and mantle convection modeling, it seems reasonable at this stage of theory development for us to use the Boussinesq approximation here, even if it is not rigorously justified.

It was noted in §7.3 that a necessary but not sufficient condition for convection in a liquid is that the temperature decrease faster upward than the adiabatic gradient. This instability is opposed by the friction action of viscosity and by thermal diffusion that tends to reduce the temperature contrast of a displaced parcel with its surroundings. The Rayleigh Number, R_a, is a nondimensional quantity giving the ratio of the thermal buoyancy to the viscosity times the thermal diffusivity, the onset of convection occurring above some critical value of this number. For convection in horizontal layers,

$$R_a = \frac{g\,\alpha_T\,(\Delta T)L^3}{\nu k_T} \quad , \tag{7.6.12}$$

where L is some typical length dimension (i.e., the thickness of the fluid layer) and ΔT is the *excess* temperature difference between the top and the bottom of the layer, i.e., the amount greater than that which would exist for an adiabatic gradient. The form of R_a varies depending on the circumstances; for example, there would be a different relationship for R_a if there were sources of internal heating or if other sources of buoyancy were present, such as might occur in compositional driven convection.

To understand instabilities better it is desirable to depart momentarily from the discussion of convection to discuss another important quantity, the Brunt–Väisälä frequency. Consider a small parcel of fluid with density ρ_0 displaced upward (which is in the negative Z direction) an amount ΔZ. The density of the surrounding fluid is $\rho_0 + (d\rho_0/dZ)\Delta Z$ and the gravitational force per unit density acting on it is $g(d\rho_0/dZ)\Delta Z$. Its equation of motion is therefore

$$\rho_0 \frac{d^2 \Delta Z}{dt^2} = g\frac{d\rho_0}{dZ}\Delta Z \quad . \tag{7.6.13}$$

If the fluid is stably stratified, this parcel will oscillate about its initial position with a frequency f_b, the Brunt–Väisälä frequency, which is seen from (7.6.13) to be

$$f_b = \sqrt{-\frac{g}{\rho_0}\frac{d\rho_0}{dZ}} \quad . \tag{7.6.14}$$

If the density variations are caused by temperature variations,

$$\frac{d\rho_0}{dZ} = \frac{\partial \rho_0}{\partial T}\frac{dT}{dZ} = -\rho_0 \alpha_T \frac{dT}{dZ} \quad , \tag{7.6.15}$$

where α_T is the thermal expansion. Substituting into (7.6.14), one obtains,

$$f_b = \sqrt{g\alpha_T \frac{dT}{dZ}} \quad . \tag{7.6.16}$$

Of course in the Earth an oscillation occurs only if the temperature gradient differs from the adiabatic gradient (see §7.3 and Eq. (7.3.4)). Consequently one must subtract the adiabatic gradient from dT/dZ to obtain f_b in the Earth:

$$f_b = \sqrt{g\alpha_T \left(\frac{dT}{dZ} - \frac{g\gamma_T T}{\phi_p}\right)} \quad . \tag{7.6.17}$$

If f_b is real, then displacement leads to oscillations, i.e., internal waves. However, if the temperature gradient decreases fast enough as Z increases, then f_b will be imaginary:

$$f_b \equiv i\gamma_b \quad . \tag{7.6.18}$$

In this case the amplitude of the disturbance will grow exponentially, or will be damped exponentially in time, depending on the sign of γ_b. Linear instability theory is usually applied to determine the sign of γ_b. This is done by assuming a perturbation in the velocity field of the form

$$v' = v(Z)\exp\left[i(k_x x + k_y y) + \gamma_b t\right] \quad , \tag{7.6.19}$$

where rectangular coordinates are used with Z vertical. Similar perturbations are made in the temperature and pressure fields. If γ_b is negative the effect of the perturbations decrease with time; while if γ_b is positive the perturbations grow exponentially with time. From symmetry considerations it is clear that the results can depend only on $k = (k_x^2 + k_y^2)^{0.5}$ and not on k_x or k_y individually. Often one plots the Rayleigh number versus some nondimensional wave number k_R defined by

$$k_R \equiv kL \quad , \tag{7.6.20}$$

where L is some characteristic length. It is found (e.g., Tritton, 1988) that for R_a below some value called the critical Rayleigh number, R_{ac}, γ_b is less than zero and convection is impossible. γ_b is greater than zero above R_{ac} for some values of k_R. In more sophisticated analyses one seeks out that value of k_R for which γ_b

is maximum and this provides valuable information on the nature of the spacing of the physical instabilities that initiate convection. In the case of classical Bernard convection in a horizontal layer heated from below, the critical Rayleigh number is 1708. However, Bernard convection does not exist in the mantle or the core. Nevertheless, the basic principles given above are still applicable to the Earth, although in the mantle one needs to consider that effective viscosity varies with temperature and there are internal heat sources. In the core compositional buoyancy needs to be considered and internal heat sources probably also are required. Instabilities (Rayleigh–Taylor) of thermal boundary layers can also be considered using the principles outlined above (e.g., Turner, 1979).

7.6.4 Turbulence

Although the concept of "turbulence" was used in §7.6.2, its definition is difficult. Usually it is defined by describing the features that are "turbulent." Instabilities affect fluid flow and reduce one's ability to predict the flow elsewhere in space or in time. One says the fluid is turbulent when it is necessary to describe the flow statistically. Much of our knowledge of turbulence comes from experiment, and the complexities of the problem are substantially compounded if a magnetic body force is present. It is not known whether the fluid motions in the outer core are turbulent.

A statistical description of turbulence begins by dividing the velocity and pressure fields into means and fluctuating parts, e.g., $\mathbf{v}+\mathbf{v}'$, where the prime represents the fluctuation velocity field. By definition the average over time of the fluctuating part, i.e., $<\mathbf{v}'>$, vanishes. Let the rectangular components of $\mathbf{v}$ be (v_1, v_2, v_3). The divergence of $\mathbf{v}$ can then be written

$$\nabla \cdot \mathbf{v} = \frac{\partial v_1}{\partial x_1} + \frac{\partial v_2}{\partial x_2} + \frac{\partial v_3}{\partial x_3} \equiv \frac{\partial v_i}{\partial x_i} \quad , \tag{7.6.21}$$

where, in the last equality, we use the Einstein convention that *repeated* subscripts mean summation over the dummy index; here the dummy index i (and later j) ranges over 1, 2 and 3.

Consider the Navier–Stokes equation in a stationary (nonrotating) system with no body forces:

$$\frac{\partial(v_i + v'_i)}{\partial t} + (v_j + v'_j)\frac{\partial(v_i + v'_i)}{\partial x_j} = -\frac{1}{\rho}\frac{\partial(P + P')}{\partial x_i} + v\frac{\partial^2(v_i + v'_i)}{\partial x_j \partial x_j} \quad .$$

$$\tag{7.6.22}$$

Averaging over time gives

$$\frac{\partial v_i}{\partial t} + v_j \frac{\partial v_i}{\partial x_j} + \left\langle v'_j \frac{\partial v'_i}{\partial x_j} \right\rangle = -\frac{1}{\rho} \frac{\partial P}{\partial x_i} + v \frac{\partial^2 v_i}{\partial x_j \partial x_j} \quad . \tag{7.6.23}$$

It is important to notice that although the average of v'_j and of $\partial v'_i/\partial x_j$ are zero, the average of their product need not be. For an incompressible fluid, $\partial v_j/\partial x_j$ vanishes, allowing one to write:

$$v_j \frac{\partial v_i}{\partial x_j} = -\frac{1}{\rho} \frac{\partial P}{\partial x_i} + v \frac{\partial^2 v_i}{\partial x_j \partial x_j} - \frac{\partial}{\partial x_j} \left\langle v'_i v'_j \right\rangle \quad , \tag{7.6.24}$$

where we have restricted our attention to *steady mean* conditions by setting $\partial v_i/\partial t$ to zero.

The term $<v'_i v'_j>$ multiplied by the density, ρ, is called the Reynolds stress. Physical insight into the Reynolds stress can be obtained by examining a two-dimensional flow near a rigid boundary. Even though turbulent fluctuations are always three-dimensional, if the imposed conditions (e.g., flat boundary) are two-dimensional, there is no variation of mean quantities in the third direction (x_3) and terms such as $\partial<v'_1 v'_3>/\partial x_3$ are zero. In this case (7.6.24) can be written

$$v_1 \frac{\partial v_2}{\partial x_1} + v_2 \frac{\partial v_2}{\partial x_2} = \frac{1}{\rho} \frac{\partial P}{\partial x_2} + v \frac{\partial^2 v_2}{\partial x_1 \partial x_1} - \frac{\partial}{\partial x_1} \left\langle v'_1 v'_2 \right\rangle \quad . \tag{7.6.25}$$

The last two terms of (7.6.25) can be rewritten

$$\frac{1}{\rho} \frac{\partial}{\partial x_1} \left(v_N \frac{\partial v_2}{\partial x_1} \right) - \rho \left\langle v'_1 v'_2 \right\rangle \quad , \tag{7.6.26}$$

where v_N, the Newtonian viscosity, is the kinematic viscosity v times the density ρ, i.e., $v_N = \rho v$. Recall that the definition of a fluid involves relating the shear stress, σ_s, to the velocity shear it produces:

$$\sigma_s = v_N \frac{\partial v_2}{\partial x_1} \quad . \tag{7.6.27}$$

Comparing the last two relations one sees that $\rho<v'_1 v'_2>$ acts as a stress on the mean flow; this stress arises from the correlation of two components of the velocity fluctuation at the same point.

Figure 7.7 shows how the mean upward dimensional velocity decreases as it approaches a lower boundary. Consider a small fluid parcel carried upward (in the x_2 direction with positive v'_2) by the turbulence. This parcel is moving from a region of low velocity to one of higher velocity and it is therefore more likely that the fluctuating velocity in the x_1 direction will be negative as shown.

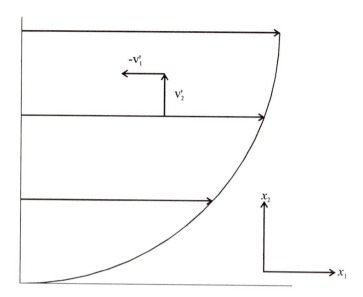

Fig. 7.7. Effective viscosity caused by correlation of two components of the velocity fluctuation at the same point.

Similarly, negative v'_2 is more likely to be associated with a positive v'_1. This produces an effective viscosity. This analogy has led to the definition of an eddy viscosity, v_E, defined by

$$-\left\langle v'_1 \, v'_2 \right\rangle \equiv v_E \frac{\partial v_1}{\partial x_2} \quad . \tag{7.6.28}$$

It is important to recognize that, unlike v, v_E is not a physical property of the fluid, but is a consequence of the Reynolds stress. This eddy viscosity may increase the magnitude of the Ekman boundary layer in the core, but to estimate this increase one needs to know if turbulence in the core exists and if so, what are the appropriate parameters to use in (7.6.28) to obtain v_E. The problem of obtaining core velocities is discussed further in §11.1. However, estimates of v_E are usually obtained empirically. Although this cannot yet be done for core conditions, there does seem to be agreement that core Ekman layers are probably less than 100 m, an agreement reached from extrapolation of the behavior of more accessible fluids.

No discussion of turbulence can be complete without discussing the concept of cascading. Consider isotropic homogeneous turbulence in which the energy spectrum, E_k, is a function of the wave number k (2π divided by the wavelength) of the eddies present and time, t. It is known (e.g., Tritton, 1988) that the change of energy with time is given by

$$\frac{\partial E_k(k,t)}{\partial t} = F(k,t) - 2\nu k^2 G(k,t) \quad , \tag{7.6.29}$$

where F and G are functions of k and t. The first term on the right side does not involve viscosity and so does not involve dissipation. It can be shown that

$$\int_0^\infty F(k,t)\,dk = 0 \quad . \tag{7.6.30}$$

Thus the term $F(k,t)$ represents the transfer of energy between wave numbers. In contrast, the second term, with $G(k,t)$, depends linearly on viscosity, and describes the dissipation of energy. Empirically it is found that when energy is fed into large eddies, these eddies break into smaller eddies, which then break into still smaller eddies and so on, until the ratio of the "surface areas" of the eddies (where shear is maximum) to the volume becomes sufficiently large for dissipation to be effective. This process, where energy is supplied to larger eddies that divide into smaller and smaller ones until a balance is achieved between energy supply (to large eddies) and removal (through heat in small eddies), is called an "energy cascade."

Using simplifying assumptions and dimensional analysis one commonly quoted cascade law is the Kolmogorov -5/3 law, which gives (e.g., Tritton, 1988)

$$E_k = A\,(\varepsilon_d)^{\frac{2}{3}} k^{-\frac{5}{3}} \quad , \tag{7.6.31}$$

where A is a constant and ε_d is the energy dissipation. This law appears to be a good first-order approximation for many flows observed in nature that have high Reynolds numbers (Tritton, 1988). Further discussion of the question of whether an energy cascade occurs in the Earth's core will be given in §9.5.3.

7.7 Energy Sources

The major uncertainties in estimating the power required to sustain the geodynamo can be traced to poor knowledge of the electrical conductivity and of the magnitude of the magnetic field within the core. As will be discussed in Chapters 8 and 9, the core's magnetic field can be divided into a poloidal magnetic field (which has a radial component) and a toroidal magnetic field (no radial component). The toroidal field is essentially confined to the core and its magnitude must be inferred from theory or from highly indirect observations.

One estimate for the power is obtained by substituting values for B and some effective decay time, τ, into

$$P_B = \frac{1}{\tau}\left[\int_V \left(\frac{B^2}{2\mu_0}\right)dV\right] \quad , \qquad (7.7.1)$$

where the integration is over all space. However, there are problems with this approach, the most obvious being what to use for B and τ. A more subtle problem is that the heat loss through Joule heating in the core remains inside as another source of buoyancy (Backus, 1975); estimates using (7.7.1) do not take this into account. Viscous heating is another source of dissipation per unit volume not yet discussed, but this is probably relatively small because magnetic viscosity (§8.2.2) is substantially greater than fluid viscosity. Using (7.7.1) one can make a very crude estimate for the power by taking B to be 20 T in the core (strong-field model; see Chapters 8 and 9), the volume of integration to be the core's volume and τ (which is related to electrical conductivity as discussed in Chapter 8) to be 10,000 yr. Doing this one obtains a value of 3×10^{10} W, a value that falls on the low side of most estimates. Although more mathematical attempts have been made to obtain estimates of the power required to sustain the Earth's magnetic field (e.g., Braginsky, 1976; Verhoogen, 1980; Stacey, 1992), the greatest uncertainties involve the assumptions and not the mathematical rigor; a value one order of magnitude larger than the 3×10^{10} W given above seems favored by most modelers. It appears the power is not known to better than one or two orders of magnitude.

Energy-source models can essentially be divided into three classes: models associated with changes in the Earth's rotation, models associated with thermal convection, and models associated with compositional convection.

(i) *Tidal friction.* This is estimated to contribute some 10^{12} W of dissipation in the core (e.g., Roberts and Gubbins, 1987), an amount sufficient to sustain the geodynamo. This has not gone unnoticed; for example, Malkus (1963, 1968) suggests differential precession drives the geodynamo. The problem is that the mechanisms for doing this are either unclear or not suitable (e.g., Rochester *et al.*, 1975; Stacey, 1977; Roberts and Gubbins, 1987). Nevertheless there is the possibility that, say, precessional torques associated with the Milankovich cycles (§3.5) can produce similar periods in the core and its magnetic field. Paleomagnetic secular variation studies have been initiated to study this possibility (e.g., Creer, personal communication, 1994), but apparently the substantial number of high-quality data required to build a convincing case for or against it are not yet available. Malkus (1994) has argued that precessional and tidal resonances may be very strong within the core, and may have a significant effect on the geodynamo.

(ii) *Thermal convection.* Thermally driven convection in the outer core was once the favored mechanism for sustaining the geodynamo. The main reasons for its decline in popularity are based on the arguments that follow. Most estimates for the power required to drive the geodynamo fall between 10^{11} and 10^{12} W. The heat flux through the core–mantle boundary is commonly estimated to be around 4×10^{12} W (see §7.4). Thermal convection is not thought to be a very efficient process for driving the geodynamo, with most estimates of the efficiency of the process varying from 10 to 20% (e.g., Verhoogen, 1980; Roberts and Gubbins, 1987).

Putting together the above arguments, one concludes that thermal convection is a marginal energy source, and may be inadequate to drive the geodynamo. However, Buffett (pers. communication) points out that in previous calculations the volume of the core has been considered invariant, but that the total volume of the core must decrease as the inner core forms because of the decrease in volume as the material freezes. There will, therefore, be significant gravitational energy, which goes to thermal energy, associated with the core contraction.

Each of the above arguments is highly vulnerable to assumptions and observational uncertainties. Our earlier estimate of the power needed to drive the geodynamo was less than that cited above; note that by increasing the toroidal magnetic field by a factor of three (well within the range of uncertainty), our previous estimate for the power to drive the geodynamo becomes compatible with conventional estimates. Furthermore, the models of heat flux out of the core neglect the possibility of radioactive sources there (see §7.4). Finally, it is not a simple matter to estimate efficiency because the ohmic heat associated with magnetic field decay is not lost from the system, but remains within as a source of buoyancy (Backus, 1975; Roberts and Gubbins, 1987). In conclusion, it is premature to rule out the possibility that thermal convection plays a significant, and possibly even a dominating, role in sustaining the Earth's magnetic field.

(iii) *Latent heat of crystallization.* Motivated by Verhoogen's (1961) arguments concerning latent heat of crystallization, Braginsky (1963, 1991) suggests that the prime source of convection buoyancy involves the release of elements less dense than iron upon freezing near the inner core boundary (see §7.4). Subsequently, Gubbins (1976, 1977), Loper (1978a,b), Loper and Roberts (1978, 1979), and Roberts and Gubbins (1987) have carried out various analyses of the problem and demonstrated that compositional buoyancy is probably the dominant source of convection sustaining the geodynamo.

(iv) *Compositional buoyancy.* The power available from the gravitational energy associated with compositional buoyancy, P_g (force × momentum), is

$$P_g = \int (\nabla \phi_g) \cdot \rho \mathbf{v} \; dV \quad , \tag{7.7.2}$$

where ϕ_g is the gravitational potential, ρ is the local density of core material, and the integration is carried out over the volume of the core. Integrating by parts and using Stokes' theorem yields

$$P_g = \int \phi_g \rho \mathbf{v} \cdot d\mathbf{S} - \int \phi_g \nabla \cdot (\rho \mathbf{v}) dV \quad , \qquad (7.7.3)$$

where $\mathbf{S}$ is the surface of the core. Gubbins (1977) assumes that there is no net volume change in the core and that the first term on the right can be neglected. Following Gubbins, and using (7.7.3) and the continuity equation, one obtains

$$P_g = \int \phi_g \frac{\partial \rho}{\partial t} dV \quad . \qquad (7.7.4)$$

The available power depends critically on the rate of growth of the inner core. Gubbins takes a freezing rate of 25 m^3 s^{-1} of material frozen at the inner core boundary (a value used by Verhoogen (1961), in calculating latent heat energy) and finds that P_g is given by

$$P_g \approx \Delta\rho \times 2 \times 10^{11} \text{ W} \quad . \qquad (7.7.5)$$

Gubbins *et al.* (1979) analyzed eigenvibration data to conclude that $\Delta\rho$, the density contrast across the inner core boundary, lies between 250 and 1250 kg m^{-3}. More recently, Shearer and Masters (1990) obtained an estimate near 550 kg m^{-3}. However, as pointed out by Buffet *et al.* (1992), only part of the density contrast of 500 kg m^{-3} can be used to sustain the geodynamo, since there is roughly a 100 to 200 kg m^{-3} increase in density associated with the volume change on freezing iron.

Buffet *et al.* (1992) argue that thermal buoyancy, associated with removal of heat at the outer core surface, is presently approximately equal to compositional buoyancy due to freezing of the inner core. However, compositional buoyancy is believed to be more efficient since its energy is lost only through viscous ohmic heating or decay of the magnetic field. In contrast, much of the heat associated with thermal convection simply leaves the core. Thus, compositional buoyancy is believed to be a more efficient process than thermal buoyancy for sustaining the geodynamo, an argument that can be placed on a more rigorous foundation than done here (e.g., Roberts and Gubbins, 1987).

A major challenge to geophysicists is to find additional constraints on the energy-source problem. Some of these may come from secular variation, as already mentioned with regard to thermal convection. Braginsky (1993) suggests that compositional driven convection leads to a roughly 80-km-thick less-dense layer at the top of the core. In contrast, there is some seismic evidence for a thicker layer (see §7.1). Planetary hydromagnetic waves are sensitive to layer thickness, and consequently future analyses of paleomagnetic secular variation data may provide valuable constraints on this hypothetical layer. It is also possible that a less-dense outer-core layer does not exist simply because the less-dense material is mixed into the outercore by convection. If so, the efficiency of compositional convection is reduced somewhat.

Stevenson (1983) suggests that analyses of terrestrial planetary magnetic fields also provide insight into the problem; he argues that Venus does not have an active dynamo primarily because it has no inner core and thus compositional buoyancy cannot sustain a dynamo.

Although the conclusion is that compositional buoyancy seems to be dominant in sustaining the geodynamo, thermal buoyancy is also likely to be important. However, all calculations and relevant observational data on this subject contain a sufficient number of problems that this "conclusion" requires continual examination. Both interdisciplinary studies and appeals to parsimony will probably be required for further progress.

Chapter Eight

Introduction to Dynamo Theory

8.1 The Dynamo Problem

8.1.1 Disc Dynamos

A disc dynamo (or homopolar or Hertzenberg dynamo) is illustrated in Fig. 8.1a. It was this type of dynamo that was suggested by Larmor (1919a,b) to explain the origin of the Earth's magnetic field. A torque, usually assumed constant, must be applied to rotate the disc with angular velocity ω in the presence of a field $\mathbf{H}$ (see Fig. 8.1a). A Lorentz force will produce an electric current, I, out from the center of the disc that can be tapped by a wire that has a "brush" contact with the disc. This wire is wound around the axis of the disc in such a way as to reinforce the initial field. Once the dynamo process starts working the initial seed field can be removed.

Many of the essentials in dynamo theory are incorporated in this simple disc dynamo picture: magnetic field energy is produced through the conversion of mechanical energy. The mechanical energy in this case is that energy required to rotate the disc. As the magnetic field increases, the torque required to rotate the disc at the same rate also increases, because the Lorentz force feeds back into the system. An initial seed field is required at the onset but not thereafter.

Because of its historical importance, a brief outline of the disc dynamo theory will be given here. Details are given in Rikitake (1966). The appropriate equations governing the system are

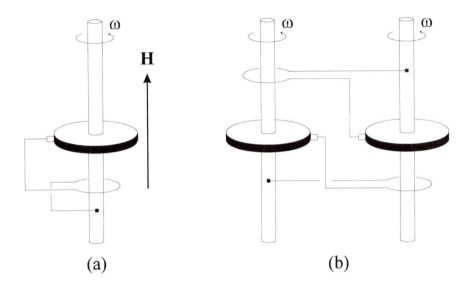

Fig. 8.1. The disc dynamo. (a) Single-disc dynamo. A torque Υ is applied to rotate a conducting disc at angular speed ω in a magnetic field $\mathbf{H}$ aligned along the axis of the disc. An electric current, induced in the rotating disc, flows outward to the edge of the disc where it is tapped by a brush attached to a wire. The wire is wound back around the axis of the disc in such a way as to reinforce the initial field. (b) Double-disc dynamo. The current induced in one disc is circulated back around the axis of the other. The winding of the wires around the rods is such that reversals of polarity can occur (modified after Rikitake, 1966).

$$L\frac{\mathrm{d}I}{\mathrm{d}t} + RI = M\omega I \qquad \text{[electrical part]} \qquad\qquad (8.1.1)$$

$$C\frac{\mathrm{d}\omega}{\mathrm{d}t} = \Upsilon \qquad\qquad \text{[mechanical part] ,} \qquad\qquad (8.1.2)$$

where L and M are derived from proper combinations of the self- and mutual-inductances of the system, t is time, R is the electrical resistance of the circuit, C is the moment of inertia of the disc and Υ is the torque. The induced e.m.f. for the disc is $M\omega I$ and the power is $M\omega I^2$. Because the power must be equal to $\Upsilon\omega$ in equilibrium, (8.1.2) becomes

$$C\frac{\mathrm{d}\omega}{\mathrm{d}t} = \Upsilon - MI^2 \quad . \qquad\qquad (8.1.3)$$

Equations (8.1.1) and (8.1.2) can be solved simultaneously for I and ω (Rikitake, 1966). The magnitude of the resulting magnetic field can easily be solved once I is obtained for the particular geometry chosen.

Calculations show that a single-disc dynamo will not produce reversals in **H** unless the rotation direction is reversed by applying an external torque of the opposite sign. In contrast, the coupled double-disc dynamo shown in Fig. 8.1b overcomes this problem. Constant torques applied to the discs result in highly varying fields, including reversals (Rikitake, 1966). Moreover, one can use disc dynamo models to produce deterministic chaotic reversals that exhibit many of the properties observed in the Earth's reversal record. However, the analogies of the solid disc, the wire, and the brushes are poor in the core. Moreover, as will be seen, realistic kinematic dynamos have been constructed that also exhibit many of the properties of the Earth's magnetic field. In the case of the Earth, the mechanical energy is that associated with fluid motions. However, the feedback system in the core operates on very different principles from the wires and brushes of disc dynamos. Therefore, in spite of the important role they played in the development of dynamo theory, no further discussion of disc dynamos will be given. Additional discussion of other laboratory models is given in §8.2.3.

8.1.2 Magnetohydrodynamics and Plasma Physics

A critical assumption in dynamo theory is that $\partial \mathbf{D}/\partial t$, the time derivative of the electric displacement current density, can be neglected (this is often referred to as the MHD or magnetohydrodynamic assumption). This contrasts with plasma physics where $\partial \mathbf{D}/\partial t$ cannot be neglected. Valuable insight justifying this assumption can be obtained by considering the microscopic derivation of Ohm's law in a conducting fluid. Let v_d be the drift velocity of electrons, μ_e the electron mass, and τ a constant (which, as will later be apparent, is a relaxation time). Let e be the charge of the electron and E the magnitude of the electric field.

The equation of motion for an electron in the presence of the force eE is

$$\mu_e \left(\frac{dv_d}{dt} + \frac{v_d}{\tau} \right) = eE \quad . \tag{8.1.4}$$

The second term on the left side is a friction term originating from collisions with other electrons. The solution of this equation is easily found to be

$$v = v_0 e^{-t/\tau} + \frac{e\tau E}{\mu_e} \quad . \tag{8.1.5}$$

Note that as $t \to \infty$ (steady state) one obtains

$$v_d = \frac{e\tau E}{\mu_e} \quad . \tag{8.1.6}$$

If N is the electron density, then the steady state current density is (Ohm's law)

$$J = Nev_d = \frac{Ne^2\tau E}{\mu_e} \equiv \sigma E \quad . \tag{8.1.7}$$

Setting $f{=}1/\tau$, f can then be interpreted as representing some mean collision frequency.

Suppose an electric field of frequency v is applied; three cases then occur:

(i) $v{\gg}f$ (the plasma domain). Here electrons (and ions) move independently. In effect, the high frequency of the applied electric field causes charge separation and $\partial \mathbf{D}/\partial t$ cannot be neglected.

(ii) $v{\ll}f$ (the magnetohydrodynamic (MHD) domain). There is no charge separation. Electrons move according to Ohm's law (steady state). In this case the magnetohydrodynamic assumption, $\partial \mathbf{D}/\partial t = \mathbf{0}$, is justified.

(iii) $v{\approx}f$ is a complicated situation that will not be considered here.

Pressures exceeding 10^{12} Pa would be required to strip electrons from their nuclei to produce plasma conditions in the Earth's core. Thus it appears justified to use the MHD approximation there, i.e.,

$$\frac{\partial \mathbf{D}}{\partial t} = \mathbf{0} \tag{8.1.8}$$

$$\nabla \times \mathbf{H} = \mathbf{J} \quad .$$

8.1.3 The Earth Dynamo Problem

The magnitude of the problem involved in obtaining a solution to the Earth's dynamo problem can be appreciated by listing the equations that must be solved simultaneously, as is done in Table 8.1. The magnetohydrodynamic assumption, that $\partial \mathbf{D}/\partial t{=}\mathbf{0}$, is always made and the fluid is usually assumed incompressible, i.e., $\nabla{\cdot}\mathbf{v}{=}0$. Buoyancy, which requires compressibility, still drives convection, but this is incorporated into the theory through the Boussinesq approximation: the fluid is assumed incompressible except for the thermal expansion (§7.6).

Simultaneous solution of (8.1.9) to (8.1.18) along with the appropriate boundary and initial conditions is referred to as the *Earth dynamo problem*. This is a formidable problem, not only because of the number of linked equations but also because even the method used to solve a particular partial differential equation depends on the boundary and initial conditions. As might be expected

Table 8.1
Equations for the Dynamo Problem

$$\nabla \times \mathbf{H} = \mathbf{J} + \frac{\partial \mathbf{D}}{\partial t}$$

(8.1.9)

$$\nabla \times \mathbf{E} = -\frac{\partial \mathbf{B}}{\partial t}$$

Maxwell's equations (8.1.10)

(8.1.11)

$$\nabla \cdot \mathbf{B} = 0$$

(8.1.12)

$$\nabla \cdot \mathbf{D} = \rho_e$$

$$\mathbf{J} = \sigma \mathbf{E} + \sigma(\mathbf{v} \times \mathbf{B})$$

Ohm's Law (8.1.13)

Navier–Stokes' equation (8.1.14)

$$\rho\left(\frac{\partial}{\partial t} + \mathbf{v} \cdot \nabla\right)\mathbf{v} + 2\rho(\Omega \times \mathbf{v})$$

$$= -\nabla P + \eta \nabla^2 \mathbf{v} + \tfrac{1}{3}\eta \nabla(\nabla \cdot \mathbf{v}) - \rho \nabla \phi_g + \mathbf{J} \times \mathbf{B}$$

$$\nabla \cdot (\rho \mathbf{v}) + \frac{\partial \rho}{\partial t} = 0$$

Continuity equation (8.1.15)

$$\nabla^2 \phi_g = -4\pi G\rho$$

Poisson's equation (8.1.16)

$$\frac{\partial T}{\partial t} = k_T \nabla^2 T + (\nabla k_T \cdot \nabla T) - \mathbf{v} \cdot \nabla T + \varepsilon$$

Generalized heat equation (8.1.17)

$$\rho = \text{Function}(P, T, H)$$

Equation of state (8.1.18)

Notation:

$\mathbf{H} \equiv$ magnetic field		$\rho \equiv$ material density
$\mathbf{B} \equiv$ magnetic induction		$\sigma \equiv$ conductivity
$\mathbf{J} \equiv$ electric current		$T \equiv$ temperature
$\mathbf{E} \equiv$ electric field		$P \equiv$ pressure
$\mathbf{D} \equiv$ electric displacement vector		$G \equiv$ gravitational constant
$\mathbf{v} \equiv$ velocity		$\phi_g \equiv$ gravitational potential
$\eta \equiv$ viscosity		$\varepsilon \equiv$ heat source term
$\rho_e \equiv$ electric charge density		$k_T \equiv$ thermal diffusivity
$\Omega \equiv$ angular velocity of rotation		

from the complexity of the problem, substantial simplifications are made to solve a subset of this problem, but even then the solution usually requires considerable mathematical analysis and intensive numerical computations.

8.2 The Magnetic Induction Equation

8.2.1 Introduction

One of the most important equations in dynamo theory, the magnetic induction equation, can be derived by combining (8.1.9) with (8.1.13) to obtain

$$\nabla \times \mathbf{H} = \sigma \mathbf{E} + \sigma(\mathbf{v} \times \mathbf{B}) \quad . \tag{8.2.1}$$

Then taking the curl of both sides of this equation and using (8.1.12), (8.1.13) and $\mathbf{B}=\mu_0\mathbf{H}$, one obtains the *magnetic induction equation*:

$$\frac{\partial \mathbf{H}}{\partial t} = \frac{1}{\sigma\mu_0} \nabla^2 \mathbf{H} + \nabla \times (\mathbf{v} \times \mathbf{H}) \quad . \tag{8.2.2}$$

Note that this equation also holds if $\mathbf{B}$ is substituted for $\mathbf{H}$, since $\mathbf{B}=\mu\mathbf{H}$ and μ, the magnetic permeability, is a constant in the Earth's core (μ_0). When $\mathbf{v} = \mathbf{0}$ this equation reduces to the vector diffusion equation for $\mathbf{H}$. In the absence of a velocity field $\mathbf{v}$, a given magnetic field will decay according to the first term on the right side. The last term in this equation gives the interaction of the velocity field with the magnetic field. This interaction can cause build-up, or breakdown, of the magnetic field, depending on the nature of the velocity field.

Use of the magnetic induction equation bypasses the need for explicitly considering the electrical currents in the core. Note, however, that the currents can always be explicitly obtained (if desired) when the magnetic field is known by taking the curl of that field. Equation (8.2.2) has never been solved in closed form, necessitating the need for other approaches, such as the use of expansions and numerical calculations.

In *kinematic dynamo models*, $\mathbf{v}$ is specified in some reasonable way (this usually means that $\mathbf{v}$ has no sources or sinks and is continuously differentiable), along with some initial magnetic field, $\mathbf{H}_0$. The problem then is to determine whether this $\mathbf{v}$ field can support an $\mathbf{H}$ field that does not decay to zero as time goes to infinity. A subset of the kinematic dynamo problems involves the search for *steady-state dynamo* solutions, i.e., solutions for which $\partial\mathbf{H}/\partial t=0$. The kinematic dynamo problem does not require that the $\mathbf{v}$ field satisfy the Navier–Stokes equation. There is, therefore, no feedback from the magnetic field to the velocity field. This means that a magnetic field that approaches infinity as time approaches infinity is an acceptable solution to a kinematic dynamo problem. Another subset of kinematic dynamo models involves *fast dynamos*, in which diffusion of the magnetic field is neglected to investigate the maximum rate of build-up of the magnetic field through dynamo processes (Soward, 1989). This is discussed further in §9.2.3.

The problem of simultaneously solving (8.1.14) and (8.2.2) represents the *hydrodynamic dynamo* problem. This problem is sometimes solved by assuming that the body force, proportional to $(\mathbf{J}\times\mathbf{B})$ occurring in the Navier–Stokes equation (8.1.14), is a perturbation. One can then solve the Navier–Stokes equation for some assumed geometry, boundary conditions, and initial conditions by neglecting the $(\mathbf{J}\times\mathbf{B})$ term. The solution for the velocity field, $\mathbf{v}$, is then substituted into the magnetic induction equation (8.2.2), which in turn is solved for some assumed boundary and initial conditions pertaining to the magnetic field. This last solution provides an estimate for **B**. It is possible to return to the Navier–Stokes equation and use this **B** to include the magnetic body force term and to solve for a new **v**. Such an iteration procedure can be continued *ad infinitum* or until computer time becomes too expensive. For this to be a valid procedure, it must be demonstrated that the iteration procedure leads to some convergence of **v** and **B**.

The above approach assumes a small Lorentz force (**J**×**B**) relative to the Coriolis force (2Ω×**v**), and the resulting dynamo is referred to as a *weak-field model*. The term *strong-field model* implies that the magnitude of the Lorentz force is equal to, or greater than, the Coriolis force. Using (8.1.8) and (8.1.9), the relative magnitude of the Lorentz force to the Coriolis force, commonly referred to as the *Elsasser number* Λ_E, is given by

$$\Lambda_E = \frac{\mu_0 H^2}{\rho k_m \Omega} \quad , \tag{8.2.3}$$

where $k_m = 1/(\sigma \mu_0)$. Most advocates of dynamo theory now argue that strong-field dynamo models are required for the Earth and, therefore, that Λ_E is equal to or greater than 1. Λ_E is equal to unity when H is approximately 3 times the dipole field at the core–mantle boundary.

Turbulent dynamo models, or *mean-field electromagnetic dynamos,* were independently developed by Steenbeck, Krause, and Rädler (Steenbeck *et al.*, 1966; Krause and Steenbeck, 1967; Rädler, 1968; Steenbeck and Krause, 1969) and by Moffatt (1961, 1970). These are discussed in detail in Chapter 9, but as the name suggests, they require that the core's velocity field be turbulent.

The possibility that *magnetohydrodynamic waves* play an important role in dynamo theory was first suggested by Braginsky (1964a,b; 1967). He assumed a strong-field model in which the instability that led to waves involved a balance between the magnetic (*M*, now called the Lorentz) force, the Archimedean force (*A*, the buoyancy force), and the Coriolis (*C*) force. These waves are commonly called *MAC waves.* Hide (1966) first developed the possibility that *MAC* waves could be responsible for much of the observed geomagnetic secular variation. He used a thin-plane approximation (the β-plane approximation), as discussed further in §8.4, and such waves are often referred to as *magnetic Rossby waves,* in analogy with Rossby waves in the atmosphere.

The above introduces much of the dynamo model terminology. However, the difficulties involved in solving the geodynamo problem should not be underestimated; all sophisticated models involve extensive computer calculations. It will be shown in §9.3 that a magnetic field that is symmetric about some axis cannot be sustained by dynamo action (note, however, that this does not necessarily preclude fluid motions with an axis of symmetry). This is the famous *Cowling Theorem*. In particular, one cannot have a perfectly symmetric magnetic field about the Earth's rotational axis. This means that three-dimensional computer modeling is required. Today there are many dynamo models, all of which make numerous simplifications and hence, in some way or another, inaccurately model the Earth's actual dynamo. Paleomagnetists and geomagnetists who wish to combine dynamo theory with observations would, therefore, be well advised to use only robust results from dynamo theory, i.e., not to use results that depend on the details of a particular model. Emphasis in the remainder of this chapter will be on developing physical concepts that might be useful to both the observationalist and the theoretician who wish to gain insight into dynamo theory without expending enormous effort to understand the complex mathematics involved. Chapter 9 continues with this goal, but also develops further the mathematics required to read most modern dynamo theory papers.

8.2.2 Physical Insight

To gain insight into the meaning of (8.2.2), the role of each of the two terms on the right side of the equation is considered separately. The first term on the right side is zero for the hypothetical case of infinite conductivity. In this case it can be shown that no induced e.m.f. occurs in a perfect conductor moving in a magnetic field. This theorem is referred to as *the frozen-in-field theorem*. It provides important insight into how a moving conductor can build up or break down a magnetic field. In the case of infinite conductivity, (8.2.2) reduces to

$$\frac{\partial \mathbf{H}}{\partial t} = \nabla \times (\mathbf{v} \times \mathbf{H}) \quad . \tag{8.2.4}$$

Consider any area, S, bounded by a line $\mathcal{L}$ in a fluid (see Fig. 8.2). Let $\mathbf{n}$ be the normal to S. Then

$$\int_S \frac{\partial \mathbf{H}}{\partial t} \cdot \mathbf{n} dS = \int_S \nabla \times (\mathbf{v} \times \mathbf{H}) \cdot \mathbf{n} dS = \int_{\mathcal{L}} (\mathbf{v} \times \mathbf{H}) \cdot d\mathcal{L} = -\int_{\mathcal{L}} \mathbf{H} \cdot (\mathbf{v} \times d\mathcal{L}) \quad .$$
$$\tag{8.2.5}$$

Now $(\mathbf{v} \times d\mathcal{L})$ is the increment of area perpendicular to $\mathcal{L}$ that is swept out in time dt. Therefore,

$$\int_S \frac{\partial \mathbf{H}}{\partial t} \cdot \mathbf{n} dS + \int_{\mathcal{L}} \mathbf{H} \cdot (\mathbf{v} \times d\mathcal{L}) = \frac{d}{dt} \int_S \mathbf{H} \cdot \mathbf{n} dS = 0 \quad . \tag{8.2.6}$$

Since the magnetic flux, ϕ, is defined as

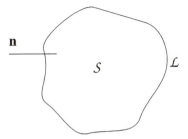

Fig. 8.2. The geometry used in the proof of the frozen-in-field theorem involves any surface with area S, bounded by $\mathcal{L}$. The unit normal to the surface is **n**.

$$\phi = \int_S \mathbf{B} \cdot \mathbf{n} \, dS = \mu_0 \int_S \mathbf{H} \cdot \hat{\mathbf{n}} \, dS \quad , \tag{8.2.7}$$

and the induced e.m.f. is equal to $-d\phi/dt$, the above equations imply that

$$\frac{d\phi}{dt} = 0 \quad . \tag{8.2.8}$$

This means there is no e.m.f. induced in the perfect conductor, and therefore no changes in the magnetic field can ever occur inside a perfect conductor. It is because of this that the magnetic field is sometimes described as being "frozen into" the conductor. Further development of the frozen flux concept and of null-flux curves is given in §11.1.2. Figure 8.3 depicts what will happen when a conductor is moved from a field-free space into a magnetic field. Because $d\phi/dt=0$, no field lines can pass through the conductor. Therefore, the field lines will be compressed in front of the moving conductor, and thus the field will be magnified there. Physically, electric current eddies are created on the infinitely small thin surface of the conductor (by Lenz's Law), resulting in magnetic flux being excluded from inside the conductor.

If only the first term on the right side of (8.2.2) is present, the magnetic induction equation is reduced to a vector diffusion equation. This explains why $(\sigma\mu_0)^{-1} \equiv k_m$ is often referred to as the *magnetic diffusivity* (or magnetic viscosity as discussed below). This could occur if the velocity field were zero, in which case the magnetic field would decay with time. For the Earth's core one anticipates that both terms on the right-hand side of the magnetic induction equation are usually present. This means there will usually be both diffusion of the magnetic field and build-up (or breakdown) of the field due to the interaction of **v** with **B**. The ratio of the second term to the first term on the right side of (8.2.2) gives an estimate of the rate of build-up of the field to decay. A dimensionless number R_m (called the *magnetic Reynolds number* by analogy

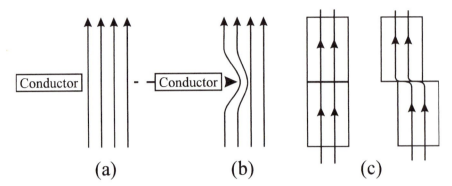

Fig. 8.3. Illustration of the frozen-in-field theorem. (a) and (b) show the effects of moving a perfect electrical conductor from a field-free region into a magnetic field indicated by vertical lines. The magnetic field lines are compressed in front of the conductor and no magnetic field lines ever pass through it as in (b). If the conductivity is finite, the magnetic field can diffuse back into the conductor. In (c) suppose there are two blocks of finite conductivity and initially the magnetic field has diffused back into the blocks. If the lower block is instantaneously moved to the right, it will carry the magnetic field lines with it as shown. Note that the magnetic field lines are concentrated in the *shear zone* at the boundary between the two blocks.

with the Reynolds number in fluid mechanics discussed later) characterizes that estimate,

$$\frac{\nabla \times (\mathbf{v} \times \mathbf{H})}{k_m \nabla^2 \mathbf{H}} \sim \frac{vL}{k_m} \equiv R_m \quad , \tag{8.2.9}$$

where L is some appropriate length dimension. It can be seen that $R_m > 1$ is a necessary, but not a sufficient, condition for a dynamo to be self-sustaining. For small magnetic Reynolds numbers the decay term will dominate and no self-sustaining dynamo is possible. The question of what to use for v and L is difficult because often more than one length scale is present. For example, a treatment given by Parker (1979), applicable to those dynamo models in which convection and nonuniform rotation of the core are assumed, requires that a so-called *dynamo number*, D, be on the order of unity. D is the product of two magnetic Reynolds numbers associated with the nonuniform rotation and convective parts of the Earth's core and is discussed further in §8.3.

An equation similar to (8.2.2) can be derived for fluid vorticity by taking the curl of the Navier–Stokes equation (8.1.14). Since the magnetic and gravity forces are conservative (their curl is zero), this operation yields the following equation for *vorticity* ($\Omega = \nabla \times \mathbf{v}$):

$$\frac{\partial \Omega}{\partial t} = \frac{\eta}{\rho} \nabla^2 \Omega + \nabla \times (\mathbf{v} \times \Omega) \quad . \tag{8.2.10}$$

This equation is of the same form as (8.2.2) with η/ρ set equal to k_m and Ω to **H**. Indeed, a theorem similar to the frozen-in-field theorem, sometimes referred to as the Helmholtz theorem for vorticity, can be derived from this equation. This theorem proves that the vorticity is carried with the fluid in a homogeneous inviscid fluid. Several important insights into the origin of the Earth's magnetic field can be gained from considering fluid mechanical analogues. One example is provided by the analogy of the magnetic induction equation with the above equation. This analogy also shows why k_m in (8.2.2) is sometimes referred to as *magnetic viscosity*. However, the analogies can only be carried so far and even experts in fluid mechanics have been led astray by the similarities of *some* of the equations. An example of one important difference between fluid mechanics and magnetic theory is that the boundary conditions at the core–mantle interface are entirely different for **H** and Ω.

In the absence of a velocity field, the magnetic induction equation (8.2.2) becomes

$$\frac{\partial \mathbf{H}}{\partial t} = k_m \nabla^2 \mathbf{H} \quad . \tag{8.2.11}$$

As has already been noted, this is a vector diffusion equation for the magnetic field. The field will decay away to $1/e$ of its initial intensity in some time τ. A dimensional analysis can be used to obtain a rough estimate of that time. Let $t \sim \tau$ and $\nabla^2 \sim 1/L^2$, where L can be taken to be roughly 3×10^6 m for the radius of the Earth's core. Then, from (8.2.10),

$$\tau \cong \frac{L^2}{k_m} \cong \frac{(3 \times 10^6)^2}{2} \text{ s.} \tag{8.2.12}$$

The free decay time estimated by this method is on the order of 100,000 yr, a time that will be shown to be too large by one order of magnitude.

It is useful to use this approximate dimensional analysis to see if laboratory dynamo models can be constructed to model the conditions of the Earth's core adequately. The subscript L will be used to denote laboratory conditions and the subscript c to denote core conditions. The above dimensional analysis yields

$$\frac{\tau_L}{\tau_c} = \frac{L_L^2 \sigma_L}{L_c^2 \sigma_c} \quad , \tag{8.2.13}$$

where the relationship $k_m = (\sigma \mu)^{-1}$ has been used.

Suppose that the dimension of a laboratory model is 1 m, and it contains liquid mercury; then, for a reversal of the Earth's magnetic field, $\tau_c = 3 \times 10^{10}$ to 3×10^{11} s, giving a value of $\tau_L = 3 \times 10^{-2}$ to 3×10^{-3} s. It is difficult to examine in detail the properties of a reversal that occurs so quickly. As mercury has one of the highest conductivities readily available for a liquid model, it is doubtful that one will be able to model in a laboratory some of the interesting phenomena occurring in the core. However, other phenomena might be modeled with more success. For example, Lowes and Wilkinson (1963, 1967 1968) have used rotating solid magnetic cylinders to increase μ_0 and therefore τ_L. Their model achieves reasonable scaling for the time factor, but at the expense of eliminating the fluid aspects of the problem. Recently, there has been renewed interest in the modeling problem with consideration, among other conductor candidates, of liquid Na. Although, as discussed above, some features of the dynamo may not be accurately modeled because of scaling problems, it is important to obtain experimental data on the effects that strong magnetic fields have on conducting fluids to test other aspects of MHD theory.

The methods for solving (8.2.11) to obtain better estimates of free decay depend on the initial and boundary conditions. Roberts (1967) suggested a problem that provides valuable insight into the free decay of the Earth's magnetic field. Figure 8.4 illustrates the geometry of the problem.

At time $t = 0$, the field **H** can be represented by $\mathbf{H}_0 = (H(z,0),0,0)$. The boundary conditions are $\mathbf{H} = \mathbf{0}$, at $z = 0$ and $z = d$ (for all time t). $H(z,t)$ can be represented by

$$H(z,t) = \sum_{n=1}^{\infty} H_n(t)\sin\frac{n\pi z}{d} \quad , \qquad (8.2.14)$$

which clearly satisfies the boundary conditions. Substitution of (8.2.14) into (8.2.11) gives

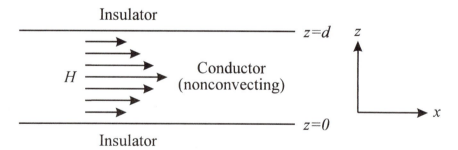

Fig. 8.4. Magnetic field decay in a conductor. A magnetic field $H(z,t)$ in the x direction is confined to a planar conducting region between two insulators. The boundaries of the insulators are at $z = 0$ and $z = d$. If there is no convection in the conductor, the magnetic field will decay away exponentially with time.

$$H_n(t) = H_n(0)\exp\left[-k_m t\left(\frac{n\pi}{d}\right)^2\right] . \tag{8.2.15}$$

This shows that the smaller wavelength features (larger n) decay faster than the longer wavelength features. Moreover, the fundamental mode, $n = 1$, decays to $1/e$ of its initial value in a time τ:

$$\tau = \left(\frac{d}{\pi}\right)^2 \frac{1}{k_m} . \tag{8.2.16}$$

The time τ is π^2 smaller, or roughly one order of magnitude less, than obtained by the previous dimensional analysis. Calculations involving spherical geometry provide similar results and indicate that the free decay time of the dipole part of the Earth's magnetic field is probably on the order of 10,000 to 30,000 yr (see §9.6). Usually the free decay times of the nondipole components are still shorter. With reasonable assumptions, the higher the degree of the harmonic representing the nondipole component, the more rapid its decay (§9.1.4). The relaxation time is related to the square of the dimensions of the feature involved.

The problem is more complicated than these simplifications make it seem, primarily because one is dealing with a vector diffusion equation in the general case. It will be shown in §9.1.4 that this complication means one cannot calculate the dipole field decay time uniquely, even if the core's electrical conductivity is perfectly known. To be able to do this one needs to know the precise nature of the magnetic field sources, which cannot be uniquely determined (Chapter 2). This problem is discussed further in §9.1.4.

8.3 The α- and ω-Effects of Dynamo Theory

8.3.1 α-Effect

Both the observational and the theoretical evidence suggest that the core motions are reasonably complex, so correlations in the velocity field over moderate dimensions may be small or negligible. If the length scale over which correlation occurs is very small, then the fluid is said to be turbulent (§7.6). However, the use of the word "turbulent" often causes unnecessary confusion. The important step in the so-called "turbulent dynamos" is simply to apply statistical approaches to obtain a mean-field approximation of the Earth's magnetic field.

Turbulent dynamics have become fashionable based on early fundamental contributions from Steenbeck *et al.* (1966), Krause and Steenbeck (1967), Rädler (1968), Steenbeck and Krause (1969), Roberts (1971), Gubbins (1974), Krause (1977), and Moffatt (1961, 1970, 1978). All turbulent models depend to some degree on the so-called α-*effect* to magnify the magnetic field.

Ohm's law can be written in terms of some current density, **J**,

$$\mathbf{J} = \sigma\mathbf{E} + \sigma(\mathbf{v} \times \mathbf{B}) \quad , \tag{8.3.1}$$

where **E** is an externally applied electric field. An effective "internal electric field" is defined by

$$\mathbf{E}_i \equiv \mathbf{v} \times \mathbf{B} \quad . \tag{8.3.2}$$

Suppose that a turbulent system is partially described by $\mathbf{v} = \mathbf{v}_0 + \mathbf{v}'$, and $\mathbf{B} = \mathbf{B}_0 + \mathbf{B}'$, where the subscript 0 denotes a steady part of the **v** or **B** field and the prime is used to indicate the fluctuating part of that field about the steady value. The average of the internal field is then

$$\langle \mathbf{E}_i \rangle = \mathbf{v}_0 \times \mathbf{B}_0 + \langle \mathbf{v}' \times \mathbf{B}' \rangle \tag{8.3.3}$$

since $\langle\mathbf{v}'\rangle = \langle\mathbf{B}'\rangle = \mathbf{0}$.

An additional e.m.f. associated with $\mathbf{v}' \times \mathbf{B}'$ (= **E**') occurs when **v**' and **B**' are correlated. Krause (1977) shows one simple example in which **E**' (or effectively, an e.m.f.) can be written as

$$\mathbf{E}' = \alpha\mathbf{B}_0 \quad . \tag{8.3.4}$$

In this case the internal electric field produced from the fluctuating parts of the **v** and **B** fields is directly related through the constant α to the steady field $\mathbf{B}_0$. This is the so-called α-effect. α need not be a scalar, but can be a second-order tensor (see §9.4).

Figure 8.5 helps illustrate (heuristically) one way in which the α-effect given by (8.3.4) can occur. Suppose that $\mathbf{v}' = \mathbf{v}_1 + \mathbf{v}_2$, where $\mathbf{v}_2$ is along the z axis and $\mathbf{v}_1$ represents a right-hand rotation about z. In fluid mechanics, **v**' is referred to as a right-handed *helical velocity*. Take $\mathbf{B}_0$ to be uniform in the x direction. $(\mathbf{v}_1 \times \mathbf{B}_0)$ will produce a current density, $\mathbf{J}_1$, flowing in planes perpendicular to the y axis in Fig. 8.5. Associated with $\mathbf{J}_1$ will be a magnetic field **B**' directed in the y direction. The interaction of **B**' with $\mathbf{v}_2$, $(\mathbf{v}_2 \times \mathbf{B}')$, will produce an electric field **E**' parallel to $\mathbf{B}_0$, as given by (8.3.4). §9.4.1 shows rigorously how the α-effect arises from statistical averaging processes not discussed here.

Equation (8.3.4) represents the α-effect in the theories of turbulent dynamos. Because a mean electric field is produced in a turbulent system, analyses of this sort are often referred to as *mean-field electrodynamics* (see §8.3.1). The

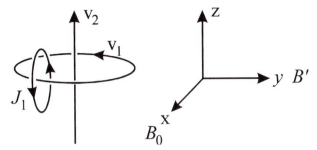

Fig. 8.5. Illustration of the α-effect. Consider a right-handed helical velocity field depicted by $\mathbf{v}_1$ and $\mathbf{v}_2$ in the presence of a field $\mathbf{B}_0$ aligned along the x axis. This will produce current loops such as $\mathbf{J}_1$, lying in the x–z plane. Associated with the current loop $\mathbf{J}_1$, is a field $\mathbf{B'}$ aligned parallel to the y axis. This new field, $\mathbf{B'}$, interacts with $\mathbf{v}_2$ to produce an electric field parallel to the x axis.

existence of the α-effect was verified experimentally by Steenbeck *et al.* (1967) in a rectangular box of liquid sodium. The internal electric field (or equivalently, the e.m.f.) can drive a current, and if the current has the proper geometry it could reinforce an initial magnetic field. Figure 8.6 illustrates how this could occur.

The experimentally measured quantity used to verify the α-effect was the e.m.f. (Steenbeck *et al.*, 1967; Krause, 1977). There is no evidence that reinforcement of the initial **B** field occurred. Even though wires, such as shown in Fig. 8.6, could be used to obtain reinforcement of the initial **B** field, the analogue in the Earth is not clear, since this "reinforcement current" would then have to be close to the core–mantle boundary. More sophisticated treatment than given above is thus required to examine the α-effect in the Earth (§9.4). Note

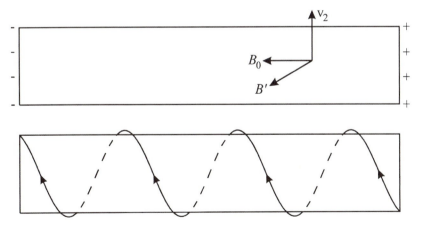

Fig. 8.6. Magnetic field reinforcement by the α-effect. $\mathbf{v}_2$, $\mathbf{B}_0$, and $\mathbf{B'}$ of Fig. 8.5 are shown in a large rectangular box. An electric field resulting from ($\mathbf{v}_2 \times \mathbf{B'}$) will produce a charge separation as shown. If wires are added around the box as shown, a current will flow in such a manner to magnify the initial **B** field.

that a necessary condition for the presence of an α-effect is a correlation between vorticity, $(\nabla \times \mathbf{v})$, and the velocity (i.e., there is some helicity, $\mathbf{v} \cdot (\nabla \times \mathbf{v}) \neq 0$). The α-effect is essentially a mechanism by which turbulent energy is converted to electrical energy.

The α-effect is fundamental to two of the most fashionable dynamo types, the $\alpha\omega$- and α^2-dynamos, as discussed below.

8.3.2 The ω-Effect and a Heuristic $\alpha\omega$-Dynamo

Parker (1955a) has presented a dynamo model that helps to provide a visual picture of the dynamo process. This concept of the dynamo is reproduced without the mathematics.

The basic idea of a dynamo is that some initial magnetic field is altered through interactions with an assumed (kinematic) or derived (hydromagnetic) velocity field in such a way that the magnetic field is reinforced. In the absence of a velocity field in the core, the magnetic field would decay with time (§8.2.3). The velocity field interacts with the magnetic field through Lenz's law. Any attempt to move an electrically conducting material into a magnetic field will induce in the conductor currents that oppose its motion. These induced currents will alter the initial magnetic field. Figure 8.3 shows what happens in one simplified case. The magnetic field lines are compressed in front of a moving conductor and the field is thus intensified there. Figure 8.3 also illustrates the frozen-in-field concept that any magnetic field lines already in the conductor will be carried with it. If the conductor is not perfect, the field can diffuse into and out of the conductor. In fact, some of this must be occurring because it would be impossible to get the field lines into a perfect conductor in the first place. The process of moving a conductor to intensify the magnetic field illustrates how kinetic energy can be converted into magnetic field energy.

Figure 8.7 illustrates how some initial magnetic field, represented by a single field line, can be magnified by moving (conductive) core fluid. The initial magnetic field has a radial component and is referred to as an S_1^0 poloidal magnetic field. The notation used here refers to the vector spherical harmonic notation developed in §9.1. Consider a simple toroidal velocity field, T_1^0, as shown in Fig. 8.7a (this velocity field produces a simple shear; the definition of a *toroidal field* implies that there is no radial component). Figure 8.7b shows that this toroidal velocity field, T_1^0, interacts with the magnetic field, S_1^0, to generate a more complex toroidal magnetic field, called a T_2^0 field. Note that this latter toroidal field has opposite signs in the two hemispheres. In simple terms,

$$S_1^0 \text{magnetic field} + T_1^0 \text{velocity field} \rightarrow T_2^0 \text{magnetic field}$$

This process is referred to as an ω-*effect.*

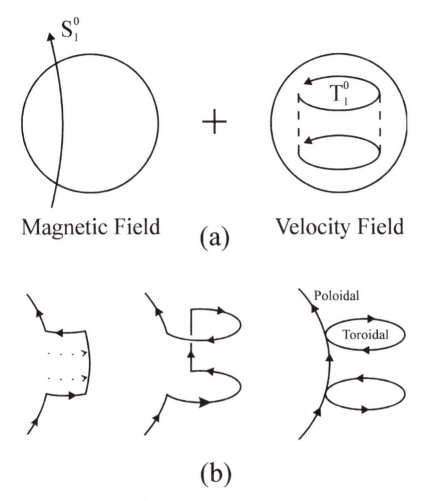

Fig. 8.7. Production of a toroidal magnetic field in the core. (a) An initial poloidal magnetic field passing through the Earth's core is shown on the left, and an initial cylindrical shear velocity field, T_1^0, is shown on the right. (b) The interaction between the velocity and the magnetic field in (a) is shown at three successive times moving from left to right. The velocity field is only shown on the left by dotted lines. After one complete circuit two new toroidal magnetic field loops of opposite sign (T_2^0) have been produced. After Parker (1955a).

Another velocity field with a radial component (poloidal) must be added in order to reinforce the original field. This is done by assuming that convection exists in the core. Figure 8.8 illustrates how a toroidal field would be affected by an upwelling of fluid in the northern hemisphere. A bulge occurs in the field line, taken to represent a segment of a toroidal field line as it moves with the fluid. The Coriolis force in the northern hemisphere will act to produce a

Fig. 8.8. Production of a poloidal magnetic field. A region of fluid upwelling, illustrated by dotted lines on the left, interacts with a toroidal magnetic field (solid line). Because of the Coriolis effect (northern hemisphere) the fluid exhibits helicity, rotating as it moves upward (thin lines, center). The magnetic field line is carried with the conducting liquid and is twisted to produce a poloidal loop as on the right. After Parker (1955a.)

counter-clockwise rotation in the region of fluid upwelling. The field lines will be twisted with this rotation and a poloidal magnetic loop (Fig. 8.8) will be produced after every 90° rotation. Figure 8.9 sketches how various poloidal loops of this sort could come together to produce a large poloidal field. Because downwelling regions produce poloidal loops of opposite sign to upwelling regions, it must be supposed that some inhomogeneity is present. The above heuristic dynamo, combining the α- and ω-processes, is an example of an $\alpha\omega$-dynamo.

The amount of twisting required to form a loop is not obvious. Note, however, that a loop (perpendicular to the plane of the paper) formed after a 90° twist will have the correct current associated with it to produce the α effect of the last section. Further discussion of this is also given in §9.4

8.3.3 Dynamo Numbers and α^2- and $\alpha^2\omega$-Dynamos

It can be shown (see §9.3) that the large-scale rotational shear that produces the ω-effect cannot produce dynamo action by itself (as may be intuitively clear from the above heuristic dynamo). Of course, to have an ω-effect the frozen-in-field component must dominate the diffusion component; i.e., the magnetic Reynolds number R_ω associated with the process must exceed one. Similar to the arguments given in (8.2.3), R_ω is given by

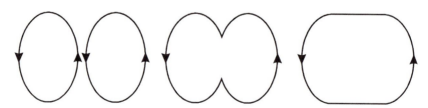

Fig. 8.9. Convergence of two poloidal loops as produced in Fig. 8.8 results in a larger poloidal loop. After Parker (1955a.)

$$R_\omega \equiv \frac{vL}{k_m} \quad , \tag{8.3.5}$$

where v is the shear velocity and L is a typical length dimension. Similarly the α-effect has an associated Reynolds number given by

$$R_\alpha \equiv \frac{\alpha L}{k_m} \quad , \tag{8.3.6}$$

where L is again a typical length dimension and α has the dimensions of velocity. The L in (8.3.6) is usually smaller than that in (8.3.5). However, neither dimension is rigorously known and so a common practice has been not to distinguish the two L's. The dynamo number, first devised by Parker, is the product of the two Reynolds numbers, i.e.,

$$D \equiv R_\omega R_\alpha \quad . \tag{8.3.7}$$

D must exceed unity for positive dynamo action in an $\alpha\omega$-dynamo. These numbers, along with other important dimensionless numbers, are given in Table 8.2.

The toroidal field in dynamos may be formed by small-scale motions rather than large-scale velocity shear. That is, the toroidal field may form via an α-effect acting on the poloidal field. When a dynamo occurs by this process it is referred to as an α^2-dynamo. In some dynamo models both an α- and an ω-effect are argued to contribute to the toroidal magnetic field generation. In this case the dynamo is sometimes referred to as an $\alpha^2\omega$-dynamo. The product of the appropriate magnetic Reynolds numbers can be obtained by using Eqs (8.3.5) and (8.3.6) to obtain the dynamo numbers for the α^2- or $\alpha^2\omega$-dynamos.

8.4 Waves in Dynamo Theory

8.4.1 MHD Waves

Waves were introduced into dynamo theory in an attempt to explain the origin of the magnetic field (Parker, 1955a; Braginsky, 1964a,b), and it may yet be that they contribute significantly to positive dynamo action, particularly if the outer core is stratified (§7.1 and §8.6). However, most current interest in waves appears to be centered around their possible role in producing secular variation, a possibility suggested by Hide (1966) (see also §4.2.7).

TABLE 8.2
Some Common Dimensionless Magnetic Numbers Used in Dynamo Theories

$R_m \equiv \dfrac{uL}{k_m}$	*Magnetic Reynolds number*. This gives the relative importance of the fluid modification of the magnetic field to the diffusion of that field. R_m must be large for the frozen-in-field concept to be a reasonable approximation. See §8.2 for further discussion.
$A \equiv \dfrac{B}{u(\rho\mu_0)^{\frac{1}{2}}}$	*Alfvén number*. The magnetic field has a minor effect on the fluid flow for $A \ll 1$. The magnetic field has a major effect on the fluid flow for $A \gg 1$. In particular, turbulence will be inhibited if A is very large.
$Q \equiv \dfrac{v_A K}{2\Omega}$	*Magnetic Rossby number*. This describes the relative importance of rotation on hydromagnetic waves; if Q is large then rotational effects are negligible. Probably $Q \ll 1$ in the Earth's core and rotational effects are very important there.
$\Lambda_E \equiv \dfrac{B^2\sigma}{\rho\Omega}$	*Elsasser number*. Importance of the magnetic force relative to the Coriolis force.
$P_m \equiv \dfrac{v}{k_m}$	*Magnetic Prandtl number*. Importance of fluid viscosity relative to magnetic viscosity (or diffusivity).
$R_R \equiv \dfrac{k}{k_m}$	*Roberts number*. Importance of thermal to magnetic diffusivity.
$D \equiv R_\omega R_\alpha$	*Dynamo number*. Product of the magnetic Reynolds numbers associated with the ω- and α-effects.

Notation:
u = estimate of the magnitude of some characteristic velocity
L = some representative length scale
σ = electrical conductivity
v = kinematic viscosity
ρ = density
k = thermal diffusivity
v_A = Alfvén velocity
K = wave number $(2\pi/\lambda$, where λ = wavelength)
Ω = angular rotation of the core
k_m = magnetic diffusivity = $1/(\mu_0\sigma)$; see text
B = magnetic induction
μ_0 = magnetic permeability of free space

Similar to waves on a vibrating string, MHD waves can travel along magnetic field lines because of a magnetic stress. Consider an electric current density **J** in the presence of a field **B**. The magnetic force **F** on the current density is

$$\mathbf{F} = \mathbf{J} \times \mathbf{B} = (\nabla \times \mathbf{H}) \times \mathbf{B}$$

$$= -\frac{\nabla(\mathbf{B} \cdot \mathbf{B})}{2\mu_0} + \nabla \cdot \left(\frac{\mathbf{BB}}{\mu_0}\right) \quad . \tag{8.4.1}$$

In the fluid regime of the core, $(\mathbf{B} \cdot \mathbf{B}/2\mu_0)$ can be interpreted as a magnetic pressure, while $(\mathbf{BB}/\mu_0)$, a second-order tensor represented in dyadic form (written as $B_i B_j$ in Einstein notation), can be interpreted as a stress (tension) along the field line, and is a reduced form of the Maxwell stress tensor.

The speed of waves on a vibrating string of density ρ and tension T_e is given by $(T_e/\rho)^{-1/2}$. By analogy it might be expected (correctly) that a wave along a magnetic field line would have a velocity $\mathbf{B}(\mu_0\rho)^{-1/2}$. This velocity is called the *Alfvén velocity* and the corresponding wave an *Alfvén wave*.

We now justify this expectation for a simplified case. Assume there are waves in a fluid that is perfectly conducting, incompressible, and inviscid. Suppose also that the initial magnetic field is uniform and denoted by $\mathbf{B}_0$. Under these conditions the appropriate equations are the magnetic induction equation for a perfect conductor

$$\frac{\partial \mathbf{B}}{\partial t} = \nabla \times (\mathbf{v} \times \mathbf{B}) \quad ; \tag{8.4.2}$$

the Navier–Stokes equation,

$$\rho \frac{\partial \mathbf{v}}{\partial t} = -\nabla P + \mathbf{J} \times \mathbf{B} - \rho \nabla \phi_g$$

$$= -\nabla \left(P + \frac{\mathbf{B}^2}{2\mu_0} + \rho \phi_g \right) + \frac{1}{\mu_0} \nabla \cdot (\mathbf{BB}) \quad , \tag{8.4.3}$$

where ϕ_g is the gravitational potential and P is pressure;

$$\nabla \cdot \mathbf{v} = 0 \quad , \tag{8.4.4}$$

indicating incompressibility of the fluid; and one of Maxwell's equations,

$$\nabla \cdot \mathbf{B} = 0 \quad . \tag{8.4.5}$$

A further simplification is made by assuming that the pressure field is always in equilibrium. This means that the first term on the right side of (8.4.3) can be taken equal to zero so the equation becomes

$$\rho \frac{\partial \mathbf{v}}{\partial t} = \frac{1}{\mu_0} \nabla \cdot (\mathbf{BB}) \quad . \tag{8.4.6}$$

This initial system, denoted by subscript 0, is now perturbed by a small amount and the perturbation values are denoted by the subscript 1, so that

$$\mathbf{B} \rightarrow \mathbf{B}_0 + \mathbf{B}_1$$

$$\rho \rightarrow \rho_0 + \rho_1$$

$$\mathbf{v} \rightarrow \mathbf{v}_1 \quad .$$

with the assumption of zero initial velocity. Substitution of these values into (8.4.2) and (8.4.6) gives for $\mathbf{B}_0$, taken in the z direction,

$$\frac{\partial \mathbf{B}_1}{\partial t} = \nabla \times (\mathbf{v} \times \mathbf{B}_0) = (\mathbf{B}_0 \cdot \nabla)\mathbf{v} \tag{8.4.7}$$

and

$$\frac{\partial \mathbf{v}}{\partial t} = (\mathbf{B}_0 \cdot \nabla)\frac{\mathbf{B}_1}{\mu_0 \rho} = \frac{\mathbf{B}_0}{\mu_0 \rho}\frac{\partial \mathbf{B}_1}{\partial z} \quad . \tag{8.4.8}$$

Differentiating (8.4.7) with respect to time and using (8.4.6),

$$\frac{\partial^2 \mathbf{B}_1}{\partial t^2} = \frac{\mathbf{B}_0^{\,2}}{\mu_0 \rho}\frac{\partial^2 \mathbf{B}_1}{\partial z^2} \quad . \tag{8.4.9}$$

This is the wave equation and shows that there are (Alfvén) waves propagated along field lines with a speed $B_0(\mu_0\rho)^{-1/2}$. For laboratory conditions, the Alfvén velocity is usually less than the sound velocity. However, great variability is found when solar and planetary bodies are considered. For example, the density of the Sun's photosphere is very low (there are approximately 6×10^{16} hydrogen atoms per cubic metre there) and the Alfvén velocity (v_A) is approximately $10^5 \, B_0$. The general dipole field of the Sun is about 10^{-4} T, while that in sunspots is roughly two orders of magnitude larger. v_A varies roughly from 10 m s^{-1} to 10^3 m s^{-1} as compared with 10^4 m s^{-1} for the estimated velocity of sound in the Sun's photosphere.

A more general case for Alfvén waves can be considered in which the field is not a perfect conductor, viscous effects are present, the first term on the right side of (8.4.3) is retained, and the field is compressible. For the same perturbations used in the last problem, it is relatively easy to show (Jackson, 1975) that (8.4.7) and (8.4.8) are replaced by

$$\frac{\partial \mathbf{B}_1}{\partial t} = \nabla \times (\mathbf{v}_1 \times \mathbf{B}_0) + \frac{\nabla^2 \mathbf{B}_1}{\sigma \mu_0} \tag{8.4.10}$$

and

$$\rho_0 \frac{\partial \mathbf{v}_1}{\partial t} = -s^2 \nabla \rho_1 - \frac{\mathbf{B}_0}{\mu_0}\times(\nabla \times \mathbf{B}_1) + \eta \nabla^2 \mathbf{v}_1 \quad , \tag{8.4.11}$$

where s is the speed of sound and η is the viscosity. Consider the possibility of plane wave solutions described by

$$\mathbf{v}_1(\mathbf{r},t) = \mathbf{v}_1 e^{i(\mathbf{k}\cdot\mathbf{r}-\omega t)} \quad . \tag{8.4.12}$$

Using this with (8.4.10) and (8.4.12) gives the following dispersion relationship for Alfvén waves that propagate parallel to the field:

$$k^2 v_A{}^2 = \omega^2 \left(1 + i \frac{k^2}{\sigma\mu_0\omega}\right)\left(1 + i \frac{\eta k^2}{\rho_0\omega}\right) \quad . \tag{8.4.13}$$

If the conductivity is large and the viscosity small, then the wave number is approximately

$$k \approx \frac{\omega}{v_A} + i \frac{\omega^2}{2 v_A{}^3}\left(\frac{1}{\mu_0\sigma} + \frac{\eta}{\rho_0}\right) \quad . \tag{8.4.14}$$

The imaginary part of k is associated with attenuation. This can be seen directly for any plane wave, since

$$e^{i(\mathbf{k}_1 \cdot \mathbf{r} + i\mathbf{k}_2 \cdot \mathbf{r} - \omega t)} = e^{i(\mathbf{k}_1 \cdot \mathbf{r} - \omega t)} e^{-\mathbf{k}_2 \cdot \mathbf{r}} \quad , \tag{8.4.15}$$

where the wave vector $\mathbf{k}$ has been written in terms of a real part, $\mathbf{k}_1$, and an imaginary part, $\mathbf{k}_2$. The factor $\exp(-\mathbf{k}_2 \cdot \mathbf{r})$ gives the decrease in amplitude of the wave with distance $\mathbf{r}$. Equation (8.4.14) shows that the attenuation increases rapidly with frequency and decreases with increases in v_A (or equivalently with B) and with increases in s. This attenuation is clearly different from that obtained from the skin-depth calculations for a solid conductor.

The real part of the hydromagnetic wave is nondispersive. That is, $\omega/k = \partial\omega/\partial k = v_A$. The phase and group velocities are both equal to the Alfvén wave velocity. In the case under consideration, *longitudinal modes* (perpendicular to the field) are also allowed. The real parts of these longitudinal modes are also nondispersive and the phase velocity is given by

$$\frac{\omega}{k} = (s^2 + v_A{}^2)^{\frac{1}{2}} \quad . \tag{8.4.16}$$

In the case when the magnetic pressure goes to zero, $B \to 0$ and $v_A \to 0$. Thus, the velocity reduces to ordinary sound velocity. In the more general case, the longitudinal wave velocity is seen to depend on the sum of the hydrostatic and magnetic pressures.

Note that (8.4.16) can be used to estimate the effect strong magnetic fields have on the speed of compressional (elastic) waves (P waves) in the outer core. The speed v is roughly 0.1 to 0.01 m s^{-1} for the core, a value roughly 10^{-5} to 10^{-6} of the speed of P waves there. This shows that the speed of compressional waves is virtually unaffected by the presence of the core's magnetic field. A more detailed analysis confirms this result (Knopoff and MacDonald, 1958).

Alfvén waves were once suggested as the cause of secular variations (Skiles, 1972a,b). However, this is now recognized to be incorrect since straightforward estimates of the Alfvén velocities are far too high (of order 0.1 to 0.01 m s^{-1}) to explain drift velocities of the order 10^{-4} m s^{-1}.

8.4.2 Planetary Waves

Although pure Alfvén waves should be considered poor candidates for secular
variation (§9.1.3) there are other hydromagnetic waves that prove to be more
promising, such as the MAC waves of Braginsky (1964a,b). We consider a
subset of such waves, called Rossby waves and magnetic Rossby waves (see also
§8.6), that are easier to treat mathematically.

Assume that there is some initial flow of a fluid that is in geostrophic balance
(pressure gradient balances the Coriolis force) and perturb this flow by a small
amount. To simplify the problem further, assume that all particle motion is
horizontal and the viscosity is zero. Rectangular coordinates are used with the
unit vectors $\hat{y}$ (north), $\hat{x}$ (east), and $\hat{z}$ (upward). Let v_1, v_2, v_3 represent
velocities in the $\hat{x}, \hat{y}, \hat{z}$ directions, respectively.

For these conditions, the appropriate Navier–Stokes equations for v_1 and v_2 are

$$\frac{Dv_1}{Dt} - f\,v_2 = -\frac{\partial P'}{\partial x} \tag{8.4.17}$$

$$\frac{Dv_2}{Dt} - f\,v_1 = -\frac{\partial P'}{\partial y} \quad , \tag{8.4.18}$$

where D/Dt is the operator $(\partial/\partial t + \mathbf{v}\cdot\nabla)$ and $P' = P/\rho$. The function f is the
Coriolis force, equal to $2\Omega\cos\theta$, where Ω is the angular velocity vector and θ is
the colatitude. Assume that $f(\theta)$ varies slowly as a function of θ and can be
approximated by

$$f(\theta + \Delta\theta) = f(\theta) + \beta y \quad . \tag{8.4.19}$$

This approximation is called the β-plane approximation, where $\beta = 2\Omega\,\sin\theta\,/R$
and R is the radius of the outer core (note that y increases northward).
Technically, this approximation is only applicable to a thin plane. It is definitely
not applicable to the entire outer core.

Differentiate (8.4.17) with respect to y, (8.4.18) with respect to x, and subtract
using the β-plane approximation to get

$$\frac{D}{Dt}\left[\frac{\partial v_1}{\partial y} - \frac{\partial v_2}{\partial x}\right] - \beta v_2 = 0 \quad , \tag{8.4.20}$$

where the fact that $\partial v_1/\partial x = \partial v_2/\partial y$ has been used (the motions are purely
horizontal). Now $(\partial v_1/\partial y - \partial v_2/\partial x) = \nabla^2\psi$, where ψ is the stream function and
$v_2 = -\partial\psi/\partial x$. Physically, the lines of constant ψ are the paths the fluid particles
follow. Substitution of these relationships into (8.4.20) gives

$$\frac{D}{Dt}(\nabla^2\psi) + \beta\frac{\partial\psi}{\partial x} = 0 \quad . \tag{8.4.21}$$

Let the initial velocity be given by $v_o\hat{x}$ and search for plane wave solutions of the form

$$\psi = \psi_0\, e^{i(kx-\omega t)} \quad , \tag{8.4.22}$$

which implies that $\nabla^2\psi = -k^2\psi$. In the perturbation approximation we also have $D/Dt \approx \partial/\partial t + v_0(\partial/\partial x)$. By substituting these last two equations into (8.4.21), solutions exist provided that

$$\omega = v_0 k - \frac{\beta}{k} \quad . \tag{8.4.23}$$

Thus there are dispersive plane waves, called Rossby or planetary waves, which satisfy (8.4.17) and (8.4.18). Since $\beta > 0$ the phase velocity, ω/k, in the reference frame of the fluid ($v_0 = 0$) is westward. However, the group velocity, $d\omega/dk$, in this reference frame is to the east. If no rotation is present, $\beta = 0$ and the plane wave solutions exhibit no dispersion.

8.5 Symmetries in Dynamo Theory

8.5.1 The Importance of Symmetries

Symmetry is important in dynamo theory because its role is often not dependent on the detail of the dynamo model. For example, consider any vector $\mathbf{X}$. The reflection of this vector through the origin yields $-\mathbf{X}$. Now consider any vector $\mathbf{H}$ which is a function of $\mathbf{X}$. $\mathbf{H}$ can always be separated into its even (symmetric), $\mathbf{H}_s$, or odd (antisymmetric), $\mathbf{H}_a$, parts since

$$\begin{aligned}
\mathbf{H}(\mathbf{X}) &= \tfrac{1}{2}\big(\mathbf{H}(\mathbf{X}) + \mathbf{H}(-\mathbf{X})\big) + \tfrac{1}{2}\big(\mathbf{H}(\mathbf{X}) - \mathbf{H}(-\mathbf{X})\big) \\
&\equiv \quad\ \mathbf{H}_s \qquad\quad + \qquad\quad \mathbf{H}_a \quad .
\end{aligned} \tag{8.5.1}$$

Let $\mathbf{H}$ be a magnetic field that satisfies (8.6.3), the magnetic induction equation. Similarly, we can separate $\mathbf{v}$ into its symmetric and antisymmetric parts:

$$\mathbf{v} \equiv \mathbf{v}_s + \mathbf{v}_a \quad . \tag{8.5.2}$$

It is interesting to note that $\mathbf{H}_s$ and $\mathbf{H}_a$ are uncoupled (i.e., changing either $\mathbf{H}_s$ or $\mathbf{H}_a$ has no effect on the other) in the magnetic induction equation if the velocity field satisfies certain symmetry conditions. To see this, remember that the curl is antisymmetric on reflection through the origin. Hence, on reflection

$$\nabla \times (\mathbf{v}_a \times \mathbf{H}_s) \quad \rightarrow \quad -\nabla \times (-\mathbf{v}_a \times \mathbf{H}_s) \quad ; \qquad (8.5.3)$$

that is, it is unchanged. Similarly, $\nabla^2 \mathbf{H}_s$ is also unchanged by this reflection and so $\partial \mathbf{H}/\partial t$ is invariant on reflection. Using similar arguments for $\mathbf{H}_a$ one finds that $\partial \mathbf{H}_d/\partial t$ in the magnetic induction equation goes to $-\partial \mathbf{H}_d/\partial t$ on reflection. Thus, providing $\mathbf{v}$ is purely symmetric with respect to the origin, a change in $\mathbf{H}_a$ in the magnetic induction equation has no effect on $\mathbf{H}_s$ and vice versa.

8.5.2 The Dynamo Families

In general one can investigate various symmetries embedded in equations by using standard group theory methods. Gubbins and Zhang (1993) have recently done this in the context of dynamo theory. A given symmetry is naturally most useful when it can be associated with a physical process, and an important example of this in the context of $\alpha\omega$-dynamos was given by Roberts (1971) and Roberts and Stix (1972). Roberts points out that if the large-scale shear that causes the ω-effect is symmetric with respect to the equator and if the α-effect is antisymmetric with respect to the equator (as might be expected since the Coriolis force changes sign across the equator), then the dynamo can be separated into two noninteracting systems made up of specific families of spherical harmonics. Roberts refers to these two families as the *dipole family* and the *quadrupole family*, a terminology generally continued in many dynamo theory papers. Members of the dipole family are characterized by spherical harmonics with $(m + l)$ being odd, where m is the order of the harmonic and l is the degree (see Chapter 2). The quadrupole family is characterized by $(m + l)$ being even. Clearly the description is complete, since all harmonics can be placed into one or the other of these two dynamo families. Note that the geocentric axial dipole field, characterized by the term with Gauss coefficient g_1^0, is in the dipole family, while the two equatorial dipole components, characterized by g_1^1 and h_1^1, are in the quadrupole family. Because this produced some confusion with the terminology Merrill and McFadden (1990) referred to the dipole family as the *primary family* and the quadrupole family as the *secondary family*. Subsequently, Gubbins and Zhang (1993) referred to these respectively as the *antisymmetric* and *symmetric families*. The latter terminology appears to be the most descriptive, providing that one remembers that the symmetry is with respect to the equator and not with respect to, say, the origin (as used in the simple example given at the beginning of the section). Hereafter we will refer to the two families as either the dipole, or antisymmetric, family

TABLE 8.3

Antisymmetric (Dipole) and Symmetric (Quadrupole) Family Terms

	Antisymmetric (dipole) family	Symmetric (quadrupole) family
Dipole	g_1^0	g_1^1, h_1^1
Quadrupole	g_2^1, h_2^1	g_2^0 g_2^2, h_2^2
Octupole	g_3^0 g_3^2, h_3^2	g_3^1, h_3^1 g_3^3, h_3^3

and the quadrupole, or symmetric, family. Table 8.3 shows how some of the low-degree Gauss coefficients are separated into the two families.

Insight into the origin of the two dynamo families, and why they are noninteracting if the symmetries are obeyed (ω even, α odd), is provided by Fig. 8.10. Roberts (1971) showed that there is interaction between the two families if these symmetries are violated. McFadden *et al.* (1991) showed how one could use paleosecular variation of lava (PSVL) data to determine the relative contributions in the past of these two families to the paleosecular variation (Chapter 6). Merrill and McFadden (1988) developed a geomagnetic field reversal model that involves interactions between the two families. This model contains two independent parts. The first is that the probability for reversal increases as the ratio of the symmetric family to antisymmetric family increases, and the second is the speculation that a polarity transition is characterized by increased interactions between the two families. Merrill and McFadden (1988) predicted that during a superchron, such as the Cretaceous Superchron, there should be a lower symmetric family to antisymmetric family ratio. This test has since been performed (McFadden *et al.*, 1991) and the prediction is clearly confirmed. This test should also be made for PSVL data during the Permo-Carboniferous (Kiaman) reverse polarity superchron. Merrill and McFadden (1988, 1990) and McFadden *et al.* (1991) argue that the long-term changes in the ratio of the two dynamo families reflect change in the core–mantle boundary conditions. Merrill and McFadden (1990) and McFadden and Merrill (1993) point out that such changes could reflect either long-term variations in the lateral make up of this boundary (e.g., topography) or possibly slow changes in the temperature that affect the overall convection pattern.

The above "reversal model" is an example of a phenomenological reversal model, one based on statistical analyses of paleomagnetic data. The first such model was that of Cox (1969) and was based on the premise that the nondipole field intensity varied randomly and on rare occasions dominated a sinusoidally

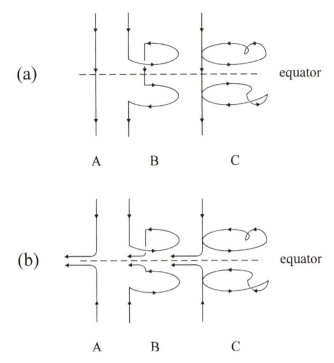

Fig. 8.10. A, initial poloidal magnetic field, antisymmetric for (a) and symmetric for (b). B, magnetic field after it has been acted on by the ω-effect. C, magnetic field after it has been acted on by both the ω-effect and the α-effect. Note that in both instances it is the same velocity field (see Fig. 8.7) but the initially antisymmetric magnetic field is converted into another antisymmetric magnetic field, and the initially symmetric magnetic field is converted into another symmetric magnetic field.

varying dipole field intensity to cause reversals. This model ignored core dynamics. However, it had the advantage that it was testable, and subsequently it was shown that the observed distribution of paleointensity data for the past five million years is incompatible with the sinusoidal variation posited by Cox (Kono, 1972; McFadden and McElhinny, 1982). Moreover, dipole sources appear dynamically coupled to nondipole sources (e.g., §9.1) and Cox's model did not consider any role for the toroidal magnetic field. The latter is implicitly included in the model by Merrill and McFadden because their model was based on the dynamo theory of Roberts. However, their model is incomplete, as evidenced by the fact that it does not include a dynamic description of the transition process.

8.6 Theories for Geomagnetic Secular Variations and Magnetic Field Reversals

8.6.1 Secular Variation

It is well known that an electromagnetic wave entering a conductor is damped to $1/e$ of its initial amplitude in a distance that is inversely proportional to the square root of the electrical conductivity. Because of this, Bullard *et al.* (1950) reasoned that the secular variation of the nondipole field originates from magnetic eddies close to the core–mantle boundary. Magnetic eddies that lie deeper in the core could not produce the secular variation since they would be screened out by the outer part of the conducting core.

It was argued that the observed westward drift of the nondipole field was caused by the outer part of the core rotating slower than the inner part. This shear in rotation rate was believed to be the result of convection. Consider a core that is initially rotating uniformly, i.e., the angular velocity ω is independent of the distance R from the rotation axis. Constant ω requires that the linear velocity, $v = \omega R$, is smaller for smaller values of R. Now consider a convection cell, as indicated in Fig. 8.11. Material that was initially close to the rotation axis, and with small relative v, is moved out into a region of high relative v, where (statistically) it will act to retard the rotation of the outer region. Material moving inward will tend to increase the rotation rate of the inner region and the total angular momentum of the system is conserved. The net effect is that the outer part of the core rotates more slowly than the inner part of the (outer) core. Because the varying part of magnetic fields originating in the inner part of the core is screened out, a westward drift of magnetic eddies in the outer part then produces westward drift of the nondipole field.

Although intuitively appealing, the relative slowing down of the outer part of the core due to convection is not required for rotating fluid systems in which convection occurs (e.g., Busse, 1975). Furthermore, it is now known that waves in a stratified MHD fluid can produce an α-effect (e.g., Gubbins and Roberts, 1987) and that MHD problems are notoriously sensitive to boundary conditions. Consequently one must be very careful of "screening arguments". Regions of electrical conductivity in the mantle will of course screen some of the core's magnetic field. However, a stratified outer core, unlike the mantle, can still have mechanical energy in the form of fluid motions that is converted to magnetic field energy. Perhaps it is best not to think of the outermost core as screening magnetic signals generated by deeper sources, but simply to consider it an integral part of the geodynamo (regardless of whether it is stratified).

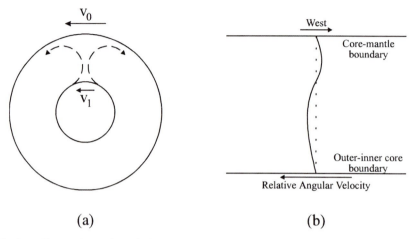

Fig. 8.11. Changes in angular velocity in the outer core produced by convection. (a) An equatorial section through the Earth shows convection in the outer core (dashed lines). The linear velocity v_0 at the core–mantle interface is larger than that, v_1, at the outer–inner core boundary because the angular velocities are equal. (b) The movement of material with smaller values of v outward and material with larger values of v inward causes a slowing of the rotation of the outer part of the core relative to the inner part. The dashed line shows the line of constant angular velocity, with higher angular velocity to the left (not drawn to scale).

Hide (1966) was the first to suggest that MHD waves similar to Rossby-type waves might explain the geomagnetic secular variation. He showed that valuable insight into the problem can be gained by considering the dispersion relation for a perfectly conducting, incompressible, inviscid, homogeneous fluid of indefinite extent rotating uniformly in a magnetic field. The dispersion relationship for the case of plane waves in a *uniformly* magnetized field that is parallel to the rotation axis (Hide, 1966) is

$$\omega = \pm[\Omega \pm (v_A^2 k^2 + \Omega^2)^{\frac{1}{2}}] \quad . \tag{8.6.1}$$

If rotation is negligible ($\Omega=0$), (8.6.1) reduces to the case of simple nondispersive Alfvén waves, $\omega_+=\omega_-=v_A\cdot\mathbf{k}$, where ω_+ denotes the positive root of (8.6.1) and ω_- the negative root. A convenient way to determine the effects of rotation is to use the Rossby number $Q=v_A k/2\Omega$ (§7.5.1). When Q is small, rotational effects dominate and the four roots of (8.6.1) are approximately

$$\left(\frac{\omega}{k}\right)_i \approx \frac{2\Omega(1+Q^2)}{k} \qquad \text{(phase velocity)}$$

$$\left(\frac{\omega}{k}\right)_m \approx \frac{\pm v_A^2 k}{2\Omega} \qquad \text{(phase velocity)}$$

$$\left(\frac{d\omega}{dk}\right)_i \approx \frac{\pm v_A^2 k}{\Omega} \qquad \text{(group velocity)}$$

$$\left(\frac{d\omega}{dk}\right)_m \approx \frac{\pm v_A^2 k}{\Omega} \qquad \text{(group velocity)} \quad .$$

(8.6.2)

The subscript i refers to the root corresponding to choosing the inner plus sign in (8.6.1), while m refers to the choice of the minus sign. The (i) subscript is associated with the normal inertial waves while the (m) subscript is associated with waves referred to as hydromagnetic inertial waves, for which there is a magnetostrophic balance of forces (i.e., when the Lorentz force is on the order of the Coriolis force). In the absence of a magnetic field, $Q = 0$ and $\omega_i = \pm 2\Omega$. This result corresponds to a semidiurnal oscillation involving no energy propagation. The presence of a weak field ($0 < Q \ll 1$) allows energy to propagate slowly along the magnetic field lines. Hide (1966) showed that reasonable wave velocities can be expected for Magnetic Rossby waves, providing that Q is not too large, unlike the case for pure Alfvén waves.

Hide (1966) constructed a model for hydromagnetic waves that used the β-plane approximation discussed in §8.4.2. His analysis leads to slow magnetohydrodynamic waves that propagate eastward (phase velocity), a result confirmed by the more rigorous mathematical treatment of Stewartson (1967). Because the β-plane approximation is only applicable to a thin shell, it was argued by Hide that westward traveling waves would exist in the outer core (thick liquid shell). The argument presented for this is intuitive rather than mathematical, and has been the subject of substantial debate. Even Hide once expressed doubts concerning its validity (Hide and Stewartson, 1972). Nevertheless, subsequent work by Acheson and Hide (1973) supported Hide's initial point of view. Acheson and Hide (1973) have also considered far more general conditions for magnetic Rossby waves than assumed to derive (8.4.18). Waves are further discussed in Chapter 9.

The Bullard *et al.* (1950) mechanism discussed above attributes secular variation to the drift of magnetic eddies near the top of the core and it was the first mechanism that gained wide support. Subsequently other very different types of mechanisms, including planetary Rossby waves (§8.4) and the drift of convection rolls (§9.6.2), also gained wide support in their time. It now seems likely that several different mechanisms, exhibiting different characteristic time scales, contribute to secular variation. A couple of examples may help illustrate this point. MHD waves in the core have periods ranging from less than a year (Alfvén waves) to more than a 100 yr (e.g., magnetic Rossby waves and dynamo waves; §8.4 and §9.4). Free-decay times could range from around 100 yr for the higher degree harmonics (e.g., degree 8 or higher) to as much as about 50,000 yr (i.e., times less than, or equal to, the decay of the dipole term with the lowest radial mode number: see Table 9.1). Changes associated with variations in the boundary conditions of the outer core provide the longest characteristic times for secular variation, and these can exceed 10^8 yr.

Although the proposed mechanisms for secular variation are many and none is well understood, there are probably significant contributions from several different mechanisms with quite different characteristic times. Processes with shorter characteristic times (less than a few hundred years) will be best resolved from analyses of historical data (e.g., from magnetic observatories), while processes with longer characteristic times will be best resolved through the analyses of paleomagnetic data.

8.6.2 Reversals

Insight into magnetic field reversals can be obtained by considering the magnetic induction equation:

$$\frac{\partial \mathbf{H}}{\partial t} = k\nabla^2 \mathbf{H} + \nabla \times (\mathbf{v} \times \mathbf{H}) \quad . \tag{8.6.3}$$

The velocity field $\mathbf{v}$ can be obtained by solving the Navier–Stokes equation, in which the magnetic field enters only through the body force ($\mathbf{J} \times \mathbf{B}$). In the MHD approximation (§8.1.2) this body force can be rewritten for core conditions as $(\nabla \times \mathbf{H}) \times \mathbf{H}/\mu_0$, where μ_0 is the permeability of free space (a constant, independent of $\mathbf{H}$). Therefore any solution $\mathbf{v}(\mathbf{H})$ of the Navier–Stokes equation must be an even function of $\mathbf{H}$, and thus the sign of the solution is independent of the sign of $\mathbf{H}$. Substitution of this even function into the last term of (8.6.3) shows that this term must be an odd function of $\mathbf{H}$. So is every other term in (8.6.3) and this implies that for any solution $\mathbf{H}$, of (8.6.3), there is an equally acceptable solution, $-\mathbf{H}$. Thus, provided there is a transition mechanism, reverse and normal states are expected from dynamo theory and the time-averaged properties are expected to be identical (Merrill *et al.*, 1979,1990).

The remaining problem dealt with in this section is the mechanism of polarity transition. An example of a simple kinematic dynamo explanation for transition is that convection ceases for a sufficient time for free decay to substantially reduce the size of the magnetic field. Meantime, the temperature gradient of the outer core would increase until convection resumes. The polarity of the new poloidal field would then depend on the sign of the residual poloidal field at the location where the resumption of convection occurred. This is an oversimplified example of one general class of reversal models, referred to here as the class of *free-decay models*. Free-decay models can be either kinematic or hydromagnetic, but they have the common property that the velocity field vanishes over sufficient time for the magnetic field to decay to a small value. The models of Parker (1969) and Levy (1972a,b,c) are examples of models that require a significant convection hiatus to initiate the reversal process. This can be contrasted with *dynamic-reversing models,* which also can be either kinematic or hydromagnetic, but which have the common property that the velocity is rarely, if ever, zero.

In modern research, computers are used to sweep through the parameter space of dynamo theory. Doing this, one finds solutions that are either steady dynamos or oscillatory dynamos (waves and reversals), a subset of which are deterministic chaotic dynamos (e.g., Olson and Hagee, 1990; Soward, 1991a; Weiss, 1991). Thus there seems to be little difficulty in constructing particular models that reverse polarity (see §9.6 and Glatzmaier and Roberts (1995b)). Because the paleomagnetic data clearly indicate that reversals have occurred (Chapter 5), theoreticians can safely conclude that the parameter space in which steady dynamos occur seems inapplicable to the Earth, except possibly during superchrons. The onset of a superchron may mark a transition from an oscillatory state to a steady one (McFadden and Merrill, 1995a).

Paleomagnetists are naturally interested in those aspects of dynamo theory that might help with their research. Thus, for example, the $\alpha\omega$ kinematic models for reversals produced by Parker (1969) and Levy (1972a,b,c) helped stimulate research into reversal transition systematics (e.g., Hoffman, 1979). Olson (1983) suggested the possibility that toroidal and poloidal magnetic fields need not reverse polarity at the same time. This suggestion is strongly supported by the three-dimensional dynamo of Glatzmaier and Roberts (1995b): the toroidal field reverses before the poloidal field. Moreover, in this model the toroidal field is typically antisymmetric with respect to the equator before and after a reversal, but symmetric midway through the reversal (see §9.6). Such studies provide a stimulus to paleomagnetic research. However, paleomagnetists should recognize that *all* dynamo models involve significant simplifications and assumptions, even when they include numerical simulations on a battery of supercomputers. At the current stage of development, such models seem too primitive for them to be testable at the level of detailed measurements of the magnetic field. The role the inner core may have on reversals is discussed in §9.6.

Conversely, the observational data can help the theoretician. For example, if it turns out that the quadrupole (symmetric) dynamo family (see §8.5) dominates the middle of a polarity transition, as seems suggested by some of the polarity transition data (Chapter 5), this would place an added constraint on dynamo theory. In addition, paleomagnetic data already appear sufficient to rule out free-decay models for reversals. The first half of the observed transition paths are not explained by free-decay models and the time for transition also seems too short. One consequence of this is the elimination of the possibility that reversals of the field are initiated when the field becomes symmetric with respect to the rotation axis (which would follow from Cowling's theorem; see §9.3).

Before leaving our broad discussion of reversals, it is worth commenting on deterministic chaos. Deterministic chaotic models exist both for disc dynamo models (e.g., Kono, 1987) and for magnetoconvection models (e.g., Weiss, 1991). The Earth's reversal chronology has become fashionable as an example where deterministic chaotic processes are "clearly operating." The subject of deterministic chaos has now been covered in many books and will not be reviewed here; an introduction to the subject and how it might apply to geomagnetic reversals can be found in Tritton (1988). Chaotic behavior requires nonlinearity and is characterized in part by extreme sensitivity to the initial conditions. An infinitesimal difference in an initial condition will eventually lead to solutions that are not correlated. Given the uncertainties in dynamo theory and the limited number of observations there should be no doubt that geomagnetic reversals themselves (e.g., Tritton, 1988; Weiss, 1991), their short-term irregular occurrence (Cortini and Barton, 1994), the long-term variations in their mean rate of occurrence, and rapid ($\approx$100-yr period) geomagnetic variations (Malinetskii *et al.*, 1990) *might* all be manifestations of deterministic chaos within the geodynamo (see also §11.4.2). It is also important to emphasize that they might not. Many magnetoconvective regimes show solutions that are steady for some parameters, oscillatory for others, and chaotic for still others (e.g., see Weiss (1991) for an excellent discussion in which the chaotic regime represents a small portion of parameter space in a class of magnetoconvection models). Furthermore, the number of observations in the chronology appears to be too few by many orders of magnitude to resolve this question empirically. The conclusion is that we do not know if geomagnetic reversals are a consequence of a deterministic chaotic process. A challenge for the future is to find some way to distinguish between reversing dynamos in the nonchaotic regime from reversing dynamos in the chaotic regime.

A related problem is that it remains unclear whether geomagnetic reversals are triggered by some process such as boundary layer instabilities or are associated with natural oscillations in the solutions that are independent of the boundary conditions, such as appears to be the case for the Sun (e.g., McFadden and Merrill, 1993). Removing this uncertainty is another challenging problem for the future.

Chapter Nine

Dynamo Theory

9.1 Vector Spherical Harmonics

9.1.1 The Helmholtz Theorem

A theorem often referred to as the Helmholtz theorem states that a vector field $\mathbf{H}$ is uniquely specified within a region $\mathcal{V}$ by its divergence and curl throughout that region and by its normal component H_n over the boundary $\mathcal{S}$ of the region. That is, given

$$\nabla \cdot \mathbf{H} = M$$
$$\nabla \times \mathbf{H} = \mathbf{J}$$
(9.1.1)

within $\mathcal{V}$ together with H_n over the boundary $\mathcal{S}$ of $\mathcal{V}$, then $\mathbf{H}$ is uniquely specified in $\mathcal{V}$. To prove this, assume there are two values $\mathbf{H}_1$ and $\mathbf{H}_2$ that each satisfy the above. Define

$$\mathbf{W} \equiv \mathbf{H}_1 - \mathbf{H}_2 \quad .$$
(9.1.2)

Then

$$\nabla \cdot \mathbf{W} = 0$$
$$\nabla \times \mathbf{W} = \mathbf{0} \quad ,$$
(9.1.3)

which implies there exists a scalar potential ψ such that $\nabla^2\psi = 0$ with $\mathbf{W} = -\nabla\psi$. Substitute ψ for each of the scalars u_1 and u_2 in a generalized form of Green's theorem that states, for any two scalars, u_1 and u_2,

$$\int_S u_1(\nabla u_2 \cdot \hat{\mathbf{n}})d\mathbf{S} = \int_v u_1(\nabla \cdot \nabla)u_2\, dv + \int_v \nabla u_1 \cdot \nabla u_2\, dv \quad . \tag{9.1.4}$$

Let $u_1=u_2=\psi$. Then because H_n is uniquely specified on $\mathbf{S}$, $W_n=H_{1n}-H_{2n}=0$, so the first integral on the left side in (9.1.4) is zero. The first integral on the right side also vanishes since $\nabla^2\psi=0$. This leaves

$$\int_v \nabla\psi \cdot \nabla\psi\, dv = \int_v \mathbf{W} \cdot \mathbf{W}\, dv = \int_v W^2\, dv = 0 \quad , \tag{9.1.5}$$

which requires that $\mathbf{W}=\mathbf{H}_1-\mathbf{H}_2=\mathbf{0}$, proving the Helmholtz theorem. It is also worth noting that any vector field can be decomposed into the sum of the gradient of a scalar potential plus the curl of a vector potential.

The above, often omitted in geophysics and physics textbooks, provides a natural place to introduce vector spherical harmonics, which are elegantly and rigorously developed in mathematics textbooks such as Morse and Feshbach (1953). Here a less elegant and less complete development is given, but one we hope will provide insight into the nature of vector spherical harmonics, a subject that must be mastered to understand the mathematics of dynamo theory.

9.1.2 Helmholtz Scalar Equation

The Helmholtz vector equation,

$$\nabla^2\mathbf{H} + k^2\mathbf{H} = \mathbf{0} \quad , \tag{9.1.6}$$

commonly occurs in many branches of physics and geophysics. For example, suppose decay of the magnetic field is assumed to be represented by $\mathbf{H}e^{-t/\tau}$, where τ is a constant and t is time. Substitution of this into the magnetic diffusion equation (8.2.11) yields (9.1.6) with $k = (1/\tau k_m)^{1/2}$. Solutions of (9.1.6) are usually obtained in terms of vector spherical harmonics. These harmonics play a central role in many dynamo theories and can be directly linked with the terminology presently used to describe the Earth's magnetic field.

The development of vector spherical harmonics is conveniently done by first analyzing (9.1.6) in its scalar form,

$$\nabla^2\psi + k^2\psi = 0 \quad , \tag{9.1.7}$$

which is referred to as the *Helmholtz scalar equation.* Solutions to Laplace's equation in spherical coordinates are given in terms of surface harmonics y_l^m (§2.2). Therefore, a solution of (9.1.7) is attempted by the substitution

$$\psi = \sum_{l=0}^{\infty} \sum_{m=1}^{\infty} f_l(r)\, y_l^m(\theta,\phi) \quad, \tag{9.1.8}$$

where $f_l(r)$ is some undetermined function of r. Substitution of (9.1.8) into (9.1.7) leads to the *radial* equation

$$\left[\frac{d^2}{dr^2} + \frac{2}{r}\frac{d}{dr} + k^2 - \frac{l(l+1)}{r^2}\right] f_l(r) = 0 \tag{9.1.9}$$

in spherical coordinates. With suitable boundary conditions imposed on $f_l(r)$, (9.1.9) may be an eigenvalue equation for k^2. Note that the radial equation depends on the degree l of the associated surface harmonic y_l^m, but not on the order m. Let

$$f_l(r) = \frac{1}{\sqrt{r}} u_l(r) \tag{9.1.10}$$

and substitute this into (9.1.9) to obtain

$$\left[\frac{d^2}{dr^2} + \frac{1}{r}\frac{d}{dr} + k^2 - \frac{\left(l+\frac{1}{2}\right)^2}{r^2}\right] u_l(r) = 0 \quad. \tag{9.1.11}$$

Equation (9.1.11) is a well-known form of Bessel's equation. Solutions to it are

$$
\begin{aligned}
u_l(r) &= \xi_{l+\frac{1}{2}}(kr) \\
u_l(r) &= N_{l+\frac{1}{2}}(kr) \quad,
\end{aligned}
\tag{9.1.12}
$$

where $\xi_{l+\frac{1}{2}}(x)$ is a Bessel function given by

$$\xi_{l+\frac{1}{2}}(x) = \left(\frac{x}{2}\right)^{l+\frac{1}{2}} \sum_{n=0}^{\infty} \left[\frac{(-1)^n}{n!\,\Gamma(n+l+\frac{3}{2})} \left(\frac{x}{2}\right)^{2n}\right] \quad. \tag{9.1.13}$$

($\Gamma(z)$ is the gamma function, given by $\Gamma(z) \equiv \int_0^{\infty} e^{-t}\, t^{z-1}\, dt$), and $N_{l+\frac{1}{2}}(x)$ is a Neumann function (or Bessel function of the second kind) given by

$$N_{l+\frac{1}{2}}(x) = \frac{\xi_{l+\frac{1}{2}}(x)\cos(l+\frac{1}{2})\pi - \xi_{-(l+\frac{1}{2})}(x)}{\sin(l+\frac{1}{2})\pi} \quad . \tag{9.1.14}$$

The solution of the Helmholtz scalar equation is therefore given by (9.1.8), where $f_l(r)$ is obtained from (9.1.10) through (9.1.14).

9.1.3 Helmholtz Vector Equation

Take ψ to represent a solution of the Helmholtz scalar equation given in §9.1.2. Then direct substitution of $\nabla\psi$ into the Helmholtz vector equation (9.1.6) shows that $\nabla\psi$ is a solution. Note that $\nabla\psi$ is irrotational, or curl free (i.e., $\nabla\times\nabla\psi = \mathbf{0}$). To find a general solution, two other vector solutions must be found that are orthogonal to $\nabla\psi$ and to each other. With this objective in mind the cross product of the terms in the Helmholtz vector equation can be taken with $\mathbf{r}$ to obtain

$$\left[\nabla^2\nabla\psi\right]\times\mathbf{r} + k^2\nabla\psi\times\mathbf{r} = \mathbf{0} \quad . \tag{9.1.15}$$

Since $\nabla^2\mathbf{r} = \mathbf{0}$, the above can be written

$$\nabla^2(\nabla\psi\times\mathbf{r}) + k^2(\nabla\psi\times\mathbf{r}) = \mathbf{0} \quad . \tag{9.1.16}$$

Therefore,

$$\mathbf{T} = \nabla\psi\times\mathbf{r} \tag{9.1.17}$$

is a second solution of the Helmholtz vector equation. For future reference, note that this vector has no radial component. For completion, a third vector solution, $\mathbf{S}$, is needed, where $\mathbf{S}$ is orthogonal to both $\mathbf{T}$ and $\nabla\psi$. Taking the curl of (9.1.16) with $\mathbf{T}$ substituted for $(\nabla\psi\times\mathbf{r})$, it is apparent that $\mathbf{S}$ is given by

$$k\mathbf{S} = \nabla\times\mathbf{T} \quad . \tag{9.1.18}$$

Multiplying both sides of (9.1.17) by k and using the fact that $\nabla\cdot\mathbf{S}=0$, we have

$$k\mathbf{T} = \nabla\times\mathbf{S} \quad . \tag{9.1.19}$$

Therefore, $\mathbf{S}$ and $\mathbf{T}$ are related to each other by identical equations. Moreover, $\mathbf{S}$ and $\mathbf{T}$ are clearly orthogonal to each other and to $\nabla\psi$.

Thus the three orthogonal vectors $\mathbf{U}(r)\equiv\nabla\psi$, $\mathbf{T}$, and $\mathbf{S}$ are the solutions to the Helmholtz vector equation. The spherical coordinate representations of these vectors are

$$\mathbf{U} = \frac{\partial \psi}{\partial r}, \frac{1}{r}\frac{\partial \psi}{\partial \theta}, \frac{1}{r\sin\theta}\frac{\partial \psi}{\partial \phi}$$

$$\mathbf{T} = 0, \frac{1}{\sin\theta}\frac{\partial \psi}{\partial \phi}, -\frac{\partial \psi}{\partial \theta} \tag{9.1.20}$$

$$\mathbf{S} = k_m r \psi + \frac{1}{k_m}\frac{\partial^2(\psi r)}{\partial r^2}, \frac{1}{k_m r}\frac{\partial^2(r\psi)}{\partial r\partial \theta}, \frac{1}{k_m r\sin\theta}\frac{\partial^2(r\psi)}{\partial r\partial \phi} .$$

Any solenoidal (i.e., divergence free) vector such as **B** can be written in terms of a vector potential **A** as

$$\mathbf{B} = \nabla \times \mathbf{A} , \tag{9.1.21}$$

where

$$\mathbf{A} = \psi_T \mathbf{r} + \nabla \psi_P \times \mathbf{r} = \psi_T \mathbf{r} + \nabla \times \psi_P \mathbf{r} , \tag{9.1.22}$$

where ψ_T and ψ_P have been used instead of ψ since there is no requirement that these be the same scalar potentials. This then gives

$$\mathbf{B} = \nabla \times \psi_T \mathbf{r} + \nabla \times (\nabla \times \psi_P \mathbf{r}) = \mathbf{T} + \mathbf{S} . \tag{9.1.23}$$

Further, note that since the current density **J** is given by the curl of the magnetic field, (9.1.18) and (9.1.19) show that toroidal magnetic field is associated with poloidal current, and poloidal magnetic field is associated with toroidal current.

Using $d\Omega$ as the increment of solid angle (i.e., $d(\cos\theta)\,d\phi$), and defining the operator L^2 by

$$L^2 = -\left[\frac{1}{\sin\theta}\frac{\partial}{\partial\theta}\left(\sin\frac{\partial}{\partial\theta}\right) + \frac{1}{\sin^2\theta}\frac{\partial^2}{\partial\phi^2}\right] , \tag{9.1.24}$$

the orthogonality conditions for **T** and **S** can be shown to be

$$\int \mathbf{T}_a \cdot \mathbf{T}_b d\Omega = \int T_a L^2 T_b d\Omega$$

$$\int \mathbf{T}_a \cdot \mathbf{S}_b d\Omega = 0 \tag{9.1.25}$$

$$\int \mathbf{S}_a \cdot \mathbf{S}_b d\Omega = \int \nabla(rS_a) \cdot \nabla(rL^2 S_b)\frac{d\Omega}{r^2}$$

for all a and b. In summary, the spherical harmonic vector **T** has no radial component and is called a *toroidal vector*, while **S** is called a *poloidal vector*. A magnetic field represented by **T** is referred to as a *toroidal magnetic field*, and one represented by **S** is referred to as a *poloidal* (spheroidal) *field*. Since **B** is always divergence free ($\nabla \cdot \mathbf{B} = 0$), it can always be represented by the sum of a toroidal and a poloidal magnetic field.

9.1.4 Free-Decay Modes

Barnes (1973), a creation scientist, used the known decrease of the dipole field during this century (Chapter 2) to conclude that the half-life of the Earth's magnetic field was on the order of 1400 yr. He further concluded that this proved that the Earth could not be older than several thousand years. As an aside, several well-known scientists through the years have suggested the next reversal will occur roughly 2000 yr from now based essentially on the same data. If the velocity term in the magnetic induction equation could be neglected then the magnetic field would decay with time, and the conclusion reached by Barnes (for the magnetic field) would not be unreasonable. Of course the velocity is not zero and positive dynamo action occurs to produce the magnetic field. Interestingly, the prediction of a reversal one to two thousand years from now also ignores the dynamic aspects; in particular, there appear to have been numerous times in the past when the dipole intensity has decreased without leading to a reversal. This is particularly so when, as in the recent past, the intensity has been above its long-term average value.

It is straightforward to substitute (9.1.20) into the magnetic induction equation, with $\mathbf{v} = \mathbf{0}$ and (for simplicity) the assumption that k_m is constant, to obtain

$$\frac{\partial \psi_T}{\partial t} = k_m \nabla^2 \psi_T$$

$$\frac{\partial \psi_P}{\partial t} = k_m \nabla^2 \psi_P \quad . \tag{9.1.26}$$

It is seen that one obtains two separate free-decay equations, one for the toroidal field and one for the poloidal field (this occurs because $\mathbf{T}$ and $\mathbf{S}$ are orthogonal). That is, if there were a convection hiatus, the toroidal and poloidal fields would decay independently of each other.

In the following, we assume free decay occurs in a conducting sphere bounded by an insulator. Based on the above equations one expects the magnetic field to decay away exponentially with time, much as has already been found in §8.2.3. As already shown, a substitution of the form $\mathbf{H}e^{-t/\tau}$ into the magnetic induction equation with $\mathbf{v} = \mathbf{0}$ yields the Helmholtz vector equation (9.1.6), for which solutions involve the vector spherical harmonics, and therefore the radial equation (9.1.11) with Bessel function solutions. Appropriate boundary conditions at the surface of the core (Elsasser, 1946a) require that $x(=kr)$ of (9.1.11) take on specific values $x_{\nu l}$, where $x_{\nu l}$ is the νth positive root in ascending order of

$$\xi_{l-\frac{1}{2}}(x) = 0 \quad . \tag{9.1.27}$$

The subscript ν is referred to as the *radial mode number*. The relaxation, or free-decay, time, $\tau_{\nu l}$, for the lth degree surface harmonic in the νth radial mode is then given by

$$\tau_{\nu l} = \frac{R^2}{k_m (x_{\nu l})^2} = \frac{R^2 \mu_0 \sigma}{(x_{\nu l})^2} \quad , \tag{9.1.28}$$

where μ_0 is the magnetic permeability of free space, σ is the electrical conductivity, and R is the radius of the core. Thus the free-decay time of a particular spherical harmonic term involves not only its degree l (e.g., dipole, quadrupole) but also its radial mode (or modes). Note again, however, that the free-decay time does not depend on the order m of the harmonic. Solutions for the poloidal mode for two different values of electrical conductivity for a sphere with 3000 km radius are shown in Table 9.1.

Note that, unlike that given in §8.2.3, the more general theory shows that the decay of the dipole field (or, for that matter, any other harmonic) cannot be uniquely determined! The decay time still goes as the square of the dimensions of the source, but one does not know the source dimensions in the core corresponding to any given spherical harmonic term (e.g., the dipole field). The spherical Bessel function provides information on the radial dimensions of the source while the spherical harmonic term provides information on the surface dimensions.

It is often assumed that the ν=1 radial mode dominates, and this mode is

Table 9.1
Estimates of Free-Decay Relaxation Times, $\tau_{\nu l}$.[a]

	l							
ν	1	2	3	4	5	6	7	8
1	5000	2460	1483	970	737	563	443	363
	50000	24600	14833	9700	7367	5633	4433	3633
2	1250	830	600	453	363	293	243	207
	12500	8300	6000	4533	3633	2933	2433	2067
3	557	417	325	263	220	187	158	137
	5566	4167	3250	2633	2200	1867	1583	1367
4	313	250	203	173	147	127	113	100
	3133	2500	2033	1733	1467	1267	1133	1000

Notes: ν, radial-mode number. l, degree of the surface spherical harmonic. Decay times are given in years. Electrical conductivities used are 10^5 Sm^{-1} (smaller values) and 10^6 Sm^{-1} (larger values).

[a]After McFadden *et al.* (1985).

commonly used to estimate the free decay of the dipole field. Also intuitively appealing is the assumption that the dominant source dimensions of the dipole are not usually smaller than those for the higher degree harmonics (i.e., the dominant radial modes for the dipole source do not have larger values of ν than the dominant radial modes for the higher degree harmonics). This assumption implies that, on the average, higher degree harmonics decay more rapidly than lower degree ones (McFadden *et al.*, 1985). Notwithstanding the inherent appeal of these conclusions, it is critical to recognize that they require physical assumptions, as is highlighted by the strong dependence on ν shown in Table 9.1.

It is interesting to note (e.g., Gubbins and Roberts, 1987) that the free-decay time for a degree *l* toroidal field in the νth radial mode is identical to that for a degree (*l*+1) poloidal field in the νth radial mode.

9.2 Kinematic Dynamos

9.2.1 Toroidal and Poloidal Fields

To gain more insight into the toroidal magnetic field we consider the special case of a velocity field

$$\mathbf{v} = (0, R\omega, 0) \tag{9.2.1}$$

in cylindrical coordinates (R, θ, z) acting on a toroidal magnetic field $\mathbf{H} = (H_R, H_\theta, H_z)$. For illustrative purposes it is assumed that the conductivity is sufficiently high that the decay term in the magnetic induction equation can be neglected. The velocity shear given by (9.2.1) leads to a toroidal magnetic field with a θ component that grows as

$$\frac{\partial H_\theta}{\partial t} = R\left[H_R\, \frac{\partial \omega}{\partial R} + H_z\, \frac{\partial \omega}{\partial z} \right] - \omega\, \frac{\partial H_\theta}{\partial \theta} \quad . \tag{9.2.2}$$

Symmetry dictates that this magnetic field can be represented in spherical coordinates (r, θ, ψ), by the toroidal vector $(0, 0, T_\psi)$. Equation (9.2.2) shows how a toroidal magnetic field can be created from shear ($\partial \omega/\partial R$ and $\partial \omega/\partial z$ terms) of a poloidal magnetic field or rotation associated with a magnetic field that varies as a function of θ ($\partial H_\theta/\partial \theta$ term). This equation also allows one to calculate the rate of build-up of the toroidal magnetic field, a rate that can be relatively high (depending on the assumed shear, rotation rate, magnetic field strength, and field inhomogeneity). This example also provides insight into the rate of production of a toroidal magnetic field by the ω-effect as described in §8.3.2.

Additional insight into toroidal and poloidal magnetic fields can be gained from Fig. 9.1, which shows representations of a few simple toroidal and poloidal fields. The following "rules" can easily be shown to be valid.

(i) For S_l^0, the number of surface latitudinal nodal lines is given by l.

(ii) For S_l^m, the number of longitudinal nodal lines is given by $(l$-$m)$.

(iii) For T_l^0, the number of latitudinal nodal lines is given by $(l$-$1)$.

Therefore, an axial geocentric dipole can be represented by an S_1^0, an axial geocentric quadrupole by an S_2^0, etc.

Because a toroidal magnetic field in the Earth's core has no radial component, the entire field line will be in the spherical core if any part of a field line is in the core. Figure 9.2 helps show why no toroidal field could exist in the Earth's mantle if the mantle were a perfect insulator. This figure shows a toroidal field loop entirely within the mantle. Since $\nabla \times \mathbf{H} = \mathbf{J}$, a current must pass vertically through this loop, contradicting the assumption that the mantle is an insulator, so no toroidal loop is possible there. If the mantle is semiconducting, then currents can exist there, as can a toroidal field. The presence of toroidal fields in the semi-conducting mantle can be imagined as originating from outward diffusion of a toroidal field produced in the core. However, the conductivity and permeability of the mantle are such that virtually no toroidal field should be manifest at the Earth's surface. Thus the core's toroidal field must be inferred from theoretical models, and, as a consequence, its magnitude is controversial.

9.2.2 Bullard–Gellman Models

Bullard and Gellman (1954) presented a brute-force kinematic dynamo model that had a pronounced impact on the geomagnetic community, in part because it was the first model that produced numerical values. Because of this, and because it still provides valuable insight into how various poloidal and toroidal magnetic and fluid velocity fields interact, it is described here in some detail. A valuable review and analysis of this model has been given by Gibson and Roberts (1969). The approach taken by Bullard and Gellman was to convert the magnetic induction equation to an eigenvalue equation. Choose a new length scale (the radius of the core is a convenient unit) such that x_{initial}, a distance measured in the original units, becomes ax_{new}. Scale the time and velocity as

$$t_{\text{initial}} \rightarrow \sigma\mu_0 a^2 t_{\text{new}}$$
$$\mathbf{v}_{\text{initial}} \rightarrow \mathbf{v}_{\text{new}} \Big/ \sigma\mu_0 a \qquad (9.2.3)$$

and represent $\mathbf{v}_{\text{new}}$ by $U\mathbf{u}$ (a convenient value for U is the maximum of $|\mathbf{v}_{\text{new}}|$). Remembering that $\nabla_{\text{initial}} \rightarrow (1/a)\nabla_{\text{new}}$, the magnetic induction equation can now be rewritten in dimensionless form as

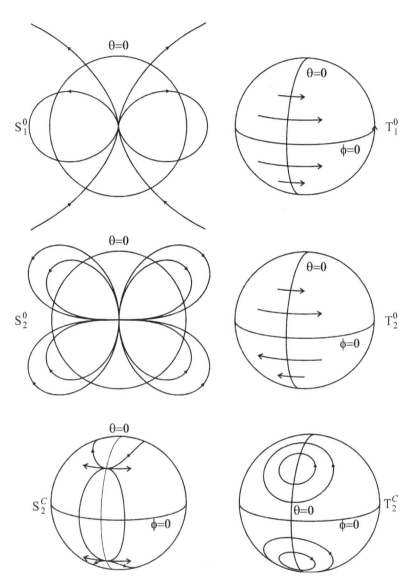

Fig. 9.1. Examples of poloidal and toroidal fields. Superscript C stand for cosine and indicates that part of the ϕ component in the surface harmonic. (\Modified after Bullard (1949a).

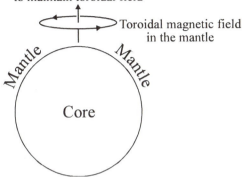

Direction of current needed
to maintain toroidal field

Toroidal magnetic field
in the mantle

Fig. 9.2. A toroidal magnetic field is assumed to exist in the Earth's electrically insulating mantle. The current needed to maintain this field must pass through the toroidal loop, contradicting the assumption of an insulating mantle.

$$\frac{\partial \mathbf{B}}{\partial t} = \nabla^2 \mathbf{B} + U\nabla \times (\mathbf{u} \times \mathbf{B}) \quad , \tag{9.2.4}$$

where the subscript has been dropped on t_{new} and ∇_{new}. For a given set of boundary conditions, (9.2.4) will have solutions for particular values (eigenvalues) of U. Because the Laplacian has mixed derivatives in spherical coordinates, it is convenient to expand the vector Laplacian and use $\nabla \cdot \mathbf{B} = 0$ to give the following form of the magnetic induction equation:

$$\frac{\partial \mathbf{B}}{\partial t} = U\nabla \times (\mathbf{u} \times \mathbf{B}) - \nabla \times (\nabla \times \mathbf{B}) \quad . \tag{9.2.5}$$

To solve this equation, Bullard and Gellman expanded both the velocity field and the magnetic field in poloidal and toroidal vectors,

$$\mathbf{u} = \sum_{\alpha} (\mathbf{s}_{\alpha} + \mathbf{t}_{\alpha})$$

$$\mathbf{B} = \sum_{\beta} (\mathbf{S}_{\beta} + \mathbf{T}_{\beta}) \quad , \tag{9.2.6}$$

where the short-hand notation of Gibson and Roberts (1969) has been used. In this notation $\Sigma_\alpha, \Sigma_\beta$, etc. implies summation of the appropriate vector spherical harmonics, y_l^m, over $\Sigma_{l=0}^\infty \Sigma_{m=0}^l$. Substituting (9.2.6) into (9.2.5) yields

$$\sum_\gamma \left[\frac{\partial \mathbf{S}_\gamma}{\partial t} + \frac{\partial \mathbf{T}_\gamma}{\partial t} \right]$$
$$= U \sum_{\alpha,\beta} \nabla \times \left[(\mathbf{s}_\alpha + \mathbf{t}_\alpha) \times (\mathbf{S}_\beta + \mathbf{T}_\beta) \right] - \sum_\gamma \nabla \times \left[\nabla \times (\mathbf{S}_\gamma + \mathbf{T}_\gamma) \right] \quad ,$$

(9.2.7)

where γ is yet another dummy index (that is, the summation is over "γ").

Now take the scalar product of $\mathbf{S}_\gamma$ with the terms in (9.2.7) and integrate over the solid angle to eliminate the θ and ϕ dependence and to make use of the orthogonality properties of the spheroidal and toroidal vectors (see (9.1.25)). This yields:

$$\left[\nabla^2 - \frac{\partial}{\partial t} \right] \mathbf{S}_\gamma = U \sum_{\alpha,\beta} \{ \mathbf{s}_\alpha \mathbf{S}_\beta \mathbf{S}_\gamma \} + \{ \mathbf{t}_\alpha \mathbf{S}_\beta \mathbf{S}_\gamma \} + \{ \mathbf{s}_\alpha \mathbf{T}_\beta \mathbf{T}_\gamma \} \quad , \qquad (9.2.8)$$

where

$$\{ \mathbf{s}_\alpha \mathbf{S}_\beta \mathbf{S}_\gamma \} = \int \int \mathbf{S}_\gamma \cdot \left[\nabla \times (\mathbf{s}_\alpha \times \mathbf{S}_\beta) \right] d\Omega$$
$$\nabla^2 \mathbf{S}_\gamma = \left[\frac{\partial}{\partial r^2} + \frac{2}{r} \frac{\partial}{\partial r} - l \frac{(l+1)}{r^2} \right] \mathbf{S}_\gamma \quad .$$

(9.2.9)

Equations similar to (9.2.9) also exist for $\{ \mathbf{t}_\alpha \mathbf{S}_\beta \mathbf{S}_\gamma \}$ and $\{ \mathbf{s}_\alpha \mathbf{T}_\beta \mathbf{T}_\gamma \}$. In addition, similar equations can be generated for $\mathbf{t}_\alpha$, by taking the scalar product of $\mathbf{T}_\gamma$ with (9.2.7) and integrating.

The basic procedure used by Bullard and Gellman (1954), as has been followed by several others (see review by Gubbins, 1974), is to assume some initial magnetic field and velocity field and to substitute these into the above equations. Bullard and Gellman assumed an initial poloidal field S_1^0, initial toroidal velocity field T_1^0 (uniform shear), and initial poloidal velocity field S_2^{2C} (C and S as superscripts refer to the cosine or sine part of the harmonic, respectively-see Fig. 9.1). They sought steady-state solutions $\partial \mathbf{S}_\gamma / \partial t = \partial \mathbf{T}_\gamma / \partial t = 0$ in (9.2.8). However, because of terms like $\{ \mathbf{s}_\alpha \mathbf{S}_\beta \mathbf{S}_\gamma \}$ in (9.2.8), this procedure leads to an infinite series of toroidal and poloidal vectors. That is, the interaction of even very simple velocity and magnetic fields leads to the generation of magnetic and velocity fields that need an infinite number of toroidal and poloidal terms for representation. One important consequence of this is that individual spherical harmonic terms should not be considered independent of other spherical harmonic terms, as is sometimes erroneously done. In particular, the assumption made by Bullard and Gellman generates terms like $T_2^0, T_2^{2C}, T_2^{2S}, S_3^0, S_3^{2C}, S_3^{2S}$, etc. They truncated the series at $l = 12$ to obtain a finite system that could be solved numerically using a computer. For $U = 2 \times 10^{-4}$ m s^{-1} (a velocity at the outer core–mantle boundary that is comparable with the westward drift velocity of the nondipole field, 0.2°/yr), and for S_1^0 at the Earth's surface around 10^{-4} T (comparable with the present field intensity), they found that the lowest order toroidal magnetic field, T_2^0, has an order of magnitude value of 10^{-2} T. It should be noted that the toroidal field in this model contains most of the magnetic energy. Because the toroidal field cannot be observed directly (§9.2.1), the Bullard–Gellman estimate for the toroidal field magnitude has been widely quoted.

The question of convergence of the infinite series used in the expansion by Bullard and Gellman was not examined adequately until Gibson and Roberts (1969) showed that it does not converge. The numerical results of Bullard and Gellman may, therefore, be radically wrong. More complicated initial flows have been assumed in attempts to circumvent this problem. An instructional dynamo model by Lilley (1970) demonstrated that the initial velocity field $T_1^0 + S_2^{2C} \rightarrow S_2^{2S}$ produces much more rapid convergence. However, even in this case the convergence still does not appear to be rapid enough (Gubbins, 1974). The most likely reason for this lack of convergence is the spherical geometry, as discussed in §9. 3.

Backus (1958) suggested one way to modify the above approach to escape from the convergence problem by noticing that convergence would occur if the velocity field were time-dependent. In particular, allowing the velocity field to be zero for long times eliminated the effects of the higher degree harmonics. Higher degree harmonics, which are associated with shorter wavelengths, decay more rapidly than lower degree harmonics. One could simply "turn off" the velocity field for as long as needed to allow the higher degree harmonics to become negligible. In this way the troublesome convergence problem could be eliminated, but at the expense of retaining terms like $\partial S_i/\partial t$ in (9.2.7). This contribution by Backus (1958) remains one of the first demonstrations that dynamos could exist, a serious concern at that time because of Cowling's theorem (§9.3).

9.2.3 Fast Dynamos

There are times when the magnetic field of the Earth (and the Sun) appears to increase dramatically in magnitude. To remind the reader that **B** and **H** are interchangeable in the MHD environment of the core, we rewrite the magnetic induction equation in terms of **B** and in a dimensionless form that emphasizes the magnetic Reynolds number, R_m:

$$\frac{\partial \mathbf{B}}{\partial t} = \frac{1}{R_m} \nabla^2 \mathbf{B} + \nabla \times (\mathbf{v} \times \mathbf{B}) \quad . \tag{9.2.10}$$

If the velocity field is steady (which precludes feedback from the Lorentz force), then (9.2.10) is linear in **B** and allows solutions of the form

$$\mathbf{B}(\mathbf{r}, t) = \mathbf{B}(\mathbf{r}) e^{Pt} \quad , \tag{9.2.11}$$

where **r** is a general position vector and P (not to be confused with Legendre polynomials) in general can have real and imaginary parts. If P is real and positive the dynamo is *steady*. If P is purely imaginary the solutions are *oscillatory*. It is important to note that one might have an infinite series of complex values for P and hence oscillatory does not necessarily imply periodic. Dynamo waves and reversals occur in oscillatory dynamos but not steady ones. For *fast dynamos*, one is interested in the case (by definition) where P approaches a positive value as R_m approaches infinity (e.g., Soward, 1989, 1994). If P remains positive but approaches zero as $R_m \rightarrow \infty$ the dynamo is said to be *slow*.

Often the steady fluid motions assumed are variations of the so called *family of ABC flows*, which in the rectangular coordinate system (x,y,z) are

$$\mathbf{v} = (A \sin z + C \cos y, B \sin x + A \cos z, C \sin y + B \cos x) \quad . \tag{9.2.12}$$

These flows exhibit helicity, are not purely toroidal (a pure toroidal velocity field cannot drive a dynamo), and played a central role in establishing the existence of dynamos (G. Roberts, 1972), see §9.3.

The study of fast dynamos has been mostly of mathematical interest (e.g., see analyses and review by Soward, 1989; 1994). A way of visualizing fast dynamo action is shown in Fig. 9.3, modified from Vainshtein and Zel'dovich (1972). This is sometimes referred to as the stretch, twist, and fold action of fast

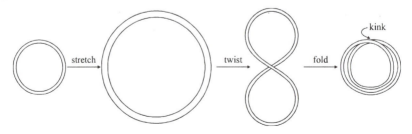

Fig. 9.3 The Alfvén dynamo mechanism. Field intensity is doubled by the stretch–twist–fold cycle. See, for example, Vainshtein and Zel'dovich (1972).

dynamos. Since we are interested in the situation $R_m \to \infty$, the effects of diffusion may be ignored. That is, the magnetic field is effectively frozen into the fluid. The magnetic field energy can by increased by first stretching the magnetic field lines (left side of Fig. 9.3), and subsequently by twisting the field lines and folding them back over to double the original field strength. Figure 9.3 is obviously contrived, but it is useful in at least two ways. First, it emphasizes to mathematicians that a geometrical approach of iterated maps may be useful in dynamo theory analyses, and second, it illustrates one way to maximize dynamo action. Indeed, in fast dynamo models, one seeks the maximum P (eigenvalues) that produces dynamo action. By doing this and by ignoring diffusion one obtains the maximum rate of build-up of the magnetic field for the velocity field being investigated.

Although fast dynamos are commonly thought to be only of mathematical interest, they are also useful for estimating the maximum rate of dynamo action in the Earth and Sun. One has already been used in Earth's case to estimate the maximum rate of build-up of the magnetic field during a polarity transition (Bogue and Merrill, 1992).

9.3 Cowling's Theorem and Other Constraints

Cowling (1934, 1957) argued that a steady poloidal magnetic field with an axis of symmetry cannot be maintained by motion symmetric about that axis. Significant effort was subsequently expended on generalizing and examining the rigor of *Cowling's theorem* (e.g., Backus and Chandrasekhar, 1956; Backus, 1957; Braginsky, 1964a,b; James *et al.*, 1980; Hide, 1981; Hide and Palmer, 1982). Backus (1957) was able to generalize the theorem to show that nonsteady axisymmetric magnetic fields cannot be maintained by symmetric fluid motions. A complete proof of Cowling's theorem using a Bullard–Gellman-type formulation was proposed by James *et al.* (1980), in which the fluid was assumed to be incompressible with no sources or sinks and the velocity was assumed to be bounded. Hide (1981) showed that the theorem is still applicable when thermoelectric and thermomagnetic effects are present, casting doubt on Hibberd's (1979) use of the Nernst–Ettinghausen effect in generating the Earth's magnetic field. Hide and Palmer (1982) have argued that Cowling's theorem is applicable even when the fluid is compressible.

Only a heuristic argument will be given here to illustrate the validity of the theorem (Cowling, 1957). Assume that there are axisymmetric magnetic fields such as shown in Fig. 9.4. At the two limiting points, O_1 and O_2, the magnetic field is zero, but $\nabla \times \mathbf{H} = \mathbf{J} \neq \mathbf{0}$ so an electric current exists. The current at O_1 and

Symmetry axis

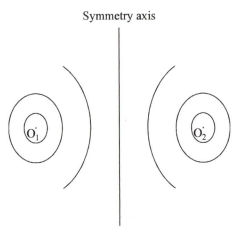

Fig. 9.4. An axisymmetric magnetic field is illustrated. The curl of this field is finite at the two limiting points O_1 and O_2. but the field itself is zero there. This leads to a contradiction and was one of the first "proofs" of Cowling's theorem. After Cowling (1957).

O_2 flows around the symmetry axis and so cannot be maintained by electrostatic forces. Also, since the magnetic field is zero at O_1 and O_2 the current cannot be maintained by the magnetic field. Therefore, the current must decay with time by ohmic dissipation, and likewise so must the associated magnetic field.

In addition to Cowling's theorem, a necessary condition for positive dynamo action is that the magnetic Reynolds number be greater than one (§8.2.2). It can also be shown by a development similar to that given in §8.2.2 that for the build-up term to dominate the decay term in the magnetic induction equation, $\lambda_M L^2/k_m$ must exceed unity, where L is the length scale involved and λ_M is the maximum eigenvalue for velocity shear (Childress, 1978). Still another necessary condition for dynamo action can be obtained by writing the field in terms of its toroidal and poloidal parts (Busse, 1975),

$$\max(\mathbf{v} \cdot \mathbf{r}) > \left[\frac{2E_s}{E_m} \right]^{\frac{1}{2}} k_m \quad , \tag{9.3.1}$$

where E_m is the total magnetic field energy, E_s is the magnetic energy in the poloidal magnetic field, and k_m is the magnetic diffusivity. Among other things, this last condition implies the physically appealing result that under steady-state conditions a poloidal magnetic field in the core requires the existence of some radial velocity motion. That is, a pure toroidal velocity field cannot produce dynamo action.

Finally, there is the intriguing question of the role of helicity in dynamo theories. Although it can be shown that no dynamo action can occur unless the volume integral of helicity over the Earth's core vanishes (i.e., mean helicity is zero), Ruzmaikin and Sokoloff (1980) have also shown that no dynamo action is possible if the helicity vanishes everywhere. So helicity and its distribution are important components of viable dynamos. Considerably more work on the necessary conditions associated with helicity will probably occur in the future, since helicity is a critical ingredient of mean-field electrodynamic approaches to dynamo theory.

Following the development by Cowling, opponents of dynamo models expended considerable effort to find a general antidynamo theorem. Proponents initially called on the Coriolis force to produce the needed asymmetry, but until the 1970s they were unable to demonstrate the existence of a single viable kinematic dynamo. Some psychological support for kinematic dynamos came from analogue models, particularly from the work of Lowes and Wilkinson (1963) and Rikitake (1966). However, these models were not totally convincing because of scaling and other problems (§8.2). Pioneering work by Childress (1969, 1970) and G. Roberts (1969, 1970, 1972) on the existence of kinematic dynamos gave a strong boost to dynamo theory. In particular, G. Roberts (1970) proved the important theorem that almost all spatially periodic motions of the *ABC* family of motions give rise to kinematic dynamos in unbounded (infinite size) conductors. ("Almost all" means that the number of spatially periodic motions that do not give rise to dynamo action is countable.) This does not assure one that the Earth has a dynamo, but it dispels doubt that any general mathematical antidynamo proof exists. Subsequently Soward (1975) gave an excellent, but mathematically involved, discussion on how random waves can produce dynamo action (also see Roberts and Gubbins, 1987). An important part of the proof by G. Roberts (1970) is to demonstrate that large-scale magnetic fields can be maintained in a system in which all fluid motions are periodic with small-scale wavelengths ("small" relative to the scale of the magnetic field). The proof of this is essentially the same as that for Bloch's Theorem dealing with electrons in a solid with a periodic potential (see Ashcroft and Merman, 1976). This result is important, not only for the existence of kinematic dynamos, but also because it provides insight into turbulent dynamos (§9.4).

The one remaining problem treated here concerns the numerical modeling problem discussed in §9.2.2. The question can be asked that if almost all periodic motions give rise to kinematic dynamos, why is there a problem in finding one that works for the Earth? This question has been investigated by Bullard and Gubbins (1977) and it appears the answer may have to do with the geometry assumed. The existence theorems usually treat periodic motions in an *infinite medium,* while the generation of the Earth's magnetic field is confined to a finite region in the Earth's outer core. Bullard and Gubbins (1977) consider two cases, a conducting fluid core bounded by an insulating mantle and a conducting fluid core bounded by a perfectly conducting mantle. Here only the first case will be discussed, in which no current can flow out of the core. This causes a concentration of current near the boundary, a concentration that can have a catastrophic effect on the nature of the magnetic field in the core. Note that this concentration of current does not occur in infinite dynamos, because nowhere is there an impediment to the current flowing between the various small-scale cells in which the velocity field is periodic. Bullard and Gubbins (1977) argue that ohmic dissipation, which depends on the square of the current density, can even be sufficient to inhibit dynamo action in some cases.

The results of G. Roberts and Childress are still encouraging because, when a large number of convection cells are present, the magnetic field in the interior will not be substantially altered by the concentration of current near the core's boundary. This will of course depend on the precise geometry of the boundary current. In any case, it appears that the core's magnetic field might be more complex than has been assumed in many numerical models. This suggestion provides a motivation for considering the possibility of turbulent dynamos.

9.4 Turbulence in the Core

9.4.1 The α- and β-Effects

In §7.6.2 the Reynolds number was defined as $R_e \equiv \rho L u / \eta$, where L is a characteristic length and u is a representative velocity. R_e for the core is around $10^{-2} L$ for velocities of order 4×10^{-4} m s^{-1} (§11.1). Thus for $L = 100 \times 10^3$ m, we have $R_e \approx 10^3$. This might suggest there is turbulence in the core, but there are two serious problems with this analysis. If turbulence is present, so that there is a significant lack of velocity coherence over distances of 100 km (i.e., L), then the analysis given in §11.1 to obtain velocities is in doubt. The second problem is that the magnetic field was neglected in the above analysis.

'Turbulence" is a difficult term to define, as can be illustrated by referring to the Reynolds number, which can be made arbitrarily small or large by the appropriate choice of L. The term "turbulent dynamo" will, therefore, be used to mean that mean-field electrodynamics can be applied. In this approach one considers some mean field value $<F>$ and some fluctuating value $F' \equiv F-<F>$, where F denotes the field of interest (e.g., the velocity or magnetic field). Clearly

$$< F' >= 0; \quad < F + G >=< F > + < G >; \quad << F > \cdot G' >= 0 \ . \quad (9.4.1)$$

Note however that $<\mathbf{F}'\cdot\mathbf{G}'>$ vanishes only if there is no correlation between $\mathbf{F}$ and $\mathbf{G}$, i.e., if $\mathbf{F}'$ and $\mathbf{G}'$ are statistically independent.

The average can be taken over space or time or both. It will usually be assumed for simplicity that the turbulence is *steady*; i.e., it does not depend on time. When the turbulence does not depend on space it is said to be *homogeneous*. The turbulence is *isotropic* if it is independent of the orientation of the coordinate system used; otherwise it is *anisotropic*.

Assume that both the velocity field and the magnetic field can be separated into a mean and a fluctuating part. Substitute these into the magnetic induction equation (8.2.2) for $\mathbf{B}$ and average over space to obtain

$$\frac{\partial <\mathbf{B}>}{\partial t} = k_m \nabla^2 <\mathbf{B}> + \nabla \times (<\mathbf{v}> \times <\mathbf{B}>) + \nabla \times (<\mathbf{v}'\times\mathbf{B}'>) \quad . \tag{9.4.2}$$

It has been shown (Childress, 1970; G. Roberts, 1972) that even when $<\mathbf{v}> = \mathbf{0}$, a steady mean fluctuating magnetic field can exist. Evidently this comes from the positive dynamo action of the last term in (9.4.2). To understand how this might occur $<\mathbf{v}' \times \mathbf{B}'>$ is expanded in a Taylor series, which in Einstein tensor notation is

$$< v'_j B'_k \varepsilon_{ijk} > = \alpha_{ij} < B_j > + \beta_{ijk} \frac{\partial B_j}{\partial x_k} + \cdots \quad . \tag{9.4.3}$$

where ε_{ijk} is the permutation tensor in which $\varepsilon_{123} = 1$ for a right-handed coordinate system, and where the tensors α_{ij} and β_{ijk} depend on the properties of the turbulence. Summation over repeated subscripts in a term is implicit. The expansion (9.4.3) needs to converge so that the power in the magnetic field does not exceed some reasonable bound, e.g., the Earth's heat output. The simplest case available to illustrate the properties of α_{ij} and β_{ijk} is one in which isotropic homogeneous turbulence is assumed in the frame of reference in which $<\mathbf{v}> = 0$. Then only δ_{ij}, the Kronecker tensor (components of which equal one when $i=j$ and are zero otherwise), and ε_{ijk} are available to construct α_{ij} and β_{ijk} Therefore

$$\begin{aligned} \alpha_{ij} &= \alpha\delta_{ij}, \\ \beta_{ijk} &= -\beta\varepsilon_{ijk} \quad . \end{aligned} \tag{9.4.4}$$

That is, to second order (9.4.3) can be written

$$< \mathbf{v}'\times\mathbf{B}'> = \alpha < \mathbf{B}> - \beta\nabla\times < \mathbf{B}> \quad . \tag{9.4.5}$$

Substitution of this back into (9.4.2) (with $<\mathbf{v}> = 0$) yields

$$\frac{\partial <\mathbf{B}>}{\partial t} = (k_m + \beta)\nabla^2 <\mathbf{B}> + \nabla\times\alpha < \mathbf{B}> \quad , \tag{9.4.6}$$

where $\nabla \times (\nabla \times \mathbf{B}) = -\nabla^2 \mathbf{B}$ has been used. One sees that β effectively adds to the diffusion (or viscosity) of the system. Therefore, if positive dynamo action occurs, it must come from the last term, referred to as the α-effect in dynamo theory. Since it comes from an effective electric field, $\mathbf{v} \times \mathbf{B}$, it is sometimes described as an effective e.m.f. in the opposite direction to the mean magnetic field (see §8.3).

It is important to note that the most general way the correlations of $\mathbf{v}'$ and $\mathbf{B}'$ can be *linearly* expressed in terms of a mean field $\mathbf{B}_0$ is through $\alpha_{ij}\beta_{j0}$ (9.4.3). Therefore it must also be possible to express a linear omega effect through α_{ij}. The way to do this mathematically is to recognize that the angular velocity, ω, is a pseudovector; i.e., it reverses sign on mirror reflection. Thus one can write α_{ij} as $v_k\varepsilon_{ijk}$, where ε_{ijk} is the permutation (pseudo) tensor, to obtain an omega effect (Phillips, 1993). This causes a blurring in the definitions of the α- and ω-effects. *Large-scale* velocity shear that takes a poloidal (S) magnetic field to a toroidal (T) field will hereafter be referred to as an omega effect and denoted ωST. *Small-scale* processes that involve the twisting of poloidal field loops to obtain toroidal field components will be referred to as an alpha effect and denoted αST. The alpha effect that takes a toroidal field to a poloidal one will be denoted αTS. Thus the combination of the processes ωST and αTS refers to a $\alpha\omega$-dynamo, αST plus αTS refers to an α^2-dynamo and if both αST and ωST are present, in addition to αTS, one has an $\alpha^2\omega$-dynamo. Strong-field dynamos and intermediate dynamos (ones that fall between weak-field and strong-field models) are sometimes approximated by allowing α_{ij} to be a function of $\mathbf{B}$ (e.g., Hagee and Olson, 1991).

Although the power to drive the geodynamo (§7.7) is not well known, it certainly cannot exceed the total heat flux at the Earth's surface. Therefore β, which increases the decay rate, cannot be too large. It seems likely that β is on the order of k_m or smaller (Roberts and Gubbins, 1987). However, it is important to recognize that if the outer core is turbulent the decay time is reduced. For example, all harmonic decay times are reduced by half if β is comparable with k_m.

The concept of an energy cascade was introduced in §7.6 where the $k^{-5/3}$ Kolmogorov law was given. Although cascading is also expected in many three-dimensional MHD systems, it is not clear whether cascading occurs in the Earth's core (even if the convection is turbulent). It is well known that small-scale motions in a rotating system can lead to large-scale magnetic fields (e.g., the α-effect). These large-scale magnetic fields then act back on the fluid in such a way as to inhibit flow across the field lines (a Maxwell stress effect), thereby possibly inhibiting a cascade. Indeed, Braginsky and Meytlis (1990) have suggested that no cascading occurs in a rotating system that produces a strong dynamo. Instead, they suggest there will be a relatively sharp maximum in the energy versus k curve. A close examination of their paper shows that the arguments against cascading are more intuitive than rigorous, so the question of cascading remains open. In contrast, Biskamp (1993) argues that cascading is common in many MHD turbulent systems. However, Braginsky and Meytlis (1990) have convincingly shown that the magnetic field will produce non-equidimensional convection cells. That is, they have shown that if there is turbulence in the core, it will be strongly anisotropic. Such anisotropy will likely play a key role in future turbulent dynamo models. That is, all the components of α_{ij} and β_{ijk} will, in general, need to be considered, but the number of independent components may be reduced by invoking symmetries.

At present it appears that one cannot safely conclude from theory whether the core is turbulent and, if it is, whether cascading occurs. Is there any observational evidence regarding this problem? Sometimes the nearly white spectrum of spherical harmonics at the CMB is used as evidence for core turbulence (e.g., Stacey, 1992). However, as we have seen, there is not a one-to-one mapping between the velocity and magnetic fields. Consequently one cannot determine the scale of the velocity field directly from these harmonics. One can, however, estimate the velocity field near the top of the core by using the secular variation, as is discussed further in §11.1.

9.4.2 α^2-Dynamos

The heuristic dynamo described in §8.3 is an example of an $\alpha\omega$-dynamo and it is this type of dynamo that seems favored for the Earth by some theoreticians (e.g., Soward, 1991b). After a large shear in the fluid generates the toroidal magnetic field from an initial poloidal field, the α-effect converts the toroidal field back to a poloidal field. In α^2-dynamo models, both the toroidal and the poloidal magnetic fields are obtained through α-effects. Two examples of α^2-dynamos are given here to illustrate them and to provide further insight into the role of helicity in dynamo theory. The notation of §8.3 is used throughout.

Example 1

This follows Childress (1978), who begins by using a first-order smoothed version of the magnetic induction equation (see discussion in §8.3):

$$\frac{\partial \mathbf{B}'}{\partial t} - k_m \nabla^2 \mathbf{B}' = \nabla \times (\mathbf{v}' \times \mathbf{B}_0) = (\mathbf{B}_0 \cdot \nabla)\mathbf{v}' - (\mathbf{v}' \cdot \nabla)\mathbf{B}_0 \quad . \qquad (9.4.7)$$

Because $\mathbf{v}'$ can be assumed to vary spatially much more rapidly than $\mathbf{B}_0$, the last term can be omitted from further discussion.

Now consider a periodic velocity field in Cartesian coordinates:

$$\mathbf{v}' \equiv u(0, \cos Kx, \sin Kx) \quad . \qquad (9.4.8)$$

Hence from (8.4.7)

$$\frac{\partial \mathbf{B'}}{\partial t} - k_m \nabla^2 \mathbf{B'} = \mathbf{B}_{0x} K u(0, -\sin Kx, \cos Kx) \quad , \tag{9.4.9}$$

where $\mathbf{B}_{0x}$ is the x component of the mean field $\mathbf{B}_0$. A steady-state solution of this last equation for the particular case that $\mathbf{B}_0$ is constant is

$$\mathbf{B'} = B_{0x} \frac{u}{k_m K} (0, \sin Kx, -\cos Kx) \quad . \tag{9.4.10}$$

In this case the effective electric field derived from the fluctuating parts (9.4.5) is

$$< \mathbf{v'} \times \mathbf{B'} > = \left(B_{0x} \frac{u^2}{k_m K}, 0, 0 \right) = \overline{\alpha} \cdot \mathbf{B}_0 \quad , \tag{9.4.11}$$

where

$$\overline{\alpha} = \frac{u^2}{k_m K} \begin{pmatrix} 1 & 0 & 0 \\ 0 & 0 & 0 \\ 0 & 0 & 0 \end{pmatrix} \quad . \tag{9.4.12}$$

Note that if $\mathbf{B}_0$ is in the x direction, then the electric current density, $\mathbf{J}_0$, produced would also be in the x direction. However, this current does not work to magnify $\mathbf{B}_{0x}$. This simple α-effect does not produce dynamo action and thus the magnetic field would decay away! In §8.3 this problem of reinforcement was bypassed by putting boundaries into the problem. Here a more satisfactory solution is obtained by examining alternative velocity fields.

The fluid velocity used above is periodic, but does not possess helicity. It is not a member of the *ABC* family of flows (see §9.2.3). A simple velocity field in the *ABC* family that does possess helicity is

$$\mathbf{v'} = u(\sin Ky, \cos Kx, \sin Ks + \cos Ky) \quad . \tag{9.4.13}$$

Substitution into (9.4.7) with similar assumptions again yields

$$< \mathbf{v'} \times \mathbf{B'} > \equiv \overline{\alpha} \cdot \mathbf{B}_0 \quad , \tag{9.4.14}$$

but now

$$\overline{\alpha} = \frac{u^2}{k_m K} \begin{pmatrix} 1 & 0 & 0 \\ 0 & 1 & 0 \\ 0 & 0 & 0 \end{pmatrix} \quad . \tag{9.4.15}$$

α_{11} interacts with a B_{0x} constant field to produce a current J_{0x} while α_{22} interacts with a B_{0y} constant field to produce J_{0y} currents. Combining J_{0x} with B_{0y} and J_{0y} with B_{0x} makes dynamo action possible in this case. This kind of interaction is called the α^2-effect. Obviously, far more complex forms of α_{ij} are possible.

Example 2

An example given by Krause and Steenbeck (1967), and repeated by Roberts (1971), provides another illustration of an α^2-dynamo. Here α is assumed (unrealistically) to be constant. Imagine a conducting sphere of radius a, bounded by an insulator. The fluid inside the sphere is homogeneous, isotropic, and turbulent with constant helicity. Under these conditions, the mean-field magnetic induction equation is

$$\frac{\partial \mathbf{B}}{\partial t} = k_m \nabla^2 \mathbf{B} + \nabla \times (\alpha \mathbf{B}) \quad . \tag{9.4.16}$$

Note that although the instantaneous motions may be far more complex than in the Bullard–Gellman dynamo (§9.2), the magnetic induction equation is easier to solve, since the $\mathbf{v} \times \mathbf{B}$ term has been replaced by $\alpha \mathbf{B}$ (a mean-field value). $\mathbf{B}$ is separated into its toroidal and poloidal magnetic fields by using the vector potentials $\mathbf{A}_T$ and $\mathbf{A}_S$ respectively for the toroidal and poloidal fields; that is

$$\mathbf{B} = \nabla \times \mathbf{A}_T + \nabla \times (\nabla \times \mathbf{A}_S) \quad . \tag{9.4.17}$$

Substitution into (9.4.16) gives

$$\frac{\partial \mathbf{A}_T}{\partial t} = k_m \nabla^2 \mathbf{A}_S + \alpha \mathbf{A}_T \quad . \tag{9.4.18}$$

The boundary conditions are that the magnetic fields must vanish when r equals a and when r approaches infinity. Solutions of this boundary value problem can be obtained by substituting the appropriate spherical harmonic expansions, such as

$$\mathbf{A}_T = \sum_{l=1}^{\infty} \mathbf{A}_T(r) Y_l(\theta,\phi) e^{-\sigma_l t} \quad , \tag{9.4.19}$$

into the above equations.

As in the Bullard–Gellman dynamo (§9.2), solutions involve linear combinations of the Bessel functions, $\xi_{l+\frac{1}{2}}$. Application of the boundary conditions necessitates that solutions exist only for certain eigenvalues of σ_l. From (9.4.9), a steady dynamo occurs if there is a real part to σ_l and if that part is less than zero. For the case at hand, σ_l turns out to be real (P. Roberts, 1971) and the marginal case ($\sigma_l = 0$) is obtained when

$$\xi_{\frac{3}{2}}(c) = 0$$

$$c = \frac{a\alpha}{k_m} \quad .$$

(9.4.20)

This is satisfied when the dimensionless number c is 4.49, and thus an α^2-dynamo occurs for all $c > 4.49$ (P. Roberts, 1971; Roberts and Stix, 1971; Krause, 1977).

9.5 Dynamo Waves

In the treatment of turbulence in §8.3 and §9.4, the velocity fields and magnetic fields were separated into their mean and fluctuating parts. In particular, in §8.3.2 the α-effect was identified with helical eddies associated with convection. However, the α-effect can also occur from waves, a point emphasized by Gubbins (1977; also Roberts and Gubbins, 1987). For example, one can divide the velocity and magnetic fields into axisymmetric and asymmetric parts,

$$\mathbf{v} = \tilde{\mathbf{v}} + \mathbf{v}'$$

$$\mathbf{B} = \tilde{\mathbf{B}} + \mathbf{B}' \quad ,$$

(9.5.1)

where the tilde refers to the field averaged over longitude and the prime refers to the asymmetric remainder. One can then proceed in a way analogous to that done in §8.3 to identify an effective e.m.f. associated with $\mathbf{v}' \times \mathbf{B}'$. However, these fluctuations may come from waves rather than convection. This introduces the possibility that a stratified core might still give rise to a dynamo, providing there is enough energy in hydromagnetic waves and if these waves are of the right character to produce amplification of some initial weak field.

Few, if any, theoreticians currently believe that waves in a completely stratified core produce the geodynamo. It is, however, possible that the outer part of the core is chemically stratified (see Chapter 7). In this case waves there could actually increase positive dynamo action or contribute to secular variation. Indeed, Braginsky (1991, 1993) argues that the outer part of the core is strongly stratified and has a very high Brunt–Väisälä frequency (see Chapter 7). He attributes the origin of secular variation with periods near 60 yr to waves traveling in this chemically stratified outermost region of the core.

Parker (1955a), in a prophetic paper, introduced the concept of dynamo waves, waves that traveled north-south rather that east–west. This has been mathematically developed further by Roberts (1972) and reviewed by Roberts and Gubbins (1987). Suppose that one has an α that depends only on r and θ such that

$$\alpha = \alpha(r)\cos\theta \quad . \tag{9.5.2}$$

The choice of $\cos\theta$ is common because it implies that α will change sign across the equator, as expected if the Coriolis force plays an important role. We further suppose that $\omega = \omega(r)$ and ω' is the fluctuating part of ω. Then it can be shown (e.g., Roberts, 1972; Roberts and Gubbins, 1987) that steady dynamos can exist only when

$$\alpha(r)\omega'(r) > 0 \quad . \tag{9.5.3}$$

Oscillatory, but not steady, solutions can exist when

$$\alpha(r)\omega'(r) < 0 \quad . \tag{9.5.4}$$

In addition, Gubbins and Roberts (1987) point out that oscillatory solutions might also exist if (9.5.3) holds. In particular, magnetic activity in the oscillatory solutions for (9.5.4) progresses as waves that move from the pole to equator. In contrast, waves that move from the equator to the pole could occur for oscillatory solutions of (9.5.3). Such waves are commonly referred to as *dynamo waves*.

Most theoreticians believe that dynamo waves are probably more applicable to the Sun than the Earth (e.g., Gubbins and Roberts, 1987). The Sun exhibits a variety of magnetic fields that migrate toward the equator (Chapter 10), whereas this does not appear to be the dominant direction of secular variation in the Earth (Chapter 2). Nevertheless, Hagee and Olson (1991) find poleward propagating dynamo waves in the quadrupole family in their calculations for some $\alpha\omega$-dynamos (see also Olson and Hagee (1987) and §4.2.7) and claim these are responsible for secular variation on the 10^3- to 10^4-yr time scale. Unfortunately, the lake sediment data they use are too few and selective to be convincing; far more data are required to test their hypothesis.

9.6 Dynamics of the Geodynamo

9.6.1 Taylor-State versus Model-Z Dynamos

In hydromagnetic dynamos one solves for the velocity and magnetic fields simultaneously. However, as before, a simplified set of equations and parameters is investigated. One approach is to neglect those terms in the Navier–Stokes equation that are thought to be the least significant. For example, the Taylor number (which gives the relative importance of the square of the Coriolis to viscous terms, i.e., the inverse square of the Ekman number) is of order 10^{30} for the Earth, which suggests that the Earth's core can be considered inviscid (except at the boundaries). The Rossby number (providing an estimate of the relative magnitude of the inertial to Coriolis forces) is around 4×10^{-7}, which suggests that the inertial forces can also be neglected. Thus the Navier–Stokes equation should be well approximated throughout most of the core by the *magnetogeostrophic* equation:

$$2\Omega \times \mathbf{v} = -\nabla P - c\mathbf{g} + \frac{1}{\rho}\mathbf{J} \times \mathbf{B} \quad . \tag{9.6.1}$$

The value of c in the gravitational term depends on whether thermal or compositional buoyancy is used (the pressure term P includes the centrifugal force; see Chapter 7). Sometimes the curl is taken of Eq. (9.6.1) to yield a "diagnostic" equation:

$$\frac{\partial \mathbf{v}}{\partial z} = \frac{\mathbf{J} \times \mathbf{B}}{2\Omega\rho} \quad . \tag{9.6.2}$$

This equation does not provide a unique solution for $\mathbf{v}$, since any velocity field that depends on spatial coordinates other than z can be added to the final solution and the equation is still satisfied. A solution must of course also be a solution to the magnetic induction equation:

$$\frac{\partial \mathbf{B}}{\partial t} = k_m \nabla^2 \mathbf{B} + \nabla \times (\mathbf{v} \times \mathbf{B}) \quad . \tag{9.6.3}$$

It turns out that there are many situations (to be specified more precisely in a moment) for which solutions to (9.6.1) and (9.6.3) (or (9.6.2) and (9.6.3)) do not exist, and when they do exist, the solutions are nonunique (e.g., Roberts, 1987).

The Taylor constraint can be illustrated using Fig. 9.5, which shows a cylinder inside a rotating sphere containing electrically conducting fluid. Take the ϕ-component of (9.6.1) and integrate over the curved surface S of the cylinder. The axially symmetric part of ∇P has no ϕ-component and the asymmetric part makes a zero contribution to the integral. Similarly, because $\mathbf{g}$ lies in meridional planes, $c\mathbf{g}$ makes no contribution. This then leaves

$$2\Omega\rho \int_S \mathbf{v} \cdot \hat{\mathbf{n}} \, dS = \int_S (\mathbf{J} \times \mathbf{B})_\phi \, dS \quad , \tag{9.6.4}$$

where $\hat{\mathbf{n}}$ is the unit normal to the cylindrical surface S. For an incompressible fluid the continuity equation requires $\nabla \cdot \mathbf{v} = 0$. By applying the divergence

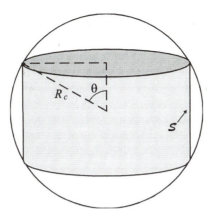

Fig. 9.5. Cylindrical surface S within the spherical core. Conservation of mass requires that there is no net flow of fluid across S. This leads to Taylor's constraint if the Coriolis and Lorentz forces dominate.

theorem, the left side of 9.6.4 vanishes. This then yields the famous Taylor constraint

$$\int_{S} (\mathbf{J} \times \mathbf{B})_{\phi} \, dS = 0 \quad , \tag{9.6.5}$$

which states that the net magnetic torque on arbitrary cylindrical surfaces aligned along the z axis must vanish.

Taylor-state dynamos, which have played a central role in many studies, involve solving (9.6.2) with the Taylor constraint, (9.6.5), simultaneously with equation (9.6.3). As we have seen, the strengths of the α- and ω-effects can be measured respectively by the magnetic Reynolds numbers R_{α} and R_{ω} (§8.3). Their product, D, is the dynamo number, which is the appropriate control parameter for $\alpha\omega$-dynamos. The control parameter is $(R_{\alpha})^{2}$ for α^{2} dynamos and $(R_{\alpha})^{2}R_{\omega}$ for $\alpha^{2}\omega$-dynamos. However, it is convenient in the following to refer to a single control parameter and D will be used to represent that parameter. Consider convection in a rotating body in which the magnetic field is weak. One expects there to be a geostrophic balance with Ekman boundary layers (§7.6), referred to as an Ekman state. As D increases, the effects of the magnetic field increase and it is found that α^{2}-dynamos almost always evolve from an Ekman state to a Taylor state (e.g., Hollerbach and Ierley, 1991; Barenghi and Jones, 1991; Barenghi, 1992, 1993; Hollerbach *et al.*, 1992; Fearn, 1994). This is not necessarily true for $\alpha\omega$- or $\alpha^{2}\omega$-dynamos. Part of the reason for this is that $\alpha^{2}\omega$-dynamos show a more complicated evolution (bifurcations of solutions) than α^{2}-dynamos. In general, the larger the ω-effect in generating toroidal magnetic fields, the more difficult it is to find a Taylor state (Fearn, 1994).

Dynamos can exist in states other than Ekman and Taylor states. For example, the Taylor state is not satisfied if viscous effects associated with boundary layers are important. Suppose that there are Ekman layers (see Fig. 7.5) near the top and bottom of the cylinder shown in Fig. 9.5 such that there is a net flow of fluid into the ends of the cylinder. To conserve mass this must be balanced by a net flow through the curved surface and the Taylor constraint is not satisfied, since the left side of (9.6.4) is not zero. In particular, suppose that the cylinder is in the Earth's core and it does not intersect the mantle or the inner core, and assume that the Ekman flow into the top and bottom is the same. Then the Taylor constraint should be replaced by

$$\int_s (\mathbf{J} \times \mathbf{B})_\phi \, d\mathbf{S} = 2\pi R_c (\nu\Omega)^{\frac{1}{2}} \frac{v_\phi \sin\theta}{|\cos\theta|^{\frac{1}{2}}} \ . \tag{9.6.6}$$

This constraint is central to the Model-Z dynamos of Braginsky (1975, 1976, 1978, 1994).

Roberts (1989b) has defined three different regimes based on the thermal wind strength, the size of the α-effect, and the strength of core–mantle coupling. Because the strength of core–mantle coupling may vary with time (e.g., Merrill and McFadden 1990; McFadden *et al.* 1991), it is possible that the geodynamo changes over time between these regimes: Model-Z with high core–mantle coupling (relative to meridional circulation strength), an intermediate regime, and a Taylor-state regime (low core–mantle coupling).

At this stage in dynamo development it is not clear whether in the first-order approximation one should use a Taylor-state approach, an approach in which viscosity and core–mantle coupling are important, or some other approach. It appears desirable to develop a variety of dynamos in the hope that some feature emerges that characterizes the differences between them, and which is testable using observational data. Core–mantle coupling is further discussed in Chapter 11.

9.6.2 Weak-Field Hydromagnetic Models

Busse (1970, 1975, 1977a,b, 1978) constructed a weak-field hydromagnetic dynamo model for the Earth and planets that stimulated considerable discussion in geomagnetism. A major assumption in these studies is that the outer core is essentially in geostrophic balance. This means that the Coriolis force is balanced against the pressure force. All other forces, such as buoyancy, inertial, friction, and Lorentz forces, are assumed to do nothing other than perturb the basic geostrophic flow (a critical set of assumptions, any one of which could be wrong). A consequence of these assumptions is that the poloidal and toroidal magnetic fields in the core usually turn out to be of similar magnitude.

To achieve geostrophic balance, the Coriolis force is equated to the pressure gradient,

$$2\Omega \times \mathbf{v} = -\frac{1}{\rho} \nabla P \quad , \tag{9.6.7}$$

where Ω is the angular velocity of rotation, P is pressure, and ρ is density. A right-handed coordinate system (x,y,z) is used, in which the z axis coincides with the rotation axis. Taking the curl of (9.6.7) yields

$$2\rho \frac{\partial \mathbf{v}}{\partial z} = \mathbf{0} \quad , \tag{9.6.8}$$

which shows that the flow is independent of z. This is a manifestation of the well-known Taylor–Proudman theorem of fluid mechanics, which states that the flow in the above situation is a steady one about the z axis. Only if friction is inserted into the problem is there any z dependence of the flow. The importance of friction can be described by an Ekman number, E, given by

$$E = \frac{\nu}{\Omega L^2} \quad , \tag{9.6.9}$$

where ν is viscosity and L is a typical length scale in the system in the z direction. E is assumed to be small (remember that ν is one of the most poorly determined quantities in the core), allowing Busse to model convection in the outer core by a cylindrical annulus. Convection in this case will thus be columnar with the column axes parallel to z (Fig. 9.6). Interestingly, these columns are not stationary, but drift about the rotation axis with time. This indicates how secular variation naturally falls out of some hydromagnetic dynamo solutions.

Using a no-slip velocity boundary condition ($\mathbf{v} = \mathbf{0}$ at the boundary), an Ekman boundary layer will occur with thickness L given in (9.6.9), so that the thickness is proportional to $E^{1/2}$. In the boundary layer, the flow will be radially outward from the column's center in the northern hemisphere for those columns having counterclockwise rotation (looking down, along the -z direction) and radially inward for columns with clockwise rotation. This produces a secondary "Ekman" flow, the magnitude of which is crucial to the dynamo mechanism. Although the magnitude of the secondary flow can be adjusted through assumed changes in the value of E in (9.6.9), E cannot be taken too large, for then the cylindrical annulus approximation could not be used. This secondary Ekman flow, produced both by the shape of the spherical boundary and by friction, together with the primary geostrophic flow, produces a net helical flow. Periodic

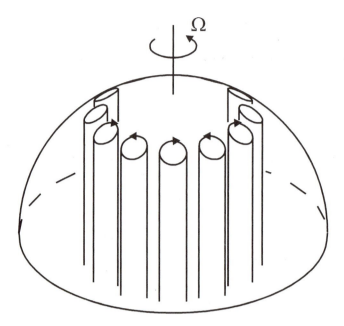

Fig. 9.6. Qualitative sketch of the onset of convection in a rotating fluid sphere following the experiments and results of Busse (1975). Convection occurs in columns parallel to the rotation axis.

helical velocity fields of sufficient magnitude and distribution can lead to viable dynamo mechanisms, as illustrated by Busse with computer calculations.

An important aspect of these models involves the convection rolls and their associated magnetic fields. These magnetic fields will be primarily toroidal ones (as can be pictured by assuming frozen flux), except near the boundaries. There the *Ekman suction* causes a poloidal field, which in principle is observable at the Earth's surface. *Ekman suction* refers to the secondary radial flow that is sucked into (or ejected from) the boundary layer from (or into) the main body of core fluid.

Indeed, it is the boundary regions near convection columns that have been invoked by Bloxham and Gubbins (1985, 1986) to explain certain observations in the historical field data. They associate two stationary pairs of magnetic foci with convection rolls in the core. Members of each pair occur in opposite hemispheres (north–south) and are approximately 120° apart. The columns that are assumed to connect members of each pair are approximately tangent to the Earth's inner core. Bloxham and Gubbins note that a third pair should be present for symmetry, but they speculate that this pair is not observed because of additional structure in the core convection. The four foci involved are stationary and so, if they are manifestations of core convection rolls, the rolls must also be stationary, unlike those of Busse. Bloxham and Gubbins (1986) argue that the rolls are pinned at the core–mantle boundary, perhaps by topography there. This subject is discussed further in §9.6.3.

One of the major departures made by Busse in the above analysis is that the core is essentially in *geostrophic balance,* implying a balance between the pressure gradient and the Coriolis force. On the other hand, if the pressure gradient is balanced by the Coriolis and magnetic forces, i.e.,

$$2\rho\Omega \times \mathbf{v} + \mathbf{J} \times \mathbf{B} = -\nabla P \quad , \tag{9.6.10}$$

then the fluid is in *magnetostrophic balance.* The consequences of assuming geostrophic or magnetostrophic balance can be severe. This is seen in the following order-of-magnitude calculation in which the Coriolis force is balanced by the magnetic (Lorentz) force (*extreme magnetostrophic balance*). The Lorentz force is

$$|\mathbf{J} \times \mathbf{B}| \approx \frac{\mathbf{B}^2}{\mu_0 l} \quad , \tag{9.6.11}$$

where l is taken as the dimension of the current loop causing the toroidal field, and the Coriolis force is $2\rho\Omega v$. Taking $l = 2\pi R_c$, where R_c is the core's radius, and v as the westward drift velocity (4×10^{-4} m s^{-1}), a value for B (toroidal field) near 130×10^{-3} T is obtained. This value is close to that obtained in Bullard–Gellman-type kinematic models (§9.2).

It is more difficult to obtain an estimate for the geostrophic balance case, but clearly the Lorentz force must (by assumption) be much smaller than that obtained above. For a very rough estimate, consider the Navier–Stokes equation (8.1.14) for an inviscid fluid. A geostrophic balance implies that the pressure term can be canceled by the Coriolis term, leaving

$$\rho\left[\frac{\partial}{\partial t} + \mathbf{v}\cdot\nabla\right]\mathbf{v} = \mathbf{J} \times \mathbf{B} - \rho\nabla\phi_g \quad . \tag{9.6.12}$$

Ignoring the gravity term and assuming steady-state conditions gives

$$|\mathbf{J} \times \mathbf{B}| \approx |\rho(\mathbf{v}\cdot\nabla)\mathbf{v}| \approx \frac{\pi v^2}{l} \quad . \tag{9.6.13}$$

Using $B^2/\mu_0 l$ as an estimate for the left side, a value for B (toroidal field) between 10^{-4} and 10^{-5} T is obtained. Thus the estimated toroidal magnetic field is two or three orders of magnitude smaller than obtained for the (extreme) magnetostrophic balance case.

One of the consequences of Busse's work is that it appears that the magnitude of the toroidal field is strongly model dependent and not well known. Also, if Busse's weak-field model is right, then the magnitude of the toroidal field is about the same as that of the poloidal field. A second important finding is that no large net velocity shear in the outer core is apparent in his convection models. The consequence of this, if confirmed, is that the Bullard mechanism for westward drift is untenable (§9.1.2).

Olson (1977) has applied a "flux-tube method" to carry out numerical estimates for the toroidal and poloidal fields. He finds that the secondary Ekman flow in Busse's model appears unable to produce a sufficient amount of helicity, a problem that might be solved by consideration of Lorentz force effects. Another objection to Busse's approach has been raised by Soward (1979), who argues that the Busse weak-field dynamos become unstable for reasonable-size magnetic fields. The controversy over balance of terms (forces in the Navier–Stokes equation) is far from resolved, but in more recent analyses it has become clear that the balance of forces can evolve in time (e.g., Fearn, 1994).

9.6.3 Strong-Field Models

If the velocity field is not affected by **B**, then the magnetic induction equation (9.6.3) is linear in **B**. In this case **B** is proportional to $\exp(p_0 t)$ as t approaches infinity, where p_0 is the real part of p, the largest positive eigenvalue for which solutions of the magnetic induction equation exist. Hence $|\mathbf{B}| \to \infty$ as $t \to \infty$, an unphysical result. McFadden *et al.* (1985) point out that this probably does not happen because the Maxwell stress affects the velocity field. Moreover, they point out this has strong support from the histogram of paleointensity data, which shows that the vast majority of absolute paleointensity values for the past five million years group around the present field intensity within a factor of two (Chapter 6).

A rough estimate of the effect the magnetic field has on the velocity field can be obtained by using the Walen Criterion (e.g., Weiss, 1991). This criterion was originally obtained by considering the buoyancy force, $g\rho\alpha\beta_s z$ (see §7.6), on a fluid element displaced upward by a distance z in an unstably stratified fluid having a superadiabatic temperature gradient β_s. There is a restoring force of magnitude $B^2 z/(\mu_0 L^2)$ associated with the Maxwell stress, B^2/μ_0, for curved magnetic field lines with semi-wavelength L. Therefore Walen (see Weiss, 1991) concludes that magnetic field would suppress convection if

$$g\alpha\beta_s < \frac{B^2}{\mu_0 \rho L^2} \ . \tag{9.6.14}$$

This illustrates that the magnetic field can have a strong effect on the velocity field. For example, in the absence of a magnetic field, strong thermal winds can be generated near the uppermost part of the core for lateral variations in temperatures on the order of only 10^{-6} of a degree (see Chapter 7). However, the present poloidal magnetic field by itself can suppress thermal winds for temperature differences on the order of one degree over several hundred kilometres.

The implication in a *strong-field dynamo* is that the Lorentz force has a significant effect on the velocity field, as is the case in the magnetogeostrophic equation (9.6.1). Usually one defines strong-field dynamos as ones in which the Elsasser number, Λ_E, is order one or larger (see §8.2). This implies that the Lorentz force is comparable to, or larger than, the Coriolis force. Examples of strong-field hydromagnetic dynamos are Taylor-state dynamos (those that obey the Taylor constraint) and Model-Z dynamos (§9.6.1). In addition, Zhang and Busse (1989) have extended the models of §9.6.2 into the strong-field regime.

In weak-field geodynamo models the velocity field is expected to be axially symmetric because of the Taylor–Proudman theorem (§9.6.2). The question naturally arises as to what happens to velocity and symmetries in strong-field models. This is of concern since the Coriolis force, the pressure gradient, and the Lorentz force are usually not colinear.

It is known that even though Cowling's theorem shows a magnetic field produced by positive dynamo action cannot be symmetric about any axis (e.g., the rotation axis), the velocity field can have such symmetry (§9.3). Intuitively one would expect that in strong-field dynamos the Lorentz force feedback into the velocity field would lead to asymmetry in the latter as well. Consequently, we conjecture that in strong-field dynamo models neither the magnetic field nor the velocity field can be symmetric about an axis. If this conjecture is proved, this might eliminate a large class of dynamo models that begin by assuming a velocity field that is symmetric about the rotation axis (but see below).

A problem with applying Cowling's theorem is that one does not know how much asymmetry is required for the dynamo to work. In particular, the present geocentric dipole field is tilted only 11° with respect to the rotation axis (Chapter 2) and large tilts throughout the past 600 million years seem rare (Chapter 6). This suggests that one might still be able to investigate dynamos by assuming velocity fields that are nearly axially symmetric, even if the conjecture made in the previous paragraph is proved. Perhaps a more crucial question involves whether the foci identified by Bloxham and Gubbins (1986) reflect convection rolls. Analyses by Zhang and Busse (1987) shed some light on this question.

Zhang and Busse (1989) extend the pioneering studies of Busse (§9.6.2) into the strong-field regime. Their investigations span a wide range of parameter space in which the Rayleigh number, the Taylor number, the Prandtl number and the magnetic Prandtl number are varied. Nevertheless, the Taylor number they use (inverse square of the Ekman number; see Chapter 7), which gives the relative importance of rotation to viscous forces, is at least 20 orders of magnitude less than that for the Earth. This suggests that some caution is required in extrapolating from their numerical dynamo models to the actual

Earth. However, they commonly find that magnetic forces lead to marked departures from the two-dimensional rolls manifested in the weak-field dynamos.

What does appear to be emerging from a variety of models is that part of the Earth's outer core is in a magnetogeostrophic state in which the magnitudes of the Lorentz and Coriolis forces are approximately equal (see §9.6.4 for elaboration). If the Lorentz force becomes too large this leads to changes in convection and the degree of positive dynamo action is reduced. In turn, this reduces the magnitude of the Lorentz force. On the other hand, if the Coriolis force is strongly dominant, then rapid growth in the magnetic field occurs relatively unimpeded until a magnetostrophic-type balance is again achieved.

Zhang and Busse (1989) also investigated whether dipole or quadrupole family symmetry was preferred, primarily by examining the axial dipole and quadrupole field stabilities. Dipole symmetry was found to be preferred in some parameter spaces while quadrupole symmetry was preferred in others. In general, they found that the strength of differential rotation is reduced by the Lorentz force. Perhaps their study will yield insight into why the dipole family appears to have dominated the quadrupole family for the Earth for the vast majority of the past 400 million years or so (Chapter 11).

Ignoring details, there does appear to be a growing consensus that the geodynamo is a strong-field dynamo (e.g., Roberts, 1988). If one assumes a magnetostrophic balance in which the magnitude of the Coriolis force ($\rho 2\Omega v$) is approximately the same as that of the Lorentz force ($B^2/\mu_0 l$), and assuming l to be around 2×10^3 km, one can use the westward drift velocity (4×10^{-4} m s^{-1}) to obtain an estimate for the toroidal magnetic field around 3×10^{-2} T. This estimate is roughly four times the dipole field downward-continued to the core–mantle boundary, and is likely the best estimate now available. Even so, given the assumptions required to make this estimate, it should be regarded as highly uncertain.

9.6.4 The Role of the Inner Core

The role of the inner core has only recently received much interest from dynamo theoreticians, an interest in part having been stimulated by the work of Bloxham and Gubbins (1985, 1986). To understand what effects the inner-core may have on the geodynamo, it is useful to define the tangential inner core cylinder (or for brevity, the tangential cylinder) as the region in the outer core contained within a cylinder that is parallel to the rotation axis and tangent to the inner-core boundary. The convection rolls of Busse (e.g., 1975; also see §9.2) do not occur in the tangential cylinder region, but they could be tangent to the inner core, a point emphasized by Bloxham and Gubbins. In particular, they point out that such tangential rolls might explain four magnetic foci that have been stationary over the past 30 yr. Individual members of these foci pairs occur opposite each other in the northern and southern hemispheres and can be connected with cylindrical columns drawn parallel to the rotation axis and tangent to the inner core. They reasonably posit that variations in the core–mantle boundary conditions, e.g., topography, might impede the movement of these convection roles, which would otherwise drift westward or eastward, depending on the circumstances (see §9.2). If all of this is true, then the observational data are providing important information on convection deep within the outer core; moreover, the inner core is seen as having an important affect on this convection. A problem with these conjectures is that the convection rolls found by Busse are for a weak-field model; they require a strong energy source and they do not appear to be stable when there is a moderately strong magnetic field present (e.g., Zhang and Jones, 1994).

Hollerbach and Jones (1993a) use an $\alpha\omega$-dynamo model to predict that the stability of the geodynamo changes with the growth of the inner core. They showed for their kinematic model that axisymmetric poloidal magnetic fields were expelled from the tangential cylinder. Diffusion of the toroidal magnetic field through the inner core also plays a central role in the Hollerbach and Jones model. Although reversals were not observed in their model, they argue that a larger inner core would stabilize the field such that reversals would occur less frequently. In contrast, the influence of the inner core on the magnetic field was found to be small for Model-Z-type dynamos (Braginsky, 1989).

Similar effects to those seen by Hollerbach and Jones are emerging from hydromagnetic dynamos (Zhang and Jones, 1994; Olson and Glatzmaier, 1995; Glatzmaier and Roberts, 1995a,b) lending support to the notion that the inner core might have a pronounced effect on outer-core dynamics. Figure 9.7 shows the convection patterns obtained by Glatzmaier and Roberts (1995a) for the first three-dimensional, time-dependent, self-consistent numerical solutions for convection and magnetic field generation in a rapidly rotating spherical MHD fluid shell with a solid conducting inner core.

The Glatzmaier–Roberts (1995a,b) dynamo is one of the most impressive model dynamos produced, and has exhibited full reversal. It is, therefore, worth describing some of its results further. The longitudinally averaged poloidal magnetic field in the outer part of the outer core typically exhibited polarity opposite to that closer to the Earth's center. The helicity of this field in the northern hemisphere is predominantly right-handed within the tangential cylinder and left-handed outside. There are two regions of toroidal magnetic field concentration, both of which lie predominantly within the tangential cylinder and which generally exhibit antisymmetry with respect to the equator. Interactions between these fields (and their corresponding velocity fields) can produce very complicated behavior, including excursions and reversals. In particular, it is found that the outer core's poloidal field frequently tries to

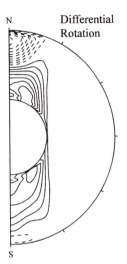
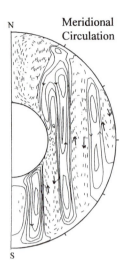

Fig. 9.7. A snapshot of the longitudinally averaged differential rotation and meridional circulation for the Glatzmaier and Roberts (1995a) model, plotted in a meridian plane showing the inner and outer cores. For the streamlines of the meridional circulation, solid (broken) contours represent counterclockwise (clockwise) directed flow. The maximum meridional flow is 10^{-3} m s^{-1}.

reverse (excursions) but is unsuccessful because diffusion of the magnetic field out of the inner core (which requires more time than the typical interval between reversal attempts) stabilizes the outer core polarity. Only one reversal was observed during the simulation. During this reversal the intensity fell to about a quarter of its usual value and the toroidal magnetic field became symmetric with respect to the equator midway through the transition. The latter occurred because the toroidal field reversed in one hemisphere before the other. It was also observed that the toroidal and poloidal magnetic fields did not reverse at the same time.

As with all dynamo models, assumptions were required in the Glatzmaier–Roberts dynamo. Perhaps the most serious of these is the assumption that a layer, similar in size and location to D" in the Earth, has very high (metallic) electrical conductivity. Such a layer can, in particular, have a dramatic effect on toroidal magnetic field generation; it is doubtful that a layer with these properties is actually present within the Earth (§7.2 and §7.3). Other assumptions include: convection driven by a specified heat flux at the inner–outer core boundary; rotation that is only about 0.002 that for the Earth; and viscosity that is nearly an order of magnitude too high. Finally, the integration time corresponds to 40,000 yr, which is at most only a small multiple of the free-decay time of the core (depending on assumptions made for the electrical conductivity of the core; see Table 9.1).

Because of such assumptions many of the details of the Glatzmaier–Roberts dynamo may not stand the test of time. An additional assumption that may cause concern is that the viscosity ν is given by $\nu=\nu_0(1+0.075l^3)$, where l is the spherical harmonic degree of the variable being acted on. This is referred to as a hyperdiffusivity and its effect is to damp out strongly small-scale features relative to large-scale ones. Nevertheless, this impressive model is one of the first that might be useful in the interpretation of paleomagnetic data. For example, some of the symmetries imparted by the inner core, such as geographical location of the toroidal and poloidal convection shear shown in Fig. 9.7, are likely robust. We make the latter conjecture based on the fact that there are remarkable similarities in the flow patterns for very different models (e.g., Hollerbach and Jones, 1993a; Zhang and Jones, 1994; Olson and Glatzmaier, 1995; Glatzmaier and Roberts, 1995a,b) and because some aspects of the flow are physically intuitive. The appropriate conditions for the fluid flow at both boundaries of the outer core are the "no-slip" boundary conditions (see §7.6) and these were used in the above models. If one assumes, reasonably, that the largest rotational shear will on the average be symmetric about the rotational axis, then this shear is expected to be confined primarily to the tangential cylinder. A consequence is that toroidal magnetic fields with rotational symmetry will also be preferentially produced within the tangential cylinder (by the ω-effect, or by some similar process). In this intuitive picture, upwellings and downwellings will occur outside the tangential cylinder where they are not resisted by a strong toroidal magnetic field. Of course, such an intuitive picture (which neglects such processes as diffusion and nonlinear coupling between the velocity and magnetic fields) needs quantitative support and this is precisely what appears to be emerging. For instance, consider the numerical calculations of Olson and Glatzmaier (1995), who impose a toroidal magnetic field to determine the structure of thermal convection in a three-dimensional rotating spherical shell of conducting fluid. They find that in the strong-field regime (large Elsasser number) the flow is essentially magnetostrophic (i.e., the Lorentz force plays a crucial role in the force balance) within the tangential cylinder while it is essentially geostrophic (i.e., the Lorentz force plays a negligible role) outside. In their model this leads to substantially different convection structures within and without the tangential cylinder; in particular, columnar-like geostrophic upwellings induce localized patches of magnetic flux on the core–mantle boundary outside the tangential cylinder, but not within it. They argue that the dipole part of the poloidal field comes from the spatial averaging of the fields within these patches. These results are generally consistent with those of the Glatzmaier–Roberts dynamo.

The likely possibility that the inner core affects the convection and magnetic field pattern is important to the observational scientist. Not only is the size of the inner core increasing with time, but so is the ratio of chemical to thermal buoyancy (see §7.4). If so, one would surmise such changes would be manifested in a variety of paleomagnetic data, including long-term drift in the intensity, in the rate of reversals, and in secular variation. As seen in Chapters 4, 5, and 6, the highest quality paleomagnetic data are mostly confined to the past few hundred million years; these data do show evidence of long-term changes, but these observed changes are probably associated with variations in the CMB boundary conditions. For example, the reversal rate (§5.3) has both increased and decreased dramatically on a time scale of about 200 million years, and it seems unlikely that both the increase and the decrease could be explained by monotonic growth of the inner core. Also, the time scale of these variations is that appropriate to the time scale expected for mantle convection. Any change produced by the small amount of growth of the inner core during this same time is likely to have been completely swamped by effects associated with changes at the CMB (§11.3). Unfortunately, a substantial amount of better quality data representing a much larger slice of geological time seems required to evaluate whether there are any very long-term trends in the properties of the Earth's magnetic field, and this will be difficult to obtain. As a point of interest, if the inner-core does affect the convection and magnetic field pattern, then one would expect the secular variation at high latitudes to have a character different from that at low latitudes, a possibility not yet adequately explored.

Chapter Ten

The Magnetic Fields of the Sun, Moon, and Planets

10.1 Origin of the Solar System

There are many observations that any comprehensive theory for the origin of the solar system must explain. Seven of the most obvious are listed in Table 10.1. No theory for the origin of the solar system has as yet won general acceptance, for no existing theory adequately explains even a majority of the observations. Many models have of course been proposed, most to be discarded later, and on occasion some have been resurrected in a modified form. A history of the subject has been provided by Brush (1990) and an overview of the science has been provided by Taylor (1992).

It is useful to mention here some aspects of the models. All models have the solar system forming around 4.6 Ga, but it is not clear whether the Sun formed earlier than, synchronous with, or later than, the planets. The chronology of the planetary accretion process is not well constrained, and separate models have been advanced to explain the accretion of the inner and of the outer planets. Nevertheless there does appear to be consensus that the Sun and the planets formed from a rotating flattened disc of gas, dust, and planetesimals. Models as to how the planetesimals themselves formed from an earlier rotating disc of gas and dust are highly speculative, but presumably the earliest stages involved clumping of material. Eventually a hierarchy of sizes, ranging from meteors to planetesimals with dimensions of hundreds to thousands of kilometres, was formed prior to the final assembly of the terrestrial planets. For example, Wetherhill (1985) developed an often-cited computer model that produced the four terrestrial (inner) planets (Mercury, Venus, Earth, and Mars) by starting with 500 planetesimals – each with mass approximately equal to one-third that of the Moon. Direct evidence for the existence of planetesimals can be seen in the history of the moon (which clearly shows an early bombardment by planetesimals; §10.3), and indirect evidence comes from the inferred cratering histories of other planets such as Mercury and Mars. Moreover, the slow rotation of Venus and the large tilt of the rotation axis of Uranus can most easily be explained if these planets were impacted by large planetesimals. In addition, one of the most popular theories for the origin of the Moon involves an impact of a giant planetesimal (possibly of the size of Mars) with the Earth (Benz and Cameron, 1990). As an important aside, the composition of the Earth's mantle and the Moon are remarkably similar, suggesting that the Earth's core had already separated out before this collision occurred. This separation probably occurred within the first 100 million years after the Earth's accretion, since there are lunar samples with radiometric ages of 4.5 Ga. Hence it is possible that the Earth had a magnetic field as early as 4.5 Ga.

Geochemical evidence appears sufficient now to rule out the once popular notion that the primitive solar nebula from which the solar system condensed was homogeneous. The homogeneous accretion model was very attractive because it was both parsimonious and explained observations such as the chemical similarities of the Sun, Moon, and Earth with certain meteorites (see observation (v) Table 10.1). There is now abundant evidence indicating that this model is inadequate. Instead it appears that planetesimals had significantly different compositions and, in addition, that some of them underwent early stages of fractionation (e.g., Taylor, 1992). The first indication that a model of perfectly homogeneous accretion might not be appropriate came from oxygen isotope analysis (Clayton *et al.*, 1973; Clayton and Mayeda, 1975, 1978). Samples that are related to one another by fractionation processes depending only on the masses of the isotopes have predictable $^{17}O/^{16}O$ trends that will be seen on plots of $^{17}O/^{16}O$ versus $^{18}O/^{16}O$. In addition, variations in composition produced by mixing of phases will fall along a straight line between the initial unmixed phases. One can, therefore, determine from such plots an isotope "fingerprint" that remains with the material regardless of how much chemical processing or mass fractionation occurs. The origin of the differences in meteorite fingerprints, particularly the famous Allende meteorite (which exhibited a very distinct oxygen isotope fingerprint), is sometimes interpreted as indicating input from some exotic source, such as a supernova (e.g., Papanastassiou and Wasserburg, 1978). Inhomogeneity of the original solar nebula is an obvious alternative, and other meteorite anomalies have convinced most investigators that there was some inhomogeneity in the primitive solar nebula (e.g., Ott, 1993).

It is now widely believed that the Sun went through FU Orionis and T-Tauri phases of evolution (e.g., Taylor, 1992). The T-Tauri phase is named after a class of stars known as T-Tauri, the best known member being a star that lies in the constellation of Taurus, the bull. Stars in this class are pre-main sequence objects and are believed to be very young on the solar system time scale ($\sim 10^6$ yr; e.g., Soderblom and Baliunas (1988)). T-Tauri stars are unusually bright and exhibit strong magnetic fields that are typically several orders of magnitude larger than that of the present Sun. An FU Orionis stage of a T-Tauri star is partially characterized by violent outbursts of material and an increased luminosity of about six magnitudes. This is thought to have produced a "snow line" at 4 to 5 AU (Astronomical Unit, equal to the mean distance of the Earth from the Sun, 150 million kilometres); it is argued that the volatiles, including water, were ejected from the region within the "snow line" by the early violent star activity. If confirmed, this would help to explain some of the compositional differences between the inner and the outer planets. Whereas all the planets have a solid core, shells of H_2O, CH_4 and NH_3 ices surround the cores of Saturn, Uranus and Neptune, and the outer shell of all the Jovian planets (Jupiter, Saturn, Uranus, Neptune) consists primarily of liquid H_2–He (see Fig. 10.1).

Observation (i) in Table 10.1 is of particular importance to theories of planetary magnetism because one proposed mechanism for angular momentum transfer from the primitive Sun to the planets involves magnetic braking (e.g., Alfvén, 1954; Alfvén and Arrhenius, 1975). During the T-Tauri phase of evolution of the Sun its strong magnetic field (which rotates with the Sun) might have transferred angular momentum from the Sun to distant ionized particles or any material that was a good electrical conductor (Alfvén, 1942a, 1954). The movement of a magnetic field past a good electrical conductor was considered in Chapter 8, and the (Maxwell) stress results in acceleration of the conductor while imposing a braking effect on the field (and hence on the Sun in this analogy). However, doubts about there being sufficient ionization, modeling difficulties, and uncertainties about the magnitude of the early solar magnetic field persist. Currently a turbulent viscosity model for transferring angular momentum seems preferred (e.g., Cameron, 1985). Nevertheless, given the capacity for oscillations in models for the origin of the solar system (e.g., Brush, 1990), and the simplicity with which both the turbulent viscosity model and the magnetic braking model have been treated, it is premature to discount either model. One useful input from observational planetary magnetism would be an estimate of the strength of the primitive solar field, and this is discussed further in §10.3 and §10.5.

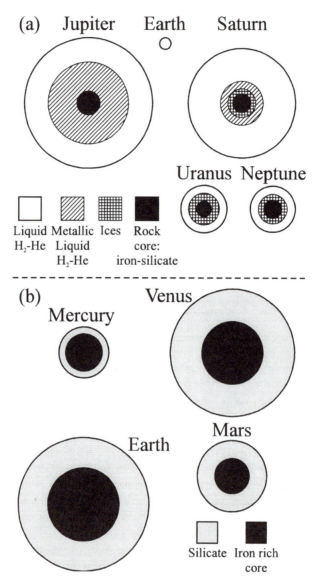

Fig. 10.1. Speculative models for the planetary interiors. (a) Models of the interiors of the Jovian planets. The Earth is shown for scale. All have a solid metallic core. Shells of H_2O, CH_4, and NH_3 ices surround the cores of Saturn, Uranus, and Neptune. The outer shell of all the Jovian planets consists primarily of liquid H_2. Jupiter and Saturn are also thought to contain shells of liquid H (after Podolak and Cameron (1974) and Connerney (1993)). (b) Models of the interiors of the terrestrial planets. All are rocky planets divided into a rocky crust (too thin to show), silicate mantle, and metallic core.

TABLE 10.1
Observations to Be Explained by Theories for the Origin of the Solar System

(i)	The Sun contains only a tiny fraction of the solar system's angular momentum; the planets, which account for only about 1/750 of the solar system mass, account for 99.5% of the angular momentum.
(ii)	There are nine planets and an asteroid belt (between Mars and Jupiter); their mean distances from the Sun follow an approximate numerical pattern known as the Titius–Bode law.
(iii)	The planets all move in the same direction about the Sun with approximately circular orbits that are nearly coplanar. The rotation of seven of the planets is prograde (i.e, their rotation is in the same sense as their orbit about the sun). Venus has a slow retrograde rotation and Uranus' rotation axis is nearly perpendicular to its orbital axis.
(iv)	There are two groups of planets separated by the asteroid belt. The inner, or terrestrial, planets consist predominantly of relatively high-density rock, and the outer, or giant, planets (excluding Pluto) have a lower density and contain much more hydrogen and helium.
(v)	Geochemical signatures of the planets, asteroids, meteors, and comets need to be explained. For example, the abundances of elements in the Sun's photosphere (excluding H, N, and noble gases) are remarkably similar to those of the Earth and to those in type 1 carbonaceous chondrites (a class of meteorites) — evidence for ^{26}Al and various isotope ratios as meteorites (e.g., Allende) also require explanation.
(vi)	The Earth's Moon is the only satellite of the terrestrial planets and is very large.
(vii)	The outer planets have many satellites, the vast majority of which exhibit prograde orbital motion and numerical regularities in their distances from their planets. There are many other numerical regularities (or resonances): as two examples, there are no asteroids with periods $1/2$, $1/3$, . . ., $1/n$, the period of Jupiter, and the three inner Galilean satellites of Jupiter have periods in the ratio 4:2:1.

10.2 The Sun

10.2.1 General Properties

The Sun is a star that currently lies on the main sequences of the Hertzsprung–Russell (temperature–luminosity) diagram. It has a mass of about 2.0×10^{30} kg and a radius of about 7.0×10^5 km. Excellent general discussion of the Sun can be found in the books by Brandt and Hodge (1964), Akasofu and Chapman (1972), and Stix (1989). According to most theoretical models, the Sun's luminosity has increased roughly 30% since the its birth 4.6 Ga ago. Thermonuclear reactions produce thermal energy in the Sun's core that is radiated out to an outer conducting shell. This outer shell has a radial dimension that is believed to be roughly 27% of the Sun's radius, yet possesses only about 1% of its mass. These dimensions and the overall spherical structure of the Sun are supported by helioseismological results (normal mode data; the restoring force believed to be the Lorentz force; e.g., Christensen-Dalsgaard (1991)). Evidence of convection in the outer shell is seen in granulation (typical dimensions on the order of 1000 km) and supergranulation (typical scale near 30,000 km). Lifetimes of the convection cells associated with the granulation are on the order of minutes and with the supergranulation on the order of an Earth day (Gilman, 1992). Large giant convection cells with radial dimensions as large as the conducting shell, and with lifetimes on the order of a month or more, are also thought to exist, although observational evidence for this is weak (Gilman, 1992). There does not appear to be any significant axisymmetric meridional circulation (Gilman, 1974, 1977, 1992). The Sun's rotation, the axis of which makes a 7° angle with the normal to the ecliptic, exerts an important influence on the convection, which is clearly turbulent (Brummell *et al.*, 1995).

Estimates of the Sun's rotation, which varies with solar latitude and with time (e.g., Howard, 1975; Gilman, 1992; Eddy, 1988), have come from measurements of the Doppler shift of spectral lines, and observations of sunspots, filaments, corona streamers, etc. Equatorial rotation rates are near 25 Earth days while estimates for the polar rotation rates are usually between 35 and 40 days. The decrease in rotation rate poleward goes roughly as $\sin^2\lambda$, where λ is solar latitude. In the lower latitudes, where estimates are thought to be more reliable, the spectroscopic measurements indicate a rotation rate that is generally one day slower than that obtained from observations of sunspots, filaments, etc., features that are associated with magnetic field. Speculations on these differences include the possibility that the magnetic fields are connected to the Sun's interior regions; for example, through the frozen-in-field theorem (see §8.2). Dicke (1970) has produced evidence on the amount of the Sun's flattening, which he interprets to demonstrate that the Sun's core is rotating nearly five times as fast as the outer regions. However, it is not clear whether such a rapid rotation of the Sun's interior would be mechanically stable. Also, even if the Dicke model were initially true, the frozen-in-field effect would eventually remove this large velocity gradient. Alternatively, hydromagnetic wave mechanisms, similar to those discussed in §9.1, may lead to preferential "drift" of the solar field.

The Sun's rotation rate varies in time with periods sometimes as low as a couple of weeks or so, and with magnitudes on the order of 5 to 10% of the mean velocity (Howard, 1975; Gilman, 1992). There is some evidence that this variation may have been larger at times in the past (Eddy *et al.*, 1977; Eddy, 1988).

Some star types appear to possess fields from 10^2 to 10^8 times the general field of the Sun (e.g., Preston, 1967; Landstreet and Angel, 1974; Angel, 1975), and fields up to 10^{12} times the Sun's general field may exist in some neutron stars (Woltjer, 1975). Moffatt (1978) suggests that many of these fields originated during the formation of the stars through compression of a galactic field, estimated to be about 10^{-9} to 10^{-10} T. Moffatt shows that the free-decay times in such bodies may easily be on the order of the lifetime of the star (§7.4.3).

However, stars with oscillating magnetic fields that originate in their cores may have substantially shorter free-decay times.

10.2.2 Solar Magnetic Field

Information on the Sun's magnetic field comes from several sources. Direct observations are available from space missions that measure the interplanetary magnetic field (IMF) originating from the Sun. Probably the most valuable indirect measurements involve utilizing the Zeeman effect. Elemental spectral lines, such as those of hydrogen, are split in the presence of the solar magnetic field and the spacing between those lines is linearly proportional to the intensity of the magnetic field component that is directed along the line of observation. On average, solar plumes will erupt parallel to magnetic field lines, so the directions of their emission also provide indirect information on the solar magnetic field. The Sun's general magnetic field is also seen in the structure of the solar corona (seen as radial plumes of electrons trapped along magnetic field lines) at times of total eclipses. Magnetic fields of varying strengths are found in the photosphere.

The magnetic field of the Sun has a strong dipole component on the order of 10^{-4} T at the Sun's photosphere. This "general" field has within it a complicated fine structure of fields, some of which often exceed 10^{-2} T. To a first-order approximation the Sun appears to have a dipolar magnetic field at high latitudes, but at lower latitudes it appears to possess a more complicated structure (Wilcox and Ness, 1965; Rosenberg and Coleman, 1969; Gilman, 1992; Wang *et al.*, 1996). In the equatorial regions the magnetic field sometimes appears to enter the Sun in two sectors and to emerge from two other sectors. Because of the Sun's rotation the nodal surfaces that separate the four apparent sectors of opposite polarity are twisted in a spiral-like fashion away from the Sun (shown in Fig. 10.2 for historical interest). The sectoral pattern for the Sun was thought to extend to solar latitudes sometimes exceeding 50° at the Sun's surface and to become restricted to a relatively narrow zone of a few degrees about the equator, at a distance of two solar radii from the Sun's surface (Schatten, 1971; Smith *et al.*, 1978). The concept of magnetic sectors is, however, rarely used in modern interpretations, the interpretations usually being made in terms of phenomena such as warped current sheets and corona holes, concepts that will be discussed shortly.

The temperatures at the photosphere are around 6000 K while in the outer corona they approach 10^6 K. Parker (1958) pioneered work showing that the heating of the corona gas is sufficient to overcome the Sun's gravitational attraction and to produce supersonic outward transport (expansion) of mainly H^+ and $^4He^{2+}$, although some heavier elements are also present. The supersonic speeds are reached at a distance of roughly two or three solar radii, beyond

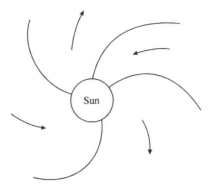

Fig. 10.2. Solar magnetic field sectors. Magnetic field entering and emerging from the Sun in an equatorial cross section sometimes appears to exhibit four sectors of alternating polarity separated by nodal surfaces. This sectoral division was thought to extend up to midlatitudes, above which the field becomes more dipolar.

which the rapidly outward moving plasma is referred to as the *solar wind* (see §2.6). This partially electrically conducting solar wind carries the Sun's magnetic field with it to distances that reach beyond Pluto (see the frozen-in-field theorem in §8.2). However, the work of Alfvén and Arrhenius (1976) suggests that strict application of the frozen-in-field theorem to the solar wind can lead to misunderstandings in some situations.

Figure 10.3 shows the interplanetary magnetic field, IMF, in a plane taken perpendicular to the plane of the ecliptic (Smith *et al.*, 1978). The field is presently outward in the northern hemisphere and inward in the southern hemisphere. There is an electric current sheet associated with the magnetic field

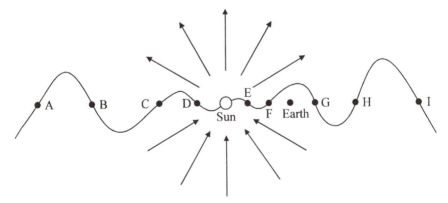

Fig. 10.3. The interplanetary magnetic field. Arrows indicate the direction of the *apparent* dipole magnetic field of the Sun. Letters A to I indicate the locations of the nodal surfaces. After Smith *et al.* (1978).

that lies close to the plane of the ecliptic (as can be determined from the curl of **H**). Observations indicate that this current sheet is warped by a few degrees and this warping can have a pronounced effect on solar particles and magnetic fields that reach the Earth via the solar wind. The nodal surfaces shown in Fig. 10.2 correspond to the points A through I, at which this warped sheet crosses the ecliptic (Smith *et al.*, 1978).

The strength of the IMF in the neighborhood of the Earth varies considerably but is around 5 nT. The component of the IMF seen by the Earth's magnetosphere, which is perpendicular to the ecliptic, changes sign when the Earth passes through a nodal surface in its orbit about the Sun. This leads to the possibility that "reconnection" of field lines from the IMF with the Earth's field lines occurs in the northern hemisphere when the IMF has a downward component (Fig. 10.4) but not in the southern hemisphere. This means that charged particles can more easily enter into the Earth's ionosphere when the Earth is in one sector of the IMF (with downward component) than in the opposite sector, because charged particles preferentially travel along field lines. Geomagnetic activity varies significantly depending on which sector the Earth is in. At present, considerable controversy exists as to whether the Earth's climate is significantly affected by this variation in the incoming particle flux. Data showing apparent correlations between the size and magnitude of low-pressure areas in the troposphere and geomagnetic activity are intriguing (e.g., see Bucha, 1980). Solar irradiance appears to vary with the 11-yr solar cycle (to be discussed shortly) by about 0.05 to 0.12%, and this is estimated to lead to mean global temperature variations of 0.03°C, which appears not to be in the

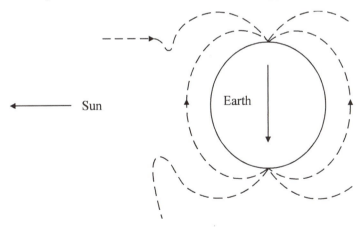

Fig. 10.4. The speculated "reconnection" of the IMF (possessing a downward field component) with the Earth's magnetic field. In this case, reconnection occurs only in the northern hemisphere. Conversely (not shown), when the IMF has an upward component, reconnection occurs only in the southern hemisphere.

detectable range (e.g., Wigley, 1988). Longer term variations in the solar "constant" may affect the Earth's climate.

In addition to the Sun's general field there is a broad hierarchy of magnetic fields. The smallest element is the individual *magnetic flux tube*, a concentration of magnetic field lines of tubular structure. At a large granular scale, small bipolar groupings and rings of flux tubes occur, followed by a network of flux tubes that follow the boundaries of supergranule cells. On still larger scale, these network fields are divided into *quiet* and *active* regions. Active regions exhibit complex magnetic structures and are where sunspots occur.

Sunspots are vortices of gas on the surface of the Sun associated with strong local magnetic fields. They appear as darker regions on the Sun's surface, but this is only in contrast with the surrounding photosphere, which is much hotter. They range in size from several times the size of the Earth to so small that telescopic observation is difficult. It was through observation of the motion of sunspots that Richard Carrington (*ca.* 1860) first discovered that the Sun rotates differentially, and not as a solid body. Although individual sunspots appear to have a lifetime of only a few weeks, they seem to occur in zones that vary latitudinally throughout the solar cycle. The spots first appear at latitudes near $\pm 40°$, reach sunspot maximum near $\pm 20°$ latitude, and die off at low latitudes of $\sim 5°$. In addition, there is a general drift of sunspots toward the equator. In 1922 Annie Maunder charted this latitude drift of spots (which was first noted by Carrington), and her chart is referred to as the Maunder butterfly diagram because of the wing-like shapes in the diagram (Fig. 10.5). The phenomenon is cyclic with a period of roughly 11 yr; it is often referred to as a quasi-periodic cycle because the length of the "period" varies between 8 and 14 yr (e.g., Gilman, 1992). In addition, fluctuations occur in this periodicity, including a period of 70 yr from 1645 to 1715, the "*Maunder Minimum*", during which time the Sun was almost without spots (Eddy, 1976, 1977; Eddy *et al.*, 1977). This Maunder Minimum is also evidenced in ^{14}C data and there appear to be other long intervals in the past when there was little or no sunspot activity (e.g., Stuiver and Quay, 1980). There have been frequent suggestions of a link between solar activity and climate, the Earth being warmer when the Sun is more active. A recent improvement in the method of categorizing sunspot activity (Hoyt *et al.*, 1994) should provide climatologists with a far more reliable tool for testing such suggestions.

Perhaps the most interesting aspect of sunspots is their very strong magnetic fields, with intensities on the order of 10^{-1} T. Although sunspot groups are often complex, about 90% of the time sunspots occur in pairs (a leader and a follower in respect to the Sun's rotation), and the spots in any given pair exhibit opposite

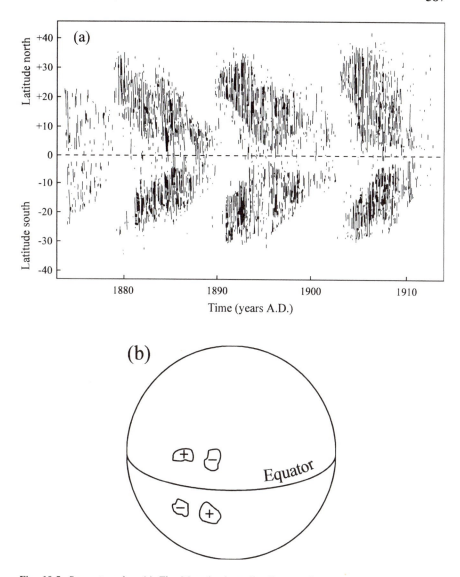

Fig. 10.5. Sunspot cycles. (a) The Maunder butterfly diagram shows the location of sunspot formation versus time in years. The beginning of a cycle is associated with sunspot formation at intermediate latitude ($\approx 30°$) and it ends about 11 yr later with the formation of sunspots near the equator. (b) The polarity of the first (left) sunspot in a pair depends on the hemisphere in which the sunspots form. The order of the polarity changes every 11 yr when sunspots begin a new 11-yr cycle. Combining this change in polarity with an 11-yr cycle shown in (a) produces a 22-yr cycle.

magnetic polarities. The sense of this polarity reverses every 11 yr, coinciding with the transition of the sunspot zone from low latitudes to high latitudes. That is, if most of the leader sunspots statistically have normal polarity (downward inclination) in the northern hemisphere during a given 11-yr cycle, then they will have a reversed polarity preference during the next 11-yr cycle. This is known as Hale's polarity law (Hale, 1924) and leads to an overall sunspot period of 22 yr (Fig. 10.5). Also, most of the leader members of sunspot groups in the southern hemisphere will have the opposite polarity to that observed in the northern hemisphere. The Sun's main dipole field also reverses polarity along with the change in polarity of the sunspots, a strong indication that the two phenomena are closely linked. The reversals of the dipole field occur near sunspot maxima, indicating (as will become apparent) that the toroidal and poloidal magnetic fields do not reverse polarity at the same time.

The origin of the magnetic fields in sunspots was first explained in magnetohydrodynamic terms by Cowling (1934). Subsequently, there have been numerous contributions to our understanding of the problem, the work of Babcock (1961) and Parker (1975, 1979) being especially stimulating. Because of the differential rotation between equatorial and polar latitudes the general poloidal magnetic field (the general dipole field) is rapidly wound up to produce a strong toroidal magnetic field (see the ω-effect in Chapters 8 and 9). Thus the resulting intense toroidal field is *antisymmetric* with respect to the solar equator and it is this strong toroidal field that is believed to be associated with sunspots. The basic idea (Cowling, 1934; Babcock, 1961; Parker, 1975) is that an instability results in an outward flow of plasma that carries the toroidal field with it. This is illustrated in Fig. 10.6, which also shows why sunspots usually occur in pairs. One sunspot is characterized by the region in which the "toroidal" magnetic field leaves the Sun's "surface", while the adjacent sunspot is characterized by the region where the "toroidal" field reenters the "surface."

An estimate of the intensity of the magnetic flux in sunspots can be obtained from the temperature estimates. The ideal gas law gives

$$\frac{P_o V_o}{T_o} = \frac{\left(P_o + \dfrac{B^2}{8\mu}\right) V_i}{T_i} \quad , \tag{10.2.1}$$

where P is pressure, V is volume, T is temperature, and the subscripts, o and i refer to conditions outside the sunspot and inside the sunspot, respectively. B is the magnetic induction and μ the magnetic permeability. Because the pressure term inside the sunspot must include that arising from the magnetic pressure (§9.1.3), the temperature inside the sunspot will be lower. Assuming quasi-

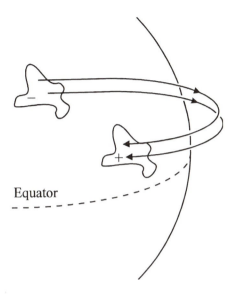

Fig. 10.6. Magnetic field lines emerge from one member of a sunspot pair and enter the Sun through the second member.

equilibrium conditions of no buoyancy ($V_o = V_i$), Eq. 10.2.1. gives the fractional decrease in temperature as

$$\frac{T_o - T_i}{T_o} = \frac{B^2}{8\mu P_o} \quad .$$

(10.2.2)

Direct substitution of measured and probable values of T and P into this equation give a value for B on the order of 10^{-1} T.

Sunspots exhibit considerable structure and better models of their magnetic fields can be constructed by using a collection of magnetic flux tubes (Parker, 1979). A flux tube can be visualized as an elongated region, often modeled with a circular cross section, in which there is a concentration of magnetic field flux. Any continuous field distribution $\mathbf{B}(\mathbf{r})$ can be decomposed into flux tubes (Parker, 1979). Each tube exerts a pressure $B^2/2\mu$ on its neighbors and strives to shorten itself with tension B^2/μ. Indeed, observational data indicate that the general magnetic field of the Sun could be better represented as consisting of many small, very intense, widely separated flux tubes rather than a broad continuous distribution. The question arises as to why the enormous magnetic pressures associated with sunspots does not cause the magnetic field to expand and disperse. Parker answers this by suggesting that nearly vertical flux tubes are brought together to form a sunspot by converging fluid flow. The flow then continues downward in the interstices of the flux tubes. This can be tested since upward traveling seismic waves will then travel slower beneath sunspots. Recent helioseismic tomography results (Duvall *et al.*, 1996) support this model by indicating that on the average there are downward flows associated with sunspots with velocities around 2 km s^{-1} that persist to depths of approximately 2000 km. Curiously, the theory of flux tubes (Parker, 1979) has been largely ignored in solar dynamo models; dynamo theory has been confined essentially to analytical or numerical solutions of equations for continuous field distributions of **B**. This may go some way to explaining why fundamental inconsistencies between theory and observation remain.

Although an analogy between sunspots and secular variation of the Earth's magnetic field is often made, the analogy has numerous weaknesses. Sunspots exist for only a few hours to a few (Earth) months and are associated with very intense magnetic fields. The ratio of a solar polarity interval (11 yr) to a terrestrial polarity interval (for the Cenozoic) is of order 10^{-4}. Hence one would expect to see in the historical record many "terrestrial spots" moving toward the equator at lower latitudes. Moreover, there should be very intense magnetic fields, some two orders of magnitude larger than the Earth's dipole field, associated with these "terrestrial spots". None of these features is seen. In addition, solar reversals are characterized by a quasi-periodicity, while terrestrial reversals have a pronounced stochastic character (Chapter 5). The absence of a good analogy is not surprising when one considers that the magnetic fields associated with sunspots probably involve instabilities in a compressible plasma interacting with a strong solar toroidal magnetic field, whereas the Earth's magnetic field originates in an electrically neutral MHD fluid whose source of convective buoyancy and whose boundary conditions are very different from those of the Sun.

Corona holes, which occur only at high latitudes and appear as dark areas in x-ray images, also vary with the solar cycle. They appear to be open magnetic field regions, that is, regions of uniform magnetic polarity in which the magnetic field lines extend outward into the interplanetary medium rather than returning into the region close to the sun to form closed loops (e.g., Wang *et al.*, 1996). Soon after the Skylab missions in 1973–1974 it became apparent that large corona holes, sometimes extending from polar regions down to 60° latitude, occur only close to sunspot minima. There are no corona holes during sunspot maxima. Unlike other regions of the sun, corona holes show no significant variation in rotation with latitude. They are believed to be a possible source of the high-speed solar winds that produce geomagnetic storms (Wang *et al.*, 1996).

10.2.3 Solar Magnetic Field Theory

Foremost among the solar phenomena that need to be explained is the solar cycle, including the reversals that occur approximately every 11 yr. (The "solar cycle" is used loosely here to refer to both the 11- and the 22-yr periodicities.) Some of the most mathematically developed dynamos (e.g., Glatzmaier, 1985; Gilman, 1992) have the solar cycle progressing in the wrong direction, leading Parker (1987) to refer to the "solar dynamo dilemma." The proposed resolution to this dilemma is that the dynamo does not originate in the convection zone in the outer part of the Sun (e.g., Stix, 1991; Brandenburg, 1994), but instead in a transition layer that separates the convection zone from the radiative transfer zone.

Unlike the Earth's outer-core fluid (which is effectively incompressible), the solar dynamo fluid is highly compressible. This is important because helicity plays a central role in dynamo theory (Chapters 8 and 9), and helicity can be greatly enhanced by the compressibility of the solar plasma. An upwelling (associated with convergence) near the bottom of the solar convection zone will lead to helicity with sign opposite to that associated with an upwelling (and divergence) near the top of the convection zone. This change in sign appears to be sufficient to make the solar cycle run in the correct direction, that is, with sunspots forming at midlatitudes at the beginning of the 11-yr cycle and near the equator at the end of the cycle.

For the above to be acceptable one must be able to explain the bipolarity of sunspots (e.g., see Fig. 10.6), which appears to provide evidence for a toroidal magnetic field. In particular, Hale's polarity rule, which appears to be violated only rarely (Stix, 1991), seems to imply there is a strong well-ordered toroidal field at some depth in the Sun. However, the turbulence in the convection zone seems sufficient to destroy such order. Only near the base of the convection zone does it appear that shear can be sufficiently large, relative to convection eddies, to produce a strong well-ordered toroidal field (Stix, 1991). Thus it seems necessary for sunspots to form near the bottom of the convection layer and then rise to the surface. Parker's (1955b) theory dealing with buoyancy of magnetic flux tubes suggests that this is feasible. Parker showed that when an isolated flux tube is in thermal and pressure equilibrium with its surroundings it is actually buoyant because the magnetic field produces some of the effective pressure and so the density of the tube is less than that of the material on either side of it. Parker (1975, 1979) estimates that a small flux tube of 10^{-2} T would rise from a depth of 10^5 km to the surface in about 2 yr. However, this ignores viscous effects and may be an order of magnitude or more too fast (e.g., Gilman, 1992). Thus it seems that the solar dynamo dilemma problem can be resolved if the dynamo originates immediately below the main convection zone in the Sun. However, as in the terrestrial case, there are many models for the solar dynamo, all of which are oversimplified and inadequate, and so this picture will almost certainly continue to undergo modifications.

One indication of the current state of dynamo theory is that in many instances essentially the same dynamo models are applied to both the Earth and the Sun, even though the conditions are very different and there are significant differences in many of the phenomena that need to be explained.

10.3 The Moon

10.3.1 General Properties

Conversion of gravitational energy associated with accretion of the Moon 4.5 to 4.6 Ga appears to have produced a magma ocean over the surface of the Moon for which depth estimates range from one hundred to many hundred kilometres. Subsequent cooling and differentiation of this magma ocean produced gabbros and anorthosites – rocks that are the major constituents of the lunar highlands. In Pre-Imbrian times (the period in which the highlands were formed), the Moon was under heavy bombardment by planetesimals and large meteorites. This bombardment decreased considerably following a period of great impacts around 3.9–4.0 Ga ago. Subsequently the lunar maria formed when basalt melt (which formed earlier from partial melting of rocks beneath the pre-Imbrian crust) flowed into large impact basins. Lunar samples returned from the maria have radiometric ages ranging from 3.1 to 3.9 Ga. Relative age relationships suggest that some maria, from which no samples were returned, may well have ages as young as 2.5 Ga (Head, 1976). It is doubtful that much volcanism, if any at all, has occurred during the past 2.5 Ga.

Although the major features of the Moon have remained essentially unchanged during the past 2.5–3.0 Ga, the absence of an atmosphere has meant that the lunar surface has been subsequently broken up into layers of debris by the constant bombardment of meteorites, micrometeorites, cosmic rays, and solar wind particles. There are some exceptions – the Tycho crater, for example, was probably formed by a meteorite impact only 300 million years ago. The surface layer of debris, which exists almost everywhere and is typically a few metres thick, is referred to as the lunar regolith. Essentially all returned samples come from this regolith and can be assumed to have been subjected to shock and radiation processes. In addition to forming the regolith, the constant rain of particles slowly mixes the upper parts of the regolith, a process known as "gardening".

The Moon has a mean density of 3.34×10^3 kg m^{-3} and a moment of inertia of 0.391 ± 0.002 $M_a R^2$ (M_a, total mass, R, radius (Blackshear and Gapcynski (1977)). Together with the geological evidence, these values are consistent with the belief that some differentiation has occurred. However, neither the extent of this differentiation nor the present lunar structure are at all well known; our knowledge remains rudimentary.

As with the Earth, seismic evidence provides the major information regarding the lunar interior. Four seismometers, and a gravity meter sometimes used as a seismometer, sent data back to the Earth for several years until they were turned off in the latter half of 1977. The seismic sources are moonquakes, meteorite impacts, and controlled impacts from various Apollo spacecraft stages that were crashed onto the lunar surface. There are several criteria for distinguishing meteorite impacts from moonquakes, including the observation that the signal from meteorite impacts has a much smaller shear wave component. Most moonquakes have Richter magnitudes less than 3 and are located at a depth between 700 and 1000 km. A few near-surface moonquakes also occur and the highest recorded lunar magnitudes (between 3 and 4) have been associated with these near-surface events. Because of the locations of the instruments (all Apollo sites were on the near side so that radio contact could be maintained with the astronauts) and the low magnitudes of the moonquakes, the rudimentary seismic structure is essentially determined for the near side.

The seismic data suggest that there is a crust roughly 70 km thick, and an upper mantle that extends to a depth of 300 km (Nakamura *et al.*, 1976) or 500 km (Dainty *et al.*, 1976), below which there is a lower mantle with a decrease in the velocity of shear waves (and probably also for compressional waves) and an increase in their attenuation. The different estimates of the extent of the upper mantle provide an indication of the problems of data resolution. Very little seismic evidence exists concerning the region extending from 1000 km to the lunar center at 1738 km depth (pressure ~4.7 GPa). These very limited seismic data (from a single event), the moment of inertia data for the Moon, and nonunique estimates of magnetic susceptibility, suggest that the Moon may have a solid metallic core with a maximum radius of 500 km (e.g., Nakamura *et al.*, 1974; Solomon and Head, 1979; Dyal and Parkin, 1971).

There are some important asymmetries regarding the Moon. These include a predominance of maria on the near side of the Moon and their associated mascon gravitational anomalies, a 2-km offset of the center of mass from the lunar center toward the Earth, and possibly a thicker crust on the far side of the Moon (Kaula *et al.*, 1974). For more detailed account of what is known about the lunar structure and interior, readers are referred to Taylor (1975, 1982).

10.3.2 The Lunar Magnetic Field

Information concerning the lunar magnetic field has been obtained from direct measurements using surface magnetometers, and satellite and subsatellite magnetometers, from indirect measurements such as the electron reflection method, and from measurements made on samples returned to the Earth.

Early spacecraft magnetometer measurements indicated that the Moon possesses only a very small magnetic field (Dolginov *et al.*, 1961, 1966; Sonett *et al.*, 1967), and it was clear that it does not possess an active dynamo. Thus it came as somewhat of a surprise to find high values of remanent magnetizations from samples returned from the Moon (see §10.3.3). The surface, satellite, and subsatellite magnetometer data clearly indicate that surface magnetic fields range up to values that occasionally exceed 300 nT (Dyal *et al.*, 1970; Barnes *et al.*, 1971; Coleman *et al.*, 1972; Sharp *et al.*, 1973; Hood, 1981). At present, magnetic measurements are available for roughly 15% of the lunar surface. The observed fields are known to originate primarily from magnetization in the lunar crust, and it varies considerably both in intensity and in direction. As a result, only an upper limit for a lunar magnetic dipole moment on the order of 1.3×10^{15} Am^2 can be obtained, compared with 8×10^{22} Am^2 for the present dipole moment of the Earth (Russell *et al.*, 1974; Russell, 1978).

Reflection of electrons by a magnetic field is not well known to most geomagnetists, but is a valuable method for estimating surface magnetic fields from satellite observations. It was first used by Howe *et al.* (1974). The solar wind and the tail of the Earth's magnetosphere provide the Moon with an abundant supply of electrons, and the absence of an atmosphere means that those electrons do not interact with atmospheric gases. Thus the Moon is an excellent candidate for the electron reflection method.

Unless its velocity **v** happens to be parallel to the magnetic induction **B**, an incident electron will travel downward in a spiral. In equilibrium, the centrifugal and Lorentz forces balance, so

$$|e\mathbf{v} \times \mathbf{B}| = \frac{m(\mathbf{v}_p)^2}{r} \quad , \tag{10.3.1}$$

where $\mathbf{v}_p$ is the component of the velocity perpendicular to **B**, and r is the radius of the spiral. If the magnitude of **B** increases on approaching the surface, $\mathbf{v}_p$ will increase and so, to conserve energy $\mathbf{v}_z$, the component of **v** toward the surface (downward) must decrease. Under appropriate conditions for the energy and angle of incidence of the electrons and the strength of the magnetic field, $\mathbf{v}_z$ can be reduced to zero for some value of **B**, say $\mathbf{B}_{ref}$. At this stage **v** is perpendicular to **B** and there is an upward force, so the electron is reflected (e.g., Parks, 1991).

The process is illustrated in Fig. 10.7. Given a reference observation $\mathbf{B}_o$ of the field together with the reflection coefficient for electrons of a known energy and angle of incidence, $\mathbf{B}_{ref}$ can be determined. The method is described by Anderson *et al.* (1975) and some details are given by Merrill and McElhinny (1983).

The most significant anomaly data are those for Rima Sirsalis, because the anomaly associated with Rima Sirsalis is probably the strongest indication that the Moon once had a moderately large magnetic field. Rima Sirsalis is a structural rille, a graben-like feature a few kilometre in width and depth and extending a few hundred kilometre in length (Fig. 10.8b). The magnetic anomaly is barely resolvable using satellite magnetometer data, but is reasonably well determined by the electron reflection method. Figure 10.8a gives some of the data used by Anderson *et al.* (1977) showing the large increase in electron reflections associated with Rima Sirsalis. Figure 10.8b shows the least-squares fit to the peak in electron reflection data along with the geographical location of Rima Sirsalis; the fit is clearly very good. The surface magnetic field strength is greater than 100 nT over a region on the order of 10 km in width and about 300 km in length (Anderson *et al.*, 1977). Anderson *et al.* (1977) show that these results can be modeled to a first approximation by assuming a uniformly

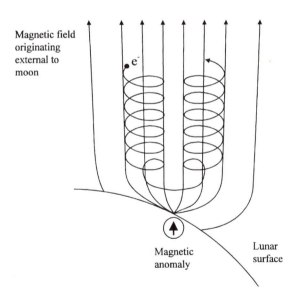

Fig. 10.7. The electron reflection method. An electron spirals in along magnetic field lines and is reflected from a magnetic crustal source lying beneath the lunar surface. After Anderson *et al.* (1975).

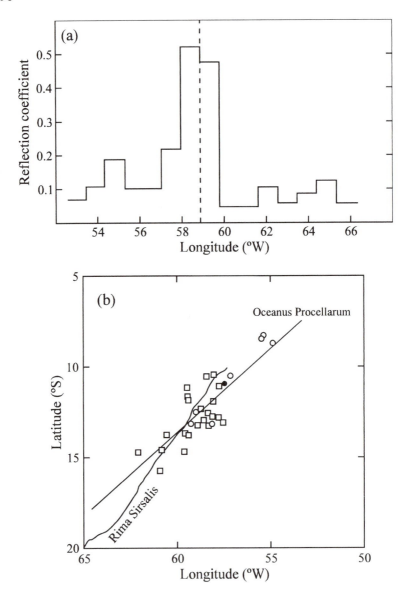

Fig. 10.8. The Rima Sirsalis magnetic anomaly. (a) Electron reflection data (14 keV) plotted versus longitude for a crossing of the Rima Sirsalis rille (located in longitude by the dashed line) at 12.5°S. (b) Electron maximum reflection peaks: 0.5 keV (□) and 14 keV (O). A direct magnetometer measurement (●) is consistent with the electron reflection data. The least-squares linear fit to the data is shown by the straight line. The location of the Rima Sirsalis rille is shown by the wavy line. Data from an Apollo 16 subsatellite flown in 1972. After Anderson *et al.* (1977).

magnetized lunar surface from which magnetized rock coinciding with a rille with a width of 10 km and depth of a few kilometre has been removed. These authors recognize the nonuniqueness of their modeling and point out that an equally acceptable model at the other extreme is one in which a magnetic dyke coincides with the trace of the rille. The authors actually prefer this latter model because the magnetic anomaly continues on after the structural manifestation of the rille disappears under Oceanus Procelarum.

It is important to point out that although the magnetic models that fit the Rima Sirsalis data are very nonunique, all the models seem to require coherence in magnetization over a few hundred kilometre. This, and other evidence, means that any hypothesis of lunar magnetism that exclusively invokes random magnetization processes is difficult to sustain.

10.3.3 Lunar Rock Magnetism

Ever since the first lunar samples returned to Earth were found to carry a remanent magnetization (Doell *et al.*, 1970; Nagata *et al.*, 1970; Runcorn *et al.*, 1970; Strangway *et al.*, 1970), considerable effort has been made to determine the origin and properties of that remanence (see reviews by Fuller (1974) and Fuller and Cisowski (1987)). Early concern that the remanence might be secondary in origin, particularly the possibility that the NRM was acquired after the collection of the sample on the lunar surface, was removed after the Apollo 12 mission carried an Apollo 11 basalt on a round trip to the Moon and back through the entire receiving and sampling process. No significant secondary component was found for this basalt.

Extensive and successful efforts to determine the magnetic properties of lunar samples and to utilize this information to gain insight into the paleolunar environments has been reviewed by Fuller and Cisowski (1987). Most work has been completed on mare basalts and highland melt rocks, since these are the most likely to have acquired a primary TRM and hence provide information on the origin and properties of the ancient lunar field. Even then, the rock magnetic problems are formidable and the majority of NRM contains secondary components.

Attempts to determine lunar paleointensities have been of most interest to planetary magnetists since, if successful, such information might be valuable in determining the origin and properties of the magnetizing field. Both absolute and relative paleointensities estimates have been obtained. Figure 10.9 compares the absolute paleointensity data trends with time to those associated with the relative paleointensity data. None of the "absolute" paleointensities is of sufficient quality to be an acceptable estimate using the much higher standards applied to terrestrial samples (see Merrill and McElhinny, 1983).

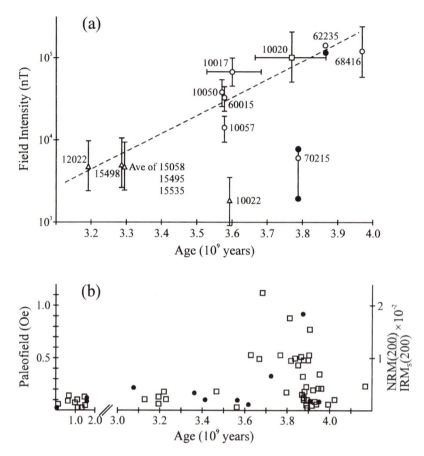

Fig. 10.9. Lunar paleointensities (a) after Stephenson *et al.* (1977) and (b) after Cisowski *et al.* (1983).

The "relative" paleointensities are obtained from the ratio of NRM to IRM$_S$, after AF demagnetization to 20 mT to remove the soft component. However, TRM and IRM are affected differently when the grain size, mineralogy, defects, etc., change, so there are problems even if the NRM is a TRM. Moreover, if the NRM is not a TRM, there is no justification at all for making relative paleointensity estimates. Consequently every single point in Figs 10.9a and 10.9b is in doubt. Yet collectively they seem to represent a strong suggestion that the intensity of the lunar magnetic field at its surface was comparable in intensity to the present-day terrestrial surface field about 3.8 Ga and was very small, or negligible, by 3.0 Ga. The rate of decrease is less clear, but it does seem that the field had a moderate (but declining) intensity for a few hundred million years after 3.9 Ga. There is also a hint of an indication that the intensity of the field might have been significantly smaller before 4.0 Ga.

10.3.4 Origin of the Ancient Lunar Magnetic Field

Various suggestions for the origin of the ancient field that produced the lunar magnetism can be grouped in three broad categories.

(i)   The first category involves external magnetic field models, such as strong fields from the Sun or the Earth (e.g., Runcorn and Urey, 1973; Strangway and Sharp, 1974; Banerjee and Mellema, 1976). These suggestions do not appear to be supported by interpretations of the paleointensity data. For example, a strong field associated with a T-Tauri phase (§10.1) would be too short-lived to explain the apparent high intensity that occurred between 3.9 and 3.6 Ga.

(ii)   The second category involves local sources, such as effects associated with shock from impacting objects (e.g., Srnka, 1977; Martelli and Newton, 1977; Srnka *et al.*, 1979), thermoelectric processes (Dyal *et al.*, 1977), and local dynamos (Murthy and Banerjee, 1973). The large spatial scale of some magnetic anomalies (§10.3.2), the larger intensity of some anomalies (§10.3.3), and the lack of systematic magnetic fields associated with craters (Fuller and Cisowski, 1987) seem to indicate that these mechanisms are unacceptable.

(iii) The final category is that of an ancient dynamo acting in the (now-solidified) central core (e.g., Runcorn, 1978, 1979, 1994). Runcorn suggests that heating by decay of superheavy radioactive elements no longer present in the Moon was sufficient to power a dynamo in a once-molten 500-km-radius metallic core. Alternatively, crystallization-produced buoyancy might have driven the hypothesized lunar dynamo (Fuller and Cisowski, 1987).

Although there are numerous uncertainties associated with the suggestion of a central dynamo, it appears to be the only one that cannot be dismissed by reasonable considerations of the data. Alternatively, one might take the conservative attitude that the origins of lunar magnetism remain a mystery.

10.4 Meteorites

Many meteorites represent remains of the primitive solar nebula (see §10.1), and so are some of the most primitive bulk matter from the solar system available for study. They are iron or stoney bodies that have fallen to Earth from elliptical orbits, suggesting that most of them originated within the solar system in the asteroid belt between Mars and Jupiter. There are exceptions, including the likely possibility that a few meteorites (such as the Shergotty Meteorite) have even come from Mars following a large impact. (There are now 10 meteorites believed to have come from Mars, possibly following a collision with a massive asteroid. These meteorites are believed to be of Martian origin because mass spectroscopic study of occluded gas shows a close match with the atmospheric composition obtained by the Viking Mars landers.)

Although numerous subclassification schemes exist, some including the use of magnetic properties (see, e.g., Cisowski, 1987), meteorites are broadly classified as metallic-rich "irons" (NiFe) and silicate-rich "stones". The iron meteorites consist mostly of the iron–nickel minerals kamacite and taenite. The stoney meteorites are divided into chondrites (named after the rounded or partially rounded polyhedral inclusions called chondrules that they contain), which are dissimilar to any terrestrial rocks, or achondrites, which are similar to terrestrial rocks consisting of ultramafic plutonic rocks and basalts. Stoney-iron meteorites have compositions intermediate between the achondrites and irons. The majority of meteorite finds are olivine-rich chondrites and iron–nickel meteorites.

Although the origin of meteorites is obscure and much debated, the formation of the vast majority of meteorites is often attributed to preexisting planetesimals, typically a few tens of kilometres to a few hundreds of kilometres in diameter. Larger sizes seem improbable because of the lack of high-pressure phases (Wood, 1967, 1979; Wood and Pellas, 1991). The interiors of some of these planetesimals became sufficiently hot to metamorphose some primitive planetary material (producing ordinary chondrites) and causing melting elsewhere (Wood, 1979). Components of the melted material separated gravitationally, forming iron-rich cores (the source of iron meteorites) and pure silicates (achondrites).

Reliable meteorite paleointensity information might shed light on magnetic field strengths early in the solar system's development, and so there has been substantial interest in obtaining such information. However, all paleointensity studies on meteorites have so far been equivocal with no convincing patterns emerging (e.g., see review by Cisowski, 1987). Consequently, meteorite paleomagnetism has not, for example, been able to determine whether there was an early T-Tauri solar phase with a strong magnetic field. Although clearly there must be mechanisms by which meteorites became magnetized, all current explanations seem to involve implausible assumptions. Thus the major use of meteorite magnetism so far has been within classification schemes.

10.5 Magnetic Fields of the Planets

10.5.1 Planetary Magnetism

The study of the magnetic fields of planets is an interesting subject in its own right. For example, if a planet has a global field that is sufficiently strong to deflect the solar wind then a magnetosphere will form (see §2.5.1 for the Earth) and this has a significant effect on the planet's environment. From our present perspective it is useful to see if the magnetic parameters of the planets might lend insight into the origin of the Earth's magnetic field.

Until recently, information about the magnetic fields of planets came mostly from indirect measurements, pioneered by Burke and Franklin (1955), who discovered strong bursts of polarized emissions from Jupiter at long radio wavelengths. These nonthermal microwave emissions were interpreted as being due to a global Jovian magnetic field and generated by trapped and precipitating electrons in the radiation belt. Decimetre and decametre emissions from Jupiter and Saturn still provide valuable information on their magnetic fields (e.g., Stannard, 1975; Brown, 1975). Studies of planetary magnetic fields have now been enhanced by direct measurements received from various planetary space probes.

Table 10.2 summarizes some of the known physical properties of the planets. Useful recent reviews of planetary magnetism include Stevenson (1983), Russell (1987), Connerney (1993), and Ness (1994). In spite of such indirect methods as mentioned above, almost all of our present knowledge of planetary magnetic fields comes from flyby missions. The measurements are generally sparse in both spatial and temporal distribution and sometimes measurements have been made within the planet's magnetosphere, where the presence of electric currents implies that the magnetic field cannot be strictly described in terms of the gradient of a scalar potential (see Backus (1986) for further discussion of this problem). Nevertheless, these measurements have provided us with a first-order picture of all the planetary magnetic fields, except for Pluto, about which virtually nothing is known. Mercury is globally magnetized, Venus is unmagnetized, and the existence of an intrinsic Martian magnetic field is controversial (if any global field does exist, it is quite small). Jupiter has a very strong, distinctly nondipolar field while Saturn has a strong field (but much weaker than Jupiter's) that, to degree 3, appears axisymmetric with respect to its spin axis. The 1986 and 1989 Voyager fly-bys of Uranus and Neptune showed global magnetic fields, but both have large angular offset of their magnetic axes

TABLE 10.2

Some Physical Properties of the Planets in the Solar System

	(M) Mass (10^{24} kg)	Planet radius (km)	Core radius (km)	Q^a	Orbital inclination	Spin period ($\times 10^5$ s)[b]	Magnetic dipole moment	Dipole tilt to rotation axis
Mercury	0.33	2,439	~1,800	?	0	+51	1.3 to 6×10^{-4} M_e	2° to 15°
Venus	4.87	6,055	~3,000	?	2°	-210	$< 5 \times 10^{-5}$ M_e	–
Earth	5.97	6,371	3,485	0.3308	23.5°	+0.864	M_e	10.7°
Mars	0.642	3,398	~1,700	0.375	24.0°	+0.886	$< 2.5 \times 10^{-4}$ M_e	uncertain
Jupiter	1899	71,600	~52,000	0.264	3.1°	+0.354	2.0×10^4 M_e	9.6°
Saturn	568	60,000	~28,000	0.207	29.0°	+0.368	5.9×10^2 M_e	0°
Uranus	87.2	25,600	–	0.26	97.9°	+0.621	47.5 M_e	58.6°
Neptune	102	24,765	–	0.26	17.5°	+0.576	25 M_e	46.8°
Moon	0.07	1738	<500	0.391	–	+23.6	$< 1.3 \times 10^{-7}$ M_e	–

[a]$Q = $ (Moment of Inertia)/(Mass × radius^2).

[b]Positive is prograde; negative is retrograde.

(59° for Uranus and 47° for Saturn) from their rotation axes and both have their magnetic centers displaced by substantial fractions (0.31 for Uranus and 0.55 for Saturn) of their planetary radius.

The global magnetic field of Mercury is usually estimated from solar wind parameters and the position of the magnetopause. (Recall from Chapter 2 that the magnetopause represents an equilibrium between the pressure from the solar wind and that from the planet's magnetic field.) For example, using a solar wind velocity of 550 km s^{-1}, a solar wind density of 14 particles cm^{-3}, and a subsolar radius (i.e., the distance of the magnetopause from the planet in the direction of the Sun) of 1.4, Mercury's magnetic moment is estimated at about 2.3×10^{18} Am^2. However, because of uncertainties in these parameters and because there are limited magnetic field measurements, Mercury's magnetic moment is not well known. Russell (1987) has recently summarized available estimates, and most fall in the range of 0.8 to 5×10^{18} Am^2; also see Table 10.2. Although many planetary scientists (e.g., Stevenson, 1983; Russell, 1987; Ness, 1994) interpret these results from Mercury as showing that it contains an active dynamo, a remanent magnetization explanation may also be possible (e.g., Strangway, 1977). Hence, it may be premature to assume (as sometimes done) that there must be a significant liquid region in the large core (extending out to ¾ of a radius) of Mercury.

With the exception of the Earth, and possibly Mercury, no other terrestrial planet has an active dynamo. The lack of a strong magnetic field for Venus is sometimes attributed to its slow rotation rate. However, Stevenson (1979) argues against this view on the basis that the estimate for Venus' Rossby number (see Table 7.4 for a definition of Rossby number) is larger than that for the Earth. It should be recognized, though, that in estimating this Rossby number the velocity used for the Venusian core fluid is obtained by taking the cube root of the estimate for the internally generated heat flux and dividing this by the planet's density. Clearly this is a very crude estimate, so it remains difficult to reject the possibility that the relatively slow rotation of Venus is a major reason for the absence of an active dynamo in the planet. Alternatively, Stevenson (1983) proposes that Venus has no dynamo because its fluid core is stably stratified (i.e., there is insufficient energy to generate convection and to power a Venusian dynamo). Mars appears to have a very weak magnetic field that is probably due to remanence (Russell, 1978), possibly acquired at a time when Mars had an active dynamo.

In contrast, all the outer planets, with the possible exception of the small planet Pluto, exhibit active dynamos and magnetospheres. It is interesting that Saturn has a magnetic moment that is $\frac{1}{30}$ of Jupiter's and yet its radius is only 17% smaller. Moreover, Saturn's magnetic field appears to be axially symmetric, in apparent conflict with Cowlings antidynamo theorem (§9.3).

Interpretation of these results requires knowledge of the planets' interiors (see Fig. 10.1). Both Saturn and Jupiter consist primarily of hydrogen and helium and are believed to have small (approximately 0.2 radius) rocky cores. The outer regions consist of insulating molecular H_2 and He, while the dynamo is believed to originate in an interior region of convecting metallic H and He (Kirk and Stevenson, 1987). This metallic region is argued to range from 0.2 to 0.8 of a radius in Jupiter and from 0.4 to 0.5 of a radius in Saturn. However, Kirk and Stevenson (1987) argue that, because of differences in the pressure and temperatures within the two planets, there is a region of immiscibility of H and He in Saturn that is not present in Jupiter. This region occurs as a conducting shell about midway to the center of Saturn and divides the insulating region of H_2 and He from the convecting dynamo region of metallic H-He. Kirk and Stevenson (1987) suggest that this conducting shell is stably stratified due to a He gradient that results from the immiscibility of H and He. They further suggest that this layer is rotating differentially with respect to the interior and therefore screens the nonaxially symmetric part of the field. Alternatively, the smaller dynamo region of Saturn may produce a smaller field that contains sufficient (undetected) asymmetry such that Cowling's theorem is not violated.

Rotation appears to play an important role in dynamo theory and so it came as a surprise to learn that the magnetic axes of Uranus and Neptune are at large angles (see Table 10.2) to their rotation axes. (One should note that the observational evidence for rotational symmetry for the Earth's average magnetic field throughout time is very strong; see Chapters 4, 5, and 6.) The spherical harmonic decompositions shown in Table 10.3 illustrate the relatively very large contributions of the second (quadrupole) and third (octupole) degree harmonics to the magnetic fields of Uranus and Neptune. The latter suggests that the source of the field is far removed from the planet's center and, more speculatively, originates in a relatively thin shell. Because characteristic times for most dynamo processes scale as the square of the characteristic dimension it is conceivable that reversal rates are much higher in Uranus and Neptune than for the Earth. Thus it might be speculated that both Uranus and Neptune have been caught at the time of a polarity transition (Schultz and Paulikas, 1990). Another possible explanation for the relatively large nondipole fields of Uranus and Neptune is that their sources are actually at some distance from the planet's center and are not radially symmetric. This could produce the observed fields, including the large deviation of the dipole axis from the rotation axis. (Note that even if a field is strongly axially symmetric, if it is not also *radially* symmetric, the spherical harmonic description of that field will exhibit significant nonzonal harmonics.) Undoubtedly there are other speculations that can "account" for the unusual properties of Uranus' and Neptune's magnetic fields, but without more data it will be difficult to evaluate these speculations.

TABLE 10.3
Spherical Harmonic Coefficients for Earth and the Four Giant Planets. After Ness (1994).

	Earth IGRF85	Jupiter O6	Saturn Z3	Uranus O3	Neptune O8
g_1^0	-29,877	+424,202	+21,535	+11,893	+9,732
g_1^1	-1,903	-65,929	0	+11,579	+3,220
h_1^1	+5,197	+24,116	0	-15,685	-9,889
g_2^0	-2,073	-2,181	+1,642	-6,030	+7448
g_2^1	+3,045	-71,106	0	-12,587	+664
h_2^1	-2,191	-40,304	0	+6,116	+11,230
g_2^2	+1,691	+48,714	0	+196	+4,499
h_2^2	-309	+7,179	0	+4,759	-70
g_3^0	+300	+7,565	+2,743	+2,705	-6,592
g_3^1	-2,208	-15,493	0	+1,188	+4,098
h_3^1	-312	-38,824	0	-7,095	-3,669
g_3^2	+1,244	+19,775	0	-4,808	-3,581
h_3^2	+284	+34,243	0	-1,616	+1,791
g_3^3	+835	-17,958	0	-2,412	+484
h_3^3	-296	-22,439	0	-2,608	-770

10.5.2 Solar System Dynamos

The Sun, Earth, Jupiter, Saturn, Uranus, and Neptune all have active dynamos today, and it is possible that Mercury does as well. The question then arises as to whether there is some fundamental scaling law that relates the magnetic fields of these bodies to each other. Considering the large number of variables available (e.g., energy flux, angular momentum, inner core size, outer core size) and the small number of bodies involved, it is not surprising that several "laws" have been suggested. One of these, the Magnetic Bode's law for planetary magnetism (Russell, 1978), is shown in Fig. 10.10. This "law" initially appeared to be supported by the data, but measurements of the magnetic fields of Uranus and Neptune have subsequently been made (Table 10.2) and they do not fit this "law".

Examples of other laws that have been suggested are those by Brecher and Brecher (1978, relating angular momentum to magnetic moment), Dolginov (1976, 1978, claiming a correlation between magnetic field intensity and the Coriolis force), Busse (1976, using an earlier weak-field model that he now appears to reject), and Stevenson (1979, who used a heuristic scaling law between heat flux and magnetic moment). Some of the most promising recent

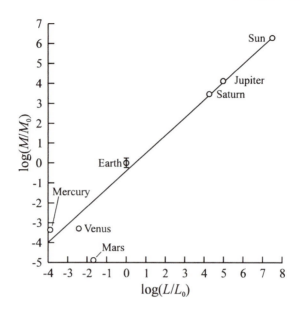

Fig. 10.10. The Magnetic Bode's "law". The log of magnetic moment M, normalized to the terrestrial moment M_0, plotted against the log of angular momentum L, normalized to the terrestrial angular momentum L_0. After Russell (1978.)

attempts at developing a scaling law presume that rough estimates of the magnetic moments produced in a planetary dynamo can be obtained by assuming magnetostrophic balance, in which the Lorentz force is balanced by the Coriolis force (Curtis and Ness, 1986; Mizutani *et al.*, 1992; Sano, 1993).

Although planetary magnetic fields may turn out to be approximately in magnetostrophic balance, there are many other factors that must also be considered, such as the nature of buoyancy and boundary conditions, factors that appear to affect the magnetic field strength significantly. Perhaps it is because of these additional factors that simple magnetostrophic balance arguments predict much stronger magnetic fields for Venus and Mars than observed (e.g., Sano, 1993). Moreover, even when they appear to work, it may be only coincidental. For example, Saturn's magnetic field strength may be significantly screened by a stratified conducting shell above the dynamo-producing region, a factor not considered in magnetostrophic balance models. For such reasons we remain skeptical that simple scaling laws will ever be sufficient to produce a convincing estimate of the moment of some astronomical body. However, they may ultimately be useful for statistical testing of broad concepts such as magnetostrophic balance.

10.6 Geomagnetic Relevance

In spite of the different conditions present in the Sun and the planets, it appears that a dynamo occurs whenever there is a relatively large conducting region that is rotating and convecting. MHD solutions are notoriously sensitive to boundary conditions, variations in the source of buoyancy, and variations in compressibility, yet the large variability in these and in the chemical compositions of the dynamo-producing regions in the Sun and planets do not stop dynamos from occurring. Despite our inability so far to produce realistic dynamo models, it would appear that the dynamo process is in fact remarkably robust, and we can but speculate on the reasons for this robustness. Perhaps, for example, the transfer of thermal energy in a large convecting system is more efficient if large-scale magnetic fields of moderate intensity are present (as opposed to many small-scale magnetic features). Clearly this is an area in which more research is needed.

Although dynamos are robust, it is not clear whether reversals are. Moreover, in the Sun and the Earth, the two bodies for which reversals have been demonstrated to occur, the properties of the reversals (such as the degree of regularity) are very different. However, the fact that solar reversals occur illustrates that an external source is not necessarily required to trigger reversals.

There are several intriguing questions that have been raised in this chapter, some of which are worth repeating. Do flux tubes play a significant role in the solar or terrestrial dynamos? Was there an ancient lunar dynamo? Does the solar dynamo originate in a transition region between the main radiative transfer region and the outer turbulent convecting region? Are there any planetary analogues? What is the reason for the high degree of asymmetry in the fields of Uranus and Neptune? In contrast, why does Saturn's field possess a very high degree of symmetry? The clear lesson is that the physics of dynamo processes is still not well understood, and that solar system dynamos (including the geodynamo) will be an active area of research for many more years to come.

Chapter 11

Examples of Synthesis

11.1 Fluid Velocities in the Core

11.1.1 Overview

Dimensionless numbers (e.g., Table 7.4) often play an important role in the fluid mechanics of the core. Several of these numbers require knowledge of the magnitude of the fluid velocity, $\mathbf{v}$, which is typically given as 10^{-4} m s^{-1}. This value originated from the estimate of the westward drift rate of the nondipole field (0.18° per year) made by Bullard *et al.* (1950; see §2.4). Shortly thereafter, Bullard and Gellman (1954; see §9.2) developed a kinematic dynamo model that appeared to produce the correct order of magnitude for the poloidal magnetic field when the fluid velocity was comparable to the drift velocity. Ever since then the 10^{-4} m s^{-1} value appears to have been accepted as scientific dogma. However, the apparently self-consistent view developed in the 1950s encountered difficulty when it was found that there were numerical convergence problems with the Bullard–Gellman dynamo.

The number of dynamo models has increased dramatically during the past few decades but it appears unlikely they will provide useful constraints on the magnitude of the core's velocity field, even though they may eventually provide some useful constraints on the directions of the velocity field (e.g., see §9.6). It appears that the best way at present to obtain estimates for the velocity field near the top of the core is by using geomagnetic secular variation data. Roberts and

Scott (1965) formalized the Bullard *et al.* (1950) approach: they assumed that the frozen flux approximation is valid (i.e., that the diffusion term in the magnetic induction equation can be ignored; see §8.2) on a time scale of tens of years to obtain the first picture of the outermost core fluid velocity. That is, providing that frozen flux applies, the secular variation reflects the fluid velocity just below the very thin boundary layer at the top of the core (§7.6).

Backus (1968) pointed out that the determination of velocities near the top of the core is nonunique, even if the frozen flux assumption is valid. The nonuniqueness stems from the fact that only the velocity component perpendicular to a magnetic field line changes that field line (§8.2). Backus defined a *null-flux curve* as the locus of points along which the radial component of the magnetic field, B_r, vanishes. He demonstrated that one component of the velocity can be uniquely determined along null-flux curves, if frozen flux applies and if the secular variation data are perfect. Although these statements are proved mathematically in the next subsection, they are also seen to be true physically by noting that when B_r is zero (i.e., the magnetic field is parallel to the CMB), any horizontal velocity perpendicular to a null-flux curve will carry the curve with it. In particular, no fluid can cross the $B_r = 0$ contour when the frozen flux condition is satisfied. Therefore the magnetic flux within a closed null-flux curve must remain invariant, another constraint first found by Backus (and formally treated in §11.1.2). With enough time, diffusion effects will of course become significant and this flux invariance no longer holds. Table 9.1 illustrates that it would, for example, be unreasonable to assume (as is sometimes done) that frozen flux applies on the time scale of a reversal transition. Booker (1969) first applied the null-flux curve concept to the geomagnetic field to show that only very limited knowledge of the outer core's velocity field could be obtained uniquely.

The subject remained dormant until Whaler (1980) and Benton (1979,1981) independently reopened it by making additional assumptions. In particular, Whaler (1980) showed that it is possible to obtain one component of the flow just below the core's top boundary layer by assuming that the fluid flow at the top of the core is toroidal (no radial component). Gubbins (1982) recognized that unique solutions could be obtained by assuming the core flow to be steady over a finite interval of time. Building on the work of Benton (1981), Voorhies and Backus (1985) formalized this idea by showing that the flow can be obtained uniquely (except for a few degenerate cases) if the flow is assumed to be the same at three different points (and providing there is perfect geomagnetic secular variation data for these different times). Le Mouël (1984) approached the problem in yet a different way by assuming that a geostrophic balance occurred near the top of the core. Braginsky and Le Mouël (1993) subsequently showed that if the outer core is also stratified, and if this stratified layer is no more than a few hundred kilometres thick, then the velocity obtained by the above methods represents the average velocity in the stratified layer. Bloxham and Jackson (1991) present an excellent overview of the subject, including important evaluations of error analyses. Hereafter, when we refer to the velocity at the top of the core, we are following the accepted practice of meaning the velocity just below the very thin uppermost boundary layer in the core.

11.1.2 The Uniqueness Problem

It is useful to work through some of the mathematics developed by Backus (1968) showing that the flux through a null-flux curve is conserved. The flux through a patch S of the surface of the core is given by

$$\int_S \mathbf{B} \cdot \mathbf{n} dS = \int_S B_r dS \quad .$$

At the CMB the radial component of the velocity, v_r, vanishes and the radial component of the magnetic induction equation, with the diffusion term dropped, (see §8.2) is

$$\frac{\partial B_r}{\partial t} + \nabla_h \cdot (\mathbf{v} B_r) = 0 \quad , \tag{11.1.1}$$

where ∇_h, the horizontal divergence operator, is

$$\nabla_h = \left(0, \frac{\partial}{R \partial \theta}, \frac{1}{R \sin \theta} \frac{\partial}{\partial \phi} \right)$$

in spherical coordinates and R is the core's radius. Naturally, the observations will not match (11.1.1) if diffusion is a significant component of the secular variation. However, a set of necessary conditions for consistency with the frozen flux hypothesis can be obtained by integrating (11.1.1) over a patch S of the core surface bounded by $\mathscr{L}$ and then applying the divergence theorem in the plane to give

$$\int_S \frac{\partial B_r}{\partial t} dS = -\oint_{\mathscr{L}} B_r v_n \, d\mathscr{L} \quad .$$

If $\mathscr{L}$ is a contour (level line) of B_r then

$$\int_S \frac{\partial B_r}{\partial t} dS = -B_r \oint_{\mathscr{L}} v_n \, d\mathscr{L} \quad .$$

Finally, if S is chosen such that $\mathscr{L}$ is a null-flux curve (i.e., $B_r = 0$) then

$$\frac{\partial}{\partial t} \int_S B_r d\mathbf{S} = 0 \quad . \tag{11.1.2}$$

This last equation shows that, for consistency with the frozen flux hypothesis, the flux within a null-flux curve must be conserved.

There is only a small portion of the core's surface associated with null-flux curves, as shown in Fig. 11.1. Hence, one can obtain only very limited knowledge of the outermost core velocity field. The nonuniqueness of the problem can be expressed in terms of generalized inverse theory. From (11.1.1) we see that any velocity field satisfying

$$\nabla_h \cdot (\mathbf{v} B_r) = 0$$

is not manifested in the secular variation, $\partial B_r/\partial t$. This implies that for any well behaved $\psi(\theta, \phi)$, if $\mathbf{v}'$ is given by

$$\mathbf{v}' = \frac{1}{B_r} \mathbf{n} \times \nabla_h \psi \tag{11.1.3}$$

(where $\mathbf{n}$ is the unit normal to the CMB), then $\mathbf{v}'$ generates no secular variation from B_r. Thus the annihilator, $\mathbf{v}'$, can be added to any velocity field satisfying (11.1.1) without any information on $\mathbf{v}'$ appearing in the secular variation. The velocity can, therefore, be determined only to within the ambiguity of ψ, which is itself arbitrary except on null-flux curves. This means that all components of $\mathbf{v}$

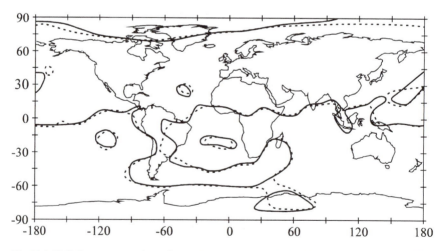

Fig. 11.1. Null-flux curves on the surface of the core for the 1975 field by Barraclough *et al.* (1975). Dashed curves are for the model up to degree 11, and solid curves are for the model up to degree 12. Diagram from E.R. Benton.

are ambiguous, except on null-flux curves where the component normal to the curve may be determined.

Reconsider (11.1.1), which assumes frozen flux, with the supposition that B_r and $\partial B_r/\partial t$ are known perfectly. Since v_r is zero, there is only one scalar equation for the two unknowns, v_θ and v_ϕ. A further scalar equation is needed to remove the nonuniqueness. One possible equation is the magnetic induction equation for the tangential component of **B**. However, it can be shown (e.g., Bloxham and Jackson, 1991) that this introduces two more unknowns. Consequently some additional assumption is required. As already noted in §11.1.1, several different assumptions have been made and these are further discussed in §11.1.3. It should be pointed out that these assumptions are usually incompatible with each other, so the velocity system would usually be overdetermined if more than one of these assumptions were made at the same time.

11.1.3 Models and Results

There are three basic models (and some hybrids that will not be discussed; see Bloxham and Jackson (1991)): in each it is assumed that frozen flux applies and that the data are perfect, and then one additional assumption is made to define the model.

(i) The Steady Motion Assumption

The additional assumption in this model is that the fluid motion remains steady over some interval of time. Change in the geometry of the magnetic field is then used to resolve the nonuniqueness in the velocity field. Bloxham and Jackson (1991) address the question of how much change in the magnetic field is needed. Discussions of error analysis are given by Voorhies (1986, 1993) and Whaler and Clarke, 1988). It is interesting to note that the existence of a magnetic jerk (§2.4) does not by itself invalidate the steady motion assumption. To see this, take the partial derivative with respect to time of (11.1.1) to obtain

$$\frac{\partial^2 B_r}{\partial t^2} + \nabla_h \cdot (\frac{\partial v}{\partial t} B_r) + \nabla_h \cdot (v \frac{\partial B_r}{\partial t}) = 0 \quad . \tag{11.1.4}$$

The second term must be zero if steady flow is assumed, but second derivatives in **B**, such as the jerk, can still exist because of the last term. The velocity field obtained under the assumption of steady motion is shown in Fig. 11.2a.

(ii) The Geostrophic Balance Assumption

The additional assumption in this model is that the horizontal pressure gradient is balanced dominantly by the horizontal Coriolis force

$$(\nabla P)_h = 2\rho(\Omega \times v)_h \quad , \tag{11.1.5}$$

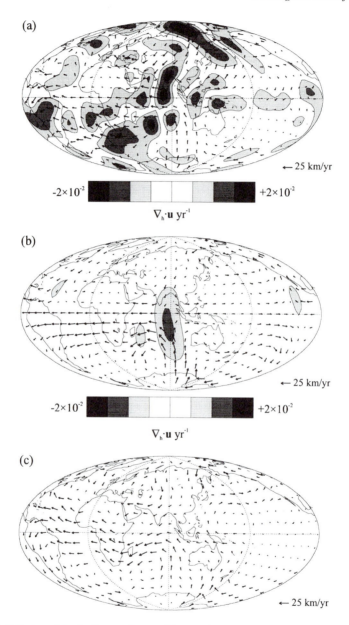

Fig. 11.2. Estimates of steady flows at the top of the core for the interval 1840–1990 under different assumptions: (a) steady unconstrained flow; (b) geostrophic flow; (c) toroidal flow. After Figs 2a, 2c, and 2e of Bloxham (1992).

where Ω is the Earth's rotation, P is pressure, and ρ is density. To eliminate the pressure term, take the curl of (11.1.5) to obtain

$$(\Omega \cdot \nabla)v = 0 \quad . \tag{11.1.6}$$

The radial component of this equation gives the so-called geostrophic constraint,

$$\nabla_h \cdot (v \cos\theta) = 0 \quad , \tag{11.1.7}$$

where θ is colatitude. This constraint along with (11.1.1) gives estimates of the velocity field such as shown in Fig. 11.2b. The horizontal component of the Coriolis force vanishes at the equator and so the above model cannot, strictly, give equatorial velocities. Thus, although the nonuniqueness is decreased, some nonuniqueness remains even with the additional assumption. However, the continuity equation still applies and so one can extrapolate for the velocities in the equatorial region. Examples of error analyses can be found in Le Mouël *et al.* (1985) and Gire and Le Mouël (1990).

(iii) The Toroidal Velocity Assumption

In this model it is assumed that $\partial v_r/\partial r$ vanishes at the CMB (v_r always does). This is then combined with the continuity equation for an incompressible fluid, $\nabla \cdot v = 0$, to obtain a second equation to be solved simultaneously with (11.1.1). Examples of error analyses can be found in Bloxham (1989) and Lloyd and Gubbins (1990). Application of this particular constraint gives estimates of the velocity field such as shown in Fig. 11.2c.

As already noted, these additional assumptions cannot be combined and, because they lead to differing results, there is naturally some controversy over which is most appropriate. For example, different models give different velocities under parts of the Americas and, as discussed in §5.4.3, this may be important in interpreting reversal transition data. Furthermore, each of the assumptions adds considerable "information" relative to the information provided by the data. In particular, there has been no significant secular variation in the Pacific region for the past few hundred years and so the velocities shown in Fig. 11.2 for that region are a direct consequence of the particular assumption made (i.e., they are not based on any data in the Pacific)!

The frozen flux assumption is common to all the models, and appears to be reasonable (e.g., Constable *et al.*, 1993) for the relevant time scales. Bloxham and Gubbins (1985) provided evidence of flux diffusion but, as argued by Bloxham (1989), this does not invalidate the assumption for the purpose of mapping the core flow: it is merely a question of whether the inevitable diffusion is detectable. Each of the models also assumes perfect data at the CMB, whereas the data are far from perfect: the poloidal magnetic field at the CMB is very poorly known above about degree 14, and its secular variation is not well determined above about degree 8. This introduces uncertainties in all of the above models, uncertainties that are not included in formal error analyses and are not even mentioned in many of the relevant papers.

In addition, there are questions that must be asked concerning the additional assumption made in each of the above models.

(i)   The core's viscosity is commonly estimated to be only a few times that of water and some dynamo models assume core convection to be turbulent. Thus it is questionable whether there is any a priori justification for assuming the motions in the outermost part of the core to be steady over time spans of a few tens of years, as assumed in model (*i*). Moreover, because of changes in core–mantle coupling that occur on a decade time scale (decade changes are discussed in §11.2), at least some variation in the fluid flow seems likely on this time scale (Bloxham and Jackson, 1991; but see Voorhies, 1993). However, as noted by Bloxham and Jackson (1991), it may be that the steady part of the flow is all we can reasonably expect to obtain.

(ii) The size of the toroidal magnetic field near the CMB is unknown. Its estimated value depends on several assumptions and on any large regional structure within the D" layer at the base of the mantle. One of the few areas of consensus in geodynamo theory is that strong-field models (those in which the Lorentz force plays a significant role) are likely appropriate for the Earth. However, it is not clear how close to the CMB the Lorentz force may still play a significant role. Indeed, Weiss (1966) concludes that much of the core's magnetic field might be concentrated in boundary layers. It is not, therefore, at all clear that the fluid motions in the outermost core are in geostrophic balance, as assumed in model (*ii*).

(iii) The concept of toroidal flow was initially motivated by the possibility that the core is stably stratified in a layer beneath the CMB, based on a suggestion first made by Higgins and Kennedy (1971). Although the assertions of Higgins and Kennedy are now in doubt, the concept of a stably stratified layer persists, even though controversial (§7.1). A criticism of the assumption of toroidal velocity at the CMB, as in model (*iii*), is that by definition it excludes the possibility of upwellings or downwellings near the CMB.

Although there are several concerns expressed above, it must be recognized that these models currently provide us with our best estimates of the core's velocity field. The order of magnitude of the velocity obtained from each of the above models is the same as that obtained by Bullard *et al.* (1950) from westward drift of the nondipole field. This similarity can be traced to the frozen flux assumption and does not appear to depend critically on the detailed assumptions made for the individual models. As frozen flux is probably an excellent first-order assumption on the time scales involved in the above analyses, we conclude that a velocity magnitude near 10^{-4} m s^{-1} is reasonably well established. Those who require more information on the velocity field (e.g., to investigate core–mantle coupling) may choose to use one of the above models, but in so doing they should be aware of the large amount of "information" on the velocity field they are adding by assumption.

11.2 Core–Mantle Coupling: Length of Day

11.2.1 Nutation and Wobble

Variations in rotation rate cause perturbations of the *length of day* (LOD) and, therefore, of universal time (UT) as determined by observations of celestial objects. Variations in the Earth's rotation rate occur over a time scale ranging from a few days to hundreds of millions of years. Short period variations are defined as less than about 5 yr, *decade variations* range from 5 yr to many tens of years, and the remainder are referred to as long-term variations. Variations of the position of the rotation axis with respect to inertial space are called *nutations* and variations with respect to geographical locations on the Earth are called *wobble*. They are easily distinguished in practice: nutation causes changes in the angle between the celestial pole and any star (e.g., Polaris, the "north star") while polar motion changes the angle between the celestial pole and the local horizontal (the observer's latitude). There will always be some wobble that accompanies nutation and vice versa. However, one can characterize the Earth's response to an excitation as predominantly either wobble or nutation.

Excitations of variations in LOD can be broadly divided into three groups.

(i)   Exchange of angular momentum between the solid earth and the fluid parts of the Earth: the atmosphere, oceans, and liquid core.

(ii) Changes in the moment of inertia tensor because of internal effects associated with the "solid" part of the Earth (e.g., earthquakes).

(iii) Effects of external torques (e.g., lunar and solar tides).

The short period variations turn out to be predominantly nutations that are caused by variations in the atmosphere (e.g., Wahr, 1986; Jault and Le Mouël, 1991) . It is also well known that there is a long-term decrease in the Earth's rotation rate associated with tidal damping. For example, there is an approximately 2-ms-per-century decrease in the LOD caused by frictional effects associated with lunar tides. At the same time the distance between the Earth and Moon is slowly increasing so that the total angular momentum of the Earth–Moon system is conserved. This long-term change in the LOD must be removed from the total change to obtain the short period and decade variations.

Modern measurements of the LOD include satellite Doppler tracking, lunar laser ranging, satellite laser ranging, and long baseline interferometry. There are many general reviews of the subject (e.g., Lambeck, 1980; Wahr, 1986; Dickey and Eubanks, 1986; Roberts, 1989a).

In this section we are primarily interested in the decade variations since they are widely believed to involve core–mantle coupling. Decade variations typically have magnitudes in the range of a few milliseconds. The ratio of the moment of inertia (about the Earth's rotation axis) of the atmosphere to the mantle is of order 10^{-6}, while that of the core to the mantle is roughly 0.13. It is commonly accepted that the fluid core is the only part of the Earth with enough inertia to cause the decade variations in the LOD. If we assume for the moment that the liquid part of the core moves as a unit, then the torque between the core and the mantle is given by

$$\Gamma = I_c \frac{d\Omega_c}{dt} = I_m \frac{d\Omega_m}{dt} = I_m \frac{\partial\Omega_m}{\partial T_d}\frac{dT_d}{dt} = -I_m \frac{2\pi}{T_d^2}\frac{dT_d}{dt} \quad , \qquad (11.2.1)$$

where subscripts c and m refer to core and mantle, respectively, I is the moment of inertia about the rotation axis, Ω represents the magnitude of the angular velocity about the rotation axis, and T_d is the mean LOD, i.e., $T_d = 2\pi/\Omega_m$. One can measure T_d over some time interval (e.g., 20 yr) and, after filtering out the short and long period variations, obtain dT_d/dt — the decade variation. This was recently done by Jault and Le Mouël (1991) using LOD data for the 1965–1985 interval. They find the magnitude of the torque required to explain the resulting decade variation to be 4×10^{18} Nm. This is the same order of magnitude of torque as has been previously found for decade variations (e.g., Roberts, 1989a).

There are four broad categories of core–mantle coupling: viscous, gravitational, electromagnetic, and topographical. It is generally agreed that the core's viscosity is probably many orders of magnitude too small to explain the decade variations in the LOD (e.g., Lambeck, 1980). Gravitational coupling occurs if there are lateral variations in density near the base of the mantle and the top of the core (e.g., Buffet *et al.*, 1993). Although model dependent, values of 10^{-9} in $\Delta\rho/\rho$ (where ρ is density) near the top of the core may be sufficient to produce significant coupling (J. Parks, personal communication). However, we have seen in §7.6 that thermal winds can occur with temperature differences as small as 10^{-6} degrees. Considering that the thermal expansion is of order 10^{-5}, it seems it would be difficult to maintain the required variations in the density in the core: density differences as small as 10^{-11} could induce flow to reduce those differences. Nevertheless, strong magnetic fields can have a significant effect on stabilizing variations in density, as evidenced by their role in sunspots (§10.2). Moreover, the above arguments can be inverted: there almost certainly is convection in the outer core, reflecting the fact that there are density variations.

Hence the gravitational mechanism deserves a closer examination than it has thus far received. In the following two subsections we examine the electromagnetic and topographical coupling mechanisms.

11.2.2 Electromagnetic coupling

The torque associated with electromagnetic coupling is $\mathbf{r} \times \mathbf{F_L}$ where $\mathbf{r}$ is the core's radius and $\mathbf{F_L}$ is the Lorentz force, $\mathbf{J} \times \mathbf{B}$. As discussed in §8.4, $\mathbf{J} \times \mathbf{B}$ can be written in terms of an effective pressure and tension:

$$\mathbf{J} \times \mathbf{B} = \frac{\nabla(\mathbf{B} \cdot \mathbf{B})}{2\mu_0} + \frac{\nabla \cdot (\mathbf{B}\mathbf{B})}{\mu_0} \ . \tag{11.2.2}$$

The effective pressure (first term on right side of Eq. (11.2.2)) does not contribute to the electromagnetic coupling at the CMB (this can be seen physically since it can be incorporated into the pressure term in the Navier–Stokes equation, or it can be seen mathematically by showing that its integral over the CMB is zero). Applying the divergence theorem to the last term gives the relevant force as $(1/\mu_0)\int(\mathbf{B} \cdot \mathbf{n})d\mathcal{S}$, where $\mathcal{S}$ is the surface area of the core–mantle boundary and $\mathbf{n}$ is the unit normal vector to this boundary. This shows that magnetic field lines must cross the CMB for there to be electromagnetic coupling. If the CMB were spherical and had no topography and if the magnetic field were purely toroidal, there would be no electromagnetic coupling across the boundary. However, this is not the case and the toroidal field can play a significant role in coupling. This can be seen by substituting the second term on the right side of (11.2.2) into the relationship for torque. Then by integrating over the core's volume and by applying the divergence theorem one obtains the total electromagnetic torque, Γ, at the CMB

$$\Gamma = \frac{1}{\mu_0} \int \mathbf{r} \times \nabla \cdot \frac{\mathbf{B}\mathbf{B}}{\mu_0} d\mathcal{V} = \frac{1}{\mu_0} \int (\mathbf{r} \times \mathbf{B}) B_r d\mathcal{S} \ ,$$

where $\mathcal{V}$ indicates the core volume. It is a common practice to separate $\mathbf{B}$ into its toroidal and poloidal parts (see toroidal and poloidal fields in §8.3 and §9.1) in the last integral to obtain the toroidal and poloidal torques:

$$\Gamma_T = \frac{1}{\mu_0} \int (\mathbf{r} \times \mathbf{B}_T) B_r d\mathcal{S} \tag{11.2.3}$$

and

$$\Gamma_p = \frac{1}{\mu_0} \int (\mathbf{r} \times \mathbf{B}_p) \mathbf{B}_r d\mathcal{S} \quad . \tag{11.2.4}$$

These equations illustrate that if the toroidal magnetic field is greater than the poloidal magnetic field at the core mantle boundary, then it will produce the larger associated torque. Our perception of the magnitude of the toroidal magnetic field is model dependent, but in the popular strong-field models it is the dominant field within the core. If the mantle were an insulator, the toroidal magnetic field would vanish at the CMB. This follows since the toroidal magnetic field is continuous across the CMB, it has no radial component, and by assumption no current can cross the CMB. However, the lowermost mantle is not a perfect insulator and may even contain regions of very high electrical conductivity. This emphasizes the crucial requirement for knowledge of the magnitude and distribution of electrical conductivity in the lowermost mantle when estimating electromagnetic coupling at the CMB. Notwithstanding these points, even when a strong toroidal magnetic field is assumed, most estimates for the electromagnetic coupling (assuming that the mantle is a semiconductor all the way to the CMB) are an order of magnitude too small to account for the LOD variations (e.g., Roberts, 1989a; Jault and Le Mouël, 1991).

Buffett (1992) has reexamined the electromagnetic coupling problem by assuming topography on the CMB up to roughly 5 km in amplitude (e.g., Morelli and Dziewonski, 1987) and by considering that there may be regions of high electrical conductivity (e.g., Knittle and Jeanloz, 1989, 1991) within the seismic D" layer. $\mathbf{B}_r$ in (11.2.3) and (11.2.4) should then be replaced by the normal component of $\mathbf{B}$, and this would no longer necessarily be in a radial direction. However, even for the large topography assumed (which is at the upper end of model values; e.g., see Creager and Jordan (1986) or Doornbos and Hilton (1989) for alternative interpretations), the increase in CMB *electromagnetic* coupling is insufficient. A different conclusion is obtained by Buffett when there are sufficiently large regions of high electrical conductivity within D". Such regions can increase the CMB coupling to the required level and beyond. Figure 11.3 helps to illustrate this type of coupling and to provide insight into the nature of electromagnetic coupling in general.

One might think that the question of electromagnetic coupling could be resolved by seeing if there is a correlation between secular variation and the LOD, as was first proposed by Vestine (1953) on the basis of an apparent correlation between a 58-yr period in the secular variation and in the LOD. No correlation would be perfect, but it is not even clear whether the secular variation should lead or lag changes in the LOD in a causal correlation. Indeed, Le Mouël *et al.* (1981) claim there is a significant correlation between secular variation and LOD where the magnetic field variation leads the LOD variation, while Langel

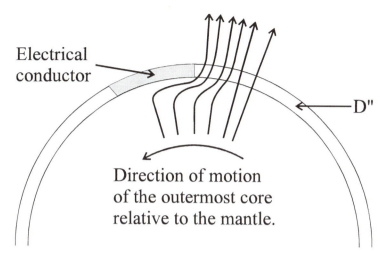

Fig. 11.3. Good electrical conductors in D" will produce an increase in core–mantle coupling because the magnetic field lines will be forced together.

et al. (1986) claim that the magnetic field changes lag those in the LOD. Similarly, Hide and Malin (1970, 1971a,b) suggested there was a correlation between small changes in the geoid and the geomagnetic secular variation after a phase shift of 160°, but this correlation has not been supported by most subsequent analyses (e.g., Khan, 1971; Lowes, 1971; Lambeck, 1980). Thus, neither calculations nor observations provide certain estimates for the magnitude of electromagnetic coupling across the CMB.

11.2.3 Topographical Coupling

Hide (1969; also see Hide, 1989, 1995) was the first to suggest that topography might exist on the CMB and that it might have important consequences for core–mantle coupling. The calculation of the effects of topography might seem relatively easy at first glance. The magnitude of the tangential shear stress should be directly related to hdP/dr, where h is the amplitude of the topography and dP/dr is the pressure gradient (e.g., Lambeck, 1980). The pressure gradient can be obtained since it is balanced by the Coriolis force in a geostrophic balance approximation. Thus it might seem straightforward, at least in principle, to obtain the magnitude of topographic coupling. However, as pointed out by Jault and Le Mouël (1991), there are some serious problems with this approach. They examine the (rotational) axially symmetric problem and point out that the Taylor constraint (§9.6) is applicable to the situation. This implies that the total zonal torque (resulting from a pure geostrophic model) on any cylinder with an axis along the Earth's rotation axis must vanish, so there will be no change in the

LOD in a pure geostrophic model. Thus the acceleration term, $d\mathbf{v}/dt$, where $\mathbf{v}$ is velocity, in the Navier–Stokes equation must be included in the force balance. Jault and Le Mouël do this and use the (tangential geostrophic) velocity model of Gire and Le Mouël (1990) and the topography model of Morelli and Dziewonski (1987) to obtain a simple order of magnitude estimate for the torque of around 2×10^{19} Nm. This value is approximately an order of magnitude too high.

It should be clear that the orientation of the topography to the core flow is important as well as the amplitudes of the flow and topography: flow parallel to an (inverted) core ridge produces no topographic coupling, while flow perpendicular to a ridge does. Jault and Le Mouël (1991) show that with minor adjustments to the topographic model of Morelli and Dziewonski they can obtain an acceptable value for the torque.

The CMB region is complex and our knowledge about it is highly speculative. It may be that core–mantle topography is negligible. Hence we arrive at the conclusion that the Jault–Le Mouël analysis shows that topographical coupling is a plausible model that depends rather critically on the right core–mantle topography. Similarly, it is possible to conclude that electromagnetic coupling might be the prime agent of core–mantle coupling, given the appropriate electrical conductivity structure within D".

11.3 Paleomagnetism and Dynamo Theory

11.3.1 Testing Geodynamo Theory

Paleomagnetism is the subfield within geophysics that appears to have the best chance of eventually providing effective tests of geodynamo theory. This is mainly because of the time constants involved. Consider the hypothetical situation in which the geodynamo is in a (quasi-) steady state, i.e., $\partial \mathbf{B}/\partial t = 0$. In this case the magnetic induction equation reduces to

$$k\nabla^2 \mathbf{B} = -\nabla \times (\mathbf{v} \times \mathbf{B}) \quad , \tag{11.3.1}$$

which illustrates that, in the steady state, diffusion effects are balanced by dynamic effects. Since free decay of the dipole field (in the lowest radial mode; see §9.1) is of order 10^4 yr (the characteristic time), at least this much time is needed to obtain any reasonable view of the behavior of the geodynamo. Moreover, many of the interesting phenomena that a viable theory should explain (e.g., magnetic reversals) have similar, or even longer (e.g., the interval of time between reversals), characteristic times.

Unfortunately, proposed tests of geodynamo theory are few and results of those tests ambiguous. For example, consider possible effects resulting from the presence of an inner core.

(i) As the inner core grows over time there will be a change in the ratio of chemical buoyancy (generated at the inner core through freezing) to thermal buoyancy (generated at the CMB) (see §7.4). This might imply a change in intensity of the Earth's magnetic field with the evolution, which could constitute a test. It has been claimed that this has been supported by paleointensity data ranging back to 3.1 billion years (Hale, 1987). The limited Pre-Cambrian paleointensities indicate a dipole moment similar to the present field, except for the 6 oldest estimates obtained by Hale, which suggest a much lower value (§6.3.4). However, there are fewer than 100 estimates for the entire Pre-Cambrian and, as noted by Merrill (1987), many of these estimates involve questionable techniques and results. The 6 new values provided by Hale are far too few to demonstrate a trend. In summary, it is such a formidable task to obtain reliable intensities for the Pre-Cambrian in the abundances needed that a convincing test will probably not be possible in the near future.

(ii) The convection and magnetic fields within the tangential cylinder containing the inner core are likely to be substantially different from those outside the cylinder. Pursuant to discussion of this point in §9.6, one might expect that some aspects of the surface geomagnetic field (e.g., secular variation) would be different within the tangential cylinder (when extended to the Earth's surface) than outside it. Although this may be one of the easier predictions of dynamo theory to test, we are not aware of any published attempts.

(iii) Magnetic fields generated in the outer core will diffuse into the inner core. This field in the inner core could have a stabilizing effect on the magnetic field in the outer core by diffusing back out into the outer core when there are rapid changes in the field near the bottom of the outer core (Hollerbach and Jones, 1993a,b). This was suggested by Hollerbach and Jones using an $\alpha\omega$-dynamo model. Diffusion times of magnetic fields into the inner core will depend on their dimensions in the outer core (similar to calculations discussed in §9.1). Hence these times will be model dependent, as evidenced by significant differences in the fields calculated by Hollerbach and Jones and those calculated by Glatzmaier and Roberts (1995a,b). Nevertheless, any marked changes in the magnetic field near the base of the outer core will, on the average, be resisted by diffusion of magnetic fields out of the inner core: the only question is by how much and over what time interval. Thus the basic concept appears sound. A problem is that it does not appear readily testable by paleomagnetic data. The rate at which reversals have occurred over the past few hundred million years is highly variable (both increasing and decreasing; see §5.3), presumably because of slow changes in CMB boundary conditions. The small growth of the inner core during this same time interval would, according to the arguments of Hollerbach and Jones, lead to a small monotonic decrease in reversal frequency. But it appears unlikely that such a change could be discerned from the more dramatic changes in the reversal rate that evidently stem from other mechanisms.

(iv) The inner core exhibits seismic anisotropy in which the P-wave velocity is a few percent faster parallel to the rotation axis than perpendicular to it (§7.1). This does not appear to be able to provide a test of dynamo theory (but see Clement and Stixrude, 1995). However, it does appear to be providing evidence that there is a strong toroidal magnetic field in the core (Karato, 1993; McSweeney *et al.*, 1996). One problem with the inner-core anisotropy studies is that the observed anisotropy (McSweeney *et al.*, 1996) is larger than that allowed by condensed matter physics calculations, even assuming perfect crystal alignment (Stixrude and Cohen, 1995b)!

The reader might think that there must be far better tests of geodynamo theory than those given above, such as the existence of magnetic field reversals. However, the prediction of the existence of field reversals was made before there was any dynamo theory (§5.1) and, perhaps surprising to many, their existence is not in fact a test of dynamo theory. It is possible to construct dynamo models that reverse and ones that do not, and the geodynamo appears to have both reversing and nonreversing regimes (§5.3.5). Paleomagnetism has demonstrated the existence of two quasi-stable polarity states and that there are transitions between those states (§5.1). Rather than a test of geodynamo theory, this information serves as a constraint on viable geodynamo models. In summary, there are few, if any, actual tests of geodynamo theory, and this disconcerting fact should serve as a challenge to theoreticians in the future.

11.3.2 Constraints on Dynamo Theory

Although, in principle, virtually any measurement of the geomagnetic field is a constraint on geodynamo theory, most measurements do not currently provide any useful constraint in the sense of having an impact on the theory. The most obvious and general constraint involves magnetic reversals, which has been discussed above and throughout Chapter 5.

The spectrum of the core's magnetic field might provide another constraint. For example, Stacey (1992) uses the observation that the spherical harmonic spectrum is nearly white (pink) at the CMB (§2.2) to argue that there is turbulence there. However, if cascading occurs in the MHD core fluid, as claimed by some (e.g., Biskamp, 1993), then the spectrum should not be white. Moreover, even if cascading does not occur, there should probably be a peak in the spectrum (Braginsky and Meytlis, 1990), i.e., again it should not be white. Thus it is not clear whether the observed geomagnetic spectrum is typical for the Earth and, even if it is, what this means in terms of a constraint on dynamo processes.

Similarly, there is the question of the strength of any constraints that stem from estimates of the fluid velocity at the top of the core (§11.1). Although there are differences between velocity models, there are also similarities, such as the northward directed flow beneath the Southern Indian Ocean (Fig. 11.3). Unfortunately, even though these velocity fields may be providing us with valuable information (e.g., on various symmetries in the dynamo), they appear to relate to too short an interval of time and to represent too small a volume of the core (without making still more additional assumptions) to provide strong constraints on geodynamo theory.

Further possible constraints do stem from paleomagnetic data (Chapters 4, 5, and 6). The following examples have not been explicitly discussed elsewhere in this book and they illustrate the usefulness of the growing symbiotic relationship between paleomagnetism and dynamo theory.

The first two examples involve the so-called dynamo families (§8.5). Symmetry arguments allowed dynamo theoreticians to define two dynamo families: the symmetric and antisymmetric families (Roberts, 1971; Roberts and Stix, 1972). These families are independent of each other when certain idealized conditions involving symmetries in the core's velocity field apply. Under different symmetry conditions they interact strongly with each other (§8.5). Based on paleosecular variation analyses of lavas that cooled during the past five million years, Merrill and McFadden (1988) suggested that reversals occur more frequently when there is increased interaction between the two dynamo families. This led to the prediction of a positive correlation between the ratio of the symmetric to antisymmetric families and the rate of reversals. In particular, they predicted that the contribution of the symmetric family to the geodynamo would have been at a minimum during superchrons, such as the Cretaceous Superchron (§5.3).

The symmetry is such that the scatter of virtual geomagnetic poles at the equator can only be caused by the symmetric family. Hence it is possible to test some of these predictions without requiring a particular model or theory for secular variation. Unfortunately, some model dependency is required to estimate relative variations in the antisymmetric family and model G was used (§6.4). These predictions were tested by McFadden *et al.* (1991) and appear confirmed, as shown in Fig. 11.4 (see Fig. 6.15 and §6.4.6). Note that this is not actually a prediction from dynamo theory: it has not been shown from theory that there

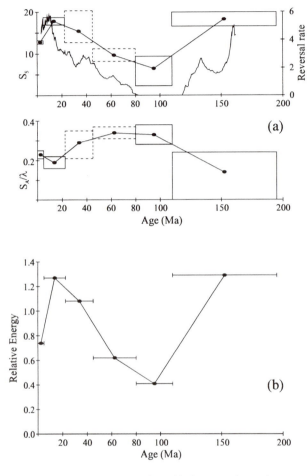

Fig. 11.4. (a) Variation of S_S and (S_A/λ) together with the reversal rate, in reversals per million years, from 190 Ma to the present. S_S (in degrees) is the contribution to PSVL scatter made by the symmetric family and (S_A/λ) is the contribution made by the antisymmetric family, where λ is the paleolatitude in degrees. The range of ages contributing to the estimated parameter and the 95% confidence limits on the estimate are indicated by the boxes around the plotted points. A solid box indicates a good fit of the data to Model G; a dotted box indicates a poor fit. (b) Estimate of the variation in the relative energy in the secular variation observed at the Earth's surface. Horizontal lines show the age range over which each estimate was performed. See also Fig. 6.15 and §6.4.6.

will be fewer reversals when the symmetric family is relatively low. Nevertheless, it is unlikely the speculation would have been made had theoreticians not developed the underlying theory for the two families and their interactions. This then represents an illustration of the growing symbiosis between practitioners of dynamo theory and paleomagnetism.

In spite of the success of the above predictions, there may be other explanations compatible with the data. For example, consider the possibility that the nondipole part of the field originates preferentially in the outermost part of the core, while most of the dipole part originates deeper in the core. Such a possibility was first suggested by Bullard *et al.* (1950) to explain secular variation (§8.6). Gubbins and Kelly (1995) develop theory and analyses based on recent secular variation data to argue that it may be the equatorial dipole that strongly dominates the variations in the symmetric family in the past. If so, one can speculate that reversals are more likely when dipole wobble (deviations from the rotation axis) is largest. This speculation is also compatible with the data illustrated in Fig. 11.4, because dipole wobble is a restricted version of the Merrill–McFadden suggestion (the equatorial dipole is part of the symmetric family). Perhaps the best way to interpret the above is simply that there appears to be a correlation between variations in the symmetric dynamo family (and the antisymmetric family; see Fig. 11.4) and reversal frequency. This appears to be a viable constraint on the theory for geomagnetic reversals. The selection rules developed by Bullard and Gellman (1954: see §9.1) indicate that it would be difficult to have dipole wobble without simultaneously generating higher degree harmonics, so we prefer our original suggestion but cannot rule out the mechanism we have constructed here. Nevertheless, the examples given above illustrate that some tuning of the "mechanisms" is possible.

There appears to be another constraint imposed on geodynamo theory that involves the two dynamo families. Thus far theory (e.g., Zhang and Busse, 1989) does not indicate any basis for preference for the antisymmetric family over the symmetric family. There is consensus that a dynamo produces the majority of the magnetic field in five of the planets. Three of these (Earth, Saturn, and Jupiter) are currently dominated by the antisymmetric family while the data are ambiguous for Uranus and Neptune (§10.5). This does not by itself constitute evidence for a preference of one family over another, but is an interesting observation. However, there is a strong case that the Earth's magnetic field has been dominated by the antisymmetric family. During the past 150 million years there have been several hundred field reversals and the quasi-stable magnetic field between these reversals has been dominated by the geocentric axial dipole term (Merrill and McFadden, 1995).

The last point deserves some elaboration. Downward continuation of the present magnetic field to the CMB indicates that the antisymmetric family is dominant in the core (almost entirely because of the axial dipole term; see §2.2).

Paleomagnetism has been very successful as a global tectonic tool because the *average* field is close to that of a geocentric axial dipole field (§6.1). We cannot rule out the possibility that contributions to the symmetric family vary more rapidly with time and thus average out faster than those associated with the antisymmetric family. However, if this is true it would still suggest that the antisymmetric family has been more stable. Regardless, it seems more likely that harmonics of the same degree have similar time constants and that the dominance of the axial dipole field at the CMB is a property of the Earth's magnetic field. In particular, if the average harmonic spectrum were white at the core–mantle boundary, the dipole field would only constitute roughly 50% of the field at the Earth's surface (Bogue and Merrill, 1992). Had this been the case in the past, the scatter would have been significantly larger than at present (which is not evident in the data, e.g., Fig. 11.4), and global paleomagnetic tectonic studies would not have been viable. We conclude that the Earth's magnetic field has been dominated by the antisymmetric dynamo family for at least the past 150 million years and probably much longer. It is not clear whether this is a fundamental property of planetary dynamos or a specific property of the Earth as a consequence of particular boundary conditions.

The geodynamo appears to have two basic states: a reversing state and a nonreversing state (Merrill and McFadden, 1994, 1995; see §5.3.5). There appears to be some threshold separating these two states that may reflect a fundamental change in dynamo processes. McFadden and Merrill (1995a) summarize the evidence for this conclusion and, based on the dynamo theory of Hagee and Olson (1991) and Phillips (1993), speculate that the change may be the result of something akin to α^2 processes (which tend to be steady) dominating the dynamo during superchrons, with $\alpha\omega$ processes (which tend to be oscillatory) dominating the rest of the time. Although it is not clear whether such speculations can be extended to more realistic strong-field dynamos, the constraint provided by the paleomagnetic data should limit the parameter space available to dynamo theoreticians.

The above list of examples is certainly not intended to be complete. For example, in §5.3.5 it is pointed out that following a reversal there may be a time interval of a few tens of thousands of years during which the probability of another reversal is decreased, possibly because the intensity of the magnetic field is higher during that time interval. Is this compatible with dynamo models that exhibit reversals? There are many other questions that can be asked and possible constraints imposed on the basis of robust analyses of paleomagnetic data as given in Chapters 4, 5, and 6.

11.4 Variations at the Core–Mantle Boundary and the Earth's Surface

11.4.1 Spatial Variation at the CMB

Magnetohydrodynamics processes are well known to be sensitive to boundary conditions. Hence it seems plausible that observations of the magnetic field might provide information on any spatial and temporal variations that occur at the CMB.

Hide (1969) appears to have been the first to suggest the existence of lateral (topographic) variations at the CMB boundary and, as already discussed in §11.2, to suggest that this could be tested using gravity data. Several properties other than topography might also vary laterally at the CMB.

Although the magnetic field can have a pronounced affect on buoyancy and hence on thermal wind generation at the CMB (as evidenced by the discussion on the stability of sunspots in §10.2), variations of less than a degree at the CMB can still lead to strong thermal winds (§7.6). The CMB should, therefore, be regarded as an isothermal boundary for many calculations. This might be thought (incorrectly) to imply that there can be no lateral temperature variations at the CMB, when it should instead be concluded that extremely small temperature differences at the CMB can have dramatic effects on the fluid velocities in the outermost core. There probably are such temperature differences at the CMB because of the chemical and physical variations that appear to be present within D" (§7.2).

There are also likely to be lateral variations in the temperature gradient (associated with chemical and physical variations within D", including variations in its thickness; see §7.2 and §7.3) and thus in the rate at which heat is conducted away from the CMB. Additionally, chemical variations might be associated with large (several orders of magnitude) variations in electrical conductivity (§7.2), which in turn could have a significant affect on the magnetic fields near the boundary (e.g., Busse and Wicht, 1992; Buffett, 1992). As can be appreciated from Fig. 9.2, this effect will be largest for toroidal magnetic fields. That figure shows there would be no toroidal magnetic fields in the mantle if D" were an insulator, but if it were a good electrical conductor there could be very strong magnetic fields in D". In summary, there are several properties that could vary laterally along the CMB and which could strongly affect the geodynamo: these include topography, temperature, temperature gradient, thermal conductivity, and electrical conductivity.

There are claims in both the geomagnetic and the paleomagnetic literature that lateral variations at the CMB have been documented. However, we suggest that all such claims appear equivocal.

As discussed in §9.6, Bloxham and Gubbins (1985) identified two pairs of magnetic foci that have remained stationary over the past few hundred years. Members of each pair are in the opposite (north–south) hemispheres and are separated in latitude by approximately 120°. Bloxham and Gubbins interpret these foci as manifestations of convection rolls that are aligned with the rotation axis in the outer core (see Fig. 9.6). They further argue that these rolls represent the main convection that produces the geodynamo and that they also represent evidence of lateral heterogeneity at the CMB. Clearly there are several assumptions involved with these assertions. It is not clear whether there are convection rolls aligned with the rotation axis in strong-field dynamos (e.g., Zhang and Busse, 1989). In those dynamo models in which rolls are present (§9.6), they drift in longitude and so there would have to be something to pin them to particular spots on the CMB. This is where the lateral variation enters, since Bloxham and Gubbins speculate that CMB topography or some other laterally varying property at the CMB does the "pinning." Finally, it is assumed that the stationary foci, which come from analyses of measurements made over the past few hundred years, have remained stationary over several million years, as they must have done if they are to be taken as evidence of heterogeneity on the CMB.

The last assumption is testable, at least in principle. If there are lateral variations in the properties of the CMB, they might be manifested in paleomagnetic data for rocks that formed during the past few million years. Gubbins and Kelly (1993) first attempted this test using a restricted paleomagnetic data set for the past 2.5 million years and claimed support for the Bloxham–Gubbins hypothesis. Recently, Johnson and Constable (1995b) have applied spherical harmonic analyses to a much more extensive paleomagnetic data set and concluded that there are significant longitudinal variations (nonzonal terms) in the time-averaged field for the past five million years. These variations neither refute nor confirm the Bloxham–Gubbins hypothesis. McElhinny *et al.* (1996a) have also carried out analyses of paleomagnetic data for rocks with ages of five million years or less and concluded that there is no convincing evidence for the existence of any nonzonal terms in the *time-averaged* field (§6.2.3). A primary difference between the last two studies is the question of the number of data required to obtain a representative value at a site that averages out secular variation. Many lava flows can erupt over a very short time interval, as shown by the nearly continuous volcanic activity on the flanks of Kilauea volcano on Hawaii since 1973. Therefore, to obtain a reasonable value of the time-averaged magnetic field it is necessary to sample rocks that clearly represent "many" distinct time intervals. When the global data are subdivided into regional averages, one has the additional problem of deciding whether there are sufficient data within each region such that each one independently averages out the field to the same extent as any other one. Our view is that there are insufficient data globally to determine anything other than the zonal quadrupole term as a fraction of the zonal dipole term (see full discussion in §6.2.3). Hence, we conclude that an adequate test of the Bloxham–Gubbins hypothesis has not been made.

Systematic changes during polarity reversals, such as preferred paths for virtual geomagnetic poles, would require some inhomogeneity in boundary conditions. Although one cannot rule out that such systematics might be attributable to properties of the inner core (Clement and Stixrude, 1995), observed "systematics" are usually attributed to variations in the CMB boundary conditions. However, even though there may be certain systematic properties involving directions during magnetic field reversals, the evidence for their existence is not convincing (McFadden and Merrill, 1995b; see also §5.4).

Differences in the time-averaged reverse and normal polarity states would provide an indirect argument for the presence of lateral variations within D". As discussed in §8.6, the dynamo equations indicate that the time-averaged normal and reverse magnetic fields should be identical, so any differences would have to be attributed to either initial or boundary conditions. However, it is unlikely that initial conditions could produce observable asymmetry in the polarity states. Thus any asymmetries in the time-averaged polarity states would require some magnetic field bias sourced from outside the core (e.g., Merrill *et al.*, 1990).

There are theoretical reasons why small biasing fields should exist. Large variations of thermal or electrical conductivity within D" would create thermal-electrically driven currents that flow, entirely or partially, within D". Chemical reactions could also give rise to electric potential differences and hence electric currents within D". Such effects have been discussed most completely in Merrill *et al.* (1990), who argue that there should be some bias from fields that originate outside the Earth's core. However, the biasing effects would be very small, so manifestation at the Earth's surface would require dramatic amplification by dynamo processes.

The existence of such internal biasing fields would also help explain the origin of the initial magnetic field required to seed the geodynamo. The alternative to having an internal field as the initiating field is to call on some external field, such as the solar field during a T-Tauri phase (§10.2). Once the dynamo is started, the initiating field can be removed and the dynamo can be self-sustaining. However, any external magnetic field lying in the ecliptic plane (e.g., from the Sun) would have been electrically screened out by the mantle early in the Earth's history, a time when the Earth's rotation rate was but a few hours. It is possible for fields that are aligned *along* the Earth's rotation axis to reach the core and serve as the initial field to start the geodynamo, but it is not clear where moderately strong fields with such orientations might come from (the galactic field, for example, appears too small).

There have been numerous papers that claim to have demonstrated the existence of significant differences between some average property of the normal and reverse polarity states. All such claims were shown to be nonrobust by McFadden *et al.* (1989), with the exception of those that claimed a significant difference between the time-averaged reverse and normal polarity states. Every paper published that carried out a spherical harmonic analysis of the time-averaged data for the two polarity states has found a significant difference (at the 95% confidence level) between the data for the two states (e.g., Merrill and McElhinny, 1977; Schneider and Kent, 1990; Merrill *et al.*, 1990; Johnson and Constable, 1995b). However, the findings have been inconsistent in that the studies did not attribute the differences to the same spherical harmonic terms (e.g., the quadrupole or octupole terms). McElhinny *et al.* (1996a) have recently made use of the new global paleomagnetic databases to reanalyze all available data. They again find apparent differences between the data from the two polarity states, but they find that the differences are a function of several parameters, including rock type, incomplete magnetic cleaning of reversely magnetized rocks, nonorthogonality of spherical harmonics in a disconnected space, and aliasing problems. They conclude that the apparent differences claimed in the past are almost certainly a consequence of factors other than genuine differences in the properties of the time-averaged magnetic fields (see §6.2).

In conclusion, theoretical considerations indicate that there should be variations on the CMB that affect the geomagnetic field. However, the magnitudes of these effects are model dependent and appear to be very small. Finally, there is no robust observational evidence from direct or indirect measurements of the magnetic field that demonstrate the existence of lateral variations in properties at the CMB.

11.4.2 Variations of the CMB Boundary Conditions with Time

After magnetic reversals, the next best established paleomagnetic conclusion about the Earth's magnetic field is the long-term nonstationarity in the mean rate at which reversals occur (Fig. 11.4). It is important to emphasize that only the existence of this nonstationarity, not its details, is well established (§5.3).

Several of the (partial differential) dynamo equations are coupled and include nonlinear terms. A full description of the mantle (including D″) and crust would involve many more coupled nonlinear partial differential equations. Evidently it is possible to consider all geomagnetic variations as simply being deterministic chaos resulting from the interaction of all these equations. However, it is more instructive to consider the core as being the system of interest and the mantle as providing varying boundary conditions for that system. One can then ponder whether the nonstationarity in reversal rate during the Phanerozoic (with a time constant of approximately 2×10^8 yr) simply reflects the nonlinearities in the core system (i.e., still deterministic chaos) or whether it is a consequence of changes in the imposed boundary conditions. Although it is not possible to rule out deterministic chaos within the core as the cause, it seems more likely that this nonstationarity reflects changes in the CMB conditions. First, the time scale is very long, for dynamo processes act primarily on time scales of 10^4 to 10^5 yr or less, as can be seen from (11.3.1). The decay processes act on this time scale (§9.1) and they must be balanced by dynamic processes to maintain the field. Moreover the change reflects a dramatic difference in reversal rate: the mean time of a polarity interval varies by more than two orders of magnitude (during the reversing state of the past 80 million years, the shortest interval was about 10^4 yr and the longest about 3×10^6 yr). Second, the Earth has without doubt undergone thermal evolution, and this has to have varied the thermal conditions at the CMB. It is recognized that MHD processes acting in the core are sensitive to boundary conditions, and so it would be most surprising were there not some manifestation in the magnetic record of this thermal evolution of the Earth. Estimates of mean rates of mantle convection of 1 to 3 cm yr^{-1} (whether whole-mantle, or layered, convection; e.g., Davies and Richards (1992)) yield mantle one-way transit times of approximately 3×10^8 and 10^8 yr, respectively. Hence we should expect to see some indication in the paleomagnetic record of changes on time scales of this order, and so it seems appropriate to associate the long-term changes in reversal rate, as well as other long-term changes in the paleomagnetic record (paleosecular variation of directions (§6.4) and intensity §6.3), with changes in conditions at the CMB (e.g., Merrill and McFadden, 1990; McFadden and Merrill, 1995b).

Unfortunately, any of the boundary conditions discussed in the previous section could also vary with time and so it is difficult to be specific about which boundary conditions might be responsible for the observed changes in the magnetic field. As noted, the characteristic time for the nonstationarity is that expected for convection processes in the mantle (e.g., Davies and Richards, 1992). The thermal boundary layer at the base of the mantle could then be connected to the thermal boundary layer at the Earth's surface (the lithosphere) either by direct (whole mantle) convection or indirectly (by layered convection or by intermediate processes, such as "flushing" that allows some subducting slabs to pass suddenly through the 670-km transition). It is to be expected that this would lead to provocative papers linking long-term variations in the magnetic field to variations in various geological phenomena at the Earth's surface (e.g., Vogt, 1975; Jacobs, 1981; Sheridan, 1983; Courtillot and Besse, 1987; Larson and Olson, 1991).

One of the earliest attempts to link processes in the core to events at the Earth's surface appears to have been made by Glass and Heezen (1967), who argued that reversals were triggered by meteorite impacts. This idea was "rediscovered" by Muller and Morris (1986). However, important geological evidence appears to be inconsistent with their model and their suggested qualitative mechanism does not appear to work when quantified (Loper, 1989). Evidence for such "external" reversal mechanisms remains particularly unconvincing; geomagnetic reversals are far more likely to be associated with internal processes, as they are for the solar dynamo (e.g., Merrill and McFadden, 1990).

Irving (1966) appears to have been the first to suggest that active tectonism might be linked to increased reversal rates. Some of the most influential earlier papers that expanded on this idea were by Vogt (1972, 1975) and it is instructive to reconstruct his evidence and reasoning. At the time of Vogt's papers it was thought that there was a sudden change in the rate of magnetic reversals around 42 Ma. Also, it appeared there was evidence for a global change in plate tectonics at the same time, as evidenced by the bend in the Hawaiian-Emperor Seamount chain (this shows a sharp change in the direction of motion of the Pacific plate, as measured relative to the Hawaiian hot spot) and because the onset of the collision of India with Eurasia was then also thought to occur near 42 Ma. It seemed reasonable to link these events via a rapid reorientation of mantlewide convection cells that simultaneously affected both the CMB (and hence the reversal rate) and the lithosphere. Because the evidence appeared solid and the mechanism was simple and appealing, these ideas were accepted by many. Subsequently, it has been shown that the sudden change in the reversal rate near 42 Ma was an error in the reversal chronology and it now appears that the onset of the collision of India with the rest of Eurasia occurred prior to 55 Ma (Beck *et al.*, 1995). Moreover, one no longer expects there to be significant simultaneous changes at the CMB and the Earth's surface.

On the other hand, and as already pointed out, we do expect some processes in the lithosphere to be related to the mantle's lowermost boundary layer. Furthermore, if we accept that long-term nonstationarities in the Earth's magnetic field are associated with changes in boundary conditions at the CMB, then this implies non-steady-state mantle convection. This was first stated by McFadden and Merrill (1984), who speculated that the nonstationarities in the reversal chronology may be associated with Rayleigh–Taylor instabilities in the thermal boundary layer at the base of the mantle. However, one could invert this speculation and argue that it is instabilities in the surface boundary layer (the lithosphere) that lead to subduction and ultimately to changes in the boundary conditions at the CMB. This then leads to enormous flexibility in potential models, for we do not even know whether linked processes occur first at the lithosphere or at the CMB. For example, if CMB changes are dominantly caused by "dregs" associated with the remains of subducted lithosphere (e.g., Gurnis and Davies, 1986), then surface changes will precede changes in MHD processes in the core. Although it would take about 10^8 yr for the lithosphere to be subducted through the mantle in mantlewide convection models, it apparently only takes about half that time for D" to experience a significant thermal anomaly associated with the subduction process (Ricard *et al.*, 1993). In contrast, if *new* hotspots originate from instabilities in D", then changes in MHD core processes would precede hotspot eruption by about 20 million years (e.g., Loper and Stacey, 1983; Griffiths and Cambell, 1991; Davies and Richards, 1992).

This flexibility has led to numerous speculations, a few of which are given in Table 11.1. In addition, there are many other indirect "causal correlations" that have been made. For example, one might expect there to be an increase in mean sea level with increase in global spreading rates, and the maximum in subduction rates (between 140 and 90 Ma the subduction and spreading rates were almost double those in the past 30 m.y. (Richards and Engebretson, 1992)) precedes the eustatic sea level curve maximum (Haq *et al.*, 1987) by about 30 m.y. (Engebretson *et al.*, 1992). An increase in volcanism associated with an increase in lithosphere production could increase the CO_2 content in the atmosphere and affect the climate through the greenhouse effect. Similarly, an increase in the area covered by shallow continental seas associated with an increase in global spreading rates could also have climatic effects. Such changes might have implications for coal and gas deposits, for the formation of shales and carbonate rocks, etc. (e.g., Haggerty, 1994).

It should be clear that the major constraint on speculation in this area is the imaginative capability of geoscientists — a weak constraint indeed. However, based on observational data (such as shown in Fig. 11.4, and elaborated on by McFadden and Merrill (1995b)) and theoretical considerations (e.g., Loper, 1992), it does appear that significant changes in D" require at least 20 million years or so. Such information might help in developing, and carrying out, tests of the many existing and conflicting speculations.

TABLE 11.1
Some Suggested Causal Correlations Between Surface Geology and Magnetic Field Reversals

True polar wander	Low occurred about 170 to 100 Ma	Courtillot and Besse (1987)
Maximum in volume of hot spot volcanism	Roughly coincides with Cretaceous Superchron	Larson (1991); Larson and Olson (1991)
Maximum in lithospheric production	Occurred about 140 to 90 Ma	Richards and Engebretson (1992)
Maxima in kimberlite magmatic activity	Roughly coincide with superchrons	Haggerty (1994)

Parsimony seems required to interpret the properties of, and the processes acting in, the Earth's deep interior. Unfortunately, parsimony is sometimes in the eye of the beholder.

SI and Gaussian CGS Units and Conversion Factors

Problems arise in the conversion of SI and CGS units in magnetism because there are two equations that are commonly used in SI units relating the magnetic induction **B** to the magnetic field strength **H** and the magnetic moment per unit volume (either **M** or **J**). These are

$$\mathbf{B} = \mu_0(\mathbf{H} + \mathbf{M}) \tag{A.1}$$

and

$$\mathbf{B} = \mu_0\mathbf{H} + \mathbf{J} \quad , \tag{A.2}$$

where μ_0 is the permeability of free space. To distinguish between these two forms of the equation **M** is usually referred to as the *magnetization* and **J** as the *magnetic polarization*.

The Kyoto General Assembly of IAGA in 1973 adopted the form (A.1) for use in geomagnetism (Anonymous, 1974). The main reason for choosing this form is that unit conversions for magnetization from CGS to SI are then simply a matter of powers of ten. Table A.1 gives SI and CGS units for the common magnetic terms used and provides conversion factors. More detailed comparisons, including a resume of the fundamental equations in electromagnetism and their form in the two systems, have been given by Payne (1981).

TABLE A.1
Common Magnetic Terms in SI and CGS Units with Conversion Factors

Magnetic term	Symbol	SI unit	CGS unit	Conversion factor
Magnetic induction	**B**	Tesla (T) = kg A^{-1} s^{-2}	Gauss	1 T = 10^4 Gauss
Magnetic field strength	**H**	A m^{-1}	Oersted (oe)	1 A m^{-1} = 4π × 10^{-3} oe
Magnetization	**M**	A m^{-1}	emu cm^{-3} = Gauss	1 A m^{-1} = 10^{-3} emu cm^{-3}
Magnetic polarization	**J**	T	emu cm^{-3} = Gauss	1T = $\dfrac{10^4}{4\pi}$ emu cm^{-3}
Magnetic dipole moment	p	A m^2	emu = Gauss cm^3	1 A m^2 = 10^3 emu
Magnetic pole strength	s_p	A m	emu = Gauss cm^3	1 A m = 10 emu
Magnetic flux	ϕ	Weber (Wb) = kg m^2 A^{-1} s^{-2}	Maxwell	1 Wb = 10^8 Maxwell
Magnetic scalar potential (for **H**)	ψ	A	emu = oe cm	1 A = 4π × 10^{-1} emu
Magnetic vector potential (for **B**)	**A**	Wb m^{-1} = kg m A^{-1} s^{-2}	emu = Gauss cm	1 Wb m^{-1} = 10^6 emu
Permeability of free space	μ_0	Henry (H) m^{-1} = kg m A^{-2} s^{-2}	1	4π × 10^{-7} H m^{-1} = 1 cgs
Permittivity of free space	ε_0	Farad (F) m^{-1} = A^2 s^4 kg m^{-3}	1	8.85 × 10^{-12} F m^{-1} = 1 cgs
Volume magnetic susceptibility	χ_m	Dimensionless	Dimensionless	4π (SI) = 1 (cgs)
Demagnetizing factor	N	Dimensionless	Dimensionless	1 (SI) = 4π (cgs)

Functions Associated with Spherical Harmonics

The purpose of this appendix is to provide further information about the Legendre functions, an explanation of the normalization used in geomagnetism, explicit forms for the lower degree functions, an explicit polynomial (rather than recursive) form for the functions that is convenient for computing purposes, and a derivation of the inclination anomaly model used in Chapter 6.

B.1 The Scalar Potential

As shown in (2.2.16), the scalar potential ψ for the magnetic field **H** of internal origin may be expanded in a spherical harmonic series as

$$\psi = \frac{a}{\mu_0} \sum_{l=1}^{\infty} \left(\frac{a}{r}\right)^{l+1} \sum_{m=0}^{l} (g_l^m \cos m\phi + h_l^m \sin m\phi) P_l^m(\cos\theta) \ , \qquad \text{(B.1.1)}$$

where a is the mean radius (6371.2 km) of the Earth, r is the radial distance from the center of the (spherical) Earth, θ is the colatitude, ϕ is the east longitude measured from Greenwich, and P_l^m is the Schmidt quasi-normalized form of the associated Legendre function $P_{l,m}$ of degree l and order m.

The components X, Y, and Z of the magnetic field **H** are then obtained (see Eq. (2.2.15)) from

$$X = \frac{1}{r}\frac{\partial \psi}{\partial \theta}; \quad Y = -\frac{1}{r\sin\theta}\frac{\partial \psi}{\partial \phi}; \quad Z = \frac{\partial \psi}{\partial r} \quad . \tag{B.1.2}$$

However, the components of the field are typically measured in nanotesla, which are units of the induction **B**. Using the superscript B to represent this, we have

$$X^B = \frac{\mu_0}{r}\frac{\partial \psi}{\partial \theta}; \quad Y^B = -\frac{\mu_0}{r\sin\theta}\frac{\partial \psi}{\partial \phi}; \quad Z^B = \mu_0\frac{\partial \psi}{\partial r} \quad . \tag{B.1.3}$$

In the special case where there is no longitudinal variation (i.e., no dependence on ϕ), (B.1.1) simplifies to the *zonal* harmonics

$$V = \frac{a}{\mu_0}\sum_{l=1}^{\infty}\left(\frac{a}{r}\right)^{l+1} g_l^0 \, P_l(\cos\theta) \quad , \tag{B.1.4}$$

where $P_l \equiv P_l^0$.

B.2 The Legendre Functions P_l

Equation (2.2.8) gives Rodriques' recursion formula for the Legendre functions as

$$P_l(\cos\theta) = P_l(\chi) = \frac{1}{2^l \, l!}\frac{d^l}{d\chi^l}\left(\chi^2 - 1\right)^l \quad . \tag{B.2.1}$$

This formula gives the first 11 functions as

$P_0(\chi) = 1$

$P_1(\chi) = \chi$

$P_2(\chi) = \frac{1}{2}(3\chi^2 - 1)$

$P_3(\chi) = \frac{1}{2}(5\chi^3 - 3\chi)$

$P_4(\chi) = \frac{1}{8}(35\chi^4 - 30\chi^2 + 3)$

$P_5(\chi) = \frac{1}{8}(63\chi^5 - 70\chi^3 + 15\chi)$

$$P_6(\chi) = \tfrac{1}{16}(231\chi^6 - 315\chi^4 + 105\chi^2 - 5)$$

$$P_7(\chi) = \tfrac{1}{16}(429\chi^7 - 693\chi^5 + 315\chi^3 - 35\chi)$$

$$P_8(\chi) = \tfrac{1}{128}(6435\chi^8 - 12012\chi^6 + 6930\chi^4 - 1260\chi^2 + 35)$$

$$P_9(\chi) = \tfrac{1}{128}(12155\chi^9 - 25740\chi^7 + 18018\chi^5 - 4620\chi^3 + 315\chi)$$

$$P_{10}(\chi) = \tfrac{1}{256}(46189\chi^{10} - 109395\chi^8 + 90090\chi^6 - 30030\chi^4 + 3465\chi^2 - 63)$$

The first 4 of these are plotted in Fig. 2.5 as functions of θ.

B.3 The Associated Legendre Functions $P_{l,m}$

Equation (2.2.11) gives the recursion formula

$$P(\cos\theta) = P_{l,m}(\chi) = \frac{1}{2^l l!}(1-\chi^2)^{\frac{m}{2}}\frac{d^{l+m}}{d\chi^{l+m}}(\chi^2 - 1)^l \qquad (B.3.1)$$

for the associated Legendre functions. From (B.3.1) and (B.2.1) it is immediately evident that $P_l = P_{l,0}$.

In many circumstances it is more convenient to use the explicit polynomial form

$$P_{l,m}(\cos\theta) = \frac{\sin^m\theta}{2^l}\sum_{t=0}^{\mathrm{Int}\left(\frac{l-m}{2}\right)}\frac{(-1)^t(2l-2t)!}{t!(l-t)!(l-m-2t)!}\cos^{(l-m-2t)}\theta \quad, \qquad (B.3.2)$$

where $\mathrm{Int}(\tfrac{1}{2}(l-m))$ is the largest integer smaller than $\tfrac{1}{2}(l-m)$.

B.4 Normalization of the Associated Legendre Functions

The associated Legendre functions (and, therefore, the Legendre functions themselves) are orthogonal, the condition being

$$\frac{1}{4\pi} \int_{\theta=0}^{\pi} \int_{\phi=0}^{2\pi} P_{l,m}(\cos\theta) \begin{Bmatrix} \cos m\phi \\ \sin m\phi \end{Bmatrix} P_{s,t}(\cos\theta) \begin{Bmatrix} \cos t\phi \\ \sin t\phi \end{Bmatrix} \sin\theta \, d\theta \, d\phi = C_{ls}^{mt} \quad , \quad \text{(B.4.1)}$$

where

$$C_{ls}^{mt} = 0 \qquad \text{unless } s = l \text{ and } t = m$$

$$C_{ll}^{mm} = \frac{1}{2(2l+1)} \frac{(l+m)!}{(l-m)!} \qquad \text{(B.4.2)}$$

and the notation

$$\begin{Bmatrix} \cos m\phi \\ \sin m\phi \end{Bmatrix}$$

implies that either $\cos m\phi$ or $\sin m\phi$ can be used.

Re-scaling the Legendre functions does not affect their orthogonality; it merely alters the numerical value of the coefficients in the expansion (B.1.1). Thus we could use *fully normalized* functions Q_l^m defined as

$$Q_l^m = \frac{P_{l,m}}{\sqrt{C_{ll}^{mm}}} \quad , \qquad \text{(B.4.3)}$$

so that

$$\frac{1}{4\pi} \int_{\theta=0}^{\pi} \int_{\phi=0}^{2\pi} \left[Q_l^m(\cos\theta) \begin{Bmatrix} \cos m\phi \\ \sin m\phi \end{Bmatrix} \right]^2 \sin\theta \, d\theta \, d\phi = 1 \quad . \qquad \text{(B.4.4)}$$

The orthonormal functions Q_l^m are used in several areas of geophysics, but in geomagnetism it has been traditional to use the Schmidt *partial normalization* (see Eq. (2.2.13)), given by

$$P_l^m = P_{l,m} \qquad\qquad \text{for } m = 0$$

$$P_l^m = \left[\frac{2(l-m)!}{(l+m)!}\right]^{\frac{1}{2}} P_{l,m} \quad \text{for } m > 0 \quad,$$

(B.4.5)

so that, from (B.4.2),

$$\frac{1}{4\pi} \int_{\theta=0}^{\pi} \int_{\phi=0}^{2\pi} \left[P_l^m(\cos\theta)\left\{\begin{matrix}\cos m\phi \\ \sin m\phi\end{matrix}\right\}\right]^2 \sin\theta\, d\theta\, d\phi = \frac{1}{2l+1} \quad.$$

(B.4.6)

Explicit functions for degrees 1 to 4 are then

$P_{1,0} = \cos\theta$

$P_{1,1} = \sin\theta; \qquad\qquad\qquad P_1^1 = P_{1,0}$

$P_{2,0} = \frac{1}{2}(3\cos^2\theta - 1);$

$P_{2,1} = 3\cos\theta\sin\theta; \qquad\qquad P_2^1 = 0.5774 P_{2,1}$

$P_{2,2} = 3\sin^2\theta; \qquad\qquad\qquad P_2^2 = 0.2887 P_{2,2}$

$P_{3,0} = \frac{1}{2}(5\cos^3\theta - 3\cos\theta)$

$P_{3,1} = \frac{3}{2}(5\cos^2\theta - 1)\sin\theta; \qquad P_3^1 = 0.4082 P_{3,1}$

$P_{3,2} = 15\cos\theta\sin^2\theta; \qquad\qquad P_3^2 = 0.1291 P_{3,2}$

$P_{3,3} = 15\sin^3\theta; \qquad\qquad\qquad P_3^3 = 0.0527 P_{3,3}$

$P_{4,0} = \frac{1}{8}(35\cos^4\theta - 30\cos^2\theta + 3)$

$P_{4,1} = \frac{5}{2}(7\cos^3\theta - 3\cos\theta)\sin\theta; \qquad P_4^1 = 0.3162 P_{4,1}$

$P_{4,2} = \frac{15}{2}(7\cos^2\theta - 1)\sin^2\theta; \qquad P_4^2 = 0.07454 P_{4,2}$

$P_{4,3} = 105\cos\theta\sin^3\theta; \qquad\qquad P_4^3 = 0.01992 P_{4,3}$

$P_{4,4} = 105\sin^4\theta; \qquad\qquad\qquad P_4^4 = 0.00704 P_{4,4}$.

B.5 Inclination Anomaly Model for Zonal Harmonics

If V is the potential for just the zonal harmonics, then from (B.1.2) and (B.1.4)

$$X = -\frac{1}{r}\frac{\partial V}{\partial \theta} = -\frac{1}{\mu_0}\sum_l \left(\frac{a}{r}\right)^{l+2} g_l^0 \frac{d}{d\theta} P_l(\cos\theta) \ , \tag{B.5.1}$$

so that at $r=a$

$$X = -\frac{1}{\mu_0}\sum_l g_l^0 \frac{d}{d\theta} P_l(\cos\theta) = -\frac{1}{\mu_0}\sum_l g_l^0 P_l'(\cos\theta) \ . \tag{B.5.2}$$

Similarly for Z we have

$$Z = -\frac{\partial V}{\partial r} = -\frac{1}{\mu_0}\sum_l -(l+1)\left(\frac{a}{r}\right)^{l+2} g_l^0 P_l(\cos\theta) \ , \tag{B.5.3}$$

so that at $r=a$

$$Z = \frac{1}{\mu_0}\sum_l (l+1) g_l^0 P_l(\cos\theta) \ . \tag{B.5.4}$$

The inclination I' of the field is given by

$$\tan I' = \frac{Z}{X} = \frac{\displaystyle\sum_l (l+1) g_l^0 P_l(\cos\theta)}{\displaystyle\sum_l -g_l^0 P_l'(\cos\theta)} \ . \tag{B.5.5}$$

Substitution of the explicit functions for $l=1$ to 4 from §B.2 then gives Eq. (6.2.6).

References

Abrahamsen, N. and Knudsen, K.L. (1979). Indication of a geomagnetic low-inclination excursion in supposed middle Weichselian interstadial marine clay at Rubjerg, Denmark. *Phys. Earth Planet. Int.*, **18**, 238-246.

Acheson, D.J. and Hide, R. (1973). Hydromagnetics of rotating fluids. *Rep. Prog. Phys.*, **36**, 159-221.

Adam, N.V., Benkova, N.P., Khramov, A.N. and Cherevko, T.N. (1975). Spherical harmonic analysis of the geomagnetic field. *Stud. Geophys. Geod.*, **19**, 141-149.

Ade-Hall, J.M., Palmer, H.C. and Hubbard, T.P. (1971). The magnetic and opaque petrological response of basalts to regional hydrothermal alteration. *Geophys. J. Roy. Astron. Soc.*, **24**, 137-174.

Ahrens, T.J. (1979). Equations of state of iron sulfide and constraints on the sulfur content of the earth. *J. Geophys. Res.*, **84**, 985-998.

Ahrens, T.J. (1987). Shock wave techniques for geophysics and planetary physics. In *Methods of Experimental Physics*, **24**, Part A, Acadamic Press, San Diego.

Aitken, M.J. (1970). Dating by archaeomagnetic and thermoluminescent methods. *Phil. Trans. Roy. Soc. London*, **A269**, 77-88.

Aitken, M.J., Alcock, P.A, Bussell, G.D. and Shaw, C.J. (1983). Palaeointensity studies on archaeological material from the near east. In *Geomagnetism of Baked Clays and Sediments* (K.M. Creer, P. Tucholka and C.E. Barton, eds), p.122-127, Elsevier, Amsterdam.

Aitken, M.J., Allsop, A.L., Bussell, G.D. and Winter, M.B. (1988a). Determination of the intensity of the Earth's magnetic field during archeological times: Reliability of the Thellier technique. *Rev. Geophys.*, **26**, 3-12.

Aitken, M.J., Allsop, A.L., Bussell, G.D. and Winter, M.B. (1988b). Comment on "The lack of reproducibility in experimentally determined intensities of the Earth's magnetic field" by D. Walton. *Rev. Geophys.*, **26**, 23-25.

Aitken, M.J., Allsop, A.L., Bussell, G.D. and Winter, M.B. (1989). Geomagnetic intensity variation during the last 4000 years. *Phys. Earth Planet. Int.*, **56**, 49-58.

Akaike, H. (1969a). Fitting autoregressive models for prediction. *Ann. Inst. Statist. Math.*, **21**, 243-247.

Akaike, H. (1969b). Power spectrum estimation through autoregressive model fitting. *Ann. Inst. Statist. Math.*, **21**, 407-419.

Akaike, H. (1970). Statistical predictor identification. *Ann. Inst. Statist. Math.*, **22**, 203-217.

Akasofu, S.I. and Chapman, S. (1961). The ring current, geomagnetic disturbance and Van Allen radiation belts. *J. Geophys. Res.*, **66**, 1321-1350.

Akasofu, S.I. and Chapman, S. (1972). *Solar-terrestrial Physics*. Clarendon Press, Oxford.

Aki, K. and Richards, P.G. (1980). *Quantitative Seismology*, Vol. 2, Freeman, San Francisco.

Alexandrescu, M., Gibert, D., Hulot, G., Le Mouël, J.-L and Saracco, G. (1995). Detection of geomagnetic jerks using wavelet analysis. *J. Geophys. Res.*, **100**, 12557-12572.

Alfvén, H. (1939). A theory of magnetic storms and of the aurorae. *K. Sven. Vetenskapad. Handl. Ser. 3*, **18**(3). (Reprinted in part with comments by A. Dessler and J. Wilcox in *EOS, Trans. Amer. Geophys. Union*, **51**, 180-194, 1970).

Alfvén, H. (1942a). Remarks on the rotation of a magnetized sphere with application to solar rotation. *Arkiv. f. Mat. Astron. Fysik*, **28A**, No.6, 2-9.

Alfvén, H. (1942b). On the existence of electromagnetic-hydromagnetic waves. *Arkiv. f. Mat. Astron. Fysik*, **29B**, No.2 (7 pp).

Alfvén, H. (1950). *Cosmical Electrodynamics*, Oxford University Press, New York, 237 pp.

Alfvén, H. (1954). *On the Origin of the Solar System*. Clarendon Press, Oxford, 194 pp.

Alfvén, H. and Arrhenius, G. (1975). *Structure and Evolutionary History of the Solar System.* D. Reidel Pub. Company, Boston, USA. (Geophys. and Astrophys. Monograph Series.) 276 pp

Alfven, H. and Arrhenius, G. (1976). Evolution of the solar system. *NASA SP-345*, 599 pp.

Allan, J.L. and Shive, P.N. (1974). Mossbauer effect observations of the "X-phase" in the ilmenite-hematite series. *J. Geomag. Geoelect.*, **26**, 329-333.

Alldredge, L.R. (1976). Effects of solar activity on geomagnetic component annual means. *J. Geophys. Res.*, **81**, 2990-2996.

Alldredge, L.R. (1977). Geomagnetic variations with periods from 13 to 30 years. *J. Geomag. Geoelect.*, **29**, 123-135.

Alldredge, L.R. (1984). A discussion of impulses and jerks in the geomagnetic field. *J. Geophys. Res.*, **89**, 4403-4412.

Alldredge, L.R. and Hurwitz, L. (1964). Radial dipoles as the sources of the earth's main magnetic field. *J. Geophys. Res.*, **69**, 2631-2640.

Alldredge, L.R. and Stearns, C.O. (1969). Dipole model of the sources of the earth's magnetic field and secular change. *J. Geophys. Res.*, **74**, 6583-6593.

Alldredge, L.R. and Stearns, C.O. (1974). A discussion of sources and description of the earth's magnetic field and its secular variations. *J. Geomag. Geoelect.*, **26**, 393-404.

Anderson, K.A., Lin, R.P. and McGuire, R.E. (1977). Linear magnetisation feature associated with Rima Sirsalis. *Earth Planet. Sci. Lett.*, **34**, 141-151.

Anderson, K.A., Lin, R.P., McGuire, R.E. and McCoy, J.E. (1975). Measurement of lunar and planetary magnetic fields by reflection of low energy electrons. *Sp. Sci. Instr.*, **1**, 439-470.

Anderson, O.L. (1979). The high temperature acoustic Grüneisen parameter in the Earth's interior. *Phys. Earth Planet. Int.*, **18**, 221-223.

Anderson, O.L. (1995). Mineral physics of iron and of the core. *Rev. Geophys. Suppl., U.S. National Report to IUGG 1991-1994*, 429-441.

Anderson, W.W., Svendsen, B. and Ahrens, T.J. (1989). Phase relations in iron-rich systems and implications for the Earths's core. *Phys. Earth Planet. Int.*, **55**, 208-220.

Angel, J.R.P. (1975). Strong magnetic fields in white dwarfs. *Ann. N.Y. Acad. Sci.*, **257**, 80-81.

Anonymous (1974). Adoption of SI units in geomagnetism. (Trans. 2nd Gen. Ass. IAGA Kyoto, 1973). *IAGA Bull.*, **35**, 148.

Anonymous (1979). Magnetostratigraphic polarity units, a supplementary chapter of the International Subcommission on Stratigraphic Classification International Stratigraphic Guide. *Geology*, 7, 578-583.

Ashcroft, N.W. and Merman, N.D. (1976). *Solid State Physics*. Holt, Rinehart and Winston, New York.

Atwater, T.M. and Mudie, J.D. (1973). Detailed near-bottom geophysical study of the Gorda rise. *J. Geophys. Res.*, **78**, 8665-8686.

Axford, W.I. (1969). Magnetospheric convection. *Rev. Geophys. Space Phys.*, **7**, 421-459.

Axford, W.I., Petschek, H.E. and Siscoe, G.L. (1965). Tail of the magnetosphere. *J. Geophys. Res.*, **70**, 1231-1236.

Baag, C. and Helsley, C.E. (1974). Geomagnetic secular variation model E. *J. Geophys. Res.*, **79**, 4918-4922.

Babcock, H.W. (1961). The topology of the Sun's magnetic field and the 22-year cycle. *Astrophys. J.*, **133**, 572-587.

Backus, G.E. (1957). The axisymmetric self-excited fluid dynamo. *Astrophys. J.*, **125**, 500-524.

Backus, G.E. (1958). A class of self sustaining dissipative spherical dynamos. *Ann. Phys., (N.Y.)*, **4**, 372-447.

Backus, G.E. (1968). Kinematics of geomagnetic secular variation in a perfectly conducting core. *Phil. Trans. R. Soc. London*, **A263**, 239-266.

Backus, G.E. (1975). Gross thermodynamics of heat engines in deep interior of earth. *Proc. Nat. Acad. Sci. USA*, **72**, 1555-1558.

Backus, G.E. (1983). Application of mantle filter theory to the magnetic jerk of 1969. *Geophys. J. Roy. Astron. Soc.*, **74**, 713-746.

Backus, G.E. (1986). Poloidal and toroidal fields in geomagnetic modeling, *Rev. Geophys.*, **24**, 75-109.

Backus, G.E. (1991). Current meters in the core of the Earth. In *Of Fluid Mechanics and Related Matters* (R. Salmon and D. Betts, eds.), *Scripps Inst. Ocean. Ref. Ser.*, **91-24**, 97.

Backus, G.E. and Chandrasekhar, S. (1956). On Cowling's theorem on the impossibility of self-maintained axisymmetric dynamos. *Proc. Nat. Acad. Sci. USA*, **42**, 105-109.

Backus, G.E. and Gilbert, F. (1967). Numerical application of a formalism for geophysical inverse problems. *Geophys. J. Roy. Astron. Soc.*, **13**, 247-276.

Backus, G.E. and Gilbert, F. (1968). The resolving power of gross earth data. *Geophys. J. Roy. Astron. Soc.*, **16**, 169-205.

Backus, G.E. and Gilbert, F. (1970). Uniqueness in the inversion of inaccurate gross earth data. *Phil. Trans. Roy. Soc. London*, **A266**, 123-192.

Backus, G., Estes, R.H., Chinn, D. and Langel, R.A. (1987). Comparing the jerk with other global models of the geomagnetic field from 1960 to 1978. *J. Geophys. Res.*, **92**, 3615-3622.

Bailey, M.E. and Dunlop, D.J. (1977). On the use of anhysteretic remanent magnetisation in paleointensity determination. *Phys. Earth Planet. Int.*, **13**, 360-362.

Baksi, A.K. (1993). A paleomagnetic polarity time scale for the period 0-17 Ma, based on $^{40}Ar/^{39}Ar$ plateau ages for selected field reversals. *Geophys. Res. Lett.*, **20**, 1607-1610.

Baksi, A.K., Hsu, V., McWilliams, M.O. and Rarrer, E. (1992). $^{40}Ar/^{39}Ar$ dating of the Brunhes-Matuyama geomagnetic field reversal. *Science*, **256**, 356-357.

Banerjee, S.K. (1977). On the origin of stable remanence in pseudo-single domain grains. *J. Geomagn. Geoelect.*, **29**, 319-329.

Banerjee, S.K. and Mellema, J.P. (1976). Early lunar magnetism. *Nature*, **260**, 230-231.

Banks, R.J. (1969). Geomagnetic variations and the electrical conductivity of the upper mantle. *Geophys. J. Roy. Astron. Soc.*, **17**, 457-487.

Banks, R.J. (1972). The overall conductivity distribution of the earth. *J. Geomag. Geoelect.*, **24**, 337-351.

Barbetti, M. (1977). Measurements of recent geomagnetic secular variation in southeast Australia and the question of dipole wobble. *Earth Planet. Sci. Lett.*, **36**, 207-218.

Barbetti, M. (1980). Geomagnetic strength over the last 50,000 years and changes in atmospheric ^{14}C concentration; emerging trends. *Radiocarbon*, **22**, 192-199.

Barbetti, M. (1983). Archaeomagnetic results from Australia. In *Geomagnetism of Baked Clays and Recent Sediments*, (K.M. Creer, P. Tucholka and C.E. Barton, eds), p.173-175, Elsevier, Amsterdam.

Barbetti, M. and McElhinny, M.W. (1972). Evidence for a geomagnetic excursion 30 000 yr B.P. *Nature*, **239**, 327-330.

Barbetti, M.F. and McElhinny, M.W. (1976). The Lake Mungo Geomagnetic Excursion. *Phil. Trans. Roy. Soc. London*, **A281**, 515-542.

Barbetti, M.F., McElhinny, M.W., Edwards, D.J. and Schmidt, P.W. (1977). Weathering processes in baked sediments and their effects on archaeomagnetic field-intensity measurements. *Phys. Earth Planet. Int.*, **13**, 346-354.

Barenghi, C.F. (1992). Nonlinear planetary dynamos in a rotating spherical shell II. The post Taylor equilibration for α^2 dynamos. *Geophys. Astrophys. Fluid Dyn.*, **67**, 27-36.

Barenghi, C.F. (1993). Nonlinear planetary dynamos in a rotating spherical shell III. $\alpha^2\omega$ dynamos and the geodynamo. *Geophys. Astrophys. Fluid Dyn.*, **67**, 163-185.

Barenghi, C.F. and Jones, C.A. (1991). Nonlinear planetary dynamos in a rotating spherical shell I. Numerical methods. *Geophys. Astrophys. Fluid Dyn.*, **60**, 211-243.

Barnes, A., Casser, P., Mihalou, J.D. and Eviatur, A. (1971). Permanent lunar surface magnetism and its deflection of the solar wind. *Science*, **172**, 716-718.

Barnes, T. (1973). *Origin and Destiny of the Earth's Magnetic Field*, Creation-Life Publishers, San Deigo, California. [See also Brush, S. (1982). Finding the age of the Earth by physics or by faith. *J. Geol. Educ.*, **30**, 34-58.

Barnett, S.J. (1933). Gyromagnetic effects; history, theory and experiments. *Physica*, **13**, 241.

Barraclough, D.R. (1974). Spherical harmonic analysis of the geomagnetic field for eight epochs between 1600 and 1910. *Geophys. J. Roy. Astron. Soc.*, **36**, 497-513.

Barraclough, D.R., Harwood, J.M., Leaton, B.R. and Malin, S.R.C. (1975). A model of the geomagnetic field at epoch 1975. *Geophys. J. Roy. Astron. Soc.*, **43**, 645-649.

Bartels, J. (1949). The standardized index, *Ks*, and the planetary index *Kp*. *IATME Bull.*, **12b**, p.97. (IUGG Publ. Office, Paris).

Bartels, J., Heck, N.H. and Johnston, H.F. (1939). The three-hour-range index measuring geomagnetic activity. *Terr. Magn. Atmos. Elec.*, **44**, 411-454.

Bartels, J., Heck, N.H. and Johnston, H.F. (1940). Geomagnetic three-hour-range indices for 1938 and 1939. *Terr. Magn. Atmos. Elec.*, **45**, 309-337.

Barton, C.E. (1978). *Magnetic Studies of Some Australian Lake Sediments*. Ph.D. thesis, Res. Sch. Earth Sci., Australian National University, Canberra.

Barton, C.E. (1982). Spectral analysis of palaeomagnetic time series and the geomagnetic spectrum. *Phil. Trans. Roy. Soc. London*, **A306**, 203-209.

Barton, C.E. (1983). Analysis of palaeomagnetic time series — techniques and applications. *Geophys. Surv.*, **5**, 335-368.

Barton, C.E. (1989). Geomagnetic secular variation: Directions and intensity. In *Encylopedia of Solid Earth Geophysics* (D.E. James, ed.), p.560-577. Van Nostrand–Reinhold, New York.

Barton, C.E. and Burden, F.R. (1979). Modifications to the Mackereth corer. *Limnol. Oceanogr.*, **24**, 977-983.

Barton, C.E. and McElhinny, M.W. (1981). A 10 000 yr geomagnetic secular variation record from three Australian maars. *Geophys. J. Roy. Astron. Soc.*, **67**, 465-485.

Barton, C.E. and McElhinny, M.W. (1982). Time series analysis of the 10 000 yr geomagnetic secular variation record from SE Australia. *Geophys. J. Roy. Astron. Soc.*, **68**, 709-724.

Barton, C.E. and McFadden, P.L. (1996). Inclination shallowing and preferred transitional VGP paths. *Earth Planet. Sci. Lett.*, **140**, 147-158.

Barton, C.E., McElhinny, M.W. and Edwards, D.J. (1980). Laboratory studies of depositional DRM. *Geophys. J. Roy. Astron. Soc.*, **61**, 355-377.

Barton, C.E., Merrill, R.T. and Barbetti, M. (1979). Intensity of the earth's magnetic field over the last 10 000 years. *Phys. Earth Planet. Int.*, **20**, 96-110.

Bauer, L.A. (1894). An extension of the Gaussian potential theory of terrestrial magnetism. *Proc. Amer. Assoc. Adv. Sci.*, **43**, 55-58.

Bauer, L.A. (1895). On the distribution and the secular variation of terrestrial magnetism, No. III. *Amer. J. Sci.*, **50**, 314-325.

Beck, R., Burbank, D., Sercombe, W., Riley, G., Barndt, J., Berry, J., Afzal, J., Khan, A., Jurgen, H., Metje, J., Cheema, A., Shafique, N., Lawrence, R. and Khan, M. (1995). Stratigraphic evidence for an early collision between northwest India and Asia. *Nature*, **373**, 55-58.

Benjamin, P. (1895). *The Intellectual Rise in Electricity.* Longmans Green, London.

Benkova, N.P. and Cherevko, T.N. (1972). Analytical representation of magnetic inclinations. *Geomag. Aeron.,* **12**, 632-634.

Benkova, N.P., Adam, N.V. and Cherevko, T.N. (1970). Application of spherical harmonic analysis to magnetic declination data. *Geomag. Aeron.,* **10**, 673-680.

Benkova, N.P., Kolomiytseva, G.I. and Cherevko, T.N. (1974). Analytical model of the geomagnetic field and its secular variations over a period of 400 years (1550-1950). *Geomag. Aeron.,* **14**, 751-755.

Benkova, N.P., Khramov, A.N., Cherevko, T. N. and Adam, N.V. (1973). Spherical harmonic analysis of the paleomagnetic field. *Earth Planet. Sci. Lett.,* **18**, 141-147.

Benkova, N.P., Kruglyakov, A.A., Khramov, A.N. and Cherevko, T.N. (1971). Spherical harmonic analysis of paleomagnetic data. *Geomag. Aeron.,* **11**, 319-321.

Benton, E.R. (1979). Magnetic probing of planetary interiors. *Phys. Earth Planet. Int.,* **20**, 111-118.

Benton, E.R. (1981). Inviscid, frozen-flux velocity components at the top of Earth's core from magnetic observations at Earth's surface. 1. A new methodology. *Phys. Earth Planet. Int.,* **18**, 157-174.

Benton, E.R., Estes, R.H. and Langel, R.A. (1987). Geomagnetic field modelling incorporating constraints from frozen flux electromagnetism. *Phys. Earth Planet. Int.,* **48**, 241-264.

Benton, E.R., Estes, R.H., Langel, R.A. and Muth, L.A. (1982). Sensitivity of selected geomagnetic properties to truncation level of spherical harmonic expansions. *Geophys. Res. Lett.,* **9**, 254-257.

Benz, W. and Cameron, A.G. (1990). Terrestrial effects of the giant impact. In *Origins of the Earth,* (H.E. Newsom and J.N. Jones, eds), p.135-150, Oxford Univ. Press, New York.

Bhargava, B.N. and Yacob, A. (1969). Solar cycle response in the horizontal force of the earth's magnetic field. *J. Geomag. Geoelect.,* **21**, 385-397.

Biermann, L. (1951). Kometenschwerfe und solare Korpuskularstrahlung. *Zeits. Astrophys.,* **29**, 274-286.

Birch, F. (1964). Density and composition of mantle and core. *J. Geophys. Res.,* **69**, 4377-4388.

Birch, F. (1965a). Speculations on the earth's thermal history. *Geol. Soc. Amer. Bull.,* **76**,133-154.

Birch, F. (1965b). Energetics of core formation. *J. Geophys. Res.,* **70**, 6217-6221.

Birch, F. (1968). On the possibility of large changes in the Earth's volume. *Phys. Earth Planet. Int.,* 1, 141-147.

Birch, F. (1972). The melting relations of iron and temperatures in the earth's core. *Geophys. J. Roy. Astron. Soc.,* **29**, 373-387.

Birkeland, K. (1908). On the cause of magnetic storms and the origin of terrestrial magnetism. *The Norweigian Aurora Polaris Expedition 1902-1903.* Aschenboug, Oslo.

Biskamp, D. (1993). *Nonlinear Magnetohydrodynamics.* Cambridge Univ. Press, New York, 378 pp.

Blackett, P.M.S. (1947). The magnetic field of massive rotating bodies. *Nature,* **159**, 658-666.

Blackett, P.M.S. (1952). A negative experiment relating to magnetism and the earth's rotation. *Phil. Trans. Roy. Soc. London,* **A245**, 309-370.

Blackett, P.M.S. (1961). Comparison of ancient climates with the ancient latitudes deduced from rock magnetic measurements. *Proc. Roy. Soc. London*, **A263**, 1-30.

Blackett, P.M.S., Clegg, J.A. and Stubbs, P.H.S. (1960). An analysis of rock magnetic data. *Proc. Roy. Soc. London*, **A256**, 291-322.

Blackshear, W.T. and Gapcynski, J.P. (1977). An improved value of the lunar moment of inertia. *J. Geophys. Res.*, **82**, 1699-1701.

Blakely, R.J. (1976). An age-dependent, two-layer model for marine magnetic anomalies. In *The Geophysics of the Pacific Ocean Basin and its Margin* (G.H. Sutton, M.H. Manghnani, R. Moberly and E.U. McAgee, eds). *Amer. Geophys. Union Monograph*, **19**, 227-234.

Blakely, R.J. and Cande, S.C. (1979). Marine magnetic anomalies. *Rev. Geophys. Space Phys.*, **17**, 204-214.

Blakely, R.J. and Cox, A. (1972). Evidence for short geomagnetic polarity intervals in the early Cenozoic. *J. Geophys. Res.*, **77**, 7065-7072.

Bloomfield, P. (1976). *Fourier Analysis of Time Series: An Introduction*, John Wiley, New York, 258 pp.

Blow, R.A. and Hamilton, N. (1978). Effect of compaction on the acquisition of a detrital remanent magnetization in fine-grained sediments. *Geophys. J. Roy. Astron. Soc.,* **52**, 13-23, 1978.

Bloxham, J. (1989). Simple models of fluid flow at the core surface derived from geomagnetic field models. *Geophys. J. Int.*, **99**, 173-182.

Bloxham, J. (1992). The steady part of the secular variation of the Earth's magnetic field. *J. Geophys. Res.*, **97**, 19565-19579.

Bloxham, J. and Gubbins, D. (1985). The secular variation of the Earth's magnetic field. *Nature*, **317**, 777-781.

Bloxham, J. and Gubbins, D. (1986). Geomagnetic field analysis. IV. Testing the frozen-flux hypothesis. *Geophys. J. Roy. Astron. Soc.*, **84**, 139-152.

Bloxham, J. and Jackson, A. (1991). Fluid flow near the surface of Earth's outer core. *Rev. Geophys.*, **29**, 97-120.

Bochev, A. (1965). The earth's magnetic field represented as dipoles. *C. R. Acad. Bulg. Sci.*, **18**, 319-322.

Bochev, A. (1969). Two and three dipoles approximating the earth's main magnetic field. *Pure Appl. Geophys.*, **74**, 29-34.

Bochev, A. (1975). Presenting the earth's magnetic field as a field of six optimal dipoles. *C. R. Acad. Bulg. Sci.*, **28**, 469-471.

Boehler, R. (1992). Melting of the Fe-FeO and Fe-FeS systems at high pressure; constraints on core temperatures. *Earth Planet. Sci. Lett.*, **111**, 217-227.

Boehler, R. (1993). Temperatures in the Earth's core from melting-point measurements of iron at high static pressures. *Nature*, **363**, 534-536,

Bogue, S.W. and Coe, R.S. (1981). Thellier paleointensity results from an R-N transition zone on Kauai, Hawaii. *EOS, Trans. Amer. Geophys. Union*, **62**, 853.

Bogue, S.W. and Coe, R.S. (1982). Successive paleomagnetic reversal records from Kauai. *Nature*, **295**, 399-401.

Bogue, S.W. and Coe, R.S. (1984). Transitional paleointensities from Kauai, Hawaii, and geomagnetic reversal models. *J. Geophys. Res.*, **89**, 10341-10354.

Bogue, S.W. and Hoffman, K.A. (1987). Morphology of geomagnetic reversals. *Rev. Geophys.*, **25**, 910-916.

Bogue, S.W. and Merrill, R.T. (1992). The character of the field during geomagnetic reversals. *Ann. Rev. Earth Planet. Sci.* **20**, 181-219.

Bogue, S.W. and Paul, H.A. (1993). Distinctive field behavior following geomagnetic reversals. *Geophys. Res. Lett.*, **20**, 2399-2402.

Bonhommet, N. and Babkine, J. (1967). Sur la présence d'aimantations inversées dans la Chaine des Puys. *C. R. Acad. Sci. Paris*, **B264**, 92-94.

Booker, J.R. (1969). Geomagnetic data and core motions. *Proc. Roy. Soc. London*, **A309**, 27-40.

Bott, M.H.P. (1967). Solution of the linear inverse problem in magnetic interpretation with application to ocean magnetic anomalies. *Geophys. J. Roy. Astron. Soc.*, **13**, 313-323.

Braginsky, S.I. (1963). Structure of the F layer and reasons for convection in the earth's core. *Dokl. Akad. Nauk SSSR Engl. Transl.*, **149**, 1311-1314.

Braginsky, S.I. (1964a). Magnetohydrodynamics of the Earth's core. *Geomag. Aeron.*, **4**, 698-711.

Braginsky, S.I. (1964b). Kinematic models of the Earth's hydromagnetic dynamo. *Geomag. Aeron.*, **4**, 572-583.

Braginsky, S.I. (1967). Magnetic waves in the Earth's core. *Geomag. Aeron.*, **7**, 851-859.

Braginsky, S.I. (1972). Spherical harmonic analyses of the main geomagnetic field in 1550-1800. *Geomag. Aeron.*, **12**, 464-468.

Braginsky, S.I. (1975). An almost axially symmetrical model of the hydromagnetic dynamo of the Earth, I. *Geomag. Aeron.*, **15**, 149-156.

Braginsky, S.I. (1976). On the nearly axially-symmetrical model of the hydromagnetic dynamo of the earth. *Phys. Earth Planet. Int.*, **11**, 191-199.

Braginsky, S.I. (1978). Nearly axially symmetrical model of the hydromagnetic dynamo of the earth. *Geomag. Aeron.*, **18**, 340-351.

Braginsky, S.I. (1989). The Z model of the geodynamo with an inner core and the oscillations of the geomagnetic dipole. *Geomag. Aeron.*, **29**, 121-126.

Braginsky, S.I. (1991). Towards a realistic theory of the geodynamo. *Geophys. Astrophys. Fluid. Dyn.*, **60**, 89-134.

Braginsky, S.I. (1993). MAC-oscillations of the hidden ocean of the core. *J. Geomag. Geoelect.*, **45**, 1517-1538.

Braginsky, S.I. (1994). The nonlinear dynamo and model-Z, 267-305. *Lectures on Solar and Planetary Dynamos* (M.R.E. Proctor and A.d. Gilbert, eds), Cambridge Univ. Press, New York, 375 pp.

Braginsky, S.I. and Kulanin, N.V. (1971). Spherical analysis of the geomagnetic field from angular data and the extrapolated g_1^0 value. III. *Geomag. Aeron.*, **11**, 786-788.

Braginsky, S.I. and Le Mouël, J.-L. (1993). Two-scale model of a geomagnetic field variation. *Geophys. J. Int.*, **112**, 147-158.

Braginsky, S.I. and Meytlis, V. (1990). Local turbulence in the earth's core. *Geophys. Astrophys.*

Fluid Dyn., **55**, 71-87.

Brandeis, G. and Marsh, B. (1989). The convective liquidus in a solidifying magma chamber; a fluid dynamic investigation. *Nature*, **339**, 613-616.

Brandenburg, A. (1994) Solar dynamos; computational background. In *Lectures on Solar and planetary dynamos*, p.117-159 (M.R.E. Proctor and A.D. Gilbert, eds), Cambridge Univ. Press, New York.

Brandt, S.O. and Hodge, P.W. (1964). *Solar System Astrophysics*. McGraw-Hill, New York, 457 pp.

Brecher, A. and Brecher, K. (1978). On the observed relation between magnetic fields and rotation in astronomical objects (abstract). In *Conference on Origins of Planetary Magnetism*, Lunar and Planetary Institute Contrib., Houston, Texas, **348**, 7-9.

Breit, G. and Tuve, M.A. (1926). A test of the existence of the conducting layer. *Phys. Rev.*, **28**, 554-575.

Brett, R. (1973). A lunar core of Fe-Ni-S. *Geochim. Cosmochim. Acta*, **37**, 165-170.

Brett, R. (1976). The current status of speculations on the composition of the core of the earth. *Rev. Geophys. Space Phys.*, **14**, 375-383.

Briden, J.C. (1968). Paleoclimatic evidence of a geocentric axial dipole field. In *History of the Earth's Crust* (R.A. Phinney, ed.), p.178-194. Princeton Univ. Press, Princeton, New Jersey.

Briden, J.C. and Irving, E. (1964). Paleoclimatic spectra of sedimentary paleoclimatic indicators. In *Problems in Paleoclimatology* (A.E.M. Nairn, ed), p.199-250. Wiley Interscience, New York.

Brock, A. (1971). An experimental study of palaeosecular variation. *Geophys. J. Roy. Astron. Soc.*, **24**, 303-317.

Brown, J.M. (1986). Interpretation of the D" zone at the base of the mantle; dependence on assumed values of thermal conductivity. *Geophys. Res. Lett.*, **13**, 1509-1512.

Brown, J.M. and McQueen, R.G. (1980). Melting of iron under core conditions. *Geophys. Res. Lett.*, **7**, 533-536.

Brown, J.M. and McQueen, R.G. (1986). Phase transitions, Grüneisen parameters and elasticity for shocked iron between 77 GPa and 400 GPa. *J. Geophys. Res.*, **91**, 7485-7494.

Brown, L.W. (1975). Saturn radio emission near 1 MHz. *Astrophys. J.*, **198**, L89-L92.

Brown, W.F. (1959). Relaxational behavior of fine magnetic particles. *J. Appl. Phys.*, **30**, 130S-132S.

Brummell, N., Cattaneo, F. and Toomre J. (1995). Turbulent dynamics in the solar convection zone. *Science*, **269**, 1370-1379.

Brunhes, B. (1906). Recherches sur le direction d'aimantation des roches volcaniques. *J. Phys.*, **5**, 705-724.

Brush, S.G. (1990). Theories of the origin of the solar system. *Rev. Mod. Phys.*, **62**, 43-112.

Bucha, V. (1967). Intensity of the earth's magnetic field during archaeological times in Czechoslovakia. *Archaeometry*, **10**, 12-22.

Bucha, V. (1969). Changes in the earth's magnetic moment and radiocarbon dating. *Nature*, **224**, 681-683.

Bucha, V. (1980). Mechanism of the relations between the changes of the geomagnetic field, solar corpuscular radiation, atmospheric circulation, and climate. *J. Geomag. Geoelect.*, **32**, 217-264.

Bucha, V. and Neustupny, E. (1967). Changes of the earth's magnetic field and radiocarbon dating. *Nature*, **215**, 261-263.

Buddington, A.F. and Lindsley, A.F. (1964). Iron-titanium oxide minerals and synthetic equivalents. *J. Petrol.*, **5**, 310-357.

Buffet, B.A. (1992). Influence of a toroidal magnetic field on the nutations of the Earth. *J. Geophys. Res.*, **98**, 2105-2118.

Buffett, B.A., Huppert, H.E., Lister, J.R. and Woods, A.W. (1992). Analytical model for solidification of the Earth's core. *Nature*, **356**, 329-331.

Buffett, B.A., Mathews, P., Herring, T. and Shapiro, I. (1993). Forced nutations of the Earth: Contributions from the effects of ellipticity and rotation on the elastic deformations. *J. Geophys. Res.* **98**, 21659-21676.

Bullard, E.C. (1949a). The magnetic field within the earth. *Proc. Roy. Soc. London*, **A197**, 433-453.

Bullard, E.C. (1949b). Electromagnetic induction in a rotating sphere. *Proc. Roy. Soc. London*, **A199**, 413-443.

Bullard, E.C. and Cooper, R.I.B. (1948). Determination of the masses required to produce a given gravitational field. *Proc. Roy. Soc. London*, **A194**, 332-347.

Bullard, E.C. and Gellman, H. (1954). Homogeneous dynamos and terrestrial magnetism. *Phil. Trans. Roy. Soc. London*, **A247**, 213-255.

Bullard, E. C. and Gubbins, D. (1977). Generation of magnetic fields by fluid motions of global scale. *Geophys. Astrophys. Fluid Dyn.*, **8**, 43-56.

Bullard, E.C., Everett, J.E. and Smith, A.G. (1965). A Symposium on Continental Drift. IV. The fit of the continents around the Atlantic. *Phil. Trans. Roy. Soc. London*, **A258**, 41-51.

Bullard, E.C., Freedman, C., Gellman, H. and Nixon, J. (1950). The westward drift of the earth's magnetic field. *Phil. Trans. Roy. Soc. London*, **A243**, 67-92.

Burakov, K.S., Gurary, G.Z., Khramov, A.N., Petrova, G.N., Rassanova, G.V. and Rodionov, G.P. (1976). Some peculiarities of the virtual geomagnetic pole positions during reversals. *J. Geomag. Geoelect.*, **28**, 295-307.

Burg, J.P. (1967). Maximum entropy spectral analysis. *37th Ann. Int. Meeting Soc. Explor. Geophys.*, Oklahoma, USA.

Burg, J.P. (1972). The relationship between maximum entropy spectra and maximum likelihood spectra. *Geophys.*, **37**, 375-376.

Burke, B.F. and Franklin, K.L. (1955). Radio emission from Jupiter. *Nature*, **175**, 1074.

Burlatskaya, S.P. (1983). Archaeomagnetic investigations in the USSR. In *Geomagnetism of Baked Clays and Sediments* (K.M. Creer, P. Tucholka and C.E. Barton, eds), p.127-137, Elsevier, Amsterdam.

Busse, F.H. (1970). Thermal instabilities in rapidly rotating systems. *J. Fluid Mech.*, **44**, 441-460.

Busse, F.H. (1975). A model of the geodynamo. *Geophys. J. Roy. Astron. Soc.*, **42**, 437-459.

Busse, F.H. (1976). Generation of planetary magnetism by convection. *Phys. Earth Planet. Int.*, **12**, 350-358.

Busse, F.H. (1977a). An example of non-linear dynamo action. *J. Geophys.*, **43**, 441-452.

Busse, F.H. (1977b). Mathematical problems of dynamo theory. In *Applications of Bifurcation*

Theory (P.H. Rabinowitz, ed.), p.175-202. Academic Press, London and New York.

Busse, F.H. (1978). Magnetohydrodynamics of the earth's dynamo. *Ann. Rev. Fluid Mech.*, **10**, 435-462.

Busse, F.H. and Wicht, J. (1992). A simple dynamo caused by conductivity variations. *Geophys. Astrophys. Fluid. Dyn.*, **64**, 135-144.

Butler, R.F. (1982). Magnetic mineralogy of continental deposits, San Juan Basin, New Mexico, and Clark's Fork Basin, Wyoming. *J. Geophys. Res.*, **87**, 7843-7852.

Butler, R.F. (1992). *Paleomagnetism*. Blackwell Scientific Publications, Boston, 319 pp.

Butler, R.F. and Banerjee, S.F. (1975). Theoretical single-domain grain size range in magnetite and titanomagnetites. *J. Geophys. Res.*, **80**, 4049-4058.

Cain, J.C. (1971). Geomagnetic models from satellite surveys. *Rev. Geophys. Space Phys.*, **9**, 259-273.

Cain, J.C. (1975). Structure and secular change of the geomagnetic field. *Rev. Geophys. Space Phys.*, **13** (Suppl.), 203-206.

Cain, J.C., Davis, W.M. and Regan, R.D. (1974). An $n = 22$ model of the geomagnetic field. *EOS, Trans. Amer. Geophys. Union*, **56**, 1108.

Cameron, A.G.W. (1985). Formation and evolution of the primitive solar nebula. In *Protostars and Planets II* (D.C. Black and M.S. Matthews, eds), p.1073-1099, Univ. Arizona, Tucson.

Cande, S. and Kent, D.V. (1976). Constraints imposed by the shape of marine magnetic anomalies on the magnetic source. *J. Geophys. Res.*, **81**, 4157-4162.

Cande, S. and Kent, D.V. (1992a). A new geomagnetic polarity time scale for the Late Cretaceous and Cenozoic. *J. Geophys. Res.*, **97**, 13917-13951.

Cande, S. and Kent, D.V. (1992b). Ultrahigh resolution marine magnetic anomaly profiles: A record of continuous paleointensity variations? *J. Geophys. Res.*, **97**, 15075-15083.

Cande, S. and Kent, D.V. (1995). Revised calibration of the geomagnetic polarity timescale for the Late Cretaceous and Cenozoic. *J. Geophys. Res.*, **100**, 6093-6095.

Carmichael, C.M. (1961). The magnetic properties of ilmenite-hematite crystals. *Proc. Roy. Soc. London*, **A263**, 508-530.

Champion, D.E. (1980). Holocene geomagnetic secular variation in the western United States: Implications for the global geomagnetic field. *Rept. Open File Series US Geol. Surv.*, **80-824**, 314 pp.

Champion, D., Lanphere, M., and Kuntz, M. (1988). Evidence for a new geomagnetic reversal from lava flows in Idaho: Discussion of short polarity reversals in the Brunhes and late Matuyama polarity chrons. *J. Geophys. Res.*, **93**, 11667-11680.

Channell, J.E.T. (1989). Paleomagnetism: Deep-sea sediments. In *The Encyclopedia of Solid Earth Geophyics* (D.E. James, ed.), p.889-891. Van Nostrand, New York.

Chapman, S. (1942). Notes on isomagnetic charts. VI. Earth-air electric currents, and the mutual consistency of the H- and D-isomagnetic charts, VII. Mathematical notes on isoporic charts and their singular points, VIII. The mutual consistency of the declination and horizontal-intensity isoporic charts. *Terr. Magn. Atmos. Elect.*, **47**, 1-13; 115-138; 139-146.

Chapman, S. (1951). Some phenomena of the upper atmosphere. *Proc. Phys. Soc. London,* **B64**,

833-844.

Chapman, S. and Bartels, J. (1940, 1962). *Geomagnetism, Vols. 1 and 2* (1940); 2nd edition (1962). Oxford University Press, Oxford, 1049pp.

Chapman, S. and Ferraro, V.C.A. (1931). A new theory of magnetic storms. *Terr. Magn. Atmosph. Elec.*, **36**, 77 and 171.

Chapman, S. and Ferraro, V.C.A. (1932). A new theory of magnetic storms. *Terr. Magn. Atmosph. Elec.*, **37**, 147 and 421.

Chapman, S. and Ferraro, V.C.A. (1933). A new theory of magnetic storms. *Terr. Magn. Atmosph. Elec.*, **38**, 79-96.

Chatterjee, J.S. (1956). The crust as the possible seat of earth's magnetism. *J. Atmosph. Terr. Phys.*, **8**, 233-239.

Chauvin, A., Roperch, P. and Duncan, R.A. (1990). Records of geomagnetic reversals from volcanic islands of French Polynesia. 2. Paleomagnetic study of a flow sequence (1.2-0.6 Ma) from the island of Tahiti and discussion of reversal models. *J. Geophys. Res.*, **95**, 2727-2752.

Chauvin, A., Duncan, R.A., Bonhommet, N. and Levi, S. (1989). Paleointensity of the Earth's magnetic field and K-Ar dating of the Louchadiere volcanic flow (Central France): New evidence for the Laschamp excursion. *Geophys. Res. Lett.*, **16**,1189-1192.

Chen, Po-Fang (1981). Geomagnetic variations with periods from 5 to 30 years at Hong Kong. *J. Geomag. Geoelect.*, **33**, 189-195.

Chevallier, R. (1925). L'aimantation des laves de l'Etna et l'orientation du champ terrestre en Sicile du XIIe and XVIIe siecle. *Ann. Phys.*, **4**, 5-162.

Chikazumi, S. (1964). *Physics of Magnetism*. Wiley, New York.

Childers, D.C. (ed.) (1978). *Modern Spectrum Analyses*. IEEE Press, New York.

Childress, S. (1969). A class of solutions of the magnetohydrodynamic dynamo problem. In *The Application of Modern Physics to the Earth and Planetary Interiors* (S. K. Runcorn, ed.), p. 629-648. Wiley, New York.

Childress, S. (1970). New solutions of the kinematic dynamo problem. *J. Math. Phys.*, **11**, 3063-3076.

Childress, S. (1978). *Lectures on Geomagnetic Dynamo Theory*. In Woods Hole Ocean Inst. Tech. Report. WHOI-78-6 Part 1, 1.

Christensen-Dalsgaard, J. (1991). Some aspects of the theory of solar oscillations. *Geophys. Astrophys. Fluid Dyn.*, **66**, 123-152.

Cisowski, S.M. (1987). Magnetism of meteorites. In *Geomagnetism Vol. 2*, (J.A. Jacobs, ed.), p.525-560, Academic Press, London.

Cisowski, S.M., Collinson, D.W., Runcorn S.K., Stephenson, A., and Fuller, M. (1983). A review of lunar paleointensity data and implications for the origin of lunar magnetism. *J. Geophys. Res.*, **88** (Suppl. A), A691-A704.

Clark, H.C. and Kennett, J.P. (1973). Paleomagnetic excursion recorded in latest Pleistocene deep-sea sediments, Gulf of Mexico. *Earth Planet. Sci. Lett.*, **19**, 267-274.

Clark, R.M. (1975). A calibration curve for radiocarbon dates. *Antiquity*, **49**, 251-266 (and comments by Suess, and reply, *Antiquity*, **50,** 61).

Clark, R.M. and Thompson, R. (1979). A new statistical approach to the alignment of time series. *Geophys. J. Roy. Astron. Soc.*, **58**, 593-607.

Clayton, R.N. and Mayeda, T.K. (1975). Genetic relations between the Moon and Meteorites. *Proc. Sixth Lunar Sci. Conf.*, 1761-1769.

Clayton, R.N. and Mayeda, T.K. (1978). Genetic relations between iron and stony meteorites. *Earth Planet. Sci. Lett.*, **40**, 168-174.

Clayton, R.N., Grossman, L. and Mayeda, T.K. (1973). A component of primitive nuclear composition in carbonaceous chondrites. *Science*, **182**, 485-488.

Clement, B. (1991). Geographical distribution of transitional VGPs: Evidence for non-zonal equatorial symmetry during the Matuyama-Brunhes geomagnetic reversal. *Earth Planet. Sci. Lett.*, **104**, 48-58.

Clement, B. and Constable C. (1991). Polarity transitions, excursions and paleosecular variation of the Earth's magnetic field. *Rev. Geophys. Suppl. to vol. 8: US Natl. Rep. to Int. Union Geodesy and Geophys. 1987-1990*, 433-442.

Clement, B. and Kent, D.V. (1991). A southern hemisphere record of the Matuyama-Brunhes polarity reversal. *Geophys. Res. Lett.*, **18**, 81-84.

Clement, B. and Martinson, D. (1992). A quantitative comparison of two paleomagnetic records of the Cobb Mountain subchron from North Atlantic deep-sea sediments. *J. Geophys. Res.*, **97**, 1735-1752.

Clement, B. and Stixrude, L. (1995). Inner core anisotropy, anomalies in the time averaged paleomagnetic field, and polarity transition paths. *Earth Planet. Sci. Lett.*, **130**, 75-85.

Coe, R.S. (1967a). The determination of paleointensities of the earth's magnetic field with emphasis on mechanisms which could cause non-ideal behaviour in Thelliers method. *J. Geomag. Geoelect.*, **19**, 157-179.

Coe, R.S. (1967b). Paleointensities of the earth's magnetic field determined from Tertiary and Quaternary rocks. *J. Geophys. Res.*, **72**, 3247-3262.

Coe, R.S. (1979). The effect of shape anisotropy on TRM direction. *Geophys. J. Roy. Astron. Soc.*, **56**, 369-383.

Coe, R.S. and Prévot, M. (1989). Evidence suggesting extremely rapid field variation during a geomagnetic reversal. *Earth Planet. Sci. Lett.*, **92**, 292-298.

Coe, R.S., Grommé, C.S. and Mankinen, E.A. (1978). Geomagnetic paleointensities from radiocarbon dated lava flows on Hawaii and the question of the Pacific non-dipole low. *J. Geophys. Res.*, **83**, 1740-1756.

Coe, R., Prévot, M. and Camps, P. (1995). New evidence for extraordinarily rapid change of the geomagnetic field during a reversal. *Nature*, **374**, 687-692.

Coleman, P.J., Jr., Lichtenstein, B.R., Russell, C.T., Sharp, L.R. and Schubert, G. (1972). Magnetic fields near the Moon. *Proc. Third Lunar Sci. Conf.*, 2271-2286.

Collinson, D.W. (1965). Origin of remanent magnetisation in certain red sandstones. *Geophys. J. Roy. Astron. Soc.*, **9**, 203-217.

Connerney, J.E.P. (1993). Magnetic fields of the outer planets. *J. Geophys. Res.*, **98**, 18659-18679.

Constable, C.G. (1985). Eastern Australian geomagnetic field intensity over the past 14000 years.

Geophys. J. Roy. Astron. Soc., **81**, 121-130.

Constable, C.G. (1992). Link between geomagnetic reversal paths and secular variation of the field over the past 5 Myr. *Nature*, **358**, 230-233.

Constable, C.G. and McElhinny, M.W. (1985). Holocene geomagnetic secular variation records from northeastern Australian lake sediments. *Geophys. J. Roy. Astron. Soc.*, **81**, 103-120.

Constable, C.G. and Parker, R.L. (1988). Statistics of the geomagnetic secular variation for the past 5 m.y. *J. Geophys. Res.*, **93**, 11569-11581.

Constable, C.G. and Tauxe, L. (1987). Paleointensity in the pelagic realm: Marine sediment data compared with archeomagnetic and lake sediments records. *Geophys. J. Roy. Astron. Soc.*, **90**, 43-59.

Constable, C.G., Parker, R.L. and Stark, P. (1993). Geomagnetic field models incorporating frozen-flux constraints. *Geophys. J. Int.*, **113**, 419-433.

Constable, S. (1993). Constraints on mantle electrical conductivity from field and laboratory measurements. *J. Geomag. Geoelect.*, **45**, 707-728.

Constable, S.C., Parker, R.L. and Constable, C.G. (1987). Occam's inversion; a practical algorithm for generating smooth models from electromagnetic sounding data. *Geophys.*, **52**, 289-300.

Cook, P.J. and McElhinny, M.W. (1979). A re-evaluation of the spatial and temporal distribution of sedimentary phosphate deposits in the light of plate tectonics. *Econ. Geol.*, **74**, 315-330.

Cook, R.M. and Belshé, J.C. (1958). Archaeomagnetism: A preliminary report from Britain. *Antiquity*, **32**, 167-178.

Cooley, J.W. and Tukey, J.W. (1965). An algorithm for the machine calculation of complex fourier series. *Math. Comp.*, **19**, 297-301.

Cortini, M, and Barton, C.C. (1994). Chaos in geomagnetic reversal records: A comparison between Earth's magnetic field data and model disk dynamo data. *J. Geophys. Res.*, **99**, 18021-18033.

Coupland, D.H. and Van der Voo, R. (1980). Long-term non-dipole components in the geomagnetic field during the last 130 Ma. *J. Geophys. Res.* **85**, 3529-3548.

Courtillot, V. and Besse, J. (1987). Magnetic field reversals, polar wander, and core-mantle coupling. *Science* **237**, 1140-1147.

Courtillot, V. and Le Mouël, J.L. (1976a). On the long-period variation of the earth's magnetic field from 2 months to 20 years. *J. Geophys. Res.*, **81**, 2941-2950.

Courtillot, V. and Le Mouël, J.L. (1976b). Time variations of the Earth's magnetic field with a period longer than two months. *Phys. Earth Planet. Int.*, **12**, 237-240.

Courtillot, V. and Le Mouël, J.-L. (1984). Geomagnetic secular variation impulses: A review of observational evidence and geophysical consequences. *Nature*, **311**, 709-716.

Courtillot, V. and Le Mouël, J.-L. (1988). Time variations of the Earth's magnetic field: From daily to secular. *Ann. Rev. Earth Planet. Sci.*, **16**, 389-476.

Courtillot, V. and Valet, J.-P. (1995). Secular variation of the Earth's magnetic field: From jerks to reversals. *C.R. Acad. Sci. Paris Ser II*, **320**, 903-922.

Courtillot, V., Ducruix, J. and Le Mouël, J.-L. (1978). Sur une accélération récente de la variation séculaire du champ magnétique terrestre. *C. R. Acad. Sci. Paris* , **D287**, 1095-1098.

Courtillot, V., Ducruix, J. and Le Mouël, J.-L. (1979). Réponse aux commentaires de L.R.Alldredge

"Sur une accélération récente de la variation séculaire du champ magnétique terrestre." *C. R. Acad. Sci. Paris*, **B289**, 173-175.

Courtillot, V., Le Mouël, J.-L. and Ducruix, J. (1984). On Backus' mantle filter theory and the 1969 geomagnetic impulse. *Geophys. J. Roy. Astron. Soc.*, **78**, 619-625.

Cowling, T.G. (1934). The magnetic field of sunspots. *Mon. Not. Roy. Astron. Soc.*, **94**, 39-48.

Cowling, T.G. (1957). *Magnetohydrodynamics*. Wiley Interscience, New York.

Cox, A. (1962). Analysis of the present geomagnetic field for comparison with paleomagnetic results. *J. Geomag. Geoelect.* **13**, 101-112.

Cox, A. (1964). Angular dispersion due to random magnetisation. *Geophys. J. Roy. Astron. Soc.*, **8**, 345-355.

Cox, A. (1968). Lengths of geomagnetic polarity intervals. *J. Geophys. Res.*, **73**, 3247-3260.

Cox, A. (1969). Geomagnetic reversals. *Science*, **163**, 237-245.

Cox, A. (1970). Latitude dependence of the angular dispersion of the geomagnetic field. *Geophys. J. Roy. Astron. Soc.*, **20**, 253-269.

Cox, A. (1975). The frequency of geomagnetic reversals and the symmetry of the non-dipole field. *Rev. Geophys. Space Phys.*, **13**, 35-51.

Cox, A. (1982). Magnetostratigraphic time scale. In *A Geologic Time Scale* (W.B. Harland, A.V. Cox, P.G. Llewellyn, C.A.G. Pickton, A.G. Smith and R. Walters, eds), p.63-84. Cambridge Univ. Press, Cambridge.

Cox, A. and Doell, R. R. (1960). Review of paleomagnetism. *Geol. Soc. Amer. Bull.*, **71**, 645-768.

Cox, A. and Doell, R.R. (1964). Long period variations of the geomagnetic field. *Bull. Seismol. Soc. Amer.*, **54**, 2243-2270.

Cox, A., Doell, R.R. and Dalrymple, G.B. (1963a). Geomagnetic polarity epochs and Pleistocene geochronometry. *Nature*, **198**, 1049-1051.

Cox, A., Doell, R.R. and Dalrymple, G.B. (1963b). Geomagnetic polarity epochs — Sierra Nevada II. *Science*, **142**, 382-385.

Cox, A., Doell, R.R. and Dalrymple, G.B. (1964a). Geomagnetic polarity epochs. *Science*, **143**, 351-352.

Cox, A., Doell, R.R. and Dalrymple, G.B. (1964b). Reversals of the earth's magnetic field. *Science*, **144**, 1537-1543.

Creager, K.C. (1992). Anisotropy of the inner core from differential travel times of the phases PKP and PKIKP. *Nature*, **356**, 309-314.

Creager, K.C. and Jordan, T.H. (1986). Aspherical structure of the core- mantle boundary from PKP travel times. *Geophys. Res. Lett.*, **13**, 1497-1500.

Creer, K.M. (1958). Preliminary palaeomagnetic measurements from South America. *Ann. Geophys.*, **14**, 373-390.

Creer, K.M. (1962). The dispersion of the geomagnetic field due to secular variation and its determination for remote times from paleomagnetic data. *J. Geophys. Res.*, **67**, 3461-3476.

Creer, K.M. (1977). Geomagnetic secular variation during the last 25 000 years: An interpretation of the data obtained from rapidly deposited sediments. *Geophys. J. Roy. Astron. Soc.*, **48**, 91-109.

Creer, K.M. (1981). Long period geomagnetic secular variation since 12000 yr B.P. *Nature,* **292,** 208-212.

Creer, K.M. (1983). Computer synthesis of geomagnetic palaeosecular variations. *Nature,* **304,** 695-699.

Creer, K.M. and Ispir, Y. (1970). An interpretation of the behaviour of the geomagnetic field during polarity transitions. *Phys. Earth Planet. Int.,* **2,** 283-293.

Creer, K.M. and Tucholka, P. (1982a). Construction of type curves of geomagnetic secular variation for dating lake sediments from east central North America. *Canad. J. Earth Sci.,* **19,** 1106-1115.

Creer, K.M. and Tucholka, P. (1982b). The shape of the geomagnetic field through the last 8,500 years over part of the northern hemipshere. *J. Geophys.,* **51,** 188-198.

Creer, K.M. and Tucholka, P. (1982c). Secular variation as recorded in lake sediments: A discussion of North American and European results. *Phil. Trans. Roy. Soc. London,* **A306,** 87-102.

Creer, K.M., Anderson, T.W. and Lewis, C.F.M. (1976b). Late Quaternary geomagnetic stratigraphy recorded in Lake Erie sediments. *Earth Planet. Sci. Lett.,* **31,** 37-47.

Creer, K. M., Georgi, D.T. and Lowrie, W. (1973). On the representation of the Quaternary and late Tertiary geomagnetic field in terms of dipoles and quadrupoles. *Geophys. J. Roy. Astron. Soc.,* **33,** 323-345.

Creer, K.M., Gross, D.L. and Lineback, J.A. (1976a). Origin of regional geomagnetic variations recorded by Wisconsinan and Holocene sediments from Lake Michigan, USA and Lake Windermere, England. *Geol. Soc. Amer. Bull.,* **87,** 531-540.

Creer, K.M., Irving, E. and Nairn, A.E.M. (1959). Palaeomagnetism of the Great Whin Sill. *Geophys. J. Roy. Astron. Soc.,* **2,** 306-323.

Creer, K.M., Irving, E. and Runcorn, S.K. (1954). The direction of the geomagnetic field in remote epochs in Great Britain. *J. Geomag. Geoelect.,* **6,** 163-168.

Creer, K.M., Thompson, R., Molyneux, L. and Mackereth, F.H. (1972). Geomagnetic secular variation recorded in the stable magnetic remanence of recent sediments. *Earth Planet. Sci. Lett.,* **14,** 115-127.

Creer, K.M., Valencio, D.A., Sinito, A.M., Tucholka, P. and Vilas, J.F.A. (1983). Geomagnetic secular variation 0-14000 yr BP as recorded by lake sediments from Argentina. *Geophys. J. Roy. Astron. Soc.,* **74,** 199-221.

Creer, K.M., Smith, G., Tucholka, P., Bonifay, E., Thouveny, N. and Truze, E. (1986). A preliminary palaeomagnetic study of the Holocene and late Würmian sediments of Lac du Bouchet (Haute Loire, France). *Geophys. J. Roy. Astron. Soc.,* **86,** 943-964.

Cullity, B.D. (1972). *Introduction to Magnetic Materials.* Addison-Wesley Publ. Co., Menlo Park, Catifornia.

Currie, R.G. (1967). Magnetic shielding properties of the earth's mantle. *J. Geophys. Res.,* **72,** 2623-2633.

Currie, R.G. (1968). Geomagnetic spectrum of internal origin and lower mantle conductivity. *J. Geophys. Res.,* **73,** 2779-2786.

Currie, R.G. (1973). Geomagnetic line spectra – 2 to 70 years. *Astrophys. Space Sci.*, **21**, 425-438.

Currie, R.G. (1974). Harmonics of the geomagnetic annual variation. *J. Geomag. Geoelect.*, **26**, 319-328.

Currie, R.G. (1976). Long period magnetic activity – 2 to 100 years. *Astrophys. Space Sci.*, **39**, 251-254.

Curtis, S. and Ness, N. (1986). Magnetostrophic balance in planetary dynamos: Predictions for Neptune's magnetosphere, *J. Geophys. Res.*, **91**, 11003-11008.

Dagley, P. and Lawley, E. (1974). Palaeomagnetic evidence for the transitional behaviour of the geomagnetic field. *Geophys. J. Roy. Astron. Soc.*, **36**, 577-598.

Dagley, P. and Wilson, R.L. (1971). Geomagnetic field reversals – A link between strength and orientation of a dipole source. *Nature Phys. Sci.*, **232**, 16.

Dainty, A.M., Toksöv, M.N. and Stein, S. (1976). Seismic investigation of the lunar interior. *Proc. Seventh Lunar Sci. Conf.*, 3057-3075.

Dalrymple, G.B. (1991). *The Age of the Earth.* Stanford Univ. Press, Stanford, California, 474 pp.

Damon, P.E. (1970). Climatic vs. magnetic perturbation of the Carbon-14 reservoir. In *Radiocarbon Variations and Absolute Chronology* (I.U. Olsson, ed.), p.571-593. Alinqvist and Wiksell, Stockholm

David, P. (1904). Sur la stabilité de la direction d'aimantation dans quelques roches volcaniques. *C. R. Acad. Sci. Paris*, **138**, 41-42.

Davies, G.F. and Gurnis, M. (1986). Interaction of mantle dregs with convection; lateral heterogeneity at the core-mantle boundary. *Geophys. Res. Lett.*, **13**, 1517-1520.

Davies, J, and Richards, M. (1992). Mantle convection. *J. Geol.*, **100**, 151-206.

Day, R. (1977). TRM and its variation with grain size. *J. Geomag. Geoelect.*, **29**, 233-265.

Day, R., Fuller, M.D. and Schmidt, V.A. (1976). Magnetic hysteresis properties of synthetic titanomagnetites. *J. Geophys. Res.,* **81**, 873-880.

Day, R., Fuller, M.D. and Schmidt, V.A. (1977). Hysteresis properties of titanomagnetites: Grain size and composition dependence. *Phys. Earth Planet. Int.*, **13**, 260-267.

Denham, C.R. (1974). Counter-clockwise motion of paleomagnetic directions 24,000 years ago at Mono Lake, California. *J. Geomag. Geoelect.*, **26**, 487-498.

Denham, C.R. (1975). Spectral analysis of paleomagnetic time series. *J. Geophys. Res.*, **80**, 1897-1901.

Denham, C.R. and Cox, A.V. (1971). Evidence that the Laschamp Polarity Event did not occur 13 300-30 400 years ago. *Earth Planet. Sci. Lett.*, **13**, 181-190.

Dicke, R.H. (1970). Internal rotation of the sun. *Ann. Rev. Astron. Astrophys.*, **8**, 297-328.

Dickey, J. and Eubanks, T. (1986). The application of space geodesy to Earth orientation studies. In *Space Geodesy and Geodynamics* (A. Anderson, and A. Cazenave, eds), Academic Press, London.

Dickson, G.O., Pitman III, W.C. and Heirtzler, J R. (1968). Magnetic anomalies in the South Atlantic and ocean floor spreading. *J. Geophys. Res.*, **73**, 2087-2100.

Dodson, R.E. (1979). Counterclockwise precession of the geomagnetic field vector and westward drift of the non-dipole field. *J. Geophys. Res.*, **84**, 637-644.

Dodson, R.E., Dunn, J.R., Fuller, M.D., Williams, I., Ito, H., Schmidt, V.A. and Wu, Yee-Ming (1978). Palaeomagnetic record of a late Tertiary field reversal. *Geophys. J. Roy. Astron. Soc.*, **53**, 373-412.

Doell, R.R. (1970). Paleomagnetic secular variation study of lavas from the Massif Central, France. *Earth Planet. Sci. Lett.*, **8**, 352-362.

Doell, R.R. (1972). Palaeomagnetic study of Icelandic lava flows. *Geophys. J. Roy. Astron. Soc.*, **26**, 459-479.

Doell, R.R. and Cox, A.V. (1963). The accuracy of the paleomagnetic method as evaluted from historic Hawaiian lava flows. *J. Geophys. Res.*, **68**, 1997-2009.

Doell, R.R. and Cox, A.V. (1965). Paleomagnetism of Hawaiian lava flows. *J. Geophys. Res.*, **70**, 3377-3405.

Doell, R.R. and Cox, A.V. (1972). The Pacific geomagnetic secular variation anomaly and the question of lateral uniformity in the lower mantle. In *The Nature of the Solid Earth* (E.C. Robertson, ed.), p.245-284. McGraw-Hill, New York

Doell, R.R., Grommé, C.S., Thorpe, A.N. and Senftle, F.E. (1970). Magnetic studies of Apollo 11 lunar samples. *Proc. Apollo 11 Lunar Sci. Conf.*, 2097-2102.

Dolginov, Sh. Sh. (1976). On the question of the energy of the precessional dynamo. In *Solar-Wind Interaction with the Planets Mercury, Venus and Mars* (N.F. Ness, ed.). *NASA Spec. Publ.*, **SP-397**, 167-170.

Dolginov, Sh. Sh. (1978). On the magnetic field of Mars: Mars 5 evidence. *Geophys. Res. Lett.*, **5**, 93-95.

Dolginov, Sh. Sh., Yeroshenko, Ye. G., Zhuzgov, L.N. and Pushkov, N.V. (1961). Investigation of the magnetic field of the Moon. *Geomag. Aeron.*, **1**, 18-25.

Dolginov, Sh. Sh., Yeroshenko, E.G., Zhuzgov, L.N. and Pushkov, N.V. (1966). Measurement of the magnetic field in the vicinity of the Moon by the artificial satellite Luna 10. *Dokl. Akad. Nauk SSSR*, **170**, 574-577.

Domen, H. (1977). A single heating method of paleomagnetic field intensity determination applied to old roof tiles and rocks. *Phys. Earth Planet. Int.*, **13**, 315-318.

Doornbos, D.J. and Hilton, T. (1989). Models of the core-mantle boundary and the travel times of internally reflected core phases. *J. Geophys. Res.*, **94**, 15741-15751.

Drewry, G.E., Ramsay, A.T.S. and Smith, A.G. (1974). Climatically controlled sediments, the geomagnetic field, and trade wind belts in Phanerozoic time. *J. Geol.*, **82**, 531-553.

Du Bois, R.L. (1974). Secular variation in southwestern United States as suggested by archeomagnetic studies. In *Proceedings of the Takesi Nagata Conference* (R.M. Fisher, M.D. Fuller, V.A. Schmidt and P.J. Wasilewski, eds), p.133. University of Pittsburgh Press, Pittsburgh.

Du Bois, R.L. (1989). Archeomagnetic results from southwest United States and Mesoamerica, and comparison with some other areas. *Phys. Earth Planet. Int.*, **56**, 18-33.

Duba, A.G. (1992). Earth's core not so hot. *Nature*, **359**, 197-198.

Duba, A.G. and Wanamaker, B.J. (1994). DAC measurement of Perovskite conductivity and implications for the distribution of mineral phases in the lower mantle. *Geophys. Res. Lett.*, **21**,

1643-1646.

Ducruix, J., Courtillot, V. and Le Mouël, J.-L. (1980). The late 1960s secular variation impulse, the eleven year magnetic variation and the electrical conductivity of the deep mantle. *Geophys. J. Roy. Astron. Soc.*, **61**, 73-94.

Duffy, T.S. and Hemley, R.J. (1995). Some like it hot: The temperature structure of the Earth. *Rev. Geophys. Suppl., U.S. National Report to IUGG 1991-1994*, 5-9.

Duncan, R.A., (1975). Palaeosecular variation at the Society Islands, French Polynesia. *Geophys. J. Roy. Astron. Soc.*, **41**, 245-254.

Dungey, J.W. (1961). Interplanetary magnetic field and the auroral zones. *Phys. Rev. Lett.*, 6,47-48.

Dungey, J.W. (1963). The structure of the exosphere, or adventures in velocity space. In *Geophysics, the Earth's Environment* (C. DeWitt, J. Hieblot and A. Lebeau, eds), p.526-537. Gordon and Breach, New York.

Dunlop, D.J. (1971). Magnetic properties of fine particle hematite. *Ann. Geophys.*, **27**, 269-293.

Dunlop, D.J. (1979). On the use of Zijderveld vector diagrams in multicomponent paleomagnetic studies. *Phys. Earth Planet. Int.*, **20**, 12-24.

Dunlop, D.J. (1990). Developments in rock magnetism. *Rep. Prog. Phys.*, **53**, 707-792.

Dunn, J.R., Fuller, M.D., Ito, H. and Schmidt, V.A. (1971). Paleomagnetic study of a reversal of the earth's magnetic field. *Science*, **172**, 840-845.

Duvall T.L., Jr., D'Silva, S., Jefferies, S., Harvey, J. and Schou, J. (1996). Downflows under sunspots detected by heliosesmic tomography. *Nature*, **379**, 235-237.

Dyal, P. and Parkin, C.W. (1971). The Apollo 12 magnetometer experiment; internal lunar properties from transient and steady magnetic field measurements. *Proc. Second Lunar Sci. Conf.*, 2391-2413.

Dyal, P., Parkin, C.W. and Daily, W.D. (1977). Global lunar crust: Electrical conductivity and thermoelectric origin of remanent magnetism. *Proc. Eighth Lunar Sci. Conf.*, 767-784.

Dyal, P., Parkin, C.W. and Sonett, C.P. (1970). Apollo 12 magnetometer – Measurement of a steady magnetic field on the surface of the Moon. *Science*, **169**, 762-764.

Dziewonski, A.M. and Anderson, D.L. (1981). Preliminary Reference Earth Model (PREM). *Phys. Earth Planet. Int.*, **25**, 297-356.

Dziewonski, A.M., and Anderson, D.L. (1984). Seismic tomography of the Earth's interior. *Amer. J. Sci.*, **72**, 483-494.

Dziewonski, A.M., and Gilbert, F. (1971). Solidity of the inner core of the Earth inferred from normal mode observations. *Nature*, **234**, 465-466.

Dziewonski, A. and Woodhouse, J. (1987). Global images of the Earth's interior. *Science*, **236**, 37-48.

Dziewonski, A., Hager, B. and O'Connell, R. (1977). Large-scale heterogeneities in the lower mantle. *J. Geophys. Res.*, **85**, 239-255.

Dzyaloshinski, I. (1958). A thermodynamic theory of "weak" ferromagnetism of antiferromagnetics. *J. Phys. Chem. Solids*, **4**, 241-255.

Eddy J.A (1976). The Maunder Minimum, *Science*, **192**, 1189-1202.

Eddy, J.A. (1977). The case of the missing sunspots. *Scient. Amer.*, **236**, 80-92.

Eddy, J.A. (1988). Variability of the present and ancient Sun; a test of solar uniformitarianism. In *Secular Solar and Geomagnetic Variations in the Last 10,000 Years* (F.R. Stephenson and A.W. Wolfendale, eds). NATO Advanced Study Series, p.1-23, Kluwer Acad. Publ., Dordrecht.

Eddy, J.A., Gilman, P.A. and Trotter, D.E. (1977). Anomalous solar rotation in the early 17th century. *Science*, **198**, 824-829.

Egbert, C.G. (1992). Sampling bias in VGP longitudes. *Geophys. Res. Lett.*, **19**, 2353-2356.

Egbert, G. and Booker, J. (1992). Very long period magnetotellurics at Tucson observatory: Implications for mantle conductivity, *J. Geophys. Res.*, **97**, 15099-15112.

Elmore, R.D. and McCabe, C. (1991). The occurrence and origin of remagnetization in the sedimentary rocks of North America. *Rev. Geophys.*, **29** (Suppl.), 377-383.

Elsasser, W.M. (1946a). Induction effects in terrestrial magnetism. 1. Theory. *Phys. Rev.*, **69**, 106-116.

Elsasser, W.M. (1946b). Induction effects in terrestrial magnetism. 2. The secular variations. *Phys. Rev.*, **70**, 202-212.

Elsasser, W.M. (1947). Induction effects in terrestrial magnetism. 3. Electric modes. *Phys. Rev.*, **72**, 821-833.

Elsasser, W.M., Ney, E.P. and Wenckler, I.R. (1956). Cosmic ray intensity and geomagnetism. *Nature*, **178**, 1226-1227.

Elston, D.P. and Purucker, M.E. (1979). Detrital magnetisation in red beds of the Moenkopi formation. *J. Geophys. Res.*, **84**, 1653-1665.

Engebretson, D., Kelley, K., Cashman, H. and Richards, M. (1992). 180 million years of subduction. *GSA Today*, **2**, 93-95,100.

Enkin, R.J. and Williams, W. (1994). Three-dimensional micromagnetic analysis of stability in fine magnetic grains, *J. Geophys. Res.*, **99**, 611-618.

Evans, M.E. (1976). Test of the dipolar nature of the geomagnetic field throughout Phanerozoic time. *Nature*, **262**, 676-677.

Evans, M.E. (1987). New archaeomagnetic evidence for the persistence of the geomagnetic westward drift. *J. Geomag. Geoelect.*, **39**, 769-772.

Evans, M.E. and McElhinny, M.W. (1966). The paleomagnetism of the Modipe gabbro. *J. Geophys. Res.*, **71**, 6053-6063.

Evans, M.E. and McElhinny, M.W. (1969). An investigation of the origin of stable remanence in magnetite-bearing igneous rocks. *J. Geomag. Geoelect.*, **21**, 757-773.

Evans, M.E. and Wayman, M.C. (1974). An investigation of small magnetic particles by means of electron microscopy. *Earth Planet. Sci. Lett.*, **9**, 365-370.

Fearn, D.R. (1994). Nonlinear Planetary Dynamos, 219-245, *Lectures on Solar and Planetary Dynamos* (M.R.E. Proctor and A.D. Gilbert, eds), Cambridge Univ. Press, New York, 375 pp

Filloux, J.H. (1980a). Observations of very low frequency electromagnetic signals in the ocean. *J. Geomag. Geoelect.*, **32**, I: 1-12 (Suppl).

Filloux, J.H. (1980b). North Pacific magnetotelluric experiments. *J. Geomag. Geoelect.*, **32**, 33-43.

Fisher, N.I., Lewis, T. and Embleton, B.J.J. (1987). *Statistical Analysis of Spherical Data*, Cambridge Univ. Press, London, 329 pp.

Fisher, R.A. (1953). Dispersion on a sphere. *Proc. Roy. Soc. London,* **A217**, 295-305.

Fisk, H.W. (1931). Isopors and isoporic movement. *Bull. Intl. Geodet. Geophys. Union,* **No. 8**, Stockholm Assembly 1930, 280-292.

Folgerhaiter, G. (1899). Sur les variations séculaires de l'inclinaison magnétique dans l'antiquité. *J. Phys. Ser. 3.,* **8**, 5-16.

Foster, J.H. and Opdyke, N.D. (1970). Upper Miocene to Recent magnetic stratigraphy in deep-sea sediments. *J. Geophys. Res.,* **75**, 4465-4473.

Fraser-Smith, A.C. (1987). Centered and eccentric geomagnetic dipoles and their poles, 1600-1985. *Rev.Geophys.,* **25**, 1-16.

Freed, W.K. (1977). The virtual geomagnetic pole path during the Brunhes-Matuyama polarity change when viewed from equatorial latitudes. *EOS, Trans. Amer. Geophys. Union,* **58**, 6, 380.

Freed, W.K. and Healy, N. (1974). Excursions of the Pleistocene geomagnetic field recorded in Gulf of Mexico sediments. *Earth Planet. Sci. Lett.,* **24**, 99-104.

Fuller, M.D. (1970). Geophysical aspects of paleomagnetism. *CRC Crit. Rev. Solid State Sci.,* **1**, 137-219.

Fuller, M.D. (1974). Lunar magnetism. *Rev. Geophys. Space Phys.,* **12**, 23-70.

Fuller, M.D. and Cisowski, S.M. (1987) Lunar paleomagnetism. In *Geomagnetism Vol. 2* (J.A. Jacobs, ed.), p.307-455, Academic Press, London.

Fuller, M.D., Williams I. and Hoffman, K.A. (1979). Paleomagnetic records of geomagnetic field reversals and the morphology of the transitional fields. *Rev. Geophys. Space Phys.,* **17**, 179-203, 1979.

Gaffin, S. (1989). Analysis of scaling in the geomagnetic polarity reversal record. *Phys. Earth Planet. Int.,* **57**, 284-289.

Gaherty, J. and Lay, T. (1992). Investigation of laterally heterogeneous shear velocity structure in D" beneath Eurasia. *J. Geophys. Res.,* **97**, 417-435.

Galliher, S.C. and Mayhew, M.A. (1982). On the possibility of detecting large-scale crustal remanent magnetisation with MAGSAT vector magnetic anomaly data. *Geophys. Res. Lett.,* **9**, 325-328.

Games, K.P. (1977). The magnitude of the paleomagnetic field: A new non-thermal, non-detrital method using sun-dried bricks. *Geophys. J. Roy. Astron. Soc.,* **48**, 315-329.

Games, K.P. (1980). The magnitude of the archaeomagnetic field in Egypt between 3000 and 0 B.C. *Geophys. J. Roy. Astron. Soc.,* **63**, 45-56.

Gans, R.F. (1972). Viscosity of the Earth's core. *J. Geophys. Res.,* **77**, 360-366.

Gauss, C.F. (1839). Allgemeine Theorie des Erdmagnetismus. In *Resultate aus den Beobachtungen magnetischen Vereins im Jahre 1838,* p.1-57. (Reprinted in *Werke,* **5**, 121-193, Gottingen, 1877; Translated by Sabine, E., in Taylor, R., *Scientific Memoirs Vol. 2,* R. & J. E. Taylor, London, 1841).

Georgi, D.T. (1974). Spherical harmonic analysis of paleomagnetic inclination data. *Geophys. J. Roy. Astron. Soc.,* **39**, 71-86.

Gibson, R.D. and Roberts, P.H. (1969). The Bullard-Gellman dynamo. In *Application of Modern Physics to the Earth and Planetary Interiors* (S.K. Runcorn, ed.), p.577-601. Wiley

Interscience, New York.

Gilman, P.A. (1974). Solar rotation. *Ann. Rev. Astron. & Astrophys.*, **12**, 47-70.

Gilman, P.A. (1977). Nonlinear dynamics of Boussinesq convection in a deep rotating spherical shell - 1. *Geophys. Astrophys. Fluid Dyn.*, **8**, 93-135.

Gilman, P.A. (1992). What can we learn about solar cycle mechanisms from observed velocity fields? In *The Solar Cycle* (K.L. Harvey, ed.), p.241-255. ASP Conf. Ser. Vol 27.

Gire, C. and Le Mouël, J.-L. (1990). Tangentially geostrophic flow at the core mantle boundary compatible with the observed geomagnetic secular variation: The large-scale component of the flow. *Phys. Earth Planet. Int.*, **59**, 259-287.

Glass, B. and Heezen, B. (1967). Tectites and geomagnetic reversals. *Nature* **214**, 372.

Glatzmaier, G.A. (1985). Numerical simulations of stellar convection dynamos. II. Field propagation in the convection zone, *Astrophys. J.*, **291**, 300-307.

Glatzmaier, G.A. and Roberts, P.H. (1995a). A three-dimensional convective dynamo solution with rotating and finitely conducting inner core and mantle. *Phys. Earth. Planet. Int.*, **91**, 63-75.

Glatzmaier, G.A. and Roberts, P.H. (1995b). A three-dimensional self-consistent computer simulation of a geomagnetic field reversal. *Nature*, **377**, 203-209.

Glen, W. 1982. *The road to Jaramillo. Critical years of the revolution in Earth science.* Stanford University Press, Stanford, California, 459 pp.

Goarant, F., Guyot, F., Peyronneau, J. and Poirier, J.P. (1992). High-pressure and high-temperature reactions between silicates and liquid iron alloys, in the diamond anvil cell, studied by analytical electron microscopy. *J. Geophys. Res.*, **97**, 4477-4487.

Gold, T. (1959). Motions in the magnetosphere of the Earth. *J. Geophys. Res.*, **64**, 1219-1224.

Gradstein, F.M, Agterberg, F.P., Ogg, J.G, Hardenbol, J., van Veen, P., Thierry, J. and Huang, Z. (1994). A Mesozoic time scale. *J. Geophys. Res.*, **99**, 24051-24074.

Graham, J.W. (1949). The stability and significance of magnetism in sedimentary rocks. *J. Geophys. Res.*, **54**, 131-167.

Graham, K.W.T. and Hales, A.L. (1957). Palaeomagnetic measurements on some Karroo dolerites. *Phil. Mag. Adv. Phys.*, **6**, 149-161.

Graham, K.W.T., Helsley, C.E. and Hales, A.L. (1964). Determination of the relative positions of continents from paleomagnetic data. *J. Geophys. Res.*, **69**, 3895-3900.

Grand, S, (1994). Mantle shear structure beneath the Americas and surrounding ocean. *J. Geophys. Res.*, **99**, 11591-11621,.

Griffiths, D.H. (1953). Remanent magnetism of varved clays from Sweden. *Nature*, **172**, 539.

Griffiths, R. and Campbell, I. (1991). On the dynamics of long-lived plume conduits in the convecting mantle. *Earth Planet. Sci. Lett.*, **103**, 214-227.

Grommé, C.S., Wright, T.L. and Peck, D.L. (1969). Magnetic properties and oxidation of iron-titanium oxide minerals in Alae and Makaopuhi lava lakes, Hawaii. *J. Geophys. Res.*, **74**, 5277-5293.

Gubbins, D. (1974). Theories of the geomagnetic and solar dynamos. *Rev. Geophys. Space Phys.*, **12**, 137-154.

Gubbins, D. (1976). Observational constraints on the generation process of the earth's magnetic field. *Geophys. J. Roy. Astron. Soc.*, **47**, 19-39.

Gubbins, D. (1977). Energetics of the earth's core. *J. Geophys.*, **43**, 453-464.

Gubbins, D. (1982). Finding core motions from magnetic observations. *Phil. Trans. Roy. Soc. London*, **A269**, 247-254.

Gubbins, D. and Coe, R.S. (1993). Longitudinally confined geomagnetic reversal paths from non-dipole transition fields. *Nature*, **362**, 51-53.

Gubbins, D. and Kelly, P. (1993). Persistent patterns in the geomagnetic field over the past 2.5 Myr. *Nature*, **365**, 829-832.

Gubbins, D. and Kelly, P. (1995). On the analysis of paleomagnetic secular variation. *J. Geophys. Res.*, **100**, 14955-14964.

Gubbins, D. and Richards, M. (1986). Coupling of the core dynamo and mantle: Thermal or topographic? *Geophys. Res. Lett.*, **13**, 1521-1524.

Gubbins, D. and Roberts, P.H. (1987). Magnetohydrodynamics of the Earth's core. In *Geomagnetism Vol. 2* (J.A. Jacobs, ed.), p.1-183, Academic Press, London.

Gubbins, D. and Zhang, K. (1993). Symmetry properties of the dynamo equations for palaeomagnetism and geomagnetism. *Phys. Earth Planet. Int.*, **75**, 225-241.

Gubbins, D., Masters, T.G. and Jacobs, J.A. (1979). Thermal evolution of the earth's core. *Geophys. J. Roy. Astron. Soc.*, **59**, 57-100.

Gurnis, M. and Davies, G. (1986). The effect of depth-dependent viscosity on convective mixing in the mantle and the possible survival of primitive mantle. *Geophys. Res. Lett.* **13**, 541-544.

Guyot, F., Peyronneau, J. and Poirier, J.-P. (1988). TEM study of high pressure reactions between iron and silicate perovskites. *Chem. Geology*, **70**, 61.

Hagee, V.L. and Olson, P. (1989). An analysis of paleomagnetic secular variation in the Holocene. *Phys. Earth Planet Int.*, **56**, 266-284.

Hagee, V.L. and Olson, P. (1991). Dynamo models with permanent dipole fields and secular variation. *J. Geophys. Res.* **96**, 11673-11687.

Haggerty, S.E. (1976). Chapters 4 and 8. In *Oxide Minerals* (Douglas Rumble III, ed.). Southern Printing Co., Blacksburg, Virginia.

Haggerty, S.E. (1978). Mineralogical constraints on Curie isotherms in deep crustal magnetic anomalies. *Geophys. Res. Lett.*, **5**, 105-108.

Haggerty, S.E. (1994). Superkimberlites: A geodynamic diamond window to the Earth's core. *Earth Planet. Sci. Lett.*, **122**, 57-69.

Haigh, G. (1958). The process of magnetisation by chemical change. *Phil. Mag.*, **3**, 267-286.

Hale, G.E. (1924). Sun-spots as magnets and the periodic reversal of their polarity. *Nature*, **113**, 105-112.

Hale, C. (1987). New Paleomagnetic data suggest a link between the Archean-Proterozoic boundary and nucleation of the inner core. *Nature*, **329**, 233-237.

Hale, C.J. and Dunlop, D.J. (1984). Evidence for an Early Archean geomagnetic field: A palaeomagnetic study of the Komati formation, Barberton greenstone belt, South Africa. *Geophys. Res. Lett.*, **11**, 97-100.

Halgedahl, S. (1989). Magnetic domains. In *The Encyclopedia of Solid Earth Geophysics* (D.E.

James, ed.), p.706-721. Van Nostrand, New York.

Halgedahl, S. (1991). Magnetic domain patterns observed on synthetic Ti-rich titanomagnetite as a function of temperature and in states of thermoremanent magnetization. *J. Geophys. Res.*, **96**, 3943-3972.

Halgedahl, S. and Fuller, M. (1981). The dependence of magnetic domain structure upon magnetization state in polycrystalline pyrrhotite. *Phys. Earth Planet. Int.*, **26**, 93-97.

Halgedahl, S. and Fuller, M. (1983). The dependence of magnetic domain structure upon magnetization state with emphasis upon nucleation as a mechanism for pseudo-single domain behavior. *J. Geophys. Res.*, **88**, 6505-6522.

Halls, H.C. (1978). The use of converging remagnetisation circles in paleomagnetism. *Phys. Earth Planet. Int.*, **16**, 1-11.

Handschumaker, D.W., Sager, W.W., Hilde, T.W.C. and Bracey, D.R. (1988). Pre-Cretaceous evolution of the Pacific plate and extension of the geomagnetic polarity reversal time scale with implications for the origin of the Jurassic "Quiet Zone." *Tectonophysics*, **155**, 365-380.

Hanna, R.L. and Verosub, K.L. (1989). A review of lacustrine paleomagnetic records from western North America: 0-40000 years BP. *Phys. Earth Planet. Int.*, **56**, 76-95.

Haq, B., Hardenbol, J. and Vail, P. (1987). Chronology of fluctuating sea levels since the Triassic. *Science*, **235**, 1156-1167.

Harland, W.B., Armstrong, R.L., Cox, A.V., Craig, L.E., Smith, A.G., and Smith, D.G. (1990). *A Geologic Time Scale 1989*. Cambridge University Press, Cambridge, 263 pp.

Harland, W.B., Cox, A.V., Llewellyn, P.G., Picton, C.A.G., Smith, A.G., and Walters, R. (1982). *A Geologic Time Scale*. Cambridge University Press, New York, 131 pp.

Harrison, C.G.A. (1968). Formation of magnetic anomaly patterns by dyke injection. *J. Geophys. Res.*, **73**, 2137-2142.

Harrison, C.G.A. (1974). The paleomagnetic record from deep-sea sediment cores. *Earth Sci. Rev.*, **10**, 1-36.

Harrison, C.G.A. (1980). Secular variation and excursions of the earth's magnetic field. *J. Geophys. Res.*, **85**, 3511-3522.

Harrison, C.G.A. (1987) The crustal field. In *Geomagnetism Vol. 1* (J.A. Jacobs, ed.), p.513-610, Academic Press, London.

Harrison, C.G.A. (1994). An alternative picture of the geomagnetic field. *J. Geomag. Geoelect.*, **46**, 127-142.

Harrison, C.G.A. and Carle, H.M. (1981). Intermediate wavelength magnetic anomalies over ocean basins. *J. Geophys. Res.*, **86**, 11585-11599.

Harrison, C.G.A. and Huang Q. (1990). Rates of change of the earth's magnetic field by recent analysis. *J. Geomag. Geoelect.*, **42**, 897-928.

Harrison, C.G.A. and Ramirez, E. (1975). Areal coverage of spurious reversals of the earth's magnetic field. *J. Geomag. Geoelect.*, **27**, 139-151.

Harrison, C.G.A. and Somayajulu, B.L.K. (1966). Behaviour of the earth's magnetic field during a reversal. *Nature*, **212**, 1193-1195.

Harwood, J.M. and Malin, S.R.C. (1976). Present trends in the Earth's magnetic field. *Nature*, **259**,

469-471.

Head, J.W. (1976). Lunar volcanism in space and time. *Rev. Geophys. Space Phys.*, **14**, 265-300.

Heirtzler, J.R., Dickson, G.O., Herron, E.M., Pitman II, W.C. and Le Pichon, X. (1968). Marine magnetic anomalies, geomagnetic field reversals, and motions of the ocean floor and continents. *J. Geophys. Res.*, **73**, 2119-2136.

Heller, F. (1980). Self-reversal of natural remanent magnetisation in the Olby-Laschamp lavas. *Nature*, **284**, 334-335.

Heller, F. and Petersen, N. (1982). Self-reversal explanation for the Laschamp/Olby geomagnetic field excursion. *Phys. Earth Planet. Int.*, **30**, 358-373.

Helsley, C.E. and Steiner, M.D. (1969). Evidence for long intervals of normal polarity during the Cretaceous period. *Earth Planet. Sci. Lett.*, **5**, 325-332.

Henshaw, P.C., Jr. and Merrill, R.T. (1980). Magnetic and chemical changes in marine sediments. *Rev. Geophys. Space Phys.*, **18**, 483-504.

Herrero-Bervera, E. and Khan, M.A. (1992). Olduvai termination; detailed palaeomagnetic analysis of a north central Pacific core. *Geophys. J. Int.*, **108**, 535-545.

Herrero-Bervera, E. and Theyer, F. (1986). Non-axisymmetric behaviour of Olduvai and Jarmillo polarity transitions recorded in north-central Pacific deep-sea sediments. *Nature*, **322**, 159-162.

Hess, H.H. (1960). Evolution of ocean basins: Report to Office of Naval Research on research supported by ONR Contract Nonr 1858 (10).

Hess, H.H. (1962). History of Ocean basins. In *Petrologic Studies: A Volume to Honor A.F. Buddington* (A.E.J. Engel, H. James and B.F. Leonard, eds), p.599-620. Geol. Soc. Amer., Boulder, Colorado.

Hibberd, F.H. (1979). The origin of the earth's magnetic field. *Proc. Roy. Soc. London*, **A369**, 31-45.

Hide, R. (1966). Free hydromagnetic oscillations of the Earth's core and the theory of the geomagnetic secular variation. *Phil. Trans. Roy. Soc. London*, **A259**, 615-647.

Hide, R. (1969). Interaction between the Earth's liquid core and solid mantle. *Nature*, **222**, 1055-1056.

Hide, R. (1981). Self-exciting dynamos and geomagnetic polarity changes. *Nature*, **293**, 728.

Hide, R. (1989). Fluctuations in the Earth's rotation and the topography of the core-mantle interface. *Phil. Trans. Roy. Soc. London*, **A328**, 351-363.

Hide, R. (1995). The topographic torque on a bounding surface of a rotating gravitating fluid and the excitation by core motions of decadal fluctuations in the Earth's rotation. *Geophys. Res. Lett.*, **22**, 961-964.

Hide, R. and Malin, S.R.C. (1970). Novel correlations between global features of the earth's gravitational and magnetic fields. *Nature*, **225**, 605.

Hide, R. and Malin, S.R.C. (1971a). Novel correlation between global features of the Earth's gravitational and magnetic fields: Further statistical considerations. *Nature Phys. Sci.*, **230**, 63.

Hide, R. and Malin, S.R.C. (1971b). Effect of rotation in latitude on correlations between Earth's gravitational and magnetic fields. *Nature Phys. Sci.*, **232**, 31.

Hide, R. and Palmer, T. (1982). Generalisation of Cowling's theorem. *Geophys. Astrophys. Fluid*

Dyn., **19**, 301-309.

Hide, R. and Stewartson, K. (1972). Hydromagnetic oscillations of the Earth's core. *Rev. Geophys. Space Phys.*, **10**, 579-598.

Higgins, G. and Kennedy, G.C. (1971). The adiabatic gradient and the melting point gradient in the core of the Earth. *J. Geophys. Res.*, **76**, 1870-1878.

Hillhouse, J. and Cox, A. (1976). Brunhes-Matuyama polarity transition. *Earth Planet. Sci. Lett.*, **29**, 51-64.

Hirooka, K. (1983). Results from Japan. In *Geomagnetism of Baked Clays and Sediments* (K.M. Creer, P. Tucholka and C.E. Barton, eds), p.150-157, Elsevier, Amsterdam.

Hoffman, K.A. (1975). Cation diffusion processes and self reversal of thermoremanent magnetization in the ilmenite-haematite solid solution series. *Geophys. J. Roy. Astron. Soc.*, **41**, 65-80.

Hoffman, K.A. (1977). Polarity transition records and the geomagnetic dynamo. *Science*, **196**, 1329-1332.

Hoffman, K.A. (1979). Behavior of the geodynamo during reversal: A phenomenological model. *Earth Planet. Sci. Lett.*, **44**, 7-17.

Hoffman, K.A. (1981). Paleomagnetic excursions, aborted reversals and transitional fields. *Nature*, **294**, 67-68.

Hoffman, K.A. (1986) Transitional field behavior from southern hemispheric lavas: Evidence for two-stage reversals of the geodynamo. *Nature*, **320**, 228-232.

Hoffman, K.A. (1989). Geomagnetic polarity reversals: Theory and models. In *The Encyclopedia of Solid Earth Geophysics* (D.E. James, ed.), p.547-555, Van Nostrand–Rheinhold, New York.

Hoffman, K.A. (1991). Long-lived transitional states of the geomagnetic field and the two dynamo families. *Nature*, **354**, 273-277.

Hoffman, K.A. (1992a). Self-reversal of thermo-remanent magnetization in the ilmenite-hematite system: Order-disorder, symmetry, and spin alignment. *J. Geophys. Res.*, **97**, 10883-10896.

Hoffman, K.A. (1992b). Dipolar reversal states of the geomagnetic field and core-mantle dynamics. *Nature*, **359**, 789-794.

Hoffman, K.A. and Day, R. (1978). Separation of multicomponent NRM: A general method. *Earth Planet. Sci. Lett.*, **40**, 433-438.

Holcomb, R.T., Champion, D.E. and McWilliams, M.O. (1986). Dating Recent Hawaiian lava flows using paleomagnetic secular variation. *Geol. Soc. Amer. Bull.*, **97**, 829-839.

Hollerbach, R. and Ierley, G.R. (1991). A model α^2-dynamo in the limit of asympototically small viscosity. *Geophys Astrophys. Fluid Dyn.*, **60**, 133- 158.

Hollerbach, R. and Jones, C.A. (1993a). Influence of the Earth's outer core on geomagnetic fluctuations and reversals. *Nature*, **365**, 541-543.

Hollerbach, R. and Jones, C.A. (1993b). A geodynamo model incorporating a finitely conducting inner core. *Phys. Earth Planet Int.*, **75**, 317-327.

Hollerbach, R., Barenghi, C. F. and Jones, C.A. (1992). Taylor's constraint in a spherical $\alpha\omega$ dynamo. *Geophys. Astrophys. Fluid Dyn.*, **67**, 37-64.

Hood, L.L. (1981). The enigma of lunar magnetism. *EOS, Trans. Amer. Geophys. Union*, **62**, 161-

163.

Hospers, J. (1953). Reversals of the main geomagnetic field I, II. *Proc. Kon. Nederl. Akad. Wetensch., B.,* **56**, 467-491.

Hospers, J. (1954a). Reversals of the main geomagnetic field III. *Proc. Kon. Nederl. Akad. Wetensch., B.,* **57**, 112-121.

Hospers, J. (1954b). Rock magnetism and polar wandering. *Nature,* **173**, 1183.

Hospers, J. (1955). Rock magnetism and polar wandering. *J. Geol.,* **63**, 59-74.

Houtermans, J. (1966). On the quantitative relationships between geophysical parameters and the natural ^{14}C inventory. *Zeits. Phys.,* **193**, 1-12.

Houtermans, J.C., Suess, H.E. and Oeschger, H. (1973). Reservoir models and production rate variations of natural radiocarbon. *J. Geophys. Res.,* **78**, 1897-1908.

Howard, R. (1975). The rotation of the sun. *Scient. Amer.,* **232**, 106-114.

Howe, H.C., Lin, R.P., McGuire, R.E. and Anderson, K.A. (1974). Energetic electron scattering from the lunar remanent magnetic field. *Geophys. Res. Lett.,* **1**, 101-104.

Hoye, G.S. (1981). Archaeomagnetic secular variation record of Mount Vesuvius. *Nature,* **291**, 216-218.

Hoyt, D.V., Schatten, K.H. and Nesmes-Ribes, E. (1994). The hundredth year of Rudolf Wolf's death: Do we have the correct reconstruction of solar activity? *Geophys. Res. Lett.,* **21**, 2067-2070.

Hulot, G. and Gallet, Y. (1996). On the interpretation of virtual geomagnetic pole (VGP) scatter curves. *Phys. Earth Planet. Int.,* submitted.

Hulot, G. and Le Mouël, J.-L. (1994). A statistical approach to the Earth's main magnetic field. *Phys. Earth Planet. Int.,* **82**, 167-183.

Hulot, G., Le Huy, M. and Le Mouël, J.-L. (1993). Secousses (jerks) de la variation séculaire et mouvements dans la noyau terrestre. *C. R. Acad. Sci. Paris Ser. II,* **317**, 333-341.

Hurwitz, L. (1960). Eccentric dipoles and spherical harmonic analysis. *J. Geophys. Res.,* **65**, 2555-2556.

Hyodo, M. (1984). Possibility of reconstruction of the past geomagnetic field from homogeneous sediments. *J. Geomag. Geoelect.,* **36**, 45-62.

Imbrie, J. and Imbrie, J.Z. (1980). Modelling the climatic response to orbital variations. *Science,* **207**, 943-953.

Inglis, D.R. (1955). Theories of the Earth's magnetism. *Rev. Mod. Phys.,* **27**, 212-248.

Irving, E. (1956). Paleomagnetic and paleoclimatological aspects of polar wandering. *Geofis. Pura. Appl.,* **33**, 23-41.

Irving, E. (1959). Palaeomagnetic pole positions: A survey and analysis. *Geophys. J. Roy. Astron. Soc.,* **2**, 51-79.

Irving, E. (1964). *Paleomagnetism and Its Application to Geological and Geophysical Problems.* Wiley, New York, 399 pp.

Irving, E. (1966). Palaeomagnetism of some Carboniferous rocks of New South Wales and its relation to geological events. *J. Geophys. Res.* **71**, 6025-6051.

Irving, E. (1977). Drift of the major continental blocks since the Devonian. *Nature,* **270**, 304-309.

Irving, E. (1979). Pole positions and continental drift since the Devonian. In *The Earth: Its Origin, Structure and Evolution* (M.W. McElhinny, ed.), p.567-593. Academic Press, London and New York.

Irving, E. and Briden, J.C. (1962). Palaeolatitude of evaporite deposits. *Nature,* **196**, 425-428.

Irving, E. and Brown, D.A. (1964). Abundance and diversity of the labyrinthodonts as a function of paleolatitude. *Amer. J. Sci.,* **262**, 689-708.

Irving, E. and Gaskell, T.F. (1962). The palaeogeographic latitude of oil fields. *Geophys. J. Roy. Astron. Soc.,* **7**, 54-64.

Irving, E. and Naldrett, A. J. (1977). Paleomagnetism in Abitibi greenstone belt, and Abitibi and Matachewan diabase dykes: Evidence of the Archean geomagnetic field. *J. Geol.,* **85**, 157-176.

Irving, E. and Parry, L.G. (1963). The magnetism of some Permian rocks from New South Wales. *Geophys. J. Roy. Astron. Soc.,* **7**, 395-411.

Irving, E. and Pullaiah, G. (1976). Reversals of the geomagnetic field, magnetostratigraphy, and relative magnitude of paleosecular variation in the Phanerozoic. *Earth. Sci. Rev.,* **12**, 35-64.

Irving, E. and Ward, M.A. (1964). A statistical model of the geomagnetic field. *Pure Appl. Geophys.,* **57**, 47-52.

Isenberg, P.A. (1991). The solar wind. In *Geomagnetism Vol. 4* (J.A. Jacobs, ed.), p.1-85. Academic Press, London.

Ishikawa, Y. and Akimoto, S. (1958). Magnetic property and crystal chemistry of ilmenite ($FeTiO_3$) and hematite (Fe_2O_3) system. 2. Magnetic property. *J. Phys. Soc. Japan,* **13**, 1298-1310.

Ishikawa, Y. and Syono, Y. (1963). Order-disorder transformation and reverse thermoremanent magnetism in the $FeTiO_3$-Fe_2O_3 system. *J. Phys. Chem. Solids,* **24**, 517-528.

Jackson, J.D. (1975). *Classical Electrodynamics.* 2nd Edition, Wiley, New York.

Jacobs, J.A. (1953). The earth's inner core. *Nature,* **172**, 297-300.

Jacobs, J.A. (1981). Heat flow and reversals of the Earth's magnetic field. *J. Geomag. Geoelect.,* **33**, 527.

Jacobs, J.A. (1984). *Reversals of the Earth's Magnetic Field.* Adam Hilger, Bristol, 230 pp.

Jacobs, J.A. (ed.) (1987). *Geomagnetism Vols 1 and 2,* Academic Press, London.

Jacobs, J.A. (ed.) (1991). *Geomagnetism Vol. 4,* 806 pp.

Jacobs, J.A. (1992). *Deep Interior of the Earth,* Chapman and Hall, London, 167 pp.

Jacobs, J.A. (1994). *Reversals of the Earth's Magnetic field,* 2nd edition, Cambridge University Press, New York, 346 pp.

James, R.W. (1971). More on secular variation. *Comm. Earth Sci. Geophys.,* **2**, 28-29.

James, R.W. (1974). The inability of latitude-dependent westward drift to account for zonal secular variation. *J. Geomag. Geoelect.,* **26**, 359-361.

James, R.W. and Winch, D.E. (1967). The eccentric dipole. *Pure Appl. Geophys.,* **66**, 77-86.

James, R.W., Roberts, P.H. and Winch, D.E. (1980). The Cowling anti-dynamo theorem. *Geophys. Astrophys. Fluid Dyn.,* **15**, 149-160.

Jault, D and Le Mouël, J.-L. (1991). Exchange of angular momentum between the core and mantle. *J. Geomag. Geoelect.,* **43**, 111-129.

Jault, D. and Le Mouël, J.-L. (1994). Does secular variation involve motions in the deep core? *Phys.*

Earth Planet. Int., **82**, 185-193.

Jeanloz, R. (1979). Properties of iron at high pressures and the state of the core. *J. Geophys. Res.*, **84**, 6059-6069.

Jeanloz, R. (1990). The nature of the Earth's core. *Ann. Rev. Earth Planet. Sci.*, **18**, 357-386

Jeanloz, R. and Thompson, A.B. (1983). Phase transitions and mantle discontinuities. *Rev. Geophys. Space Phys.*, **21**, 51-74.

Jeanloz, R. and Wenk, J-R. (1988). Convection and anisotropy of the inner core. *Geophys. Res. Lett.*, **15**, 72-75.

Johnson, C.L. and Constable, C.G. (1995a). Palaeosecular variation recorded by lava flows over the last 5 Myr. *Phil. Trans. Roy. Soc. London*, **A354**, 89-141.

Johnson, C.L. and Constable, C.G. (1995b). The time-averaged field as recorded by lava flows over the past 5 Myr years. *Geophys. J. Int.*, **122**, 489-519.

Johnson, E.A., Murphy, T. and Torrenson, O.W. (1948). Pre-history of the Earth's magnetic field. *Terr. Magn. Atmosph. Elect.*, **53**, 349-372.

Johnson, H.P. (1979). Magnetisation of the oceanic crust. *Rev. Geophys. Space Phys.*, **17**, 215-226.

Johnson, H.P. and Hall, H.M. (1978). A detailed rock magnetism and opaque mineralogy study of the basalts from the Nazca Plate. *Geophys. J. Roy. Astron. Soc.*, **52**, 45-64.

Johnson, H.P. and Merrill, R.T. (1978). A direct test of the Vine-Matthews hypothesis. *Earth Planet. Sci. Lett.*, **40**, 263-269.

Johnson, H.P., Van Patten, D., Tivey, M., and Sager, W. (1995). Geomagnetic polarity reversal rate for the Phanerozoic. *Geophys. Res. Lett.*, **22**, 231-234.

Johnson, R. (1982). Brunhes-Matuyama magnetic reversal dated at 790,000 yr. B.P. by marine astronomical correlations. *Quat. Res.*, **17**, 135-147.

Jones, D.L., Robertson, I.D.M. and McFadden, P.L. (1975). A palaeomagnetic study of the Precambrian dyke swarms associated with the Great Dyke of Rhodesia. *Trans. Geol. Soc. Sth. Africa*, **78**, 57-65.

Jones, G.M. (1977). Thermal interaction of the core and the mantle and long term behaviour of the geomagnetic field. *J. Geophys. Res.*, **82**, 1703-1709.

Kanasewich, E.R. (1973). *Time Sequence Analysis in Geophysics*. University of Alberta Press, Edmonton, Alberta

Karato, S. (1993). Inner core anisotropy due to magnetic field-induced orientation of iron. *Science*, **262**, 1708-1711.

Karlin, R. and Levi, S. (1985). Geochemical and sedimentological control of the magnetic properties of hemipelagic sediments. *J. Geophys. Res.*, **90**, 10373-10392.

Karlin, R., Lyle, M. and Heath, G.R. (1987). Authigenic magnetite formation in suboxic marine sediments. *Nature*, **326**, 490-493.

Kaula, W.M., Schubert, G., Lingenfelter, R.E., Sjogren, W.L. and Wollenhaupt, W.R. (1974). Apollo laser altimetry and inferences as to lunar structure. *Proc. Fifth Lunar Sci. Conf.*, 3049-3058.

Keefer, C.M. and Shive, P. N. (1981). Curie temperature and lattice constant reference contours for synthetic titanomaghemites. *J. Geophys. Res.*, **86**, 987-998.

Keeler, R.N. and Mitchell, A.C. (1969). Electrical conductivity demagnetization and the high pressure phase transitions in shock compressed iron. *Solid State Commun.*, 7, 271-274.

Kennett, J.P. (ed.) (1980). *Magnetic Stratigraphy of Sediments.* Dowden, Hitcbingson and Ross, Stroudsburg, Pennsylvania.

Kent, D.V. and Gradstein, F.M. (1986). A Jurassic to recent chronology, in *The Geology of North America, Vol. M,* The Western North Atlantic Region (P.R. Vogt and B.E. Tucholke, eds), p.45-50, Geol. Soc. Amer., Boulder, Colorado.

Kent, D.V. and Opdyke, N.D. (1977). Palaeomagnetic field intensity variation recorded in a Brunhes Epoch deep-sea sediment core. *Nature*, 266, 156-159.

Khan, M.A. (1971). Correlation function in geophysics. *Nature*, 230, 57.

Khodair, A.A. and Coe, R.S. (1975). Determination of geomagnetic paleointensities in vacuum. *Geophys. J. Roy. Astron. Soc.*, 42, 107-115.

Khramov, A.N. (1955). Study of remanent magnetization and the problem of stratigraphic correlation and subdivision of non-fossiliferous strata. *Akad. Nauk SSSR*, 100, 551-554 (in Russian).

Khramov, A.N. (1957). Paleomagnetism as a basis for a new technique of sedimentary rock correlation and subdivision. *Akad. Nauk SSSR*, 112, 849-852 (in Russian).

Khramov, A.N. (1958). *Palaeomagnetism and Stratigraphic Correlation.* Gostoptechizdat, Leningrad, 218 pp. (English translation by Lojkine, A.J., published by Dept. of Geophysics, Australian Natl. Univ., Canberra, 1960).

Kidd, R.G.W. (1977). The nature and shape of the sources of marine magnetic anomalies. *Earth Planet. Sci. Lett.*, 33, 310-320.

Kincaid, C. (1995). Subduction dynamics: From the trench to the core-mantle boundary. *Rev. Geophys. Suppl., U.S. National Report to IUGG 1991-1994*, 401-412.

King, J.W., Banerjee, S.K. and Marvin, J. (1983). A new rock magnetic approach to selecting sediments for geomagnetic paleointensity studies: Application to paleointensity for the last 4000 years. *J. Geophys. Res.*, 88, 5911-5921.

King, J.W., Banerjee, S.K., Marvin, J. and Özdemir, Ö. (1982). A comparison of different magnetic methods for determining the relative grain size of magnetite in different materials: Some results from lake sediments. *Earth Planet. Sci. Lett.*, 59, 404-419.

King, R.F. and Rees, A.I. (1966). Detrital magnetism in sediments: An examination of some theoretical models. *J. Geophys. Res.*, 71, 561-571.

Kirk, R.L. and Stevenson, D.J. (1987). Hydrodynamic constraints on deep zonal flow in the giant planets. *Astrophys. J.*, 316, 836-846.

Kirschvink, J.L. (1980). The least squares line and plane and the analysis of palaeomagnetic data. *Geophys. J. Roy. Astron. Soc.*, 62, 699-718.

Kirschvink, J. and Lowenstam, H. (1979). Mineralization and magnetization of Chiton teeth: Paleomagnetic, sedimentologic and biologic implications of organic magnetite. *Earth Planet. Sci. Lett.*, 44, 193-204.

Kittel, C. (1949). Physical theory of ferromagnetic domains. *Rev. Mod. Phys.*, 21, 541-583.

Knittle, E. and Jeanloz, R. (1986). High-pressure metalization of FeO and implications for Earth's core. *Geophys. Res. Lett.*, **13**, 1541-1544.

Knittle, E. and Jeanloz, R. (1987). Synthesis and equation of state of (Mg,Fe)SiO₃ perovskite to over 100 GPa. *Science*, **235**, 668-670.

Knittle, E. and Jeanloz, R. (1989). Simulating the core-mantle boundary: An experimental study of high-pressure reactions between silicates and liquid iron. *Geophys. Res. Lett.*, **16**, 609-612.

Knittle, E. and Jeanloz, R. (1991). Earth's core-mantle boundary : Results of experiments at high pressures and temperatures. *Science*, **251**, 1438-1443.

Knopoff, L. and MacDonald, G. (1958). The magnetic field and the central core of the Earth. *Geophys. J. Roy. Astron. Soc.*, **1**, 216-223.

Kobayashi, K. (1961). An experimental demonstration of the production of chemical remanent magnetisation with Cu-Co alloy. *J. Geomag. Geoelect.*, **12**, 148-163.

Koci, A. and Sibrava, V. (1976). The Brunhes-Matuyama boundary at central European localities, Report No. 3 of IGCP project 72/1/24 in *Quaternary Glaciations in the Northern Hemisphere*, p.135-160, Prague.

Kolesova, V.I. and Kropachev, E.P. (1973). Spherical harmonic analysis of the geomagnetic field for the 1965 epoch up to $n = 23$ according to ground- based data. 2. Results. *Geomag. Aeron.*, **13**, 127-131 (English translation), p.154-159 (in Russian).

Kono, M. (1972). Mathematical models of the earth's magnetic field. *Phys. Earth Planet. Int.*, **5**, 140-150.

Kono, M. (1973). Spherical harmonic analysis of the geomagnetic field from inclination and declination data. *Rock Magn. Paleogeophys. Tokyo*, **1**, 124-129.

Kono, M. (1978). Reliability of palaeointensity methods using alternating field demagnetisation and anhysteretic remanence. *Geophys. J. Roy. Astron. Soc.*, **54**, 241-261.

Kono, M. (1987). Rikitake two disc dynamo and paleomagnetism. *Geophys. Res. Lett.*, **14**, 21-24.

Kono, M. and Tanaka, H. (1995a). Mapping the Gauss coefficients to the pole and the models of paleosecular variation. *J. Geomag. Geoelect.* **47**, 115-130.

Kono, M. and Tanaka, H. (1995b). Intensity of the geomagnetic field in geological time: A statistical study. In *The Earth's Central Part: Its Structure and Dynamics* (T. Yukutake, ed.), p.75-94, Terrapub, Tokyo.

Kovacheva, M. (1980). Summarised results of the archeomagnetic investigations of the geomagnetic field variation for the last 8000 yr in southeastern Europe. *Geophys. J. Roy. Astron. Soc.*, **61**, 57-64.

Kovacheva, M. and Kanarchev, M. (1986). Revised archaeointensity data from Bulgaria. *J. Geomag. Geoelect.*, **38**, 1297-1310.

Kovacheva, M. and Veljovich, D. (1977). Geomagnetic field variations in southeastern Europe between 6500 and 100 years B.C. *Earth Planet. Sci. Lett.*, **37**, 131-137.

Krause, F. (1977). Mean-field electrodynamics and dynamo theory of the Earth's magnetic field. *J. Geophys.*, **43**, 421-440.

Krause, F. and Steenbeck, M. (1967). Some simple models of magnetic field regeneration by non-mirror-symmetric turbulence. *Z. Naturforsch.*, **22a**, 671-675. [English translation: Roberts & Stix (1971) p.81-95.]

Kristjansson, L. (1985). Some statistical properties of palaeomagnetic directions in Icelandic lava flows. *Geophys. J. Roy. Astron. Soc.*, **80**, 57-71.

Kristjansson, L. (1995). New palaeomagnetic results from Icelandic Neogene lavas. *Geophys. J. Int.*, **121**, 435-443.

Kristjansson, L. and Gudmundsson, A. (1980). Geomagnetic excursions in late-glacial basalt outcrops in south-western Iceland. *Geophys. Res. Lett.*, **7**, 337-340.

Kristjansson, L. and McDougall, I. (1982). Some aspects of the late Tertiary geomagnetic field in Iceland. *Geophys. J. Roy. Astron. Soc.*, **68**, 273-294.

LaBrecque, J.L., Kent, D.V. and Cande, S.C. (1977). Revised magnetic polarity time scale for Late Cretaceous and Cenozoic time. *Geology*, **5**, 330-335.

Lahiri, B.N. and Price, A.T. (1939). Electromagnetic induction in non- uniform conductors, and the determination of the conductivity of the earth from terrestrial magnetic variations. *Phil. Trans. Roy. Soc. London*, **A237**, 509-540.

Laj, C. (1989). Geomagnetic polarity reversals: Observations. In *The Encyclopedia of Solid Earth Geophysics* (D.E. James, ed.), p.535-547, Van Nostrand–Rheinhold, New York.

Laj, C., Guitton, S., Kissel, C. and Mazaud, A. (1988). Complex behavior of the geomagnetic field during three successive polarity reversals 11-12 m.y. B.P. *J. Geophys. Res.*, **93**, 11655-11666.

Laj, C., Mazaud, A., Weeks, R., Fuller, M., and Herrero-Bervera, E. (1991). Geomagnetic reversal paths. *Nature*, **351**, 447.

Laj, C., Mazaud, A., Weeks, R., Fuller, M., and Herrero- Bervera, E. (1992a). Geomagnetic reversal paths [discussion]. *Nature*, **359**, 111-112.

Laj, C., Mazaud, A., Weeks, R., Fuller, M., and Herrero-Bervera, E. (1992b). Statistical assessement of the preferred longitudinal bands for recent geomagnetic field reversal records. *Geophys. Res. Lett.*, **19**, 2003-2006.

Lamb, H. (1883). On electrical motions in a spherical conductor. *Phil. Trans. Roy. Soc. London*, **A174**, 519.

Lambeck, K. (1980). *The Earth's Variable Rotation: Geophysical Causes and Consequences.* Cambridge University Press, Cambridge.

Landstreet, J.D. and Angel, J.R.P. (1974). The polarisation spectrum and magnetic field strength of the White Dwarf Grw + 70°8247. *Astrophys. J.*, **196**, 819-825.

Langel, R.A. (1987). The main field. In *Geomagnetism Vol.1* (J.A. Jacobs, ed.), p.249-512. Academic Press, London.

Langel, R.A. (1990a). Global magnetic anomaly maps derived from POGO spacecraft data. *Phys. Earth Planet. Int.*, **62**, 208-230.

Langel, R.A. (1990b). Study of the crust and mantle using magnetic surveys by Magsat and other satellites. *Proc. Indian Acad. Sci. (Earth Planet. Sci)*, **99**, 581-618.

Langel, R.A. (1993). The use of low altitude satellite data bases for modelling of core and crustal fields and the separation of external and internal fields. *Surv. Geophys.*, **14**, 31-87.

Langel, R.A. and Estes, R.H. (1982). A geomagnetic field spectrum. *Geophys. Res. Lett.*, **9**, 250-253.

Langel, R.A., Baldwin, R.T. and Green, A.W. (1995). Toward an improved distribution of magnetic observatories for modeling of the main geomagnetic field and its temporal change. *J. Geomag. Geoelect.*, **47**, 475-508.

Langel, R.A., Purucker, M. and Rajaram, M. (1993). The equatorial electrojet and associated currents as seen in Magsat data. *J. Atmosph. Terr. Phys.*, **55**, 1233-1269.

Langel, R.A., Kerridge, D.J., Barraclough, D.R. and Malin, S.R.C. (1986). Geomagnetic temporal change: 1903-1982, a spline representation. *J. Geomag. Geoelectr.*, **38**, 573-597.

Langereis, C.G, van Hoof, A.A.M. and Rochette, P. (1992). Longitudinal confinement of geomagnetic reversal paths as a possible sedimentary artefact. *Nature*, **358**, 226-230.

Langereis, C.G., Linssen, J.H., Mullender, T. and Zijderveld, J. (1989). Demagnetization. In *The Encylcopedia of Solid Earth Geophysics* (D.E. James, ed.), p.201-211. Van Nostrand, New York.

Larmor, J. (1919a). Possible rotational origin of magnetic fields of sun and earth. *Elec. Rev.*, **85**, 412.

Larmor, J. (1919b). How could a rotating body such as the sun become a magnet? *Rept. Brit. Assoc. Adv. Sci. 1919*, 159-160.

Larsen, J.C. (1975). Low frequency (0.1-6.0 cpd) electromagnetic study of deep mantle electrical conductivity beneath the Hawaiian Islands. *Geophys. J. Roy. Astron. Soc.*, **43**, 17-46.

Larson, E.E. and Walker, T.R. (1975). Development of chemical remanent magnetisation during early stages of red bed formation in late Cenozoic sediments, Baja, California. *Geol. Soc. Amer. Bull.*, **86**, 639-650.

Larson, E.E., Watson, D.E. and Jennings, W. (1971). Regional comparison of a Miocene geomagnetic transition in Oregon and Nevada. *Earth Planet. Sci. Lett.*, **11**, 391-400.

Larson, E.E., Ozima, M., Nagata, T. and Strangway, D. (1969). Stability of remanent magnetisation of rocks. *Geophys. J. Roy. Astron. Soc.*, **17**, 263-292.

Larson, R.L. (1991). Latest pulse of Earth: Evidence for a mid-Cretaceous superplume. *Geology*, **19**, 963-966.

Larson, R.L. and Olson, P. (1991). Mantle plumes control magnetic reversal frequency. *Earth Planet. Sci. Lett.*, **107**, 437-447.

Lawley, E. (1970). The intensity of the geomagnetic field in Iceland during Neogene polarity transitions and systematic deviations. *Earth Planet. Sci. Lett.*, **10**, 145-149.

Lay, T. (1989). Structure of the core-mantle transition zone; a chemical and thermal boundary layer. *EOS, Trans. Amer. Geophys. Union*, **70**, 49, 54-55, 58-59.

Lay, T. (1995). Seismology of the lower mantle and core–mantle boundary. *Rev. Geophys. Suppl., U.S. National Report to IUGG 1991-1994*, 325-328.

Lay, T. and Young, C. (1990). The stably-stratified outermost core revisited. *Geophys. Res. Lett.*, **17**, 2001-2004.

Le Mouël, J.-L. (1984). Outer core geostrophic flow and secular variation of Earth's magnetic field. *Nature*, **311**, 734-735.

Le Mouël, J.L., Ducruix, J. and Duyen, C. (1982). The world-wide character of the 1969-1970 impulse of the secular acceleration rate. *Phys. Earth Planet. Int.*, **28**, 337-350.

Le Mouël, J.-L., Gire, C. and Madden, T. (1985). Motions at the core surface in geostrophic approximation. *Phys. Earth Planet. Int.*, **39**, 270-287.

Le Mouël, J.-L., Madden, T., Ducruix, J. and Courtillot, V. (1981). Decade fluctuations in the geomagnetic westward drift and the earth's rotation. *Nature*, **290**, 763-765.

Le Pichon, X. (1968). Sea-floor spreading and continental drift. *J. Geophys. Res.*, **73**, 3661-3697.

Le Pichon, X. and Heirtzler, J.R. (1968). Magnetic anomalies in the Indian Ocean and seafloor spreading. *J. Geophys. Res.*, **73**, 2101.

Leaton, B.R. and Malin, S.R.C. (1967). Recent changes in the magnetic dipole moment of the earth. *Nature*, **213**, 1110.

Lee, S. and Lilley, F.E.M. (1986). On paleomagnetic data and dynamo theory. *J. Geomag. Geoeclect.*, **38**, 797-806.

Levi, S. (1975). Comparison of two methods of performing the Thellier experiment (or, how the Thellier experiment should not be done). *J. Geomag. Geoelect.*, **27**, 245-255.

Levi, S. (1989). Chemical remanent magnetization. In *Encyclopedia of Solid Earth Geophysics* (D.E. James, ed.), p.49-58. Van Nostrand–Reinholdt, New York.

Levi, S. and Banerjee, S.K. (1976). On the possibility of obtaining relative paleointensities from lake sediments. *Earth Planet. Sci. Lett.*, **29**, 219-226.

Levi, S. and Karlin, R. (1989). A sixty thousand year paleomagnetic record from Gulf of California sediments: Secular variation, late Quaternary excursions and geomagnetic implications. *Earth Planet. Sci. Lett.*, **92**, 219-226.

Levi, S. and Merrill, R.T. (1976). A comparison of ARM and TRM in magnetite. *Earth Planet. Sci. Lett.*, **32**, 171-184.

Levi, S. and Merrill, R.T. (1978). Properties of single-domain, pseudo-single-domain, and multidomain magnetite. *J. Geophys. Res.*, **83**, 309-323.

Levi, S., Audunsson, H., Duncan, R.A., Kristjansson, L., Gillot, P.Y. and Jakobsson, S. (1990). Late Pleistocene geomagnetic excursion in Icelandic lavas: Confirmation of the Laschamp excursion. *Earth Planet. Sci. Lett.*, **96**, 443-457.

Levy, E.H. (1972a). Effectiveness of cyclonic convection for producing the geomagnetic field. *Astrophys. J.*, **171**, 621-633.

Levy, E.H. (1972b). Kinematic reversal schemes for the geomagnetic dipole. *Astrophys. J.*, **171**, 635-642.

Levy, E.H. (1972c). On the state of the geomagnetic field and its reversals. *Astrophys. J.*, **175**, 573-581.

Li, X. and Jeanloz, R. (1990). Laboratory studies of the electrical conductivity of silicate perovskites at high pressures and temperatures. *J. Geophys. Res.*, **95**, 5067-5078.

Li, Z.X., Powell, C.M., Embleton, B.J.J. and Schmidt, P.W. (1991). New palaeomagnetic results from the Amadeus Basin and their implications for stratigraphy and tectonics. In *Geological and Geophysical Studies in the Amadeus Basin* (R.J. Korsch and J.M. Kennard, eds.). *Bur. Min. Resour. Aust. Bull.*, **236**, 349-360.

Liddicoat, J.C. (1992). Mono Lake excursion in Mono Basin, California, and at Carson Sink and Pyramid Lake, Nevada. *Geophys. J. Int.*, **108**, 442-452.

Liddicoat, J.C. and Coe, R.S. (1979). Mono Lake geomagnetic excursion. *J. Geophys. Res.*, **84**,

261-271.

Liddicoat, J.C., Coe, R.S., Lambert, P.W. and Valastro, S. (1979). Palaeomagnetic record in late Pleistocene and Holocene dry lake deposits at Tlapacoya, Mexico. *Geophys. J. Roy. Astron. Soc.*, **59**, 367-377.

Lilley, F.E.M. (1970). On kinematic dynamos. *Proc. Roy. Soc. London*, **A316**, 153-167.

Lilley, F.E.M. and Day, A.A. (1993). D'Entrecasteaux, 1792: Celebrating a bicentennial in geomagnetism. *EOS, Trans. Amer. Geophys. Union*, **74**, 97, 102-103.

Lindemann, F.A. (1910). Uber die Berechnung Molekularer Eigenfrequenzen. *Physik. Zeit.*, **14** (Jarg 11), 609-612.

Lindemann, F.A. (1919). Note on the theory of magnetic storms. *Phil. Mag.*, **38**, 669-684.

Lindsley, D.H. (1976). Chapters 1 and 2. In *Oxide Minerals* (Douglas Rumble Ill, ed.). Southern Printing Co., Blacksburg, Virginia.

Lingenfelter, R.E. and Ramaty, R. (1970). Astrophysical and geophysical variations in C-14 production. In *Radiocarbon Variations and Absolute Chronology* (I.U. Olsson, ed.), p.513-535. Alinqvist and Wiksell, Stockholm.

Liu, L.-G. (1976). Orthorhombic perovskite phase observed in olivine, pyroxene and garnet at high pressures and temperatures. *Phys. Earth Planet. Int.*, **11**, 289-298.

Livermore, R.A., Vine, F.J. and Smith, A.G. (1983). Plate motions and the geomagnetic field. I. Quaternary and late Tertiary. *Geophys. J. Roy. Astron. Soc.*, **73**, 153-171.

Livermore, R.A., Vine, F.J. and Smith, A.G. (1984). Plate motions and the geomagnetic field. II. Jurassic to Tertiary. *Geophys. J. Roy. Astron. Soc.*, **79**, 939-961.

Lloyd, D. and Gubbins, D. (1990). Toroidal fluid motion at the top of Earth's core. *Geophys. J. Int.*, **100**, 455-467.

Lock, J. and McElhinny, M.W. (1991). The global paleomagnetic database: Design, installation and use with ORACLE. *Surv. Geophys.*, **12**, 317-491.

Loper, D.E. (1978a). The gravitationally powered dynamo. *Geophys. J. Roy. Astron Soc.*, **54**, 389-404.

Loper, D.E. (1978b). Some thermal consequences of a gravitationally powered dynamo. *J. Geophys. Res.*, **83**, 5961-5970.

Loper, D.E. (1989). Thermal mechanical and magnetic core-mantle interactions. *EOS, Trans. Amer. Geophys. Union*, **70**, 276.

Loper, D.E. (1992). On the correlation between mantle plume flux and the frequency of reversals of the geomagnetic field. *Geophys Res. Lett.*, **19**, 25-28.

Loper, D.E. and Roberts, P.H. (1978). On the motion of an iron-alloy core containing a slurry. *Geophys. Astrophys. Fluid Dyn.*, **9**, 289-321.

Loper, D.E. and Roberts, P.H. (1979). Are planetary dynamos driven by gravitational settling? *Phys. Earth Planet. Int.*, **20**, 192-193.

Loper, D.E. and Roberts, P.H. (1980). On the motion of an iron-alloy core containing a slurry. II. *Geophys. Astrophys. Fluid Dyn.*, **16**, 83-127.

Loper, D.E. and Stacey, F.D. (1983). The dynamical and thermal structure of deep-mantle plumes. *Phys. Earth Planet. Int.*, **33**, 304-317.

Lowes, F.J. (1955). Secular variation and the non-dipole field. *Ann. Geophys.*, **11**, 91-94.

Lowes, F.J. (1971). Significance of the correlation between spherical harmonic fields. *Nature*, **230**, 61.

Lowes, F.J. (1974). Spatial power spectrum of the main geomagnetic field and extrapolation to the core. *Geophys. J. Roy. Astron. Soc.*, **36**, 717-730.

Lowes, F.J. (1975). Vector errors in spherical harmonic analysis of scalar data. *Geophys. J. Roy. Astron. Soc.*, **42**, 637-651.

Lowes, F.J. (1976). The effect of a field of external origin on spherical harmonic analysis using only internal coefficients. *J. Geomag. Geoelect.*, **28**, 515-516.

Lowes, F.J. and Runcorn, S.K. (1951). The analysis of the geomagnetic secular variation. *Phil. Trans. Roy. Soc. London*, **A243**, 525-546.

Lowes, F.J. and Wilkinson, I. (1963). Geomagnetic dynamo: A laboratory model. *Nature*, **198**, 1158-1160.

Lowes, F.J. and Wilkinson, I. (1967). A laboratory self-exciting dynamo. In *Magnetism and the Cosmos* (NATO Advanced Study Inst. on Planetary and Steller Magnetism, University of Newcastle Upon Tyne, 1965.), p.121-125, Elsevier, Amsterdam.

Lowes, F.J. and Wilkinson, I. (1968). Geomagnetic dynamo: An improved laboratory model. *Nature*, **219**, 717-718.

Lowrie W., and Kent, D.V. (1983). Geomagnetic reversal frequency since the Late Cretaceous. *Earth Planet. Sci. Lett.*, **62**, 305-313.

Lowrie, W. and Alvarez, W. (1981). One hundred million years of geomagnetic polarity history. *Geology*, **9**, 392-397.

Lowrie, W. and Fuller, M. (1971). On the alternating field demagnetisation characteristics of multidomain thermoremanent magnetisation in magnetite. *J. Geophys. Res.*, **76**, 6339-6349.

Lund, S.P. (1994). Paleomagnetic secular variation. In *Trends in Geophysical Research* (Council of Scientific Research Integration, Trivandrum, India).

Lund, S.P. (1996). A comparison of Holocene paleomagnetic secular variation records from North America. *J. Geophys. Res.*, **101**, 8007-8024.

Lund, S.P. and Banerjee, S.K. (1979). Paleosecular variations from lake sediments. *Rev. Geophys. Space Phys.*, **17**, 244-248.

Lund, S.P. and Banerjee, S.K. (1985). Late Quaternary paleomagnetic field secular variation from two Minnesota lakes. *J. Geophys. Res.*, **90**, 803-825.

Lund, S.P. and Keigwin, L. (1994). Measurement of the degree of smoothing in sediment paleomagnetic secular variation records: An example from late Quaternary deep-sea sediments of the Bermuda Rise, NW Atlantic Ocean. *Earth Planet. Sci. Lett.*, **122**, 317-330.

Lutz, T.M. (1985). The magnetic reversal record is not periodic. *Nature*, **317**, 404-407.

Lutz, T.M. and Watson, G.S. (1988). Effects of long-term variation on the frequency spectrum of the geomagnetic reversal record. *Nature*, **334**, 240-242.

MacDonald, K.C. and Holcombe, T.L. (1978). Inversion of magnetic anomalies and seafloor spreading in the Cayman Trough. *Earth Planet. Sci. Lett.*, **40**, 407-414.

MacDonald, K.C., Haymon, R.M., Miller, S.P., Sempere, J.C., and Fox, P.J. (1988). Deep-tow and sea beam studies of dueling propagation ridges on the East Pacific rise near 20°40'S. *J. Geophys. Res.*, **93**, 2875-2898.

Mackereth, F.J.H. (1958). A portable core sampler for lake deposits. *Limnol. Oceanogr.*, **3**, 181-191.

Mackereth, F.J.H. (1971). On the variation in direction of the horizontal component of remanent magnetisation in lake sediments. *Earth Planet. Sci. Lett.*, **12**, 332-338.

Malin, S.R.C. and Hodder, B.M. (1982). Was the 1970 geomagnetic jerk of internal or external origin? *Nature*, **296**, 726-728.

Malin, S.R.C., Hodder, B.M. and Barraclough, D.R. (1983). Geomagnetic secular variation: A jerk in 1970. *75th Anniversary Volume of Ebro Observatory* (J.O. Cardus, ed.), p.239-256. Tarragona, Spain.

Malinetskii, G.G., Potapov, A.B., Gizzatulina, S.M., Rusmaikin, A.A., and V.D. Rukavishnikov, (1990). Dimension of geomagnetic attractor from data on length of day variations. *Phys. Earth Planet. Int.*, **59**, 170-181.

Malkus, W.V.R. (1963). Precessional torques as the cause of geomagnetism. *J. Geophys. Res.*, **68**, 2871-2886.

Malkus, W.V.R. (1968). Precession as the cause of geomagnetism. *Science*, **160**, 259-264.

Malkus, W.V.R. (1994). Energy sources for planetary dynamos. In *Lectures on Solar and Planetary Dynamos* (M.R.E. Proctor and A.D. Gilbert, eds), p.161-180, Cambridge Univ. Press, Cambridge.

Mankinen, E.A., Prévot, M., Grommé, C.S. and Coe, R. (1985). The Steens Mountain (Oregon) geomagnetic polarity transition. 1. Directional history, duration of episodes, and rock magnetism. *J. Geophys. Res.*, **90**, 10393-10416.

Marsh, B.D. (1981). On the crystallinity, probability of occurrence, and rheology of lava and magma. *Contrib. Mineral. Petrol.*, **78**, 85-98.

Marshall, M., Chauvin, A. and Bonhommet, N. (1988). Preliminary paleointensity measurements and detailed magnetic analyses of basalts from the Skalamaelifell excursion, southwest Iceland. *J. Geophys. Res.*, **93**, 11681-11698.

Martelli, G. and Newton, G. (1977). Hypervelocity cratering and impact magnatization of basalt. *Nature*, **269**, 478.

Marzocchi, W. and Mulargia, F. (1990). Statistical analysis of the geomagnetic reversal sequences. *Phys. Earth Planet. Int.*, **61**, 149-164.

Mason, R.G. and Raff, A.D. (1961). Magnetic survey off the west coast of North America, 32°N latitude to 42°N latitude. *Geol. Soc. Amer. Bull.*, **72**, 1259-1265.

Matassov, V.G. (1977). *The electrical conductivity of iron-silicon alloys at high pressures and Earth's core*. Ph.D. thesis, Univ. California Lawrence Livermore Lab. (UCRL-52322), 180 pp.

Matsushita, S. (1967). Solar quiet and lunar daily variation fields. In *Physics of Geomagnetic Phenomena Vol. 1* (S. Matsushita and W.H. Campbell, eds), p.302-424. Academic Press, London and New York.

Matthews, D.H. and Bath. J. (1967). Formation of magnetic anomaly pattern of mid-Atlantic Ridge. *Geophys. J. Roy. Astron. Soc.*, **13**, 349-357.

Mattis, D.G. (1965). *The Theory of Magnetism.* Harper & Row, New York.

Matuyama, M. (1929). On the direction of magnetisation of basalt in Japan, Tyosen and Manchuria. *Proc. Imp. Acad. Japan*, **5**, 203-205.

Maxwell, A.E., Von Herzen, R.P., Hsü, K.J., Andrews, J.E., Saito, T., Percival, S.F., Milow, E.D. and Boyce, R.E. (1970). Deep sea drilling in the South Atlantic. *Science*, **168**, 1047-1059.

Mayaud, P.N. (1968). Indices *Kn, Ks, Km,* 1964-1967. *Centre National de la Recherche Scientifique* (Paris).

Mayaud, P.N. (1980). Derivation, Meaning and Use of Geomagnetic Indices, *Amer. Geophys. Union Mongr. Ser.,* **22**.

Mazaud A. and Laj, C. (1991). The 15 m.y. geomagnetic reversal periodicity: A quantitative test. *Earth Planet. Sci. Lett.,* **107**, 689-696.

Mazaud A., Laj, C., de Seze, L. and Verosub, K.L. (1983). 15-Myr periodicity in the reversal frequency of geomagnetic reversals since 100 Ma. *Nature*, **304**, 328-330.

Mazaud, A., Laj, C., Bard, E., Arnold, M. and Tric, E. (1991). Geomagnetic field control of ^{14}C production over the last 80 ky: Implications for the radiocarbon time-scale. *Geophys. Res. Lett.,* **18**, 1885-1888.

McCabe, C. and Elmore, R.D. (1989). The occurrence and origin of Late Paleozoic remagnetization in the sedimentary rocks of North America. *Rev. Geophys.,* **27**, 471-494.

McClelland, E. and Goss, C. (1993). Self-reversal of chemical remanent magnetization on the transformation of maghemite to haematite. *Geophys. J. Int.,* **112**, 517-532.

McDonald, K.L. (1957). Penetration of a geomagnetic secular field through a mantle with variable conductivity. *J. Geophys. Res.,* **62**, 117-141.

McDonald, K.L. and Gunst, R.H. (1968). Recent trends in the earth's magnetic field. *J. Geophys. Res.,* **73**, 2057-2067.

McDougall, I. and Harrison, T.M. (1988). *Geochronology and Thermo-chronology by the $^{40}Ar/^{39}Ar$ Method*, Oxford University Press, New York, 212 pp.

McDougall, I. and Tarling, D.H. (1963). Dating of polarity zones in the Hawaiian Islands. *Nature*, **200**, 54-56.

McDougall, I. and Tarling, D.H. (1964). Dating geomagnetic polarity zones. *Nature*, **202**, 171-172.

McDougall, I., Watkins, N.D. and Kristjansson, L. (1976a). Geochronology and paleomagnetism of a Miocene-Pliocene lava sequence at Bessatadaá, eastern Iceland. *Amer. J. Sci.,* **276**, 1078-1095.

McDougall, I., Watkins, N.D., Walker, G.P.L. and Kristjansson, L. (1976b). Potassium-argon and paleomagnetic analysis of Icelandic lava flows: Limits on the age of anomaly 5. *J. Geophys. Res.,* **81**, 1505-1512.

McDougall, I., Saemundsson, K., Johannesson, H., Watkins, N.D. and Kristjansson, L. (1977). Extension of the geomagnetic polarity time scale to 6.5 my: K-Ar dating, geological and paleomagnetic study of a 3,500 m lava succession in western Iceland. *Geol. Soc. Amer. Bull.,* **88**, 1-15.

McElhinny, M.W. (1971). Geomagnetic reversals during the Phanerozoic. *Science*, **172**, 157-159.

McElhinny, M.W. (1973). *Palaeomagnetism and Plate Tectonics.* Cambridge University Press,

Cambridge, 358 pp.

McElhinny, M.W. and Brock, A. (1975). A new paleomagnetic result from East Africa and estimates of the Mesozoic paleoradius. *Earth Planet. Sci. Lett.*, **27**, 321-328.

McElhinny, M.W. and Evans, M.E. (1968). An investigation of the strength of the geomagnetic field in the early Precambrian. *Phys. Earth Planet. Int.*, **1**, 485-497.

McElhinny, M.W. and Lock, J. (1991). Global paleomagnetic database complete. *EOS, Trans. Amer. Geophys. Union*, **72**, 579.

McElhinny, M.W. and Lock, J. (1993a). Global paleomagnetic database updated. *EOS, Trans. Amer. Geophys. Union*, **74**, 180.

McElhinny, M.W. and Lock, J. (1993b). Global paleomagnetic database supplement number one, update to 1992. *Surv. Geophys.*, **14**, 303-329.

McElhinny, M.W. and Lock, J. (1994). Paleomagnetic databases — new updates. *EOS, Trans. Amer. Geophys. Union*, **75**, 428.

McElhinny, M.W. and Lock, J. (1995). Four IAGA databases released in one package. *EOS, Trans. Amer. Geophys. Union*, **76**, 266.

McElhinny, M.W. and Merrill, R.T. (1975). Geomagnetic secular variation over the past 5 my. *Rev. Geophys. Space Phys.*, **13**, 687-708.

McElhinny, M.W. and Senanayake, W.E. (1980). Paleomagnetic evidence for the existence of the geomagnetic field 3.5 Ga ago. *J. Geophys. Res.*, **85**, 3523-3528.

McElhinny, M.W. and Senanayake, W.E. (1982). Variations in the geomagnetic dipole. 1. The past 50 000 years. *J. Geomag. Geoelect.*, **34**, 39-51.

McElhinny, M.W., McFadden, P.L. and Merrill, R.T. (1996a). The time-averaged paleomagnetic field 0-5 Ma. *J. Geophys. Res.*, in press.

McElhinny, M.W., McFadden, P.L. and Merrill, R.T. (1996b). The myth of the Pacific dipole window. *Earth Planet. Sci. Lett.*, in press.

McFadden, P.L. (1980). The best estimate of Fisher's precision parameter κ. *Geophys. J. Roy. Astron. Soc.*, **60**, 397-407.

McFadden, P.L. (1984a). Statistical tools for the analysis of geomagnetic reversal sequences. *J. Geophys. Res.*, **89**, 3363-3372.

McFadden, P.L. (1984b). 15-Myr periodicity in the frequency of geomagnetic reversals since 100 Myr. *Nature*, **311**, 396.

McFadden, P.L (1987). Comment on "Aperiodicity of magnetic reversals?" *Nature*, **330**, 27.

McFadden, P.L. and Lowes, F.J. (1981). The discrimination of mean directions drawn from Fisher distributions. *Geophys. J. Roy. Astron. Soc.*, **67**, 19-33.

McFadden, P.L. and McElhinny, M.W. (1982). Variations in the geomagnetic dipole. 2. Statistical analysis of VDM's for the past 5 million years. *J. Geomag. Geoelect.*, **34**, 163-189.

McFadden, P.L. and McElhinny, M.W. (1984). A physical model for palaeosecular variation. *Geophys. J. Roy. Astron. Soc.*, **78**, 809-823.

McFadden, P.L. and McElhinny, M.W. (1988). The combined analysis of remagnetization circles and direct observations in palaeomagnetism. *Earth Planet. Sci. Lett.*, **87**, 161-172.

McFadden, P.L. and McElhinny, M.W. (1995). Combining groups of paleomagnetic directions or poles. *Geophys. Res. Lett.*, **22**, 2191-2194.

McFadden, P.L. and Merrill, R.T. (1984). Lower mantle convection and geomagnetism. *J. Geophys. Res.* **89**, 3354-3362.

McFadden, P.L. and Merrill, R.T. (1986). Geodynamo energy source constraints from paleomagnetic data. *Phys. Earth Planet. Int.*, **43**, 22-33.

McFadden, P.L. and Merrill, R.T. (1993). Inhibition and geomagnetic field reversals. *J. Geophys. Res.*, **98**, 6189-6199.

McFadden, P.L. and Merrill, R.T. (1995a). Fundamental transitions in the geodynamo as suggested by paleomagnetic secular variation data. *Phys. Earth Planet. Int.*, **91**, 253-260.

McFadden, P.L. and Merrill, R.T. (1995b). History of Earth's magnetic field and possible connections to core-mantle boundary processes. *J. Geophys. Res.*, **100**, 317-316.

McFadden, P.L. and Merrill, R.T. (1997). Paleointensity and reversals of the geomagnetic field, in preparation.

McFadden, P.L., McElhinny, M.W. and Merrill, R.T. (1989). Geomagnetic field: asymmetries, in *Encyclopedia of Solid Earth Geophysics* (D.E. James, ed.), p.489-495, Van Nostrand.

McFadden, P.L., McElhinny, M.W. and Merrill, R.T. (1997a). The myth of polarity bias, in preparation.

McFadden, P.L., Merrill, R.T. and Barton, C.E. (1993). Do virtual geomagnetic poles follow preferred paths during geomagnetic reversals? *Nature*, **361**, 342-344.

McFadden, P.L., Merrill, R.T. and McElhinny, M.W. (1985). Non-linear processes in the geodynamo: Palaeomagnetic evidence. *Geophys. J. Roy. Astron. Soc.*, **83**, 111-126.

McFadden, P.L., Merrill, R.T. and McElhinny, M.W. (1988). Dipole/Quadrupole family modelling of paleosecular variation. *J. Geophys. Res.*, **93**, 11583-11588.

McFadden, P.L., Merrill, R.T. and McElhinny, M.W. (1997b). The reliability of the geomagnetic polarity timescale, in preparation.

McFadden, P.L., Merrill, R.T., Lowrie, W. and Kent, D.V. (1987). The relative stabilities of the reverse and normal polarity states of the earth's magnetic field. *Earth Planet. Sci. Lett.*, **82**, 373-383.

McFadden, P.L., Merrill, R.T., McElhinny, M.W. and Lee, S. (1991). Reversals of the Earth's magnetic field and temporal variations of the dynamo families. *J. Geophys. Res.*, **96**, 3923-3933.

McKenzie, D.P. and Parker, R.L. (1967). The North Pacific – An example of tectonics on a sphere. *Nature*, **216**, 1276-1280.

McLeod, M. (1994). Magnetospheric and ionospheric signals in magnetic observatory monthly means: Electrical conductivity of the deep mantle. *J. Geophys. Res.*, **99**, 13577-13590.

McNish, A.G. (1940). Physical representations of the geomagnetic field. *Trans. Amer. Geophys. Union*, **21**, 287-291.

McPherron, R.L. (1979). Magnetospheric substorms. *Rev. Geophys. Space Phys.*, **17**, 657-681.

McPherron, R.L. (1991). Physical processes producing magnetospheric substorms and magnetic storms. In *Geomagnetism Vol. 4* (J.A. Jacobs, ed.), p.593-739, Academic Press, London.

McSweeney, T., Creager, K. and Merrill, R.T. (1996). Inner-core anisotropy and implications for the geodynamo. *Phys. Earth Planet. Int.* (submitted).

McWilliams, M.O., Holcomb, R.T. and Champion, D.E. (1982). Geomagnetic secular variation from ^{14}C dated lava flows on Hawaii and the question of the Pacific non-dipole low. *Phil. Trans. Roy. Soc. London*, **A306**, 211-222.

Menvielle, M. and Berthelier, A. (1991). The K-derived planetary indices: Description and availability. *Rev. Geophys.*, **29**, 415-432.

Mercanton, P.L. (1926). Inversion de l'inclinaison magnétique terrestre aux âges géologiques. *Terr. Magn. Atmosph. Elec.*, **31**, 187-190.

Merrill, R.T. (1975). Magnetic effects associated with chemical changes in igneous rocks. *Geophys. Surv.*, **2**, 277-311.

Merrill, R.T. (1977). The demagnetisation field of multidomain grains. *J. Geomag. Geoelect.*, **29**, 185-292.

Merrill, R.T. (1981). Towards a better theory of thermal remanent magnetisation. *J. Geophys. Res.*, **86**, 937-949.

Merrill, R.T. (1985). Correlating magnetic field polarity changes with geologic phenomena. *Geology*, **13**, 487-490.

Merrill, R.T. (1987). Use and abuse of intensity data. *Nature* **329**, 197-198.

Merrill, R.T. and McElhinny, M.W. (1977). Anomalies in the time-averaged paleomagnetic field and their implications for the lower mantle. *Rev. Geophys. Space Phys.*, **15**, 309-323.

Merrill, R.T. and McElhinny, M.W. (1983). *The Earth's Magnetic Field: Its History, Origin and Planetary Perspective*. Academic Press, London and New York, 401 pp.

Merrill, R.T. and McFadden, P.L. (1988). Secular variation and the origin of geomagnetic field reversals. *J. Geophys. Res.* **93**, 11589-11597.

Merrill, R.T. and McFadden, P.L. (1990). Paleomagnetism and the nature of the geodynamo. *Science* **248**, 345-350.

Merrill, R.T. and McFadden, P.L. (1994). Geomagnetic field stability: Reversal events and excursions. *Earth Planet. Sci. Lett.* **121**, 57-69.

Merrill, R.T. and McFadden, P.L. (1995). Dynamo theory and paleomagnetism. *J. Geophys. Res.* **100**, 317-326.

Merrill, R.T., McElhinny, M.W. and Stevenson, D.J. (1979). Evidence for long-term asymmetries in the earth's magnetic field and possible implications for dynamo theories. *Phys. Earth Planet. Int.*, **20**, 75-82.

Merrill, R.T., McFadden, P.L. and McElhinny, M.W. (1990). Paleomagnetic tomography of the core-mantle boundary. *Phys. Earth Planet. Int.*, **64**, 87-101.

Mewaldt, R.E., Cummings, A.C. and Stone, E.C. (1994). Anomalous cosmic rays: Interstellar interlopers in the heliosphere and magnetosphere. *EOS, Trans. Amer. Geophys. Union*, **75**, 185 and 193.

Meynadier, L., Valet, J.P., Weeks, R., Shackleton, N.J. and Hagee, V.L. (1992). Relative geomagnetic intensity of the field during the last 140 ka. *Earth Planet. Sci. Lett.*, **114**, 39-57.

Milankovitch, M. (1941). *Serb Akad Beogr. Spec. Publ.* 132 (Translated - Israel Progarm for Scientific Translation, Jerusalem, 1969).

Mitchell, A.C. (1939). Chapter on the history of terrestrial magnetism. *Terr. Magn. Atmosph. Elec.*, **44**, 77.

Mizutani, H., Yamamoto, T. and Fujimura, A. (1992). Prediction of magnetic field strengh of Neptune: Scaling laws of planetary magnetic fields. *Adv. Space Res.*, **12**, 265-279.

Moffatt, H.K. (1961). The amplification of a weak applied magnetic field by turbulence in fluids of moderate conductivity. *J. Fluid Mech.*, **11**, 625-635.

Moffatt, H.K. (1970). Turbulent dynamo action at low magnetic Reynolds numbers. *J. Fluid Mech.*, **41**, 435-452.

Moffatt, H.K. (1978). *Magnetic Field Generation in Electrically Conducting Fluids*. Cambridge University Press, Cambridge, 343 pp.

Momose, K. (1963). Studies on the variation of the earth's magnetic field during Pleistocene time. *Bull. Earthquake Res. Inst. Tokyo Univ.*, **41**, 487-534.

Moon, T. and Merrill, R.T. (1984). The magnetic moments of non-uniformly magnetized grains. *Phys. Earth Planet. Int.*, **34**, 186-194.

Moon, T. and Merrill, R.T. (1985). Nucleation theory and domain states in muliti-domain material. *Phys. Earth Planet. Int.*, **37**, 214-222.

Moon, T. and Merrill, R.T. (1986). A new mechanism for stable viscous remanent magnetization and overprinting during long magnetic polarity intervals. *Geophys. Res. Lett.*, **13**, 737-740.

Moon, T. and Merrill, R.T. (1988). Single domain theory of remanent magnetiz-ation. *J. Geophys. Res.*, **93**, 9202-9210.

Moore, T.E. and Delcourt, D.C. (1995). The Geopause. *Rev. Geophys.*, **33**, 175-209.

Morel, P. and Irving, E. (1978). Tentative paleocontinental maps for the early Phanerozoic and Proterozoic. *J. Geol.*, **86**, 535-561.

Morelli, A. and Dziewonski, A.M. (1987). Topography of the core-mantle boundary and lateral homogeneity of the liquid core. *Nature* **325**, 678-683.

Morelli, A., Dziewonski, A.M. and Woodhouse, J.H. (1986). Anisotropy of the inner core inferred from *PKIKP* travel times. *Geophy. Res. Lett.*, **13**, 1545-1548.

Morgan, W.J. (1968). Rises, trenches, great faults, and crustal blocks. *J. Geophys. Res.*, **73**, 1959-1982.

Morley, L.W. and Larochelle, A. (1964). Paleomagnetism as a means of dating geological events. In *Geochronology in Canada* (F.F. Osborne, ed.), *Roy. Soc. Canad. Spec. Publ.*, **8**, 39-51, University of Toronto Press, Toronto.

Morrish, A.H. (1965). *The Physical Principles of Magnetism*. Wiley, New York.

Morse, P.M. and Feshbach, H. (1953). *Methods of Theoretical Physics*, Vols 1 and 2. McGraw Hill, New York.

Moskowitz, B. (1993). Micromagnetic study of the influence of crystal defects on coercivity in magnetite. *J. Geophys. Res.*, **98**, 18011-18026.

Muller, R. and Morris, D. (1986). Geomagnetic reversals from impacts on the Earth. *Geophys. Res. Lett.* **13**, 1177.

Murrell, M.T. and Burnett, D.S. (1986). Partitioning of K, U, and Th between sulfide and silicate liquids; implications for radioactive heating of planetary cores. *J. Geophys. Res.*, **91**, 8126-

8136.

Murthy, V.R. and Banerjee, S.K. (1973). Lunar evolution: How well do we know it now? *The Moon*, 7, 149-171.

Murthy, V.R. and Hall, H.T. (1970). The chemical composition of the earth's core: Possibility of sulphur in the core. *Phys. Earth Planet. Int.*, 2, 276-282.

Nagata, T. (1961). *Rock Magnetism* (revised edition). Maruzen Co., Tokyo.

Nagata, T. (1965). Main characteristics of recent geomagnetic secular variation. *J. Geomag. Geoelect.*, 17, 263-276.

Nagata, T. (1969). Lengths of geomagnetic polarity intervals. *J. Geomag. Geoelect.*, 21, 701-704.

Nagata, T., Ishikawa, Y., Kinoshita, H., Kuno, M., Syono, Y. and Fisher, R.M. (1970). Magnetic properties and natural remanent magnetisation of lunar materials. *Proc. Apollo 11 Lunar Sci. Conf.*, 2325-2340.

Naidu, P.S. (1974). Are geomagnetic reversals independent? *J. Geomag. Geoelect.*, 26, 101-104.

Nakajimi, T., Yashkawa, K., Natsuhara, N., Kawai, H. and Hones, S. (1973). Very short geomagnetic excursion 18 000 yr B.P. *Nature*, 244, 8-10.

Nakamura, Y., Latham, G., Lammlein, D., Ewing, M., Duennebier, F.K. and Dorman, J. (1974). Deep lunar interior inferred from recent seismic data. *Geophys. Res. Lett.*, 1, 137-140.

Nakamura, Y., Latham, G.V., Dorman, J. and Duennebier, F.K. (1976). Seismic structure of the Moon: A summary of current status. *Proc. Seventh Lunar Sci. Conf.*, 3113-3121.

Needham, J. (1962). *Science and Civilisation in China, Vol. 4. Physics and Physical Technology*, part 1. *Physics*. Cambridge University Press, Cambridge.

Néel, L. (1949). Théorie du trainage magnétique des ferromagnétiques aux grains fins avec applications aux terres cuites. *Ann. Geophys.*, 5, 99-136.

Néel, L. (1955). Some theoretical aspects of rock magnetism. *Adv. Phys.*, 4, 191-243.

Negi, J.G. and Tiwari, R.K. (1983). Matching long term periodicities of geomagnetic reversals and galactic motions of the solar system. *Geophys. Res. Lett.*, 10, 713-716.

Negi, J.G. and Tiwari, R.K. (1984). Periodicities of paleomagnetic intensity and paleoclimatic variations: A Walsh spectral approach. *Earth Planet. Sci. Lett.*, 70, 139-147.

Negrini, R.M., Davis, J.O. and Verosub, K.L. (1984). Mono Lake geomagnetic excursion found at Summer Lake, Oregon. *Geology*, 12, 643-646.

Ness, G., Levi, S. and Couch, R. (1980). Marine magnetic anomaly time scales for the Cenozoic and Late Cretaceous: A précis, critique, and synthesis. *Rev. Geophys. Space Phys.*, 18, 753-770.

Ness, N.F. (1965). The Earth's magnetic tail. *J. Geophys. Res.*, 70, 2989-3005.

Ness, N.F. (1994). Intrinsic magnetic fields of the planets: Mercury to Neptune. *Phil. Trans. Roy. Soc. London*, A349, 249-260.

Newell, A.J., Dunlop, D.J and Williams, W. (1993). A two-dimensional micromagnetic model of magnetizations and fields in magnetite. *J. Geophys. Res.*, 98, 9533-9549.

Niitsuma, N. (1971). Detailed study of the sediments recording the Matuyama-Brunhes geomagnetic reversal. *Tohoku Univ. Sci. Rpt 2nd Ser.* (Geology), 43, 1-39.

Ninkovich, D., Opdyke, N.D., Heezen, B.C. and Foster, J.H. (1966). Paleomagnetic stratigraphy, rates of deposition and tephrachronology in North Pacific deep-sea sediments. *Earth Planet.*

Sci. Lett., **1**, 476-492.

Nishitani, T. (1979). Grain size effect on the low-temperature oxidation of titanomagnetites. *Zeits. Geophys.*, **50**. 137-142.

Noltimeier, H.C. and Colinvaux, P.A. (1976). Geomagnetic excursion from Imuruk Lake, Alaska. *Nature*, **259**, 197-200.

Nord, G.L. and Lawson, C.A. (1992). Magnetic properties of ilmenite70-hematite30; effect of transformation-induced twin boundaries. *J. Geophys. Res.*, **97**, 10897-10910.

O'Reilly, W. (1984). *Rock and Mineral Magnetism.* Chapman and Hall, New York, 220 pp.

O'Reilly, W. and Banerjee, S.K. (1966). Oxidation of titanomagnetites and self-reversal. *Nature*, **211**, 26-28.

Ogg, J.C. (1995) Magnetic polarity time scale of the Phanerozoic. In *Global Earth Physics: A Handbook of Physical Constants* (T.J. Ahrens, ed.), p.240-270, Amer. Geophys. Union, Washington D.C.

Ohno, M. and Hamano, Y. (1992). Geomagnetic poles over the past 10,000 years. *Geophys. Res. Lett.*, **19**, 1715-1718.

Ohno, M. and Hamano, Y. (1993). Global analysis of the geomagnetic field: Time variation of the dipole moment and the geomagnetic pole in the Holocene. *J. Geomag. Geoelect.*, **45**, 1455-1466.

Ohtani, E. (1985). The primordial terrestrial magma ocean and its implication for stratification of the mantle. *Phys. Earth Planet. Int.*, **38**, 70-80.

Okada, M., and Niitsuma, N. (1989). Detailed paleomagnetic records during the Brunhes-Matuyama geomagnetic reversal and a direct determination of depth lag for magnetization in marine sediments. *Phys. Earth Planet. Int.*, **56**, 133-150.

Olson, P.L. (1977). *Internal Waves and Hydromagnetic Induction in the Earth's Core*, Ph.D. dissertation, University of California, Berkeley.

Olson, P.L. (1983). Geomagnetic polarity reversals in a turbulent core. *Phys. Earth Planet. Int.*, **33**, 260-274.

Olson, P. and Glatzmaier, G. (1995). Magnetoconvection in a rotating spherical shell; structure of flow in the outer core. *Phys. Earth and Planet. Int.*, **92**, 109-118.

Olson, P. and Hagee, V.L. (1987). Dynamo waves and palaeomagnetic secular variation. *Geophys. J. Roy. Astron. Soc.*, **88**, 139-159.

Olson, P. and Hagee, V.L. (1990). Geomagnetic polarity reversals, transition field structure and convection in the outer core. *J. Geophys. Res.*, **95**, 4609-4620.

Onwumechilli, A. (1967). Geomagnetic variations in the equatorial zone. In *Physics of Geomagnetic Phenomena, Vol. 1* (S. Matsushita and W.H. Campbell, eds), p.425-507. Academic Press, London and New York.

Opdyke, N.D. (1961). The palaeoclimatological significance of desert sandstone. In *Descriptive Palaeoclimatology* (A.E.M. Nairn, ed.), p.45-60. Wiley Interscience, New York.

Opdyke, N.D. (1962). Palaeoclimatology and continental drift. In *Continental Drift* (S.K. Runcorn, ed.), p.41-65. Academic Press, London and New York

Opdyke, N.D. (1972). Paleo magnetism of deep-sea cores. *Rev. Geophys. Space Phys.*, **10**, 213-249.

Opdyke, N.D. and Channell, J.E.T. (1996). *Magnetic Stratigraphy*. Academic Press, San Diego, Calif.

Opdyke, N.D. and Glass, B.P. (1969). The paleomagnetism of sediment cores from the Indian Ocean. *Deep-Sea Res.*, **16**, 249-261.

Opdyke, N.D. and Henry, K.W. (1969). A test of the dipole hypothesis. *Earth Planet. Sci. Lett.*, **6**, 139-151.

Opdyke, N.D. and Runcorn, S.K. (1956). New evidence for reversal of the geomagnetic field near the Pliocene-Pleistocene boundary [Ariz.]. *Science*, **123**, 1126-1127.

Opdyke, N.D. and Runcorn, S.K. (1960). Wind direction in the western United States in the late Paleozoic. *Geol. Soc. Amer. Bull.*, **71**, 959-972.

Opdyke, N.D., Burckle, L.H. and Todd, A. (1974). The extension of the magnetic time scale in sediments of the Central Pacific Ocean. *Earth Planet. Sci. Lett.*, **22**, 300-306.

Opdyke, N.D., Kent, D.V. and Lowrie, W. (1973). Details of magnetic polarity transitions recorded in a high deposition rate deep-sea core. *Earth Planet. Sci. Lett.*, **20**, 315-324.

Ott, U. (1993). Interstellar grains in meteorites. *Nature*, **364**, 25-32.

Özdemir, Ö. and Dunlop, D.J. (1988). Crystallization remanent magnetization during the transformation of maghemite to hematite. *J. Geophys. Res.*, **93**, 6530-6544.

Özdemir, Ö. and Dunlop, D.J. (1989). Chemico-viscous remanent magnetization in the Fe_3O_4-Fe_2O_3 system. *Science*, **243**, 1043-1047.

Ozima, M. and Ozima, M. (1967). Self-reversal of remanent magnetization in some dredged submarine basalts. *Earth Planet. Sci. Lett.*, **3**, 213-215.

Ozima, M. and Sakamoto, N. (1971). Magnetic properties of synthesised titanomagnetite. *J. Geophys. Res.*, **76**, 7035-7046.

Ozima, M., Funaki, M., Hamada, N., Aramaki, S. and Fujii, T. (1992). Self-reversal of thermo-remanent magnetization in pyroclastics from the 1991 eruption of Mt. Pinatubo, Phillipines. *J. Geomag. Geoelect.*, **44**, 979-984.

Pal, P.C. and Roberts, P. (1988). Long-term polarity stability and strength of the geomagnetic dipole. *Nature*, **331**, 702-705.

Papanastassiou, D.A. and Wasserburg, G.J. (1978). Sr isotopic anomalies in the Allende meteorite. *Geophys. Res. Lett.*, **5**, 595-598.

Parker, E.N. (1955a). Hydromagnetic dynamo models. *Astrophys. J.*, **122**, 293-314.

Parker, E.N. (1955b). The formation of sunspots from the Solar toroidal field. *Astrophys. J.*, **121**, 491-507.

Parker, E.N. (1958). Dynamics of the interplanetary gas and magnetic fields. *Astrophys. J.*, **128**, 664-676.

Parker, E.N. (1969). The occasional reversal of the geomagnetic field. *Astrophys. J.*, **158**, 817-830.

Parker, E.N. (1975). The generation of magnetic fields in astrophysical bodies. X. Magnetic buoyancy and the solar dynamo. *Astrophys. J.*, **198**, 205-209.

Parker, E.N. (1979). *Cosmical Magnetic Fields*. Clarendon Press, Oxford.

Parker, E.N. (1987). The dynamo dilemma. *Solar Physics*, **110**, 11-21.

Parker, R.L. (1973). The rapid calculation of potential anomalies. *Geophys. J. Roy. Astron. Soc.*, **31**,

447-455.

Parker, R.L. (1977). Understanding inverse theory. *Ann. Rev. Earth Planet. Sci.*, **5**, 35-64.

Parker, R.L. (1994). *Geophysical Inverse Theory*, Princeton Univ. press, Princeton New Jersey, 386 pp.

Parker, R.L. and Huestis, S.P. (1974). The inversion of magnetic anomalies in the presence of topography. *J. Geophys. Res.*, **79**, 1587-1593.

Parker, R.L. and Whaler, K.A. (1981). Numerical methods for establishing solutions to the inverse problem of electromagnetic induction. *J. Geophys. Res.*, **86**, 9574-9584.

Parks, G.K. (1991). *Physics of Space Plasmas: An Introduction.* Addison-Wesley, New York, 538 pp.

Paschmann, G. (1991). The Earth's magnetopause. In *Geomagnetism Vol. 4* (J.A. Jacobs, ed.), p.295-331. Academic Press, London.

Patton, B.J. and Fitch, T.L. (1962). Anhysteretic remanent magnetisation in small steady fields. *J. Geophys. Res.*, **67**, 307-311.

Payne, M.A. (1981). SI and Gaussian CGS units, conversions and equations for use in geomagnetism. *Phys. Earth Planet. Int.*, **26**, P10-P16.

Peddie, N.W. (1979). Current loop models of the earth's magnetic field. *J. Geophys. Res.*, **84**, 4517-4523.

Peng, L. and King, J.W. (1992). A late Quaternary geomagnetic secular variation record from Lake Waiau, Hawaii and the question of the Pacific non-dipole low. *J. Geophys. Res.*, **97**, 4407-4424.

Perrin, M., Prévot, M. and Mankinen, E. (1991). Low intensity of the geomagnetic field in Early Jurassic time. *J. Geophys. Res.*, **96**, 14197-14210.

Petersen, N., Dobeneck, T.V. and Vali, H. (1986). Fossil bacterial magnetite in deep-sea sediments from the South Atlantic Ocean. *Nature,* **320**, 611-615.

Peterson, N. and Bleil, U. (1973). Self-reversal of remanent magnetisation in synthetic titanomagnetite. *Zeits. Geophys.*, **39**, 965-977.

Petherbridge, J. (1977). A magnetic coupling occurring in partial self-reversal of magnetism and its association with increased magnetic viscosity in basalts. *Geophys. J. Roy. Astron. Soc.*, **50**, 395-406.

Peyronneau, J. and Poirier, J.-P. (1989). Electrical conductivity of the material of the Earth's lower mantle. *Nature*, **342**, 537-539.

Phillips, C.G. (1993). *Mean dynamos.* Ph.D. thesis, University of Sydney, Australia.

Phillips, J.D. (1977). Time variation and asymmetry in the statistics of geomagnetic reversal sequences. *J. Geophys. Res.*, **82**, 835-843.

Phillips, J.D. and Cox, A. (1976). Spectral analysis of geomagnetic reversal time scales. *Geophys. J. Roy. Astron. Soc.*, **45**, 19-33.

Phillips, J.D., Blakely, R.J. and Cox, A. (1975). Independence of geomagnetic polarity intervals. *Geophys. J. Roy. Astron. Soc.*, **43**, 747-754.

Pick, T. and Tauxe, L. (1991). Chemical remanent magnetization in synthetic magnetite. *J. Geophys. Res.,* **96**, 9925-9936.

Pick, T. and Tauxe, L. (1993). Holocene paleointensities: Thellier experiments on submarine basaltic glass from the East Pacific Rise. *J. Geophys. Res.*, **98**, 17949-17964.

Piper, J.D.A. and Grant, S. (1989). A paleomagnetic test of the axial dipole assumption and implications for continental distribution through geological time. *Phys. Earth Planet. Int.*, **55**, 37-53.

Pitman III, W.C., Herron, E.M. and Heirtzler, J.R. (1968). Magnetic anomalies in the Pacific and sea floor spreading. *J. Geophys. Res.*, **73**, 2069-2085.

Podolak, M. and Cameron, A.G.W. (1974). Models of the giant planets. *Icarus*, **22**, 123-148.

Poirier, J.P. (1988). Transport properties of liquid metals and viscosity of the Earth's core. *Geophys. J. Roy. Astron. Soc.*, **92**, 99-105.

Poirier, J.P. (1991). *Introduction to the Physics of Earth's Interior*. Cambridge Univ. Press, New York, 264 pp.

Preston, G.W. (1967). Studies of stellar magnetism – Past, present, and future. In *The Magnetic and Related Stars* (Proc. AAS–NASA Symposium, R.C. Cameron, ed.), p.3-28, Mono Book Corp., Baltimore, Maryland.

Prévot, M. (1977). Large intensity changes of the non-dipole field during a polarity transition. *Phys. Earth Planet. Int.*, **13**, 342-345.

Prévot, M. and Camps, P. (1993). Absence of preferred longitudinal sectors for poles from volcanic records of geomagnetic reversals. *Nature*, **366**, 53-57.

Prévot, M. and Perrin, M. (1992). Intensity of the Earth's magnetic field since Precambrian time from Thellier-type paleointensity data and inferences on the thermal history of the core. *Geophys. J. Int.*, **108**, 613-620.

Prévot, M., Derder, M.E., McWilliams, M.O. and Thompson, J. (1990). Intensity of the Earth's magnetic field: Evidence for a Mesozoic dipole low. *Earth Planet. Sci. Lett.*, **97**, 129-139.

Prévot, M, Mankinen, E., Coe, R.S. and Grommé, C.S. (1985). The Steens Mountain (Oregon) geomagnetic polarity transition. 2. Field intensity variations and discussion of reversal models. *J. Geophys. Res.*, **90**, 10417-10448.

Pudovkin, I.M. and Kolesova, V.I. (1968). Dipole model of the main geomagnetic field (based on an analysis of the Z_{ao} field). *Geomag. Aeron.*, **8**, 301-302.

Pudovkin, I.M., Kolesova, V.I. and Valuyeva, G.Ye. (1968). Spatial structure of the residual Z_{ao} field. *Geomag. Aeron.*, **8**, 483-486.

Quidelleur, X., and Courtillot, V. (1996). On low-degree spherical harmonic models of paleosecular variation. *Phys. Earth Planet. Int.*, in press.

Quidelleur, X. and Valet, J.-P. (1994). Paleomagnetic records of excursions and reversals: Possible biases caused by magnetization artefacts. *Phys. Earth. Planet. Int.*, **82**, 27-48.

Quidelleur, X., Valet, J.-P., Courtillot, V. and Hulot, G. (1994). Long-term geometry of the geomagnetic field for the last 5 million years; an updated secular variation database from volcanic sequences. *Geophys. Res. Lett.*, **21**, 1639-1642.

Quidelleur X., Valet, J.-P., LeGoff, M. and Bouldoire, X. (1995). Field dependence on magnetization of laboratory-redeposited deep-sea sediments: First results. *Earth. Planet. Sci. Lett.*, **133**, 311-325.

Rädler, K.-H. (1968). On the electrodynamics of conducting fluids in turbulent motion. The principles of mean field electrodynamics. *Z. Naturforsch.*, **23a**, 1841-1851. [English translation: Roberts & Stix (1971)]

Rampino, M.R. and Caldeira, K. (1993). Major episodes of geologic change: Correlations, time structure and possible causes. *Earth Planet. Sci. Lett.*, **114**, 215-227.

Raup, D.M. (1985). Magnetic reversals and mass extinctions. *Nature*, **314**, 341-343.

Readman, P.W. and O'Reilly, W. (1970). The synthesis and inversion of non-stoichiometric titanomagnetites. *Phys. Earth Planet. Int.*, **4**, 121-128.

Readman, P.W. and O'Reilly, W. (1972). Magnetic properties of oxidised (cation deficient) titanomagnetites (Fe, Ti, □)$_3$0$_4$. *J. Geomag. Geoelect.*, **24**, 69-90.

Rekdal, T. and Doornbos, D.J. (1992). The times and amplitudes of core phases for a variable core-mantle boundary layer. *Geophys. J. Int.*, **108**, 546-556.

Ricard, Y., Richards, M., Lithgo-Bertelloni, C. and Le Stunff, Y. (1993). A geodynamic model of mantle density heterogeneity. *J. Geophys. Res.* **98**, 21895-21909.

Richards, M.A. and Engebretson, D.C. (1992). Large-scale mantle convection and the history of subduction. *Nature*, **355**, 437-440.

Rikitake, T. (1958). Oscillations of a system of disc dynamos. *Proc. Cambridge Philos. Soc.*, **54**, 89-105.

Rikitake, T. (1966). *Electromagnetism and the Earth's Interior*. Elsevier, Amsterdam.

Rimbert, F. (1959). Contribution a l'étude de l'action de champs alternatifs sur les aimantations remanentes des roches, applications géophysiques. *Rev. Inst. Fr. Pet.*, **14**, 17 and 123.

Ringwood, A.E., (1958). On the chemical evolution and densities of the planets. *Geochim. Cosmochim. Acta*, **15**, 257-283.

Ringwood, A.E. (1977). Composition of the core and implications for the origin of the earth. *Geochim. J.*, **11**, 111-135.

Ringwood, A.E. (1978). Origin of the Moon. *Lunar & Planet. Sci.*, **9**, 961-963.

Ringwood, A.E. (1979). *Origin of the Earth and Moon*. Springer-Verlag, New York.

Rivin, Yu.R. (1974). 11-year periodicity in the horizontal component of the geomagnetic field. *Geomag. Aeron.*, **14**, 97-100.

Roberts, G.O. (1969). Dynamo waves. In *The Application of Modern Physics to the Earth and Planetary Interiors* (S.K. Runcorn, ed.), p.603-628, Wiley, New York.

Roberts, G.O. (1970). Spatially periodic dynamos. *Phil. Trans. Roy. Soc. London*, **A266**, 535-558.

Roberts, G.O. (1972). Dynamo action of fluid motions with two-dimensional periodicity. *Phil. Trans. Roy. Soc. London*, **A271**, 411-454.

Roberts, P.H. (1967). *An Introduction to Magnetohydrodynamics*. Longmans. London; Elsevier, New York.

Roberts, P.H, (1971). Dynamo theory. In *Mathematical Problems in the Geophysical Sciences*. (W.H. Reid, ed.), p.129-206, Amer. Math. Soc., Providence, Rhode Island.

Roberts, P.H. (1972). Electromagnetic core-mantle coupling. *J. Geomag. Geoelect.*, **24**, 231-259.

Roberts, P.H. (1987). Origin of the Main Field: Dynamics. In *Geomagnetism Vol. 2* (J.A. Jacobs, ed.), p.251-306, Academic Press, London.

Roberts, P.H. (1988). Future of geodynamo theory. *Geophys. Astrophys. Fluid Dyn.*, **42**, 3-31.

Roberts, P.H. (1989a). Core-mantle coupling. In *Encyclopedia of Solid Earth Geophysics* (D.E. James, ed.), p.148-160, Van Nostrand–Reinhold, N.Y.

Roberts, P.H. (1989b). From Taylor state to model-Z. *Geophys. Astrophys. Fluid Dyn.*, **49**, 143-160.

Roberts, P.H. and Gubbins, D. (1987). Origin of the main field: Kinematics. In *Geomagnetism Vol. 2* (J.A. Jacobs, ed.), p.185-249, Academic Press, London.

Roberts, P.H. and Scott, S. (1965). On the analysis of the secular variation. 1. A hydromagnetic constraint: Theory. *J. Geomag. Geoelectr.*, **17**, 137-151.

Roberts, P.H., and Stix, M. (1971). The turbulent dynamo: A translation of a series of papers by F. Krause, K.-H. Rädler and M. Steenbeck. Tech. Note 60, NCAR, Boulder, Colorado.

Roberts, P.H. and Stix, M. (1972). α-effect dynamos by the Bullard-Gellman formalism. *Astron. Astrophys.*, **18**, 453-466.

Roche, A. (1951). Sur les inversions de l'aimantation remanente des roches volcaniques dans les monts d'Auvergne. *C. R. Acad. Sci. Paris*, **233**, 1132-1134.

Roche, A. (1953). Sur l'origine des inversions l'aimentation constantées dans les roches d'Auvergne. *C. R. Acad. Sci. Paris*, **236**, 107-109.

Roche, A. (1956). Sur la date de la dernière inversion due champ magnétique terrestre. *C. R. Acad. Sci. Paris*, **243**, 812-814.

Roche, A. (1958). Sur les variation de direction du champ magnétique terrestre au cours du Quaternaire. *C. R. Acad. Sci. Paris*, **246**, 3364-3366.

Rochester, M.G., Jacobs, J.A., Smytie, D.E. and Chong, K.F. (1975). Can precession power the geomagnetic dynamo? *Geophys. J. Roy. Astron. Soc.*, **43**, 661-678.

Roperch, P., Bonhommet, N. and Levi, S. (1988). Paleointensity of the Earth's magnetic field during the Laschamp excursion and its geomagnetic implications. *Earth Planet. Sci. Lett.*, **88**, 209-219.

Rosenberg, R.L. and Coleman, P.J., Jr. (1969). Heliographic latitude dependence of the dominant polarity of the interplanetary magnetic field. *J. Geophys. Res.*, **74**, 5611-5622.

Rossi, B. (1964). *Cosmic Rays*. McGraw-Hill, New York

Roy, J.L. and Park, J.K. (1972). Red beds: DRM or CRM? *Earth Planet. Sci. Lett.*, **17**, 211-216.

Runcorn, S.K. (1954). The Earth's Core. *Trans. Amer. Geophys. Union*, **35**, 49-63.

Runcorn, S.K. (1955). The electrical conductivity of the earth's mantle. *Trans. Amer. Geophys. Union*, **36**, 191-198.

Runcorn, S.K. (1956). Paleomagnetic comparisons between Europe and North America. *Proc. Geol. Assoc. Canada*, **8**, 77-85.

Runcorn, S.K. (1959). On the hypothesis that the mean geomagnetic field for parts of geological time has been that of a geocentrical axial multipole. *J. Atmosph. Terr. Phys.*, **14**, 167-174.

Runcorn, S.K. (1961). Climatic change through geological time in the light of paleomagnetic evidence for polar wandering and continental drift. *Quart. J. Roy. Met. Soc.*, **87**, 282-313.

Runcorn, S.K. (1978). The ancient lunar core dynamo. *Science*, **199**, 771-773.

Runcorn, S.K. (1979). An iron core in the Moon generating an early magnetic field? *Proc. Tenth Lunar Planet. Sci. Conf.*, 2325-2333.

Runcorn, S.K., (1992). Polar path in geomagnetic reversals, *Nature*, **356**, 654.

Runcorn, S.K. (1994). The Early magnetic field and primeval satellite system of the moon: Clues to planetary formation. *Phil. Trans. Roy. Soc. London*, **A349**, 181-196.

Runcorn, S.K. and Urey, H.C. (1973). A new theory of lunar magnetism. *Science*, **180**, 636-638.

Runcorn, S.K., Collinson, D.W., O'Reilly, W., Battey, M.H., Stephenson, A., Jones, J.M., Manson, A.J. and Readman, P.W. (1970). Magnetic properties of Apollo 11 lunar samples. *Proc. Apollo 11 Lunar Sci. Conf.*, 2369-2387.

Russell, C.T. (1978). Re-evaluating Bode's law of planetary magnetism. *Nature*, **272**, 147-148.

Russell, C.T. (1987). Planetary magnetism. In *Geomagnetism Vol. 2* (J.A. Jacobs, ed.), p.457-523, Academic Press, London.

Russell, C.T., Coleman, P.J., Jr., Lichtenstein, B.R. and Schubert, G. (1974). The permanent and induced magnetic dipole moment of the Moon. *Proc. Fifth Lunar Sci. Conf.*, 2747-2760.

Rutten, M.G. (1959). Paleomagnetic reconnaissance of mid-Italian volcanoes. *Geol. Mijnbouw*, **21**, 373-374.

Ruzmaikin, A.A. and Sokoloff, D.D. (1980). Helicity, linkage and dynamo action. *Geophys. Astrophys. Fluid Dyn.*, **16**, 73-82.

Ryall, P. and Ade-Hall, J.M. (1975). Laboratory-induced self reversal of thermoremanent magnetisation in pillow basalts. *Nature*, **257**, 117-118.

Sabatier, P.C. (1977). On geophysical inverse problems and constraints. *J. Geophys*, **43**, 115-137.

Sakai, H. and Hirooka, K. (1986). Archaeointensity determinations from western Japan. *J. Geomag. Geoelect.*, **38**, 1323-1329.

Sano, Y. (1993). The magnetic fields of the planets: A new scaling law of the dipole moments of planetary magnetism. *J. Geomagn. Geoelect.*, **45**, 65-77.

Sato, M. and Wright, T.L. (1966). Oxygen fugacities directly measured in volcanic gases. *Science*, **153**, 1103-1105.

Saxena, S.K., Shen, G. and Lazor, P. (1993). Experimental evidence for a new iron phase and implications for Earth's core. *Science*, **260**, 1312-1314.

Saxena, S.K., Shen, G. and Lazor, P. (1994). Temperature in the Earth's core based on melting and phase transformation experiments on iron. *Science*, **264**, 405-407.

Schatten, K.H. (1971). Large-scale properties of the interplanetary magnetic field. *Rev. Geophys. Space Phys.*, **9**, 773-812.

Schmidt, A. (1924). Das erdmagnetische Außenfeld. *Zeits. Geophys.*, **1**, 3-20.

Schmidt, A. (1935). *Tafeln der normierten Kugelfunktionen und ihrer Ableitungen nebst den Logarithmen dieser Zahlen sowie Formein zur Entwicklung nach Kugelfuntionen.* Engelhard-Reyler Verlag, Gotha, Germany, 52 pp.

Schmidt, A. (1939). Zur Frage der Hypothetischen die erdoberfliche durchdringenden Strome. Mit einen zusatz von J. Bartels. *Gerlands Beitr. Geophys.*, **55**, 292.

Schmitz, D., Frayser, J.B. and Cain, J.C. (1982). Application of dipole modelling to magnetic anomalies. *Geophys. Res. Lett.*, **9**, 307-310.

Schneider, D.A. and Kent, D.V. (1990). The time-averaged paleomagnetic field. *Rev. Geophys.* **28**, 71-96.

Schneider, D.A. and Mello, G.A. (1994). Pleistocene geomagnetic intensity variation from Sulu Sea sediments, *EOS, Trans. Amer. Geophys. Union*, **75** (No.44), 1994 Fall Meeting Suppl., p.193.

Schouten, H. and Denham, C.R. (1979). Modelling the oceanic magnetic source layer. In *Deep Drilling Results in the Atlantic Ocean: Ocean Crust* (M. Talwani, C.G. Harrison and D.E. Hayes, eds), p.151-159. Maurice Ewing Ser., Amer. Geophys. Union.

Schouten, H. and McCamy, K. (1972). Filtering marine magnetic anomalies. *J. Geophys. Res.*, **77**, 7089-7099.

Schult, A. (1968). Self-reversal of magnetisation and chemical composition of titanomagnetites in basalts. *Earth Planet. Sci. Lett.*, **4**, 57-63.

Schultz, A. and Larsen, J. (1987) On the electrical conductivity of the Earth's interior. 1. Mid-mantle response function computation, *Geophys. J. Roy. Astron. Soc.*, **88**, 733-761.

Schultz, M. (1991). The magnetosphere. In *Geomagnetism Vol. 4* (J.A. Jacobs, ed.), p.87-293. Academic Press, London.

Schultz, M. and Paulikas, G. (1990). Planetary magnetic fields: A comparative view. *Adv. Space Res.*, **10**, 55-64.

Schuster, A. (1898). On the investigation of hidden periodicities with application to a supposed 26 day period of meteorological phenomena. *Terr. Magn.*, **3**, 13-41.

Secco, R.A. and Schloessin, H.H. (1989). The electrical resistivity of solid and liquid Fe at pressures up to 7 GPa. *J. Geophys. Res.*, **94**, 5887-5894.

Seki, M. and Ito, K. (1993). A phase transition model for geomagnetic polarity reversals. *J. Geomag. Geoelect.*, **45**, 79-88.

Senanayake, W.E. and McElhinny, M.W. (1981). Hysteresis and susceptibility characteristics of magnetite and titanomagnetites: Interpretation of results from basaltic rocks. *Phys. Earth Planet. Int.*, **26**, 47-55.

Senanayake, W.E., McElhinny, M.W. and McFadden, P.L. (1982). Comparisons between the Thelliers' and Shaw's paleointensity methods using basalts less than 5 million years old. *J. Geomag. Geoelect.*, **34**, 141-161.

Shankland, T., Peyronneau, J. and Poirier, J.-P. (1993). Electrical conductivity of the lower mantle. *Nature*, **366**, 453-455.

Sharp, L.R., Coleman, P.J., Jr., Lichtenstein, B.R., Russell, C.T. and Schubert, G. (1973). Orbital mapping of the lunar magnetic field. *The Moon*, 7, 322-341.

Shaw, G.H. (1978). Effects of core formation. *Phys. Earth Planet. Int.*, **16**, 361-369.

Shaw, J. (1974). A new method of determining the magnitude of the palaeomagnetic field. *Geophys. J. Roy. Astron. Soc.*, **39**, 133-141.

Shaw, J. (1975). Strong geomagnetic fields during a single Icelandic polarity transition. *Geophys. J. Roy. Astron. Soc.*, **40**, 345-350.

Shaw, J. (1977). Further evidence for a strong intermediate state of the palaeomagnetic field. *Geophys. J. Roy. Astron. Soc.*, **48**, 263-269.

Shearer, P. and Masters, G. (1990). The density and shear velocity contrast at the inner core boundary. *Geophys. J. Int.*, **102**, 491-498.

Shearer, P.M., Toy, K.M. and Orcutt, J.A. (1988). Axisymmetric Earth models and inner core anisotropy. *Nature*, **333**, 228-232.

Sheridan, R. (1983). Phenomenon of pulsation tectonics related to the breakup of the Eastern North America continental margin. *Tectonophysics*, **94**, 169-185.

Shibuya, H., Cassidy, J., Smith, I.E.M. and Itaya, T. (1992). A geomagnetic excursion in the Brunhes epoch recorded in New Zealand basalts. *Earth Planet. Sci. Lett.*, **111**, 41-48.

Shure, L. and Parker, R.L. (1981). An alternative explanation for intermediate wavelength magnetic anomalies. *J. Geophys. Res.*, **86**, 11600-11608.

Shure, L., Parker, R.L. and Backus, G.E. (1982). Harmonic splines for geomagnetic field modelling. *Phys. Earth Planet. Int.*, **28**, 215-229.

Skiles, D.D. (1970). A method of inferring the direction of drift of the geomagnetic field from paleomagnetic data. *J. Geomag. Geoelect.*, **22**, 441-461.

Skiles, D.D. (1972a). The laws of reflection and refraction of incompressible magnetohydrodynamic waves at fluid-solid interface. *Phys. Earth Planet. Int.*, **5**, 90-98.

Skiles, D.D. (1972b). On the transmission of the energy in an incompressible magnetohydrodynamic wave into a conducting solid. *Phys. Earth Planet. Int.*, **5**, 99-109.

Smith, A.G., Hurley, A M. and Briden, J.C. (1981). *Phanerozoic Paleocontinental World Maps*. Cambridge University Press, Cambridge.

Smith, E.J., Tsurutani, B.T. and Rosenberg, R.L. (1978). Observations of the interplanetary sector structure up to heliographic latitudes of 16°: Pioneer 11. *J. Geophys. Res.*, **83**, 717-724.

Smith, G. and Creer, K.M. (1986). Analysis of geomagnetic secular variations 10000 to 30000 years BP, Lac du Bouchet, France. *Phys. Earth Planet. Int.*, **44**, 1-14.

Smith, P.J. (1967a). The intensity of the Tertiary geomagnetic field. *Geophys. J. Roy. Astron. Soc.*, **12**, 239-258.

Smith, P.J. (1967b). The intensity of the ancient geomagnetic field: A review and analysis. *Geophys. J. Roy. Astron. Soc.*, **12**, 321-362.

Smith, P.J. (1967c). On the suitability of igneous rocks for ancient geomagnetic field intensity determination. *Earth Planet. Sci. Lett.*, **2**, 99-105.

Smith, P.J. (1968). Pre-Gilbertian conceptions of terrestrial magnetism. *Tectonophysics*, **6**, 499-510.

Smith, P.J. (1970a). Petrus Peregrinus Epistola – The beginning of experimental studies of magnetism in Europe. *Atlas* (*News Supp. to Earth Sci. Revs.*), **6**, A11-A17.

Smith, P.J. (1970b). Do magnetic polarity-oxidation state correlations imply self-reversal? *Comm. Earth Sci. Geophys.*, **1**, 74-85.

Smith, P.J. and Needham, J. (1967). Magnetic declination in medieval China. *Nature*, **214**, 1213-1214.

Smylie, D.E., Clark, G.K.C. and Ulrych, T.J. (1973). Analysis of irregularities in the earth's rotation. In *Methods in Computational Physics Vol. 13*, p.391. Academic Press, London and New York

Soderblom, D.R. and Baliunas, S.C. (1988). The Sun among the stars; what stars indicate about solar variability. In *Secular Solar and Geomagnetic Variations in the Last 10,000 years* (F.R. Stephenson and A.W. Wolfendale, eds). NATO Advanced Study Series, p.25-48, Kluwer Acad. Publ., Dordrecht.

Solomon, S.C. and Head, J.W. (1979). Vertical movement in mare basins; relation to mare emplacement, basin tectonics, and lunar thermal history. *J. Geophys. Res.*, **84**, 1667-1682.

Sonett, C.P., Colburn, D.S. and Currie, R.G. (1967). The intrinsic magnetic field of the Moon. *J. Geophys. Res.*, **72**, 5503-5507.

Souriau, A. and Poupinet, G. (1991). A study of the outermost liquid core using differential travel times of SKS and S3KS phases. *Phys. Earth Planet. Int.*, **68**, 183-199.

Soward, A.M. (1989). On dynamo action in a steady flow at large magnetic Reynolds numbers. *Geophys. Astrophys. Fluid Dyn.*, **49**, 3-22.

Soward, A.M. (1975). Random waves and dynamo action. *J. Fluid Mech.*, **69**, 145-177.

Soward, A.M. (1979). Convection driven dynamos. *Phys. Earth Planet. Int.*, **20**,134-151.

Soward, A.M. (1991a). The earth's dynamo. *Geophys. Astrophys. Fluid Dyn.*, **62**, 191-209.

Soward, A.M. (1991b). Magnetoconvection. *Geophys. Astrophys. Fluid Dyn.*, **62**, 229-247.

Soward, A.M. (1994). Fast dynamos. In *Lectures on Solar and Planetary Dynamos* (M.R.E. Proctor and A.D. Gilbert, eds), p.181-218. Cambridge Univ. Press, New York.

Speiser, T.W. and Ness, N.F. (1967). The neutral sheet in the geomagnetic tail: Its motion, equivalent currents, and field line convection through it. *J. Geophys. Res.*, **72**, 131-141.

Spell, T.L. and McDougall, I. (1992). Revisions to the age of the Brunhes-Matuyama boundary and the Pleistocene geomagnetic polarity timescale. *Geophys. Res. Lett.*, **19**, 1181-1184.

Spiess, F.N. and Mudie, J.D. (1970). Small scale topographic and magnetic features of the seafloor. In *The Sea Vol. 4* (A.E. Maxwell, ed.), p.205-250. Wiley, New York.

Srivastava, B.J. and Abbas, H. (1977). Geomagnetic secular variation in India – Regional and local features. *J. Geomag. Geoelect.*, **29**, 51-64.

Srnka, L.J. (1977). Spontaneous magnetic field generation in hypervelocity impact. *Proc. Eighth Lunar Sci. Conf.*, 785-792.

Srnka, L.J., Martelli, G., Newton, G., Cisowski, S.M., Fuller, M.D. and Schaal, R.B. (1979). Magnetic field, shock effects and remanent magnetisation in a hypervelocity impact experiment. *Earth Planet. Sci. Lett.*, **42**, 127-137.

Stacey, F.D. (1977). *Physics of the Earth*, 2nd Ed., Wiley, New York, 414 pp.

Stacey, F.D. (1992). *Physics of the Earth*, 3rd Ed., Brookfield Press, Brisbane, 513 pp.

Stacey, F.D. and Banerjee, S.K. (1974). *The Physical Principles of Rock Magnetism*. Elsevier, Amsterdam, 195 pp.

Stacey, F.D., Brennan, B.J. and Irvine, R.D. (1981). Finite strain theories and comparisons with seismological data. *Geophys. Surv.*, **4**, 189-232.

Stacey, F.D., Spiliopoulos, S.S. and Barton, M.A. (1989). A critical re-examination of the thermodynamic basis of Lindemann's melting law. *Phys. Earth Planet. Int.*, **55**, 201-207.

Stannard, D. (1975). Observations of the magnetic field structure of Jupiter. *Geophys. J. Roy. Astron. Soc.*, **41**, 327-330.

Starchenko, S. and Shcherbakov, V. P (1994). The behavior of the magnetosphere during reversals. *Trans. (Dokl.) Russian Acad., (Earth Sci.)*, **322**, 1-6.

Steenbeck, M. and Krause, F. (1969). On the dynamo theory of stellar and planetary magnetic fields, I. A.C. dynamos of solar type. *Astron. Nachr.*, **291**, 49-84. [English translation: Roberts

& Stix (1971), p.147-220]

Steenbeck, M. Krause, F. and Rädler, K.-H. (1966). A calculation of the mean electromotive force in an electrically conducting fluid in tubulent motion, under the influence of Coriolis forces. *Z. Naturforsch.*, **21a**, 369-376. [English translation: Roberts & Stix (1971), p.29-47]

Steenbeck, M., Kirko, I.M., Gailitis, A., Klawina, A.P., Krause, F., Laumonis, I.J. and Lielausis, O.A. (1967). An experimental verification of the α-effect. *Monatsber, Deut. Akad. Wiss. Berlin*, **9**, 716-719. [English translation: Roberts & Stix (1971), p.97-102]

Stehli, F.G. (1968). A paleoclimatic test of the hypothesis of an axial dipolar magnetic field. In *History of the Earth's Crust* (R.A. Phinney, ed.), p.195-205. Princeton Univ. Press, Princeton, New Jersey.

Stephenson, A. (1975). The observed moment of a magnetized inclusion of high Curie point within a titanomagnetite particle of lower Curie point. *Geophys. J. Roy. Astron. Soc.*, **40**, 29-36.

Stephenson, A. and Collinson, D.W. (1974). Lunar magnetic field paleointensities determined by anhysteretic remanent magnetisation method. *Earth Planet. Sci. Lett.*, **23**, 220-228.

Stephenson, A., Runcorn, S.K. and Collinson, D.W. (1977). Paleointensity estimates from lunar samples 10017 and 10020. *Proc. Eighth Lunar Sci. Conf.*, 679-688.

Stern, D.P. (1977). Large scale electric fields in the earth's magnetosphere. *Rev. Geophys. Space Phys.*, **15**, 156-194.

Stern, D.P. (1989). A brief history of magnetospheric physics before the spaceflight era. *Rev. Geophys.*, **27**, 103-114.

Sternberg, R.S. (1983). Archaeomagnetism in the southwest of North America. In *Geomagnetism of Baked Clays and Recent Sediments* (K.M. Creer, P. Tochulka and C.E. Barton, eds), p.158-167, Elsevier, Amsterdam.

Sternberg, R.S. (1989). Archaeomagnetic paleointensity in the American southwest during the past 2000 years. *Phys. Earth Planet. Int.*, **56**, 1-17.

Sternberg, R.S. and McGuire, R.H. (1990). Archeomagnetic secular variation in the American southwest, A.D. 700-1450. In *Archeomagnetic Dating* (J. Eighmy and R. Sternberg, eds), p.199.

Stevenson, D.J. (1974). Planetary magnetism. *Icarus*, 22, 403-415.

Stevenson, D.J. (1979). Turbulent thermal convection in the presence of rotation and a magnetic field: A heuristic theory. *Geophys. Astrophys. Fluid Dyn.*, **12**, 139-169.

Stevenson, D.J. (1981). Models of the earth's core. *Science*, **214**, 611-619.

Stevenson, D.J. (1983). Planetary magnetic fields. *Rept. Prog. Phys.*, **46**, 555-620.

Stewartson, K. (1967). Slow oscillations of a fluid in a rotating cavity in the presence of a toroidal magnetic field. *Proc. Roy. Soc. London*, **A299**, 173-187.

Stigler, S. (1987). Aperiodicity of magnetic reversals? *Nature*, **330**, 26-27.

Stix, M. (1991) The Solar Dynamo. *Geophys. Astrophys. Fluid Dyn.*, **62**, 211-228.

Stix, M. (1989). *The Sun*. Springer-Verlag, New York, 390 pp.

Stixrude, L. and Cohen, R.E. (1995a). Constraints on the crystalline structure of the inner core; mechanical instability of bcc iron at high pressure. *Geophys. Res. Lett.*, **22**, 125-128.

Stixrude, L. and Cohen, R.E. (1995b). High-pressure elasticity of iron and anisotropy of Earth's inner core. *Science*, **267**, 1972-1975.

Stoner, E.C. and Wohlfarth, E.P. (1948). A mechanism of magnetic hysteresis in heterogeneous alloys. *Phil. Trans. Roy. Soc. London*, **A240**, 599-642.

Stothers, R.B. (1986). Periodicity of the Earth's magnetic reversals. *Nature*, **332**, 444-446.

Strangway, D.W. (1977). The magnetic fields of the terrestrial planets. *Phys. Earth Planet. Int.*, **15**, 121-130.

Strangway, D.W. and Sharpe, H.A. (1974). Lunar magnetism and an early cold Moon. *Nature*, **249**, 227-230.

Strangway, D.W., Larson, E.E. and Goldstein, M. (1968). A possible cause of high magnetic stability in volcanic rocks. *J. Geophys. Res.*, **73**, 3787-3795.

Strangway, D.W., Larson, E.E. and Pearce, G.W. (1970). Magnetic studies of lunar samples breccia and fines. *Proc. Apollo 11 Lunar Sci. Conf.*, 2435-2451.

Stuiver, M. and Quay, P. (1980). Changes in atmospheric carbon-14 attributed to a variable sun. *Science*, **207**, 11-19.

Su, W.-J., Woodward, R. and Dziewonski, A. (1994). Degree 12 model of shear velocity heterogeneity in the mantle. *J. Geophys. Res.*, **96**, 6945-6980.

Suzuki, Y. and Sato, R. (1970). Viscosity determination in the Earth's outer core from ScS and SKS phases. *J. Phys. Earth*, **18**, 157-170.

Swingler, D.N. (1979). A comparison between Burg's maximum entropy method and a non-recursive technique for the spectral analysis of deterministic signals. *J. Geophys. Res.*, **84**, 679-685.

Swingler, D.N. (1980). Burg's maximum entropy algorithm versus the discrete Fourier transform as a frequency estimator for truncated real sinosoids. *J. Geophys. Res.*, **85**, 1435-1438.

Tackley, P.J. (1995). Mantle dynamics: Influence of the transition zone. *Rev. Geophys. Suppl., U.S. National Report to IUGG 1991-1994*, 275-282.

Tackley, P.J., Stevenson, D.J., Glatzmaier, G.A., and Schubert, G. (1993). Effects of an endothermic phase transition at 670 km depth in a spherical model of convection in the Earth's mantle. *Nature*, **361**, 699-704.

Tanaka, H. and Kono, M. (1994). Paleointensity database provides new resource. *EOS, Trans. Amer. Geophys. Union*, **75**, 498.

Tanaka, H., Kono, M. and Uchimura, H. (1995). Some global features of palaeointensity in geological time. *Geophys. J. Int.*, **120**, 97-102.

Tanaka, H., Otsuka, A., Tachibana, T. and Kono, M. (1994). Paleointensities for 10-22 ka from volcanic rocks in Japan and New Zealand. *Earth Planet. Sci. Lett.*, **122**, 29-42.

Tanguy, J.C. (1970). An archaeomagnetic study of Mt. Etna: The magnetic direction recorded in lava flows subsequent to the twelfth century. *Archaeometry*, **12**, 115-128.

Tauxe, L. (1993). Sedimentary records of relative paleointensity of the geomagnetic field; theory and practice. *Rev. Geophys.*, **31**, 319-354.

Tauxe, L. and Valet, J.-P. (1989). Relative paleointensity of the Earth's magnetic field from marine sedimentary records: A global perspective. *Phys. Earth Planet. Int.*, **56**, 59-68.

Tauxe, L. and Wu, G. (1990). Normalized remanence in sediments of the western equatorial Pacific:

Relative paleointensity of the geomagnetic field? *J. Geophys. Res.*, **95**, 12337-12350.

Taylor, S.R. (1975). *Lunar Science: A Post-Apollo View.* Pergamon Press, New York.

Taylor, S.R. (1982). *Planetary Science: A Lunar Perspective.* Lunar and Planetary Institute, Houston, Texas.

Taylor, S.R. (1992). *Solar System Evolution: A New Perspective.* Cambridge Univ. Press, Cambridge U.K., 307 pp.

Thellier, E. (1937a). Aimantation des terres ciutes; application à la recherche de l'intensité du champ magnétique terrestre dans le passé. *C. R. Acad. Sci. Paris*, **204**, 184-186.

Thellier, E. (1937b). Recherche de l'intensité du champ magnétique terrestre dans le passé: Premier résultats. *Ann. Inst. Phys. Globe, Univ. Paris*, **15**, 179-184.

Thellier, E. (1981). Sur la direction du champs magnétique terrestre en France, durant les deux derniers millénaires. *Phys. Earth Planet. Int.*, **24**, 89-132.

Thellier, E. and Thellier, O. (1951). Sur la direction du champ magnétique terrestre, retrouvée sur des parois de fours des époques punique et romaine, á Carthage. *C. R. Acad. Sci. Paris*, **233** 1476-1478.

Thellier, E. and Thellier, O. (1952). Sur la direction du champ magnétique terrestre, dans la région de Trèves, vers 380 après J.-C. *C.R. Acad. Sci. Paris*, **234**, 1464-1466.

Thellier, E. and Thellier, O. (1959a). Sur l'intensité du champ magnétique terrestre dans le passé historique et géologique. *Ann. Geophys.*, **15**, 285-376.

Thellier, E. and Thellier, O. (1959b). The intensity of the geomagnetic field in the historical and geological past. *Akad. Nauk. SSR. Izv. Geophys. Ser.*, 1296-1331.

Thomas, R. (1983). Summary of prehistoric archaeointensity data from Greece and eastern Europe. In *Geomagnetism of Baked Clays and Recent Sediments* (K.M. Creer, P. Tucholka and C.E. Barton, eds), p.117-122, Elsevier, Amsterdam.

Thompson, R. and Turner, G.M. (1979). British geomagnetic master curve 10,000 yr B.P. for dating European sediments. *Geophys. Res. Lett.*, **6**, 249-252.

Thouveny, N. and Creer, K.M. (1992). Geomagnetic excursions in the past 60 ka: Ephemeral secular variation features. *Geology*, **20**, 399-402.

Thouveny, N., Creer, K.M. and Blunk, I. (1990). Extension of the Lac du Bouchet palaeomagnetic record over the last 120,000 years. *Earth Planet. Sci. Lett.*, **97**, 140-161.

Thouveny, N., Creer, K.M. and Williamson, D. (1993). Geomagnetic moment variations in the last 70,000 years, impact on production of cosmogenic isotopes. *Global Planet. Change*, **7**, 157-172.

Tric, E., Laj, C., Jehanno, C., Valet, J.-P., Kissel C., Mazaud, A. and Iaccarino, S. (1991a). High-resolution record of the upper Olduvai transition from Po Valley (Italy) sediments; support for dipolar transition geometry? *Phys. Earth Planet. Int.*, **65**, 319-336.

Tric, E., Laj, C., Valet, J-P., Tucholka, P., Paterne, M. and Guichard, F. (1991b). The Blake geomagnetic event; transition geometry, dynamical characteristics and geomagnetic significance. *Earth Planet. Sci. Lett.*, **102**, 1-13.

Tric, E., Valet, J.P., Tucholka, P., Paterne, M., Labeyrie, L., Guichard, F., Tauxe, L. and Fontugne, M. (1992). Paleointensity of the geomagnetic field during the last 80,000 years. *J. Geophys.*

Res., **97**, 9337-9351.

Tritton, D.J. (1988). Deterministic chaos, geomagnetic reversals, and the spherical pendulum. In *Proceedings of the NATO Advanced Study Institute on Geomagnetism and Palaeomagnetism* (F.J. Lowes, ι .W. Collinson, J.H. Parry, S.K. Runcorn, D.C. Tozer, and A.M. Soward, eds), p.215-226. Kluwer Acad. Publ., Dordrecht.

Tucker, P. and O'Reilly, W. (1980). Reversed thermoremanent magnetization in synthetic titanomagnetites as a consequence of high temperature oxidation. *J. Geomag. Geoelect.*, **32**, 341-355.

Turner, G.M. and Lillis, D.A. (1994). A palaeomagnetic secular variation record for New Zealand during the past 2500 years. *Phys. Earth Planet. Int.*, **83**, 265-282.

Turner, G.M. and Thompson, R. (1979). Behaviour of the earth's magnetic field as recorded in the sediment of Loch Lomond. *Earth Planet. Sci. Lett.*, **42**, 412-426.

Turner, G.M. and Thompson, R. (1981). Lake sediment record of the geomagnetic secular variation in Britain during Holocene times. *Geophys. J. R. Astr.Soc.*, **65**, 703-725.

Turner, G.M. and Thompson, R. (1982). Detransformation of the British geomagnetic secular variation record for Holocene times. *Geophys. J. R. Astr. Soc.*, **70**, 789-792.

Turner, G.M., Evans, M.E. and Hussin, I.S. (1982). A geomagnetic secular variation study (31000-19500 bp) in western Canada. *Geophys. J. Roy. Astron. Soc.*, **71**, 159-171.

Turner, J.S. (1979). *Buoyancy Effects in Fluids*. Cambridge University Press, Cambridge, 368 pp.

Ulrych, T.J. (1972). Maximum entropy power spectrum of long period geomagnetic reversals. *Nature*, **235**, 218-219.

Ulrych, T.J. and 'Bishop, T.N. (1975). Maximum entropy spectral analysis and autoregressive decomposition. *Rev. Geophys. Space Phys.*, **13**, 183-200.

Urakawa, S., Kato, K. and Kumazawa, M. (1987). Experimental study on the phase relations in the system FE-Ni-O-S up to 15 GPa. In *High-Pressure Research in Mineral Physics* (M. Manghnani and Y. Syono, eds), p.95-111, Amer. Geophys. Union, Washington D.C.

U.S.-Japan Paleomagnetic Cooperation Program in Micronesia (1975). Paleosecular variation of lavas from the Marianas in the Western Pacific Ocean. *J. Geomag. Geoelect.*, **27**, 57-66.

Uyeda, S. (1958). Thermoremanent magnetism as a medium of paleomagnetism with special references to reverse thermoremanent magnetism. *Japan J. Geophys.*, **2**, 1-123.

Vainshtein, S.I. and Zel'dovich, Ya. B. (1972). Origin of magnetic fields in astrophysics. *Usp. Fiz. SSSR*, **106**, 431-457. [English translation: *Sov. Phys. Usp.*, **15** , 159-172]

Valet, J.P. and Laj, C. (1981). Paleomagnetic record of two successive Miocene geomagnetic reversals in western Crete. *Earth Planet. Sci. Lett.*, **54**, 53-63.

Valet, J.P. and Meynadier, L. (1993). Geomagnetic field intensity and reversals during the past four million years. *Nature*, **366**, 234-238.

Valet, J.-P., Laj, C., and Langereis, C. (1988a). Sequential geomagnetic reversals recorded in Upper Tortonian marine clays in western Crete (Greece). *J. Geophys. Res.*, **93**, 1131-1151.

Valet, J.-P., Laj, C., and Tucholka, P. (1986). High-resolution sedimentary record of a geomagnetic reversal. *Nature*, **322**, 27-32.

Valet, J.-P., Tauxe, L., and Clark, D. (1988b). The Matuyama-Brunhes transition recorded in Lake

Tecopa sediments (California). *Earth Planet. Sci. Lett.*, **87**, 463-472.

Valet, J.-P., Tauxe, L., and Clement, B. (1989). Equatorial and mid-latitude records of the last geomagnetic reversal from the Atlantic Ocean. *Earth Planet. Sci.*, **94**, 371-384.

Valet, J.-P., Tucholka, P., Courtillot, V. and Meynadier, L. (1992). Palaeomagnetic constraints on the geometry of the geomagnetic field during reversals. *Nature*, **356**, 400-407.

Van Allen, J.A., Ludwig, G.H., Ray, E.C. and McIlwain, C.E. (1958). Observations of high intensity radiation by satellites 1958 Alpha and Gamma. *Jet Propulsion*, **28**, 588.

Van der Bos, A. (1971). Alternative interpretation of maximum entropy spectral analysis. *IEEE Trans. Inform. Theory*, **IT-17**, 493-494.

Van der Voo, R. (1993). *Paleomagnetism of the Atlantic, Tethys and Iapetus Oceans*, Cambridge Univ. Press, Cambridge U.K., 411 pp.

van Hoof, A.A.M. (1993). The Gilbert/Gauss sedimentary geomagnetic reversal record from southern Sicily. *Geophys. Res. Lett.*, **20**, 835-838.

van Hoof, A.A.M. and Langereis, C.G. (1992a). The upper Kaena sedimentary geomagnetic reversal record from Sicily. *J. Geophys. Res.*, **97**, 6941-6957.

van Hoof, A.A.M. and Langereis, C.G. (1992b). The upper and lower Thvera sedimentary geomagnetic reversal record from southern Sicily. *Earth Planet. Sci. Lett.*, **114**, 59-75.

van Zijl, J.S.V., Graham, K.W.T. and Hales, A.L. (1962a). The paleomagnetism of the Stormberg lavas of South Africa. I. Evidence for a genuine reversal of the earths's magnetic field in Triassic-Jurassic times. *Geophys. J. Roy. Astron. Soc.*, **7**, 23-39.

van Zijl, J.S.V., Graham, K.W.T. and Hales, A.L. (1962b). The paleomagnetism of the Stormberg lavas of South Africa. II. The behaviour of the magnetic field during a reversal. *Geophys. J. Roy. Astron. Soc.*, **7**, 169-182.

Vandamme, D. (1994). A new method to determine paleosecular variation. *Phys. Earth Planet. Int.*, **85**, 131-142.

Verhoogen, J. (1956). Ionic ordering and self-reversal of impure magnetites. *J. Geophys. Res.*, **61**, 201-209.

Verhoogen, J. (1961). Heat balance of the earth's core. *Geophys. J. Roy. Astron. Soc.*, **4**, 276-281.

Verhoogen, J. (1980). *Energetics of the Earth*. National Academy of Science, Washington, D.C.

Verosub, K.L. (1977). Depositional and postdepositional processes in the magnetisation of sediments. *Rev. Geophys. Space Phys.*, **15**, 129-143.

Verosub, K.L. (1979). Paleomagnetism of varved sediments from western New England: Secular variation. *Geophys. Res. Lett.*, **6**, 245-248.

Verosub, K.L. (1982). Geomagnetic excursions: A critical assessment of the evidence as recorded in sediments of the Brunhes epoch. *Phil. Trans. Roy. Soc. London*, **A306**, 161-168.

Verosub, K.L. and Banerjee, S.K. (1977). Geomagnetic excursions and their paleomagnetic record. *Rev. Geophys. Space Phys.*, **15**, 145-155.

Vestine, E.H. (1953). On variations of the geomagnetic field, fluid motions, and the rate of the earth's rotation. *J. Geophys. Res.* **58**, 127-145.

Vestine, E.H. (1954). The earth's core. *Trans. Amer. Geophys. Union*, **35**, 63-72.

Vestine, E.H. (1967). Main geomagnetic field. In *Physics of Geomagnetic Phenomena Vol. 1* (S.

Matsushita and W.H. Campbell, eds.), p.181-234. Academic Press, London and New York.

Vestine, E.H., Lange, I., LaPorte, L. and Scott, W.E. (1947a). The geomagnetic field, its description and analysis. *Carnegie Inst. Washington Rept. No. 580.*

Vestine, E.H., LaPorte, L., Lange, I., Cooper, C. and Hendrix, W.C. (1947b). Description of the earth's main magnetic field, 1905-1945. *Carnegie Inst. Washington Rept. No. 578.*

Vine, F.J. and Matthews, D.H. (1963). Magnetic anomalies over oceanic ridges. *Nature*, **199**, 947-949.

Vogt, P.R. (1972). Evidence for a global synchronism in mantle plume convection, and possible significance for geology. *Nature*, **240**, 338-342.

Vogt, P.R. (1975). Changes in geomagnetic frequency at times of tectonic change: Evidence for coupling between core and upper mantle processes. *Earth Planet. Sci. Lett.*, **25**, 313-321.

Voorhies, C.V. (1986). Steady flows at the top of Earth's core derived from geomagnetic field models. *J. Geophys. Res.*, **91**, 12444-12466.

Voorhies, C.V. (1993). Geomagnetic estimates of steady surficial core flow and flux diffusion: Unexpected geodynamo experiments. In *Dynamics of Earth's Deep Interior and Earth Rotation, Geophysical Monograph 72, IUGG 12*, Amer. Geophys. Union, Washington D. C.

Voorhies, C.V. and Backus, G.E. (1985). Steady flows at the top of the core from geomagnetic field models: The steady motions theorem. *Geophys. Astrophys. Fluid Dyn.*, **32**, 163-173.

Wada, M. and Inoue, A. (1966). Relation between the carbon-14 production rate and the geomagnetic moment. *J. Geomag. Geoelect.*, **18**, 485-488.

Wahr, J. (1986). Geophysical aspects of polar motion variations in the length of day, and the Luni-Solar nutations. In *Space Geodesy and Geodynamics* (A. Anderson and A. Cazenave, eds), Academic Press, London.

Walton, D. (1979). Geomagnetic intensities in Athens between 2000 BC and AD 400. *Nature*, **277**, 643-644.

Walton, D. (1988a). The lack of reproducibility in experimentally determined intensities of the Earth's magnetic field. *Rev. Geophys.*, **26**, 15-22.

Walton, D. (1988b). Comments on "Determination of the intensity of the Earth's magnetic field during archeological times: Reliability of the Thellier technique." *Rev. Geophys.*, **26**, 13-14.

Wang, Chen-To (1948). Discovery and application of magnetic phenomena in China. I. The lodestone spoon of the Han. *Chinese J. Archaeology, Acad. Sinica*, **3**, 119.

Wang, Y.-M., Hawley, S. and Sheeley, N., Jr. (1996). The magnetic nature of corona holes. *Science*, **271**, 464-469.

Wasilewski, P.J. (1973). Magnetic hysteresis of natural materials. *Earth Planet. Sci. Lett.*, **20**, 67-72.

Watkins, N.D. (1969). Non-dipole behaviour during an Upper Miocene geomagnetic polarity transition in Oregon. *Geophys. J. Roy. Astron. Soc.*, **17**, 121-149.

Watkins, N.D. (1973). Brunhes epoch geomagnetic secular variation on Reunion Island. *J. Geophys. Res.*, **78**, 7763-7768.

Webb, E.N. and Chree, C. (1925). *Australasian Antarctic Exploration, 1911-14 Scientific Reports*, Series B, Vol. 1, Terrestrial Magnetism, p.285. Govt. Printer, Sydney.

Wegener, A. (1924). *The Origin of Continents and Oceans* (English translation by J.G.A. Skerl). Methuen, London.

Wei, Q.Y., Li, T.C., Chao, G.Y., Chang, W.S., Wang, S.P. and Wei, S.F. (1983). Results from China. In *Archaeomagnetism of Baked Clays and Sediments* (K.M. Creer, P. Tucholka and C.E. Barton, eds), p.138-150, Elsevier, Amsterdam.

Weiss, N.O. (1966). The expulsion of magnetic flux by eddies. *Proc. Roy. Soc. London*, **A293**, 310-328.

Weiss, N.O. (1991). Magnetoconvection. *Geophys. Astrophys. Fluid Dyn.*, **62**, 229-247.

Weiss, N.O. (1994). Solar and stellar dynamos, In *Lectures on Solar and Planetary Dynamos* (M.R.E. Proctor and A.D. Gilbert, eds), p.59-97, Cambridge Univ. Press, New York.

Wells, J.M. (1969). *Non-linear Spherical Harmonic Analysis of Paleomagnetic Data*. Ph.D. thesis, University of California, Berkley.

Wells, J.M. (1973). Non-linear spherical harmonic analysis of paleomagnetic data. In *Methods in Computational Physics Vol. 13, Geophysics* (B.A. Bolt, ed.), p.239-269. Academic Press, New York and London.

Wescott-Lewis, M.F. and Parry, L.G. (1971). Thermoremanence in synthetic rhombohedral iron-titanium oxides. *Aust. J. Phys.*, **24**, 735-742.

Wetherhill, G.W. (1985). Occurrence of giant impacts during the growth of terrestrial planets. *Science*, **228**, 877-879.

Whaler, K.A. (1980). Does the whole of the Earth's core convect? *Nature*, **287**, 528-530.

Whaler, K.A. and Clarke, S.O. (1988). A steady velocity field at the core-mantle boundary. *Geophys. J. R. Astron. Soc.*, **86**, 453-473.

Whitney, J., Johnson, H.P., Levi, S. and Evans, B.W. (1971). Investigations of some magnetic and mineralogical properties of the Laschamp and Olby flows, France. *Quat. Res.*, **1**, 511-521.

Widmer, R., Masters, G. and Gilbert, F. (1992). Observably split multiplets; data analysis and interpretation in terms of large-scale aspherical structure. *Geophys. J. Int.*, **111**, 559-576.

Wigley, T.M.C. (1988). The climate of the past 10,000 years and the role of the Sun. In *Secular Solar and Geomagnetic Variations in the Last 10,000 Years* (NATO Advanced Study Series, F.R. Stephenson and A.W. Wolfendale, eds), p.209-224, Kluwer Acad. Publ., Dordrecht.

Wilcox, J.M. and Ness, N.F. (1965). Quasi-stationary corotating structure in the interplanetary medium. *J. Geophys. Res.*, **70**, 5793-5805.

Williams, Q., Jeanloz, R., Bass, J., Svendson, B. and Ahrens, T.J. (1987). Melting curve of iron to 250 GPa: A constraint on the temperature of the earth's center. *Science*, **236**, 181-183.

Williams, W. and Dunlop, D.J. (1989). Three-dimensional micromagnetic modelling of ferromagnetic domain structure. *Nature*, **337**, 634-637.

Wilson, R.L. (1962a). The palaeomagnetic history of a doubly-baked rock. *Geophys. J. Roy. Astron. Soc.*, **6**, 397-399.

Wilson, R.L. (1962b). The palaeomagnetism of baked contact rocks and reversals of the earth's magnetic field. *Geophys. J. Roy. Astron., Soc.*, **7**, 194-202.

Wilson, R.L. (1970). Permanent aspects of the earth's non-dipole magnetic field over Upper Tertiary times. *Geophys. J. Roy. Astron. Soc.*, **19**, 417-437.

Wilson, R.L. (1971). Dipole offset-the time-averaged paleomagnetic field over the past 25 million years. *Geophys. J. Roy. Astron. Soc.*, **22**, 491-504.

Wilson, R.L. (1972). Paleomagnetic differences between normal and reversed field sources, and the problem of far-sided and right-handed pole positions. *Geophys. J. Roy,. Astron. Soc.*, **28**, 295-304.

Wilson, R.L. and Ade-Hall, J.M. (1970). Paleomagnetic indications of a permanent aspect of the non-dipole field. In *Palaeogeophysics* (S.K. Runcorn, ed.), p.307-312, Academic Press, New York and London.

Wilson, R.L. and Haggerty, S.E. (1966). Reversals of the earth's magnetic field. *Endeavour*, **25**, 104-109.

Wilson, R.L. and McElhinny, M.W. (1974). Investigation of the large scale paleomagnetic field over the past 25 million years; eastward shift of the Icelandic spreading ridge. *Geophys. J. Roy. Astron. Soc.*, **39**, 571-586.

Wilson, R.L. and Watkins, N.D. (1967). Correlation of petrology and natural magnetic polarity in Columbia Plateau basalts. *Geophys. J. Roy. Astron. Soc.*, **12**, 405-424.

Wilson, R.L., Dagley, P. and McCormack, A.G. (1972). Palaeomagnetic evidence about the source of the geomagnetic field. *Geophys. J. Roy. Astron. Soc.*, **28**, 213-224.

Winograd, I.J., Coplen, B.J., Landwehr, J.M., Riggs, A.C., Ludwig, K.R., Szabo, B.J., Kolesar, P.T. and Revesz, K.M. (1992). Continuous 500,000-year climate record from vein calcite in Devil's Hole, Nevada. *Science*, **258**, 255-260.

Wolfendale, A.W. (1963). *Cosmic Rays*. George Mewnes Ltd., London.

Wollin, G., Ryan, W.B.F. and Ericson, D.B. (1978). Climatic changes, magnetic intensity variations and fluctuations of the eccentricity of the Earth's orbit during the past 2,000,000 years and a mechanism which may be responsible for the relationship. *Earth Planet. Sci. Lett.*, **41**, 395-397.

Woltjer, L. (1975). Astrophysical evidence for strong magnetic fields. *Ann. N. Y. Acad. Sci.*, **257**, 76-79.

Wood, B. and Nell, J. (1991). High-temperature electrical conductivity of the lower mantle phase (MgFe)O. *Nature*, **351**, 309-311.

Wood, J.A. (1967). Chondrites: Their metallic minerals, thermal histories and parent planets. *Icarus*, **6**, 1-49.

Wood, J.A. (1979). *The Solar System*. Prentice-Hall, New Jersey.

Wood, J.A. and Pellas, P. (1991). What heated the parent meteorite planets? In *The Sun in Time* (C.P. Sonnett, M. Giampapa and M. Mathews, eds), p.740-760, Univ. Arizona Press, Tucson, Arizona.

Woodward, R. and Masters, G. (1991). Lower-mantle structure from ScS-S differential travel times. *Nature*, **352**, 231-233.

Xu, S. and Merrill, R.T. (1989). Microstress and microcoercivity in multidomain grains. *J. Geophys. Res.*, **94**, 10627-10636.

Yaskawa, K., Nakajima, T., Kawai, N., Torn, M., Natsuhara, N. and Hore, S. (1973). Paleomagnetism of a core from Lake Biwa (1). *J. Geomag. Geoelect.*, **25**, 447-474.

Ye, J. and Merrill, R.T. (1991). An attempt to reconcile magnetic domain imaging observations with theory. *Geophys. Res. Lett.,* **18**, 593-596.

Ye, J. and Merrill, R.T. (1995). The use of renormalization group theory to explain the large variations of domain states observed in titanomagnetites and implications for paleomagnetism. *J. Geophys. Res.,* **100**, 17899-17907.

Yoo, C.S., Holmes, N.C., Ross, M., Webb, D.J. and Pike, C. (1993). Shock temperatures and melting of iron at Earth core conditions. *Phys. Rev. Lett.,* **70**, 3931-3934.

Young, C.J. and Lay, T. (1987). Evidence for a shear velocity discontinuity in the lower mantle beneath India and the Indian Ocean. *Phys. Earth Planet. Int.,* **49**, 37-53.

Young, C.J. and Lay, T. (1990). Multiple phase analysis of the shear velocity structure in the D" region beneath Alaska. *J. Geophys. Res.,* **95**, 17385-17402.

Yukutake, T. (1961). Archaeomagnetic study on volcanic rocks in Oshima Island, Japan. *Bull. Earthq. Res. Inst. Univ. Tokyo,* **39**, 467.

Yukutake, T. (1965). The solar cycle contribution to the secular change in the geomagnetic field. *J. Geomag. Geoelect.,* **17**, 287-309.

Yukutake, T. (1967). The westward drift of the earth's magnetic field in historic times. *J. Geomag. Geoelect.,* **19**, 103-116.

Yukutake, T. (1971). Spherical harmonic analysis of the earth's magnetic field for the 17th and 18th centuries. *J. Geomag. Geoelect.,* **23**, 11-23.

Yukutake, T. (1979). Geomagnetic secular variations on the historical time scale. *Phys. Earth Planet. Int.,* **20**, 83-95.

Yukutake, T. (1989). Geomagnetic secular variation theory. In *Encyclopedia of Solid Earth Geophysics* (D.E. James, ed.), p.578-584, Van Nostrand–Reinhold, New York.

Yukutake, T. (1993). The geomagnetic non-dipole field in the Pacific. *J. Geomag. Geoelect.,* **45**, 1441-1453.

Yukutake, T. and Tachinaka, H. (1968). The non-dipole part of the earth's magnetic field. *Bull. Earthq. Res. Inst. Univ. Tokyo,* **46**, 1027-1074.

Yukutake, T. and Tachinaka, H. (1969). Separation of the earth's magnetic field into the drifting and the standing parts. *Bull. Earthq. Res. Inst. Univ. Tokyo,* **47**, 65-97.

Zhang, K. and Busse, F.H. (1987). Magnetohydrodynamic dynamos driven by convection in rotating spherical fluid shells. *XIX General Assembly of IUGG, Abstracts Vol. 2.* **IAGA** GA1.3-8, 440.

Zhang, K. and Busse, F.H. (1989). Convection driven magnetohydrodynamic dynamos in rotating spherical shells. *Geophys. Astrophys. Fluid Dyn.,* **49**, 97-116.

Zhang, K. and Jones, C.A. (1994). Convection motions in the Earth's fluid core. *Geophys. Res. Lett.,* **21**, 1939-1942.

Zidarov, D. (1969). Geomagnetic field variations. *Geomag. Aeron.,* **9**, 856-861.

Zidarov, D. (1970). Forecasting the geomagnetic field. *Geomag. Aeron.,* **10**, 533-539.

Zidarov, D. (1974). Presentation of magnetic fields with multipolar fields and solution of the inverse magnetic problem. *C. R. Acad. Bulg. Sci.,* **27**, 1493-1496.

Zidarov, D. and Bochev, A. (1965). Analysis of the geomagnetic field and its secular variation. *Geomag. Aeron.,* **5**, 746-748.

Zidarov, D. and Bochev, A. (1969). Representation of secular geomagnetic field variation as variation of the field of optimum geomagnetic dipoles. *Geomag. Aeron.*, **9**, 252-256.

Zidarov, D.P. and Petrova, T. (1974). Representation of the earth's magnetic field as a field of a circular loop. *C. R. Acad. Bulg. Sci.*, **27**, 203-206.

Zijderveld, J.D.A. (1967). A.C. demagnetization of rocks. In *Methods in Palaeomagnetism* (D.W. Collinson, K.M. Creer and S.K. Runcorn, eds.), p.256-286, Elsevier, New York.

Author Index

Barbetti 1983...116, 121, 122.
Barbetti and McElhinny 1972...150.
Barbetti and McElhinny 1976...105, 150, 214.
Barbetti *et al.* 1977...105.
Barenghi 1992...365.
Barenghi 1993...365.
Barenghi and Jones 1991...365.
Barnes, A. *et al.* 1971 ...394.
Barnes, T. 1973...344.
Barnett 1933...288.
Barraclough 1974...57, 119.
Barraclough *et al.* 1975...412.
Bartels 1949...67.
Bartels *et al.* 1939...66, 67.
Bartels *et al.* 1940...67.
Barton 1978...157.
Barton 1982...158, 159, 161, 162.
Barton 1983...147, 158, 159, 161, 162.
Barton 1989...37, 47, 56.
Barton and Burden 1979...133.
Barton and McElhinny 1981...135, 160.
Barton and McElhinny 1982...157.
Barton and McFadden 1996...213.
Barton *et al.* 1979...122, 123, 127, 129.
Barton *et al.* 1980...88, 131.
Bauer 1894...57.
Bauer 1895...117, 137.
Beck *et al.* 1995...434.
Benjamin 1895...3.
Benkova and Cherevko 1972...228.
Benkova *et al.* 1970...56.
Benkova *et al.* 1971...228.
Benkova *et al.* 1973...228.
Benkova *et al.* 1974...57.
Benton 1979...410.
Benton 1981...410.
Benton *et al.* 1982...37, 44.
Benton *et al.* 1987...41.
Benz and Cameron 1990...378.
Bhargava and Yacob 1969...158.
Biermann 1951...16.
Birch 1964...271.
Birch 1965a...283.
Birch 1965b...283.
Birch 1968...271.
Birch 1972...271.
Birkeland 1908...58.
Biskamp 1993...359, 425.
Blackett 1947...17, 286.
Blackett 1952...12, 17.
Blackett 1961...225.
Blackett *et al.* 1960...12.
Blackshear and Gapcynski 1977...393.
Blakely 1976...176.
Blakely and Cande 1979...176.

Blakely and Cox 1972...178.
Bloomfield 1976...161.
Blow and Hamilton 1978...233.
Bloxham 1989...415.
Bloxham 1992...414.
Bloxham and Gubbins 1985...143, 19, 41, 53,
    57, 259, 368, 372, 415, 430.
Bloxham and Gubbins 1986...143, 368, 371,
    372, 41, 53.
Bloxham and Jackson 1991...411, 413, 416.
Bochev 1965...43.
Bochev 1969...43.
Bochev 1975...43.
Boehler 1992...280.
Boehler 1993...280.
Bogue and Coe 1981...205, 206.
Bogue and Coe 1982...205.
Bogue and Coe 1984...205.
Bogue and Hoffman 1987...206.
Bogue and Merrill 1992...206, 215, 353, 428.
Bogue and Paul 1993...206, 208, 209.
Bonhommet and Babkine 1967...149.
Booker 1969...410.
Bott 1967...173, 174.
Braginsky 1963...302.
Braginsky 1964a...311, 328, 353.
Braginsky 1964b...311, 323, 328, 353.
Braginsky 1967...311.
Braginsky 1972...55, 57.
Braginsky 1975...366.
Braginsky 1976...301, 366.
Braginsky 1978...366.
Braginsky 1989...373.
Braginsky 1991...302, 362.
Braginsky 1993...303, 362.
Braginsky 1994...366.
Braginsky and Kulanin 1971...57.
Braginsky and Le Mouël 1993...410.
Braginsky and Meytlis 1990...358, 359, 425.
Brandeis and Marsh 1989...275.
Brandenburg 1994...391.
Brandt and Hodge 1964...381.
Brecher and Brecher 1978...286, 405.
Breit and Tuve 1926...13.
Brett 1973...271.
Brett 1976...265, 271.
Briden 1968...225.
Briden and Irving 1964...98, 225, 226.
Brock 1971...261.
Brown, J.M. 1986...280, 282.
Brown, J.M. and McQueen 1980...280.
Brown, J.M. and McQueen 1986...280.
Brown, L.W. 1975...401.
Brown, W.F. 1959...78.
Brummell *et al.* 1995...382.

Courtillot *et al.* 1978...50.
Courtillot *et al.* 1979...50.
Courtillot *et al.* 1984...50, 51.
Cowling 1934...18, 353, 388.
Cowling 1957...353, 354.
Cox 1962...250, 255.
Cox 1964...250.
Cox 1968...122, 123, 126, 129, 186, 203, 241, 43.
Cox 1969...149, 169, 186, 331.
Cox 1970...247, 248.
Cox 1975...228, 229.
Cox 1982...183, 197.
Cox and Doell 1960...12,118, 218.
Cox and Doell 1964...239.
Cox *et al.* 1963a...169.
Cox *et al.* 1963b...169.
Cox *et al.* 1964a...169.
Cox *et al.* 1964b...169.
Creager 1992...268, 269.
Creager and Jordan 1986...268, 420.
Creer 1958...12.
Creer 1962...248, 249, 255.
Creer 1977...134, 140, 154.
Creer 1981...134, 140.
Creer 1983...140, 141, 142, 143, 145, 146, 154, 233.
Creer and Ispir 1970...209.
Creer and Tucholka 1982a...134.
Creer and Tucholka 1982b...139.
Creer and Tucholka 1982c...134, 139.
Creer *et al.* 1954...12, 217.
Creer *et al.* 1959...249.
Creer *et al.* 1972...11, 133.
Creer *et al.* 1973...228, 229.
Creer *et al.* 1976a...134.
Creer *et al.* 1976b...134.
Creer *et al.* 1983...134.
Creer *et al.* 1986...134.
Cullity 1972...75, 76.
Currie 1967...159.
Currie 1968...50, 52, 159.
Currie 1973...158.
Currie 1974...157.
Currie 1976...158.
Curtis and Ness 1986...406.
Dagley and Lawley 1974...205.
Dagley and Wilson 1971...206.
Dainty *et al.* 1976...393.
Dalrymple 1991...113.
Damon 1970...125.
David 1904...10, 164.
Davies, G.F. and Gurnis 1986...270.
Davies, J. and Richards 1992...433, 435.
Day 1977...74.

Day *et al.* 1976...74.
Day *et al.* 1977...74.
Denham 1974...135, 137, 151.
Denham 1975...157, 160.
Denham and Cox 1971...135, 151.
Dicke 1970...382.
Dickey and Eubanks 1986...418.
Dickson *et al.* 1968...181.
Dodson 1979...137, 138.
Dodson *et al.* 1978...205, 206, 208.
Doell 1970...257.
Doell 1972...257.
Doell and Cox 1963...116.
Doell and Cox 1965...257.
Doell and Cox 1972...148, 214, 227, 257, 259, 260.
Doell *et al.* 1970...397.
Dolginov 1976...405.
Dolginov 1978...405.
Dolginov *et al.* 1961...394.
Dolginov *et al.* 1966...394.
Domen 1977...103.
Doornbos and Hilton 1989...420.
Drewry *et al.* 1974...98, 225.
Du Bois 1974...116.
Du Bois 1989...116, 119.
Duba 1992...280.
Duba and Wanamaker 1994...276.
Ducruix *et al.* 1980...50.
Duffy and Hemley 1995...280.
Duncan 1975...261.
Dungey 1961...59.
Dungey 1963...59.
Dunlop 1971...87.
Dunlop 1979...91.
Dunlop 1990...69, 74, 83.
Dunn *et al.* 1971...205.
Duvall *et al.* 1996...390.
Dyal and Parkin 1971...393.
Dyal *et al.* 1970...394.
Dyal *et al.* 1977...399.
Dziewonski and Anderson 1981...267.
Dziewonski and Anderson 1984...268.
Dziewonski and Gilbert 1971...266.
Dziewonski and Woodhouse 1987...260, 268.
Dziewonski *et al.* 1977...260, 268.
Dzyaloshinski 1958...86.
Eddy 1976...386.
Eddy 1977...386.
Eddy 1988...382.
Eddy *et al.* 1977...382, 386.
Egbert 1992...221.
Egbert and Booker 1992...48,276.
Elmore and McCabe 1991...90, 92.
Elsasser 1946a...18, 344.

Hide 1966...311, 323, 334, 335.
Hide 1969...421, 429.
Hide 1981...353.
Hide 1989...421.
Hide 1995...421.
Hide and Malin 1970...421.
Hide and Malin 1971a...421.
Hide and Malin 1971b...421.
Hide and Palmer 1982...353.
Hide and Stewartson 1972...335.
Higgins and Kennedy 1971...416.
Hillhouse and Cox 1976...205, 206, 208, 210.
Hirooka 1983...116.
Hoffman 1975...164, 165.
Hoffman 1977...210.
Hoffman 1979...210, 337.
Hoffman 1981...214.
Hoffman 1986...205.
Hoffman 1989...186, 205.
Hoffman 1991...205, 212.
Hoffman 1992a...164, 165.
Hoffman 1992b...205, 212.
Hoffman and Day 1978...91.
Holcomb *et al.* 1986...116.
Hollerbach and Ierley 1991...365.
Hollerbach and Jones 1993a...373, 375, 423.
Hollerbach and Jones 1993b...423.
Hollerbach *et al.* 1992...365.
Hood 1981...394.
Hospers 1953...10, 12.
Hospers 1954a...10, 12.
Hospers 1954b...217.
Hospers 1955...10.
Houtermans 1966...127.
Houtermans *et al.* 1973...127.
Howard 1975...382.
Howe *et al.* 1974...394.
Hoye 1981...116, 117.
Hoyt *et al.* 1994...386.
Hulot and Gallet 1996...255, 258.
Hulot and Le Mouël 1994...147, 255, 258.
Hulot *et al.* 1993...52.
Hurwitz 1960...139.
Hyodo 1984...205.
Imbrie and Imbrie 1980...113, 114.
Inglis 1955...18, 286, 288.
Irving 1956...12, 101, 224, 225.
Irving 1959...12.
Irving 1964...167, 218.
Irving 1966...434.
Irving 1977...98.
Irving 1979...98.
Irving and Briden 1962...98, 225.
Irving and Brown 1964...225.
Irving and Gaskell 1962...98, 225.

Irving and Naldrett 1977...225.
Irving and Parry 1963...168, 196.
Irving and Pullaiah 1976...196, 261.
Irving and Ward 1964...249, 261.
Isenberg 1991...57, 59.
Ishikawa and Akimoto 1958...87.
Ishikawa and Syono 1963...87, 164, 165.
Jackson 1975...291.
Jacobs 1953...273.
Jacobs 1981...433.
Jacobs 1984...113.
Jacobs 1987...271.
Jacobs 1991...60.
Jacobs 1992...265, 271, 280, 284.
Jacobs 1994...163, 206.
James 1971...55.
James 1974...55.
James and Winch 1967...221.
James *et al.* 1980...353.
Jault and Le Mouël 1991...417, 418, 420, 421,
    422.
Jault and Le Mouël 1994...52.
Jeanloz 1979...271.
Jeanloz 1990...265, 280.
Jeanloz and Thompson 1983...270.
Jeanloz and Wenk 1988...268.
Johnson, C.L. and Constable 1995a...221, 258,
    259.
Johnson, C.L. and Constable 1995b...235, 236,
    430, 432.
Johnson, E.A. *et al.* 1948...11.
Johnson, H. P. and Hall 1978...86.
Johnson, H.P. 1979...176, 177.
Johnson, H.P. and Merrill 1978...175, 176.
Johnson, H.P. *et al.* 1995...197.
Johnson, R. 1982...114.
Jones 1977...199.
Jones *et al.* 1975...91.
Kanasewich 1973...157.
Karato 1993...424.
Karlin and Levi 1985...88, 131, 171.
Karlin *et al.* 1987...86.
Kaula *et al.* 1974...393.
Keefer and Shive 1981...86.
Keeler and Mitchell 1969...274.
Kennett 1980...181.
Kent and Gradstein 1986...185, 199, 200.
Kent and Opdyke 1977...131, 132.
Khan 1971...421.
Khodair and Coe 1975...108.
Khramov 1955...10.
Khramov 1958...10, 12.
Kidd 1977...176.
Kincaid 1995...281.
King, J.W. *et al.* 1983...108, 131.

Lowes 1976...37.
Lowes and Runcorn 1951...43.
Lowes and Wilkinson 1963...316, 355.
Lowes and Wilkinson 1968...316.
Lowrie and Alvarez 1981...183.
Lowrie and Fuller 1971...74.
Lowrie and Kent 1983...198.
Lund 1994...139, 147.
Lund 1996...134, 147.
Lund and Banerjee 1979...134.
Lund and Banerjee 1985...134, 145, 146.
Lund and Keigwin 1994...130.
Lutz 1985...202.
Lutz and Watson 1988...202.
Macdonald and Holcombe 1978...172.
Macdonald et al. 1988...172, 180.
Mackereth 1958...11, 133.
Mackereth 1971...11, 133.
Malin and Hodder 1982...50.
Malin et al. 1983...50.
Malinetskii et al. 1990...338.
Malkus 1963...301.
Malkus 1994...301.
Mankinen et al. 1985...205, 209, 215.
Marsh 1981...275.
Marshall et al. 1988...150.
Martelli and Newton 1977...399.
Marzocchi and Mulargia 1990...202.
Mason and Raff 1961...172.
Matassov 1977...274.
Matsushita 1967...63.
Matthews and Bath 1967...175.
Mattis 1965...17, 76.
Matuyama 1929...10, 164, 168.
Maxwell et al. 1970...181.
Mayaud 1968...67.
Mayaud 1980...66.
Mazaud and Laj 1991...202, 203.
Mazaud et al. 1983...202.
Mazaud et al. 1991...129.
McCabe and Elmore 1989...90.
McClelland and Goss 1993...164, 165, 166.
McDonald 1957...48, 49.
McDonald and Gunst 1968...55, 56.
McDougall and Harrison 1988...113.
McDougall and Tarling 1963...169.
McDougall and Tarling 1964...169.
McDougall et al. 1976a...181.
McDougall et al. 1976b...181.
McDougall et al. 1977...181.
McElhinny 1971...196.
McElhinny 1973...98, 167, 175, 218.
McElhinny and Brock 1975...98, 222.
McElhinny and Evans 1968...105, 226.
McElhinny and Lock 1991...197.

McElhinny and Lock 1993a...197, 232.
McElhinny and Lock 1993b...197.
McElhinny and Lock 1994...232.
McElhinny and Lock 1995...232.
McElhinny and Merrill 1975...247, 251, 252, 257, 258, 260.
McElhinny and Senanayake 1980...226.
McElhinny and Senanayake 1982...123, 124, 128, 129, 131, 241, 242, 243, 252.
McElhinny et al. 1996a...218, 219, 220, 229, 230, 231, 232, 233, 234, 235, 236, 237, 430, 432.
McElhinny et al. 1996b...261, 268.
McFadden 1980...100.
McFadden 1984a...188, 191, 192, 193, 194, 197, 198, 199, 214.
McFadden 1984b...202.
McFadden 1987...202.
McFadden and Lowes 1981...99.
McFadden and McElhinny 1982...110, 123, 239, 240, 241, 252, 332.
McFadden and McElhinny 1984...249, 252, 253, 258.
McFadden and McElhinny 1988...91.
McFadden and McElhinny 1995...220, 232.
McFadden and Merrill 1984...187, 192, 193, 194, 195, 198, 199, 200, 202, 434.
McFadden and Merrill 1986...188, 190, 200, 203, 214.
McFadden and Merrill 1993...188, 191, 199, 200, 201, 203, 214, 338.
McFadden and Merrill 1995a...201, 202, 214, 337, 428.
McFadden and Merrill 1995b...209, 212, 214, 260, 431, 433, 435.
McFadden and Merrill 1997...203.
McFadden et al 1993...213.
McFadden et al. 1985...116, 208, 345, 346, 370.
McFadden et al. 1987...194, 195.
McFadden et al. 1988...253, 254, 256.
McFadden et al. 1989...432.
McFadden et al. 1991...257, 258, 262, 331, 366, 425.
McFadden et al. 1993...205, 213.
McFadden et al. 1997a...193.
McFadden et al. 1997b...192, 195, 196, 198, 200.
McKenzie and Parker 1967...175.
McLeod 1994...276.
McNish 1940...43.
McPherron 1979...61.
McPherron 1991...61, 66.
McSweeney et al. 1996...424.
McWilliams et al. 1982...261.

Parker, E.N. 1987...391.
Parker, R.L. 1973...173, 174.
Parker, R.L. 1977...44, 45.
Parker, R.L. 1994...48, 173, 175, 177.
Parker, R.L. and Huestis 1974...44, 45, 173, 174, 177.
Parker, R.L. and Whaler 1981...48.
Parks 1991...60, 394.
Paschmann 1991...58.
Patton and Fitch 1962...107.
Payne 1981...437.
Peddie 1979...43, 141.
Peng and King 1992...135.
Perrin *et al.* 1991...245.
Petersen *et al.* 1986...109.
Peterson and Bliel 1973...166.
Petherbridge 1977...164, 166.
Peyronneau and Poirier 1989...276.
Phillips, C.G. 1993...358, 428.
Phillips, J.D. 1977...194, 198.
Phillips, J.D. and Cox 1976...157, 186, 187.
Phillips, J.D. *et al.* 1975...186.
Pick and Tauxe 1991...86.
Pick and Tauxe 1993...107.
Piper and Grant 1989...223, 224.
Pitman *et al.* 1968...181.
Podolak and Cameron 1974...380.
Poirier 1988...274, 275.
Poirier 1991...265, 266, 269, 270, 274, 278, 279, 280, 282, 284.
Preston 1967...382.
Prévot 1977...206.
Prévot and Camps 1993...206, 211, 212.
Prévot and Perrin 1992...245, 247.
Prévot *et al.* 1985...205, 206, 209, 215.
Prévot *et al.* 1990...108, 119, 245.
Pudovkin and Kolesova 1968...43.
Pudovkin *et al.* 1968...43.
Quidelleur and Courtillot 1996...233, 235, 255.
Quidelleur and Valet 1994...213.
Quidelleur *et al.* 1994...218, 220, 221, 233, 236.
Quidelleur *et al.* 1995...213.
Rädler 1968...311, 318.
Rampino and Caldeira 1993...202.
Raup 1985...202.
Readman and O'Reilly 1970...86, 166.
Readman and O'Reilly 1972...86.
Rekdal and Doornbos 1992...268.
Ricard *et al.* 1993...435.
Richards and Engebretson 1992...260, 435.
Rikitake 1958...241.
Rikitake 1966...306, 307, 355.
Rimbert 1959...107.
Ringwood 1958...288.

Ringwood 1977...271.
Ringwood 1978...271.
Ringwood 1979...265, 272.
Rivin 1974...158.
Roberts, G. 1969...355.
Roberts, G. 1970...18, 355.
Roberts, G. 1972...352, 355, 357.
Roberts, P.H. 1967...316.
Roberts, P.H. 1971...318, 330, 331, 361, 362, 425.
Roberts, P.H. 1972...362, 363.
Roberts, P.H. 1987...364.
Roberts, P.H. 1988...372.
Roberts, P.H. 1989a...418, 420.
Roberts, P.H. 1989b...366.
Roberts, P.H. and Gubbins 1987...301, 302, 303, 355, 358, 362, 363.
Roberts, P.H. and Scott 1965...410.
Roberts, P.H. and Stix 1971...362.
Roberts, P.H. and Stix 1972...253, 330, 425.
Roche 1951...10, 168.
Roche 1953...10.
Roche 1956...10, 168.
Roche 1958...10.
Rochester *et al.* 1975...301.
Roperch *et al.* 1988...149, 150.
Rosenberg and Coleman 1969...383.
Rossi 1964...126.
Roy and Park 1972...89.
Runcorn 1954...12.
Runcorn 1955...49.
Runcorn 1956...12.
Runcorn 1959...12, 136, 137.
Runcorn 1961...225.
Runcorn 1978...399.
Runcorn 1979...399.
Runcorn 1992...260.
Runcorn 1994...399.
Runcorn and Urey 1973...399.
Runcorn *et al.* 1970...397.
Russell 1978...394, 403, 405, 406.
Russell 1987...401, 403.
Russell *et al.* 1974...394.
Rutten 1959...169.
Ruzmaikin and Sokoloff 1980...355.
Ryall and Ade-Hall 1975...166.
Sabatier 1977...44.
Sakai and Hirooka 1986...117, 119.
Sano 1993...406.
Sato and Wright 1966...86, 88.
Saxena *et al.* 1993...268.
Saxena *et al.* 1994...280.
Schatten 1971...383.
Schmidt 1924...16.
Schmidt 1935...28.

Index

International Geophysics Series

EDITED BY

RENATA DMOWSKA

Division of Applied Science
Harvard University
Cambridge, Massachusetts

JAMES R. HOLTON

Department of Atmospheric Sciences
University of Washington
Seattle, Washington

* Out of Print

Volume 37 J. A. Jacobs. The Earth's Core, Second Edition. 1987

Volume 38 J. R. Apel. Principles of Ocean Physics. 1987

Volume 39 Martin A. Uman. The Lightning Discharge. 1987

Volume 40 David G. Andrews, James R. Holton, and Conway B. Leovy. Middle Atmosphere Dynamics. 1987

Volume 41 Peter Warneck. Chemistry of the Natural Atmosphere. 1988

Volume 42 S. Pal Arya. Introduction to Micrometeorology. 1988

Volume 43 Michael C. Kelley. The Earth's Ionosphere. 1989

Volume 44 William R. Cotton and Richard A. Anthes. Storm and Cloud Dynamics. 1989

Volume 45 William Menke. Geophysical Data Analysis: Discrete Inverse Theory, Revised Edition. 1989

Volume 46 S. George Philander. El Niño, La Niña, and the Southern Oscillation. 1990

Volume 47 Robert A. Brown. Fluid Mechanics of the Atmosphere. 1991

Volume 48 James R. Holton. An Introduction to Dynamic Meteorology, Third Edition. 1992

Volume 49 Alexander A. Kaufman. Geophysical Field Theory and Method.
Part A: Gravitational, Electric, and Magnetic Fields. 1992
Part B: Electromagnetic Fields I. 1994
Part C: Electromagnetic Fields II. 1994

Volume 50 Samuel S. Butcher, Gordon H. Orians, Robert J. Charlson, and Gordon V. Wolfe. Global Biogeochemical Cycles. 1992

Volume 51 Brian Evans and Teng-Fong Wong. Fault Mechanics and Transport Properties of Rocks. 1992

Volume 52 Robert E. Huffman. Atmospheric Ultraviolet Remote Sensing. 1992

Volume 53 Robert A. Houze, Jr. Cloud Dynamics. 1993

Volume 54 Peter V. Hobbs. Aerosol-Cloud-Climate Interactions. 1993

Volume 55 S. J. Gibowicz and A. Kijko. An Introduction to Mining Seismology. 1993

Volume 56 Dennis L. Hartmann. Global Physical Climatology. 1994

Volume 57 Michael P. Ryan (ed.). Magmatic Systems. 1994

Volume 58 Thorne Lay and Terry C. Wallace. Modern Global Seismology. 1995

Volume 59 Daniel S. Wilks. Statistical Methods in the Atmospheric Sciences. 1995

Volume 60 Frederik Nebeker. Calculating the Weather. 1995

Volume 61 Murry L. Salby. Fundamentals of Atmospheric Physics. 1996

Volume 62 James P. McCalpin (ed.). Paleoseismology. 1996

Volume 63 Ronald T. Merrill, Michael W. McElhinny, and Phillip L. McFadden. The Magnetic Field of the Earth: Paleomagnetism, the Core, and the Deep Mantle. 1996